Random Signals and Processes Primer with MATLAB

Gordana Jovanovic Dolecek

Random Signals and Processes Primer with MATLAB

Gordana Jovanovic Dolecek
Department of Electronics
Instituto Nacional de Astrofisica
Optica y Electronica (INAOE)
Tonantzintla, Puebla, Mexico

Additional material to this book can be downloaded from http://extra.springer.com

ISBN 978-1-4939-0028-2 ISBN 978-1-4614-2386-7 (eBook)
DOI 10.1007/978-1-4614-2386-7
Springer New York Heidelberg Dordrecht London

Printed on acid-free paper

Springer is part of Springer Science+Business Media (www.springer.com)

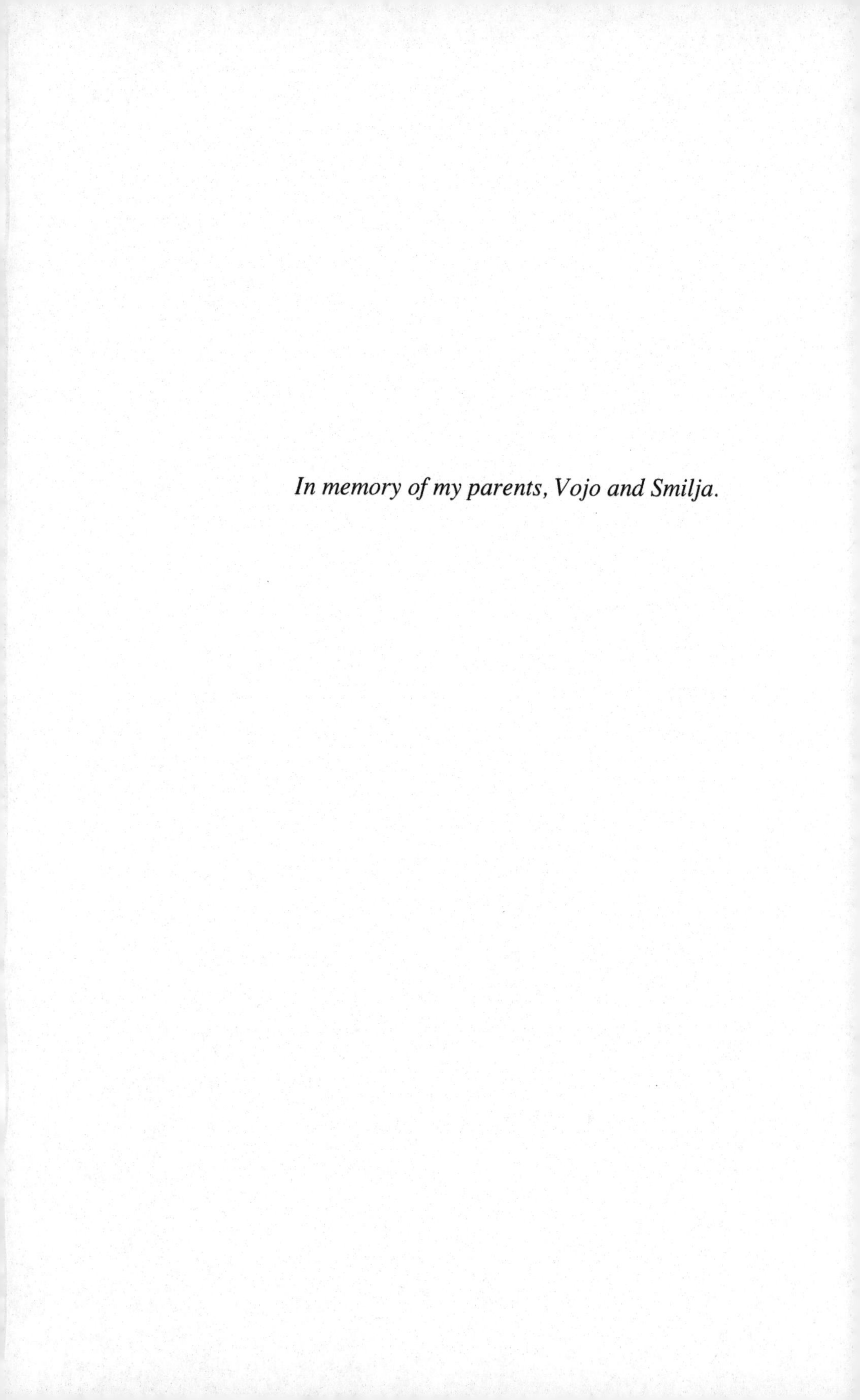

In memory of my parents, Vojo and Smilja.

Preface

The concept of randomness is important in almost all aspects of modern engineering systems and sciences. This includes computer science, biology, medicine, the social sciences, and management sciences, among others.

Whenever we are unable to know the exact future behavior of some phenomena, we say that it is *random*. As such, random phenomena can only be approximated.

Because of this, work with random phenomena calls for an understanding of probability concepts as well as the basic concepts of the theory of random signals and processes.

Despite the number of books written on this topic, there is no book which introduces the concept of random variables and processes in a simplified way.

The majority of books written on this topic are built on a strong mathematical background which sometimes overwhelms the reader with mathematical definitions and formulas, rather than focusing on a full understanding of the concepts of random variables and related terms. In this way, sometimes the reader learns the mathematical definitions without a full understanding of the meaning of the underlying concepts.

Additionally, many books which claim to be introductory are very large. This is not encouraging for readers who really desire to understand the concept first.

Taking into account that the concept of random signals and processes is also important in many fields without the strong mathematical background, one of our main purposes in writing this book is to explain the basic random concepts in a more reader-friendly way; thus making them accessible to a broader group of readers.

As the title indicates, this text is intended for theses who are studying random signals for the first time.

It can be used for self-study and also in courses of varying lengths, levels, and emphases. The only prerequisite is knowledge of elementary calculus.

Motivating students is a major challenge in introductory texts on random signals and processes. In order to achieve student motivation, this book has features that distinguish it from others in this area which we hope will attract prospective readers:

- Besides the mathematical definitions, simple explanations of different terms like density, distribution, mean value, variance, random processes, autocorrelation function, ergodicity, etc., are provided.
- In an effort to facilitate understanding, a lot of examples which illustrate concepts are followed by the numerical exercises which are already solved.
- It is widely accepted that the use of visuals reinforces better understanding. As such, MATLAB exercises are used extensively in the text, starting in Chap. 2, to further explain and to give visual and intuitive representations of different concepts.
- We chose MATLAB because MATLAB, along with its accompanying toolboxes, is the tool of choice for most educational and research applications. In order for the reader to be able to focus on understanding the subject matter and not on programming, all exercises and MATLAB codes are provided.
- Pedagogy recognizes the need for more active learning in improving the quality of learning. Making a reader active rather than passive can enhance understanding. In pursuit of this goal, each chapter provides several demo programs as a complement to the textbook. These programs allow the reader to become an active participant in the learning process. The readers can run the programs by themselves as many times as they want, choosing different parameters.
- Also, in order to encourage students to think about the concepts presented, and to motivate them to undertake further reading, we present different questions at the end of each chapter. The answers to these are also provided so that the reader can check their understanding immediately.

The text is divided into seven chapters.

The first chapter is a brief introduction to the basic concepts of random experiments, sample space, and probability, illustrated with different examples.

The next chapter explains what a random variable is and how it is described by probability distribution and density function while considering discrete, continuous, and mixed variables. However, there are many applications—when a partial description of random variable is either needed or possible—in which we need some parameters for the characterization of the random variable. The most important parameters, mean value and variance, are explained in detail and are illustrated with numerous examples.

We also show how to find the characteristics of the random variable after its transformation, if the characteristics of the random variable before transformation are known. Some other important characteristics are also described, such as characteristic function and moment generating function.

The concept of a multidimensional random variable is examined in Chap. 3 considering a two-dimensional random variable, which is described by the joint distribution and density. Expected values and moments are also introduced and discussed. The relations between random variables such as dependence and correlation are also explained. This chapter also discusses some other important topics, such as transformation and the characteristic function.

The most important random variable the Gaussian, or normal random variable is described in detail in Chap. 4 and the useful properties which make it so important in different applications are also explained.

Chapter 5 presents some important random variables and describes their useful properties. Some of these include lognormal, Rayleigh, Rician, and exponential random variables. We also discuss variables related to exponential variables such as Laplacian, Gamma, Erlang's, and Weibull random variables.

Special attention is also paid to some important discrete random variables such as Bernoulli, binomial, Poisson, and geometric random variables.

The concept of a random process is introduced and explained in Chap. 6. A description of random process is given and its important characteristics, such as stationarity and ergodicity, are explained. Special attention is also given to an autocorrelation function, and its importance and properties are explained.

Chapter 7 examines the spectral description of a random process based on the Fourier transform. Difficulties in the application of the Fourier transform to random processes and differences in applying the Fourier transform of deterministic signals are elaborated. The power spectral density is explained along with its important properties. Some important operations of processes like sum, multiplication with a sinusoidal signal, and LTI filtering are described through the lens of the spectral description of the process.

Puebla, Mexico Gordana Jovanovic Dolecek

Acknowledgements

The author would like to offer thanks for the professional assistance, technical help, and encouragement that she received from the Institute for Astrophysics, Optics, and Electronics (INAOE), Puebla, Mexico. Without this support it would not have been possible to complete this project.

She would also thank her colleague Professor Massimiliano Laddomada for his useful comments for improving the text.

Contents

Chapter 1
Introduction to Sample Space and Probability

1.1 Sample Space and Events

Probability theory is the mathematical analysis of random experiments [KLI86, p. 11]. An *experiment* is a procedure we perform that produces some result or *outcome* [MIL04, p. 8].

An experiment is considered *random* if the result of the experiment cannot be determined exactly. Although the particular outcome of the experiment is not known in advance, let us suppose that all possible outcomes are known.

The mathematical description of the random experiment is given in terms of:

- Sample space
- Events
- Probability

The set of all possible outcomes is called *sample space* and it is given the symbol S. For example, in the experiment of a coin-tossing we cannot predict exactly if "head," or "tail" will appear; but we know that all possible outcomes are the "heads," or "tails," shortly abbreviated as H and T, respectively. Thus, the sample space for this random experiment is:

$$S = \{\mathrm{H}, \mathrm{T}\}. \tag{1.1}$$

Each element in S is called a *sample point*, s_i. Each outcome is represented by a corresponding sample point. For example, the sample points in (1.1) are:

$$s_1 = \mathrm{H}, \quad s_2 = \mathrm{T}. \tag{1.2}$$

When rolling a die, the outcomes correspond to the numbers of dots (1–6). Consequently, the sample set in this experiment is:

$$S = \{1, 2, 3, 4, 5, 6\}. \tag{1.3}$$

G.J. Dolecek, *Random Signals and Processes Primer with MATLAB*,
DOI 10.1007/978-1-4614-2386-7_1,

We assume here that coins and dice are *fair* and have *no memory*, i.e., each outcome is equally likely on each toss, regardless of the results of previous tosses.

It is helpful to give a geometric representation of events using a *Venn diagram*. This is a diagram in which sample space is presented using a closed-plane figure and sample points using the corresponding dots. The sample spaces (1.1) and (1.3) are shown in Fig. 1.1a, b, respectively.

The sample sets (1.1) and (1.3) are *discrete* and *finite*. The sample set can also be *discrete* and *infinite*. If the elements of the sample set are *continuous* (i.e., not countable) thus the sample set S is continuous. For example, in an experiment which measures voltage over time T, the sample set (Fig. 1.2) is:

$$S = \{s | V_1 < s < V_2\}. \tag{1.4}$$

In most situations, we are not interested in the occurrence of specific outcomes, but rather in certain characteristics of the outcomes. For example, in the voltage measurement experiment we might be interested if the voltage is positive or less than some desired value V. To handle such situations it is useful to introduce the concept of an event.

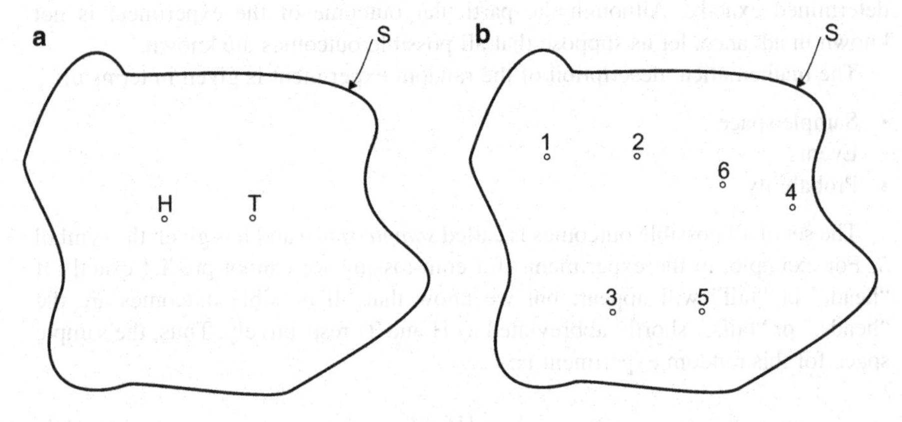

Fig. 1.1 Sample spaces for coin tossing and die rolling. (**a**) Coin tossing. (**b**) Die rolling

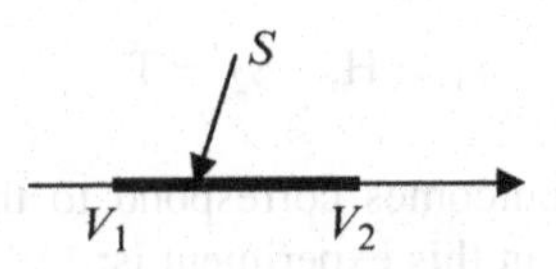

Fig. 1.2 Example of continuous space

Any subset of the sample space S is called an *event* and denoted with capital letters (A, B, etc.). Thus, the event for the voltage measurement can be $A = \{v > 0\}$. In the die rolling experiment the event can be, for example, the occurrence of even number of dots,

$$A = \{2, 4, 6\}. \tag{1.5}$$

Note that an outcome can also be an event. For example, in the sample space (1.1), the event is either H, or T (i.e., $A = \{\mathrm{H}\}$ is the event that "heads" occurred). Similarly, $B = \{\mathrm{T}\}$ is the event that "tails" occurred in the experiment.

An event that has no sample points is called a *null event*.

Figure 1.3 illustrates a Venn diagram representation of a sample space S, with the element s_i and events A_1, A_2, and A_3.

Example 1.1.1 A coin is tossed until either "heads" or "tails" appears twice in succession. Find the sample space S and its subset which corresponds to the event $A = \{$"No more than three tosses are necessary"$\}$.

Solution The sample space is formed by the outcomes of two "heads" denoted by H, or two "tails," denoted by T, independent of the number of tosses.

Two tosses:

$$\begin{aligned} s_1 &= \{\mathrm{H}, \mathrm{H}\}, \\ s_2 &= \{\mathrm{T}, \mathrm{T}\}, \end{aligned} \tag{1.6}$$

Three tosses:

$$\begin{aligned} s_3 &= \{\mathrm{T}, \mathrm{H}, \mathrm{H}\}, \\ s_4 &= \{\mathrm{H}, \mathrm{T}, \mathrm{T}\}, \end{aligned} \tag{1.7}$$

Four tosses:

$$\begin{aligned} s_5 &= \{\mathrm{H}, \mathrm{T}, \mathrm{H}, \mathrm{H}\}, \\ s_6 &= \{\mathrm{T}, \mathrm{H}, \mathrm{T}, \mathrm{T}\}, \end{aligned} \tag{1.8}$$

etc.

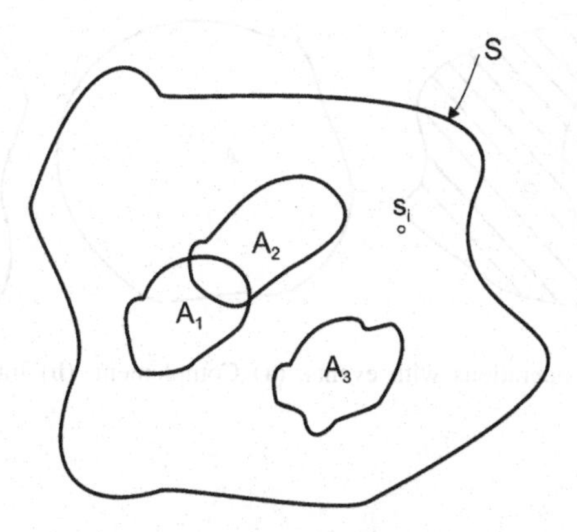

Fig. 1.3 Venn diagram illustration of sample space S, element s_i and events A_1, A_2, and A_3

The sample space is infinite but countable:

$$S = \{s_1, s_2, s_3, s_4, \ldots\}. \tag{1.9}$$

Event A is formed by (1.6) and (1.7):

$$A = \{s_1, s_2, s_3, s_4\}. \tag{1.10}$$

1.1.1 Operations with Events

The *complement* of an event A–denoted as $\overline{A}$–is an event which contains all sample points not in A. For example, the complement of the event A in Fig. 1.1a is the event $B = \overline{A} = \{\text{T}\}$.

The *union* (or *sum*) of events A and B–denoted as $A \cup B$–is an event which contains all points in A and B. For example, the union of the events A and B in Fig. 1.1a is: $A \cup B = \{\text{H, T}\}$. The union is also denoted as $A + B$.

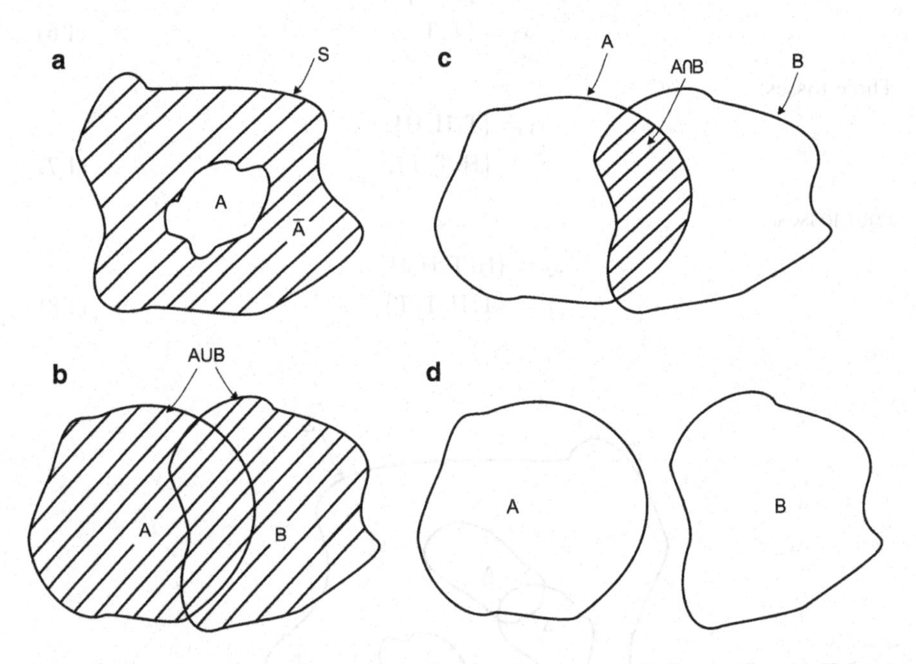

Fig. 1.4 Illustration of operations with events. (**a**) Complement. (**b**) Intersection. (**c**) Union. (**d**) Disjoint events

The *joint event* or *intersection* of events A and B–denoted as $A \cap B$–is an event which contains points common to A and B (i.e., the event "both A and B occurred"). The intersection is sometimes called the *product* of events and is denoted as AB. For example, the joint event of A and B from Fig. 1.1a is a null event, $A \cap B = \{0\}$, because the events A and B cannot occur simultaneously.

If events A and B do not have elements in common, $A \cap B = 0$, then they are said to be *disjoint*, or *mutually exclusive*. The events A and B from Fig. 1.1a are mutually exclusive.

Figure 1.4 is a Venn diagram representation of different operations with events.

1.2 Probability of Events

1.2.1 Axioms

The probability of event A is a real number assigned to event A–denoted as $P\{A\}$–which satisfies the following axioms, formulated by the Russian mathematician A.N. Kolmogorov:

Axiom I

$$P\{A\} \geq 0. \tag{1.11}$$

The axiom states that the probability of event A is a nonnegative number.

Axiom II

$$P\{S\} = 1. \tag{1.12}$$

The axiom states that the sample space is itself an event which always occurs.

Axiom III The axiom states if two events A and B are mutually exclusive (i.e., $A \cap B = 0$) then the probability of either event A, or B occurring, $P\{A \cup B\}$, is the sum of their probabilities:

$$P\{A \cup B\} = P\{A + B\} = P\{A\} + P\{B\}. \tag{1.13}$$

Since the word *axiom* means self-evident statement [MIL04, p. 12], the proof for (1.11)–(1.13) is omitted.

Example 1.2.1 In a coin tossing experiment, the events $A = \{\mathrm{H}\}$ and $B = \{\mathrm{T}\}$ are mutually exclusive (i.e., only "heads" or "tails" can occur), resulting in:

$$P\{\mathrm{H} \cup \mathrm{T}\} = P\{\mathrm{H}\} + P\{\mathrm{T}\} = 1. \tag{1.14}$$

The corresponding sample space is:

$$S = A \cup B \tag{1.15}$$

and from (1.14) we have:

$$P\{S\} = 1. \tag{1.16}$$

1.2.2 *Properties*

From the Axioms, I, II, and III, we can derive more properties of probability as shown here:

P.1 Event A and its complement $\overline{A}$ are mutually exclusive. Consequently, the union of $\overline{A}$ and A is a certain event with probability 1.

Proof From Axiom III, we have:

$$1 = P\{\overline{A} \cup A\} = P\{\overline{A}\} + P\{A\}. \tag{1.17}$$

From here, the probability of the complement of event A is:

$$P\{\overline{A}\} = 1 - P\{A\}. \tag{1.18}$$

P.2 Probability of event A is:

$$0 \leq P\{A\} \leq 1. \tag{1.19}$$

Proof From Axiom I the minimum probability of event A is equal to 0, this corresponds to the maximum value of the probability of its complement, which is equal to 1 (see (1.18)). Similarly, from Axiom I, the minimum value of the probability of the complement of A is equal to 0. In this case, from (1.18) it follows that the maximum value of $P\{A\}$ is equal to 1.

P.3 If B is a subset of event A (Fig. 1.5), $B \subset A$, then,

$$P\{B\} \leq P\{A\}. \tag{1.20}$$

Proof From Fig. 1.5, we found:

$$\begin{aligned} A &= B \cup \overline{B}, \\ P\{A\} &= P\{B\} + P\{\overline{B}\}, \\ P\{B\} &= P\{A\} - P\{\overline{B}\}. \end{aligned} \tag{1.21}$$

Given that $P\{\overline{B}\} \geq 0$, thus (1.20) follows.

P.4 Probability of null events (empty events) is zero

$$P\{0\} = 0. \tag{1.22}$$

Proof This property follows the property P.1 and Axiom II, recognizing that the null event is the complement of the certain event S and thus:

$$P\{0\} = 1 - P\{S\} = 1 - 1 = 0. \tag{1.23}$$

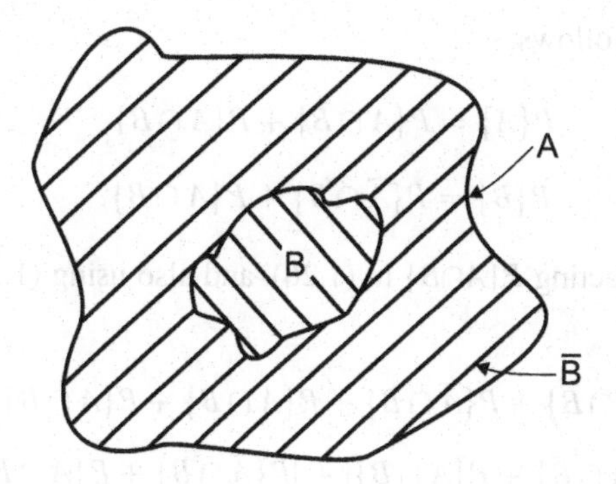

Fig. 1.5 Illustration of the subset of A

P.5 If events A and B are not mutually exclusive, Fig. 1.6 (i.e., their intersection is not an empty set) then:

$$P\{A \cup B\} = P\{A\} + P\{B\} - P\{A \cap B\}. \tag{1.24}$$

Proof To derive (1.24) we must first note that $A \cup B$ can be written as the union of three disjoint events (Fig. 1.6):

$A \cap \overline{B}$ (The points in A which are not in B),
$\overline{A} \cap B$ (The points in B which are not in A),
$A \cap B$ (The points in both, A and B).

$$A \cup B = (A \cap \overline{B}) \cup (\overline{A} \cap B) \cup (A \cap B). \tag{1.25}$$

Thus, from Axiom III, and (1.25) we have:

$$P\{A \cup B\} = P\{A \cap \overline{B}\} + P\{\overline{A} \cap B\} + P\{A \cap B\}. \tag{1.26}$$

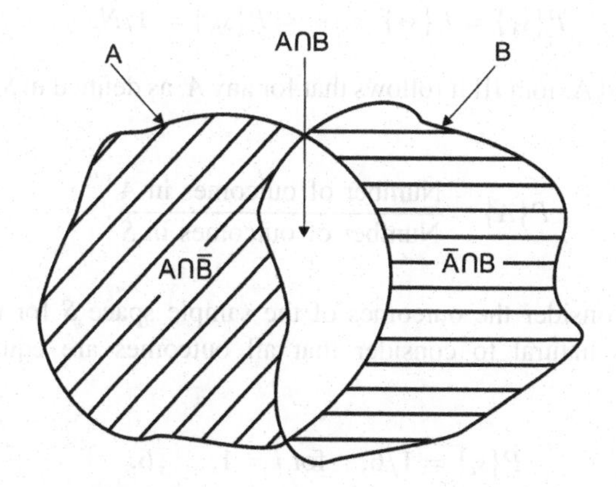

Fig. 1.6 Events A and B are not mutually exclusive

From Fig. 1.6, it follows:

$$P\{A\} = P\{A \cap \overline{B}\} + P\{A \cap B\}, \tag{1.27}$$

$$P\{B\} = P\{\overline{A} \cap B\} + P\{A \cap B\}. \tag{1.28}$$

Adding and subtracting $P\{A \cap B\}$ in (1.26)–and also using (1.27) and (1.28)–we arrive at:

$$\begin{aligned} P\{A \cup B\} &= P\{A \cap \overline{B}\} + P\{\overline{A} \cap B\} + P\{A \cap B\} + P\{A \cap B\} - P\{A \cap B\} \\ &= [P\{A \cap \overline{B}\} + P\{A \cap B\}] + [P\{\overline{A} \cap B\} + P\{A \cap B\}] - P\{A \cap B\} \\ &= P\{A\} + P\{B\} - P\{A \cap B\}. \end{aligned} \tag{1.29}$$

1.3 Equally Likely Outcomes in the Sample Space

Consider a random experiment whose sample space S is given as:

$$S = \{s_1, s_2, \ldots, s_N\}. \tag{1.30}$$

In many practical cases, it is natural to suppose that all outcomes are equally likely to occur:

$$P\{s_1\} = P\{s_2\} = \cdots = P\{s_N\}. \tag{1.31}$$

From Axioms II and III, we have:

$$P\{S\} = 1 = P\{s_1\} + P\{s_2\} + \cdots + P\{s_N\}, \tag{1.32}$$

which implies, taking (1.31) into account:

$$P\{s_1\} = P\{s_2\} = \cdots = P\{s_N\} = 1/N. \tag{1.33}$$

From (1.33) and Axiom III it follows that for any A, as defined in S, the following will be true:

$$P\{A\} = \frac{\text{Number of outcomes in } A}{\text{Number of outcomes in } S}. \tag{1.34}$$

Example 1.3.1 Consider the outcomes of the sample space S for the die rolling experiment. It is natural to consider that all outcomes are equally probable, resulting in:

$$P\{s_i\} = 1/6, \quad \text{for } i = 1, \ldots, 6. \tag{1.35}$$

If the event A is defined as in (1.5), then event A has three elements resulting in:

$$P\{A\} = 3/6 = 1/2. \tag{1.36}$$

1.4 Relative Frequency Definition of Probability

A random experiment must be, in principle, repeatable an arbitrary number of times under the same conditions [HEL91, p. 17]. Consider that the experiment has been repeated N times and the number of times that event A occurred in N trials is denoted as n_A. Define the *relative frequency* $q(A)$ as:

$$q(A) = \frac{n_A}{N}. \tag{1.37}$$

This property, called *statistical regularity*, implies that if the number of trials is increased the relative frequency varies less and less and approaches the probability of event A,

$$q(A) = \frac{n_A}{N} \to P\{A\}, \quad \text{as} \quad N \to \infty. \tag{1.38}$$

It should be noted that observations of the statistical stability of relative frequency (statistical regularity) of many phenomena served as the starting point for developing a mathematical theory of probability [GNE82, p. 45].

The probability defined in (1.38), obeys Axioms I–III described in Sect. 1.2.1, as shown in the following.

Axiom I From (1.38), it is obvious that the numbers n_A and N are always positive and thus the probability $P\{A\}$ cannot be negative.

Axiom II If in all trials the event A occurs, then $n_A = N$, and A is a certain event, and

$$P\{A\} = N/N = 1. \tag{1.39}$$

Axiom III If two events A and B are mutually exclusive $(A \cap B) = 0$, then

$$n_{A+B} = n_A + n_B, \tag{1.40}$$

and the probability is equal to:

$$P\{A \cup B\} = P\{A + B\} = n_{A+B}/N = n_A/N + n_B/N = P\{A\} + P\{B\}. \tag{1.41}$$

1.5 Conditional Probability

Often the occurrence of one event may be dependent upon the occurrence of another. If we have two events A and B, we may wish to know the probability of A if we already know that B has occurred. Given an event B with nonzero probability,

$$P\{B\} > 0, \tag{1.42}$$

we define such probability as conditional probability $P\{A|B\}$,

$$P\{A|B\} = \frac{P\{A \cap B\}}{P\{B\}} = \frac{P\{AB\}}{P\{B\}}, \tag{1.43}$$

where $P\{AB\}$ is the probability of the joint occurrence of A and B. It is called a *joint probability*.

Equation (1.43) can be interpreted, depending on the relations between A and B, as described in the following.

- The event A occurs, if and only if, the outcome s_i is in the intersection of events A and B (i.e., in $A \cap B$) as shown in Fig. 1.7a.
- If we have $A = B$ (Fig. 1.7b), then, given that event B has occurred, event A is a certain event with the probability 1. The same is confirmed from (1.43), making

$$P\{AB\} = P\{B\}. \tag{1.44}$$

- The other extreme case is when events A and B do not have elements in common (Fig. 1.7c), resulting in the conditional probability $P\{A|B\}$ being equal to zero.

The conditional probability, since it is still probability, must satisfy the three axioms (1.11)–(1.13).

Axiom I The first axiom is obviously satisfied, because $P\{B\}$ is positive, and the joint probability

$$P\{AB\} = P\{A \cap B\} = P\{A\} + P\{B\} - P\{A \cup B\} \tag{1.45}$$

is also positive because the following relation holds:

$$P\{A + B\} = P\{A \cup B\} = P\{A\} + P\{B\} - P\{A \cap B\} \leq P\{A\} + P\{B\}. \tag{1.46}$$

Axiom II The second axiom is satisfied because

$$P\{S|B\} = \frac{P\{SB\}}{P\{B\}} = \frac{P\{B\}}{P\{B\}} = 1. \tag{1.47}$$

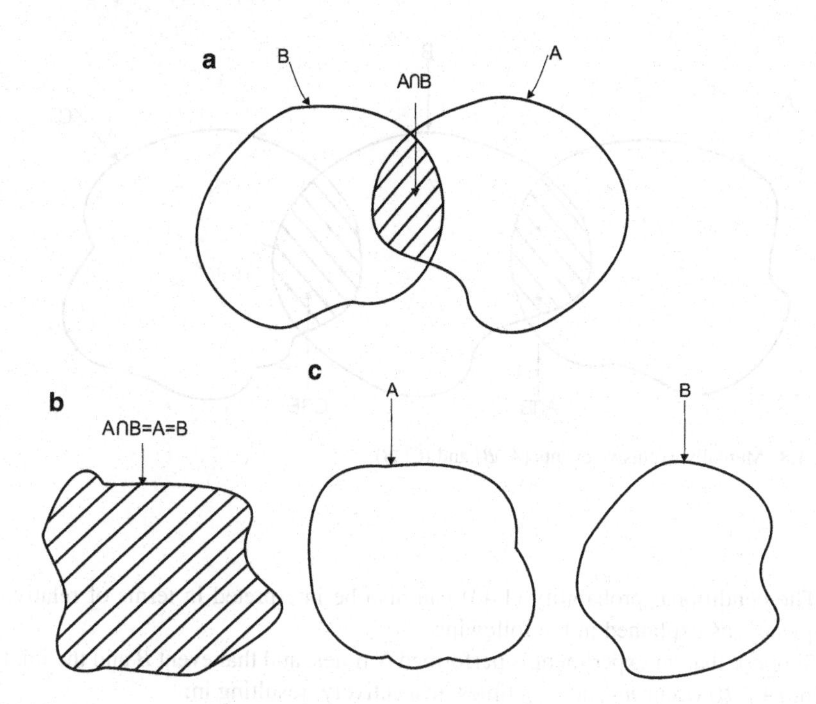

Fig. 1.7 Events A and B

Axiom III To prove Axiom III, consider the union of event A with a mutually exclusive event C in the conditional probability,

$$\begin{aligned}P\{A\cup C|B\} &= \frac{P\{A\cup C, B\}}{P\{B\}} = \frac{P\{A\cap B, C\cap B\}}{P\{B\}} = \frac{P\{(A\cap B)\cup(C\cap B)\}}{P\{B\}}\\ &= \frac{P\{A\cap B\}+P\{C\cap B\}-P\{(A\cap B)\cap(C\cap B)\}}{P\{B\}}.\end{aligned} \tag{1.48}$$

Since A and C are mutually exclusive, then events ($A\cap B$) and ($C\cap B$) are also mutually exclusive, as shown in Fig. 1.8, resulting in:

$$P\{(A\cap B)\cap(C\cap B)\} = 0. \tag{1.49}$$

Placing (1.49) into (1.48) and using the definition of conditional probability (1.43), we easily show that the Axiom III holds:

$$\begin{aligned}P\{A\cup C|B\} &= \frac{P\{A\cap B\}+P\{C\cap B\}}{P\{B\}} = \frac{P\{A\cap B\}}{P\{B\}} + \frac{P\{C\cap B\}}{P\{B\}}\\ &= P\{A|B\} + P\{C|B\}\end{aligned} \tag{1.50}$$

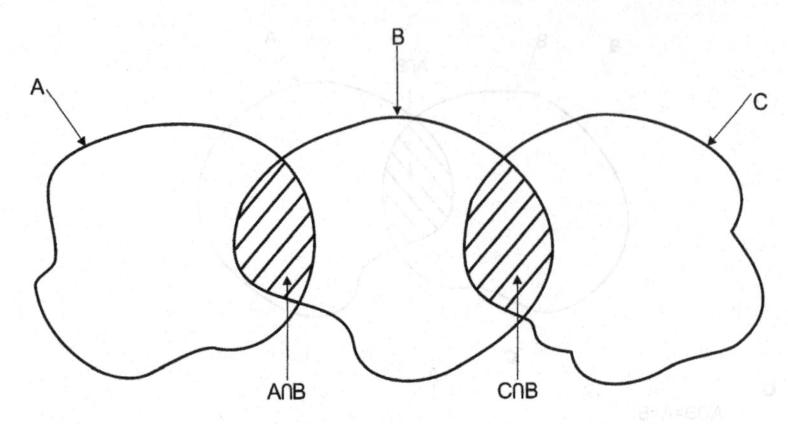

Fig. 1.8 Mutually exclusive events ($A \cap B$) and ($C \cap B$)

The conditional probability (1.43) can also be interpreted in terms of relative frequency, as explained in the following.

Suppose that an experiment is performed N times, and that event B and the joint event $(A \cap B)$ occur n_B and $n_{A,B}$ times, respectively, resulting in:

$$P\{A|B\} = \frac{P\{A \cap B\}}{P\{B\}} = \frac{P\{AB\}}{P\{B\}} = \frac{n_{A,B}/N}{n_B/N} = \frac{n_{A,B}}{n_B}. \tag{1.51}$$

Example 1.5.1 Two dice are rolled. What is the probability that the sum of both dice dots is 6, if it is already known that the sum will be an even number?

Solution Denote the following events:

$$\begin{aligned} A &= (\text{``The sum of both dice dots is 6''}), \\ B &= (\text{``An even sum of dots is rolled''}). \end{aligned} \tag{1.52}$$

All possible values of rolling two dice are shown in Table 1.1 and they form the space S; the first and second numbers in parentheses present the values of dots for the first and second dice, respectively, in each roll.

The sample space S has 36 outcomes:

$$S = \{s_1, \ldots, s_{36}\}. \tag{1.53}$$

The outcomes of event A are shown in bold and marked with an asterisk in Table 1.1. There are five total outcomes in A:

$$A = \{s_5, s_{10}, s_{15}, s_{20}, s_{25}\}. \tag{1.54}$$

Table 1.1 Outcomes of the space S in Example 1.5.1

$s_1 = (1,1)^*$	$s_7 = (2,1)$	$s_{13} = (3,1)^*$	$s_{19} = (4,1)$	$\mathbf{s_{25} = (5,1)^*}$	$s_{31} = (6,1)$
$s_2 = (1,2)$	$s_8 = (2,2)^*$	$s_{14} = (3,2)$	$\mathbf{s_{20} = (4,2)^*}$	$s_{26} = (5,2)$	$s_{32} = (6,2)^*$
$s_3 = (1,3)^*$	$s_9 = (2,3)$	$\mathbf{s_{15} = (3,3)^*}$	$s_{21} = (4,3)$	$s_{27} = (5,3)^*$	$s_{33} = (6,3)$
$s_4 = (1,4)$	$\mathbf{s_{10} = (2,4)^*}$	$s_{16} = (3,4)$	$s_{22} = (4,4)^*$	$s_{28} = (5,4)$	$s_{34} = (6,4)^*$
$\mathbf{s_5 = (1,5)^*}$	$s_{11} = (2,5)$	$s_{17} = (3,5)^*$	$s_{23} = (4,5)$	$s_{29} = (5,5)^*$	$s_{35} = (6,5)$
$s_6 = (1,6)$	$s_{12} = (2,6)^*$	$s_{18} = (3,6)$	$s_{24} = (4,6)^*$	$s_{30} = (5,6)$	$s_{36} = (6,6)^*$

From (1.34) the probability of A is given as:

$$P\{A\} = n_A/N = 5/36. \tag{1.55}$$

The outcomes of the event B are denoted with an asterisk in Table 1.1. The total number of outcomes n_B is 18:

$$B = \{s_1, s_3, s_5, s_8, s_{10}, s_{12}, s_{13}, s_{15}, s_{17}, s_{20}, s_{22}, s_{24}, s_{25}, s_{27}, s_{29}, s_{32}, s_{34}, s_{36}\}. \tag{1.56}$$

Using (1.51), we find the desired conditional probability:

$$P\{A|B\} = n_{A,B}/n_B = 5/18. \tag{1.57}$$

1.6 Total Probability and Bayes' Rule

1.6.1 Total Probability

Consider two events, A and B, which are not mutually exclusive, as shown in Fig. 1.9.

We wish to obtain information regarding the occurrence of A by exploring two cases:

- B would have occurred (event $(A \cap B)$).
- B would not have occurred (event $(A \cap \overline{B})$).

The events $(A \cap B)$ and $(A \cap \overline{B})$ are mutually exclusive, resulting in:

$$P\{A\} = P\{A \cap \overline{B}\} + P\{A \cap B\}. \tag{1.58}$$

Using (1.43) we can express (1.58) in terms of conditional probabilities:

$$P\{A\} = P\{A|B\}P\{B\} + P\{A|\overline{B}\}P\{\overline{B}\}. \tag{1.59}$$

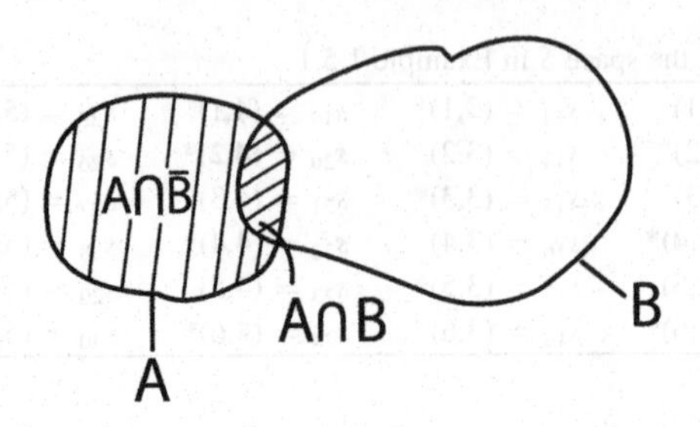

Fig. 1.9 Illustration of total probability

This result is known as *total probability* because it finds the probability of A in terms of conditional probabilities given B and given the complement of B.

Example 1.6.1 A device can operate under two kinds of conditions in time t:

$$\begin{aligned} &\text{Regular, (event } A) \\ &\text{Nonregular, (event } \bar{A}). \end{aligned} \tag{1.60}$$

The corresponding probabilities are:

$$P\{A\} = 0.85, \quad P\{\overline{A}\} = 1 - P\{A\} = 0.15. \tag{1.61}$$

The probability that the device will fail under regular conditions in certain time t, is 0.1, in nonregular conditions it may fail with a probability of 0.7.

Find the total probability $P\{F\}$ that the device will fail during the time t.

Solution The conditional probability that the device will fail, given A, is:

$$P\{F|A\} = 0.1. \tag{1.62}$$

Similarly, the conditional probability that the device will fail under nonregular conditions (given $\overline{A}$) is:

$$P\{F|\overline{A}\} = 0.7. \tag{1.63}$$

Using the formula for total probability (1.59), we find:

$$P\{F\} = P\{F|A\}P\{A\} + P\{F|\overline{A}\}P\{\overline{A}\} = 0.1 \times 0.85 + 0.7 \times 0.15 = 0.19. \tag{1.64}$$

The results (1.58) and (1.59) can be generalized as discussed in the following.

Theorem of Total Probability *Let $B_1, \ldots, B_N$ be a set of mutually exclusive events (i.e., $B_i \cap B_j = 0$, for all $i \neq j$), and event A is the union of N mutually exclusive events ($A \cap B_i$), $i = 1, \ldots, N$.*

Then,

$$P\{A\} = \sum_{i=1}^{N} P\{A|B_i\}P\{B_i\}. \tag{1.65}$$

Proof Event A can be written as:

$$A = (A\cap B_1)\cup(A\cap B_2)\cup(A\cap B_3)\cup(A\cap B_4)\cup(A\cap B_5) = \bigcup_{i=1}^{N}(A\cap B_i). \tag{1.66}$$

From here, using (1.13) and (1.43) we arrive at:

$$P\{A\} = P\left\{\bigcup_{i=1}^{N}(A\cap B_i)\right\} = \sum_{i=1}^{N} P\{A\cap B_i\} = \sum_{i=1}^{N} P\{A|B_i)P(B_i). \tag{1.67}$$

Example 1.6.2 The theorem of total probability is illustrated in Fig. 1.10 for $N = 5$. Events $B_1, \ldots, B_5$ are mutually exclusive. Event A which is the union of N mutually exclusive events $A\cap B_i$, $i = 1, \ldots, 5$, is presented by a shaded area in the figure.

$$A = \{A\cap B_1\}\cup\{A\cap B_2\}\cup\{A\cap B_3\}\cup\{A\cap B_4\}\cup\{A\cap B_5\}. \tag{1.68}$$

Note than the events ($A\cap B_i$) are mutually exclusive.

$$P\{A\} = \sum_{i=1}^{5} P\{A\cap B_i\} = \sum_{i=1}^{5} P\{AB_i\} = \sum_{i=1}^{5} P\{A|B_i\}P\{B_i\}. \tag{1.69}$$

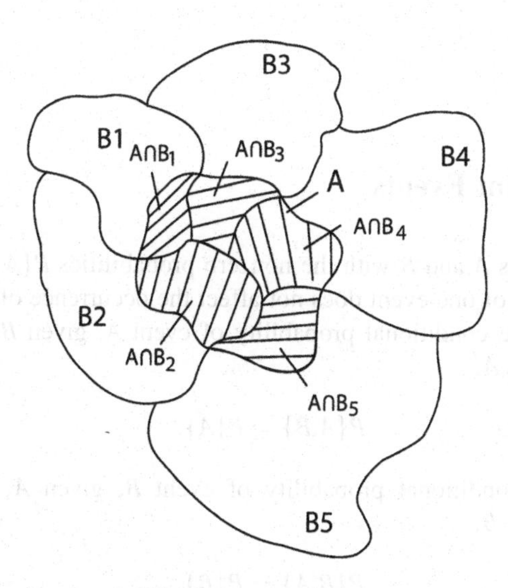

Fig. 1.10 Illustration of total probability for $i = 1, \ldots, 5$

1.6.2 Bayes' Rule

Let the space S be subdivided into N mutually exclusive events B_j, $j = 1, \ldots, N$. The probabilities of events B_j are called *a priori probabilities* because they represent the probabilities of B_j before the experiment is performed. Similarly, the probabilities $P\{A|B_j\}$ are typically known prior to conducting the experiment, and are called *transition probabilities.*

Now, suppose that the experiment is performed and, as a result, event A occurred. The probability of the occurrence of any of the events B_j—knowing that the event A has occurred–is called *a posteriori probability* $P\{B_j|A\}$. The a posteriori probability is calculated using conditional probability (1.43) and the theorem of total probability (1.65) resulting in:

$$P\{B_j|A\} = \frac{P\{AB_j\}}{P\{A\}} = \frac{P\{A|B_j\}P\{B_j\}}{\sum_{i=1}^{N} P\{A|B_i\}P\{B_i\}}. \tag{1.70a}$$

The formula in (1.70a) is known as *Bayes' rule.*

Because (1.70a) is a complicated formula, some authors like Haddad [HAD06, p. 32] consider it is better to write Bayes' rule using the following two equations:

$$P\{B_j|A\} = \frac{P\{AB_j\}}{P\{A\}} = \frac{P\{A|B_j\}P\{B_j\}}{P\{A\}}$$

$$P\{A\} = \sum_{i=1}^{N} P\{A|B_i\}P\{B_i\}. \tag{1.70b}$$

1.7 Independent Events

Consider two events A and B with the nonzero probabilities $P\{A\}$ and $P\{B\}$, and that the occurrence of one event does not affect the occurrence of the other event. That means that the conditional probability of event A, given B, is equal to the probability of event A,

$$P\{A|B\} = P\{A\}. \tag{1.71}$$

Similarly, the conditional probability of event B, given A, is equal to the probability of event B,

$$P\{B|A\} = P\{B\}. \tag{1.72}$$

If (1.71) and (1.72) are satisfied, we say that the events are *statistically independent* (i.e., the occurrence of the one event does not affect the occurrence of the other event).

From (1.67), we write:

$$P\{A|B\} = \frac{P\{AB\}}{P\{B\}} = P\{A\}, \tag{1.73}$$

$$P\{B|A\} = \frac{P\{AB\}}{P\{A\}} = P\{B\}. \tag{1.74}$$

From (1.73) and (1.74), it follows that the joint probability of two statistically independent events is equal to the product of the two corresponding probabilities:

$$P\{AB\} = P\{A\}P\{B\}. \tag{1.75}$$

This result can be generalized to N mutually independent events $A_j, j = 1, \ldots, N$. If they are independent, then any one of them is independent of any event formed by unions, intersections, or complements of the others [PAP65, p. 42].

As a consequence, it is necessary that for all combinations i, j, k, $1 \leq i \leq j \leq k \leq, \ldots, \leq N$,

$$\begin{aligned} P\{A_iA_j\} &= P\{A_i\}P\{A_j\} \\ P\{A_iA_jA_k\} &= P\{A_i\}P\{A_j\}P\{A_k\} \\ P\{A_1 \ldots A_N\} &= \prod_{i=1}^{N} P\{A_i\} \end{aligned} \tag{1.76}$$

There are $2^N - N - 1$ of these conditions. [PEE93, p. 21].

Example 1.7.1 The system shown in Fig. 1.11 has two units in serial connection such that both units must operate in order for the system to operate. The probabilities of failure-free operations of units in time t, are p_1 and p_2. The failure of one unit does not depend on the failure of another unit. Find the probability that only the first unit failed if the system has failed.

Solution The space S is subdivided into $N = 4$ mutually exclusive events B_j, $j = 1, \ldots, 4$:

$$B_1 = \{\text{Both units operate}\}, \tag{1.77}$$

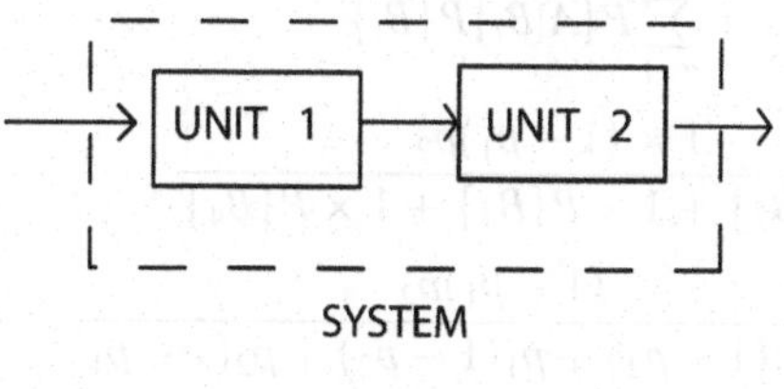

Fig. 1.11 Serial connection of units

$$B_2 = \{\text{Both units failed}\}, \tag{1.78}$$

$$B_3 = \{\text{First unit operates and second unit failed}\}, \tag{1.79}$$

$$B_4 = \{\text{First unit failed and second unit operates}\}. \tag{1.80}$$

The corresponding probabilities are a priori probabilities and using (1.73) we have:

$$P\{B_1\} = p_1 p_2, \tag{1.81}$$

$$P\{B_2\} = (1 - p_1)(1 - p_2), \tag{1.82}$$

$$P\{B_3\} = p_1(1 - p_2), \tag{1.83}$$

$$P\{B_4\} = (1 - p_1)p_2. \tag{1.84}$$

The event A is the occurrence of system failure.

$$A = \{\text{system failure}\}. \tag{1.85}$$

The corresponding transition probabilities are:

$$P\{A|B_1\} = 0. \tag{1.86}$$

(The system cannot fail if both units work properly).

$$P\{A|B_2\} = P\{A|B_3\} = P\{A|B_4\} = 1. \tag{1.87}$$

(As defined in the problem, the failure of the system occurs if either unit 1, unit 2, or both units, fail).

The desired a posteriori probability (the probability that if the system failed it was because the first unit failed, while second one worked properly), $P\{B_4|A\}$ is obtained using Bayes' formula (1.70a):

$$\begin{aligned} P\{B_4|A\} &= \frac{P\{A, B_4\}}{P\{A\}} = \frac{P\{A|B_4\}P\{B_4\}}{\sum_{i=1}^{4} P\{A|B_i\}P\{B_i\}} \\ &= \frac{1 \times (1 - p_1)p_2}{1 \times P\{B_2\} + 1 \times P\{B_3\} + 1 \times P\{B_4\}} \\ &= \frac{(1 - p_1)p_2}{(1 - p_1)(1 - p_2) + p_1(1 - p_2) + p_2(1 - p_1)} = \frac{(1 - p_1)p_2}{1 - p_1 p_2}. \end{aligned} \tag{1.88}$$

1.8 Numerical Exercises

Exercise E.1.1 The experiment consists of rolling two dice and observing the odd sum of the corresponding number of dots. Describe the sample space.

Answer All possible values of rolling two dice are shown in Table 1.2 and form the space S. The first and second numbers in parentheses represent the values of dots for the first and second dice, respectively.

The number of outcomes is 18, and the sample space is:

$$S = \{s_1, \ldots, s_{18}\}. \tag{1.89}$$

Exercise E.1.2 Let event A in the sample space S, from Exercise E.1.1., be: "the sum of dots is 5." Find the probability of event A assuming that all outcomes are equally likely.

Answer Event A is formed from the following sample points in the sample space (1.89): s_2, s_5, s_7, and s_{10}, denoted in bold in Table 1.2.

$$A = \{s_2, s_5, s_7, s_{10}\}. \tag{1.90}$$

From (1.34) the probability of event A is:

$$P\{A\} = 4/18 = 2/9. \tag{1.91}$$

Exercise E.1.3 Event B in the sampling space (1.89) denotes that the sum of the dots is less than 5. Define event B, and find its probability.

Answer From Table 1.2 we can see that the following outcomes, denoted with an asterisk, satisfy the desired property: s_1 and s_4. Thus, event B is:

$$B = \{s_1, s_4\}, \tag{1.92}$$

and according to (1.34) the probability of event B is:

$$P\{A\} = 2/18 = 1/9. \tag{1.93}$$

Table 1.2 Outcomes of the space S in Exercise E.1.3

$s_1 = (1,2)$*	$s_4 = (2,1)$*	$s_7 =$ **(3,2)**	$s_{10} =$ **(4,1)**	$s_{13} = (5,2)$	$s_{16} = (6,1)$
$s_2 =$ **(1,4)**	$s_5 =$ **(2,3)**	$s_8 = (3,4)$	$s_{11} = (4,3)$	$s_{14} = (5,4)$	$s_{17} = (6,3)$
$s_3 = (1,6)$	$s_6 = (2,5)$	$s_9 = (3,6)$	$s_{12} = (4,5)$	$s_{15} = (5,6)$	$s_{18} = (6,5)$

Exercise E.1.4 A coin is tossed three times. Find the sample space S. It is assumed that all outcomes are equally likely. Define event A–that there will be two "heads"–and find the probability of event A.

Answer All of the possible outcomes are:

$$s_1 = \{\mathrm{H,H,H}\}, s_2 = \{\mathrm{H,H,T}\}, s_3 = \{\mathrm{T,T,T}\}, s_4 = \{\mathrm{T,T,H}\}, s_5 = \{\mathrm{T,H,T}\},$$
$$s_6 = \{\mathrm{H,T,T}\}, s_7 = \{\mathrm{T,H,H}\}, s_8 = \{\mathrm{H,T,H}\}. \tag{1.94}$$

The sample space is:

$$S = \{s_1, \ldots, s_8\}. \tag{1.95}$$

From (1.94), event A is defined as:

$$A = \{s_2, s_7, s_8\}. \tag{1.96}$$

The total number of elements in S and A are 8 and 3, respectively. Therefore, using (1.34) the probability of the event A is given as:

$$P\{A\} = 3/8. \tag{1.97}$$

Exercise E.1.5 Three messages are sent and each of them can be received with a different degree of accuracy, described by the following events:

$$\begin{aligned} A &= \{\text{the received message is correct}\}, \\ B &= \{\text{the received message is partially correct}\}, \\ C &= \{\text{the received message is incorrect}\}. \end{aligned} \tag{1.98}$$

The events in (1.98) are independent and their probabilities are known:

$$\begin{aligned} P\{A\} &= p_1, \\ P\{B\} &= p_2, \\ P\{C\} &= p_3. \end{aligned} \tag{1.99}$$

(a) Find the probability of event D,
$D = \{\text{all three messages are received correctly}\}$.
(b) Find the probability of the event E,
$E = \{\text{at least one message is received incorrectly}\}$.

Answer
(a) From (1.76) the probability that all three messages are received correctly is:

$$P\{D\} = p_1{}^3. \tag{1.100}$$

(b) From (1.13) the probability that each message is received either correctly or partially correctly (mutually exclusive events) is equal to the sum of $p_1 + p_2$.

Consequently, the probability that all three messages are received either correctly or partially correctly is equal to $(p_1 + p_2)^3$. Note that this event is a complement of event E. Therefore, from (1.18) we arrive at:

$$P\{E\} = 1 - P\{\overline{E}\} = 1 - (p_1 + p_2)^3. \tag{1.101}$$

Exercise E.1.6 There are n_1 students in a group that usually obtain "excellent" grades, n_2 students that usually obtain "good" grades, and n_3 students that usually obtain "fair" grades. On a test, the students, that usually get "excellent" grades, have equal probabilities of getting "excellent" and "good" grades. Similarly, those, who usually get "good" grades, have equal probabilities of getting "excellent," "good," or "fair" marks, and those, who usually get "fair" grades, have equal probabilities of receiving "good" or "fair" grades.

A student is randomly selected. Find the probability of the event:

$$A = \{\text{student receives a ``good'' grade}\}. \tag{1.102}$$

Answer We define the following events,

$$\begin{aligned} A_1 &= \{\text{a student who usually gets ``excellent'' grades is selected}\},\\ A_2 &= \{\text{a student who usually gets ``good'' grades is selected}\},\\ A_3 &= \{\text{a student who usually gets ``fair'' grades is selected}\}. \end{aligned} \tag{1.103}$$

Knowing that the total number of students is,

$$n = n_1 + n_2 + n_3, \tag{1.104}$$

using (1.34) we find the probabilities of the events A_i, $i = 1, \ldots, 3$,

$$\begin{aligned} P\{A_1\} &= n_1/n.\\ P\{A_2\} &= n_2/n.\\ P\{A_3\} &= n_3/n. \end{aligned} \tag{1.105}$$

The corresponding conditional probabilities are:

$$P\{A|A_1\} = 1/2; \quad P\{A|A_2\} = 1/3; \quad P\{A|A_3\} = 1/2. \tag{1.106}$$

From (1.65), we obtain:

$$P\{A\} = P\{A_1\}P\{A|A_1\} + P\{A_2\}P\{A|A_2\} + P\{A_3\}P\{A|A_3\}$$
$$= \frac{0.5n_1 + \frac{1}{3}n_2 + 0.5n_3}{n}. \quad (1.107)$$

Exercise E.1.7 A given system has two units. In order for the system to be in the operational mode, each unit must be in the operational mode, as well. The failures of units are independent. The probabilities of failure – free operations for units 1 and 2 are known and denoted as p_1 and p_2, respectively.

Find the probability that the second unit fail, if it is known that system failed. (The first unit worked).

Answer We define the following events:

$$\begin{aligned} A_1 &= \{\text{both units work}\}, \\ A_2 &= \{\text{the first unit failed and the second unit works}\}, \\ A_3 &= \{\text{the first unit works and second unit failed}\}, \\ A_4 &= \{\text{ both units failed}\}. \end{aligned} \quad (1.108)$$

The corresponding probabilities are:

$$\begin{aligned} P\{A_1\} &= p_1p_2, \\ P\{A_2\} &= (1 - p_1)p_2, \\ P\{A_3\} &= p_1(1 - p_2), \\ P\{A_4\} &= (1 - p_1)(1 - p_2). \end{aligned} \quad (1.109)$$

The system failed if any or both units failed.
Defining event A as:

$$A = \{\text{the system failed}\}, \quad (1.110)$$

we write the corresponding conditional probabilities:

$$\begin{aligned} &P\{A|A_1\} = 0 \text{ (If both units work, the system cannot fail)} \\ &P\{A|A_2\} = P\{A|A_3\} = P\{A|A_4\} = 1 \text{ (The system fails if any, or both units fail).} \end{aligned} \quad (1.111)$$

Denote the event that the second unit fails, if we know that the system failed, as $(A_3|A)$.

The corresponding probability is obtained using Bayes's formula:

$$P\{A_3|A\} = \frac{P\{A|A_3\}P\{A_3\}}{P\{A\}} = \frac{p_1(1-p_2)}{P\{A\}}. \quad (1.112)$$

$$\begin{aligned} P\{A\} &= \sum_{i=1}^{4} P\{A|A_i\}P\{A_i\} = 0 + 1 \times P\{A_2\} + 1 \times P\{A_3\} + 1 \times P\{A_4\} \\ &= (1-p_1)p_2 + (1-p_2)p_1 + (1-p_1)(1-p_2). \end{aligned} \quad (1.113)$$

Placing (1.113) into (1.112), we obtain:

$$P\{A_3|A\} = \frac{p_1(1-p_2)}{(1-p_1)p_2 + (1-p_2)p_1 + (1-p_1)(1-p_2)}. \quad (1.114)$$

Exercise E.1.8 A binary communication channel transmits one of two possible symbols "1" or "0." Because of channel noise, "0" can be received as "1" and vice versa. The a priori probability that a particular symbol ("1" or "0") is sent is known:

$$P\{\text{``0''}\} = P\{\text{``1''}\} = 0.5. \quad (1.115)$$

The error probabilities are also known:

$$\begin{aligned} P\{\text{``0''}|\text{``1''}\} &= p(\text{The probability that ``0'' was received when ``1'' was sent}), \\ P\{\text{``1''}|\text{``0''}\} &= q(\text{The probability that ``1'' was received when ``0'' was sent}). \end{aligned} \quad (1.116)$$

(a) Find the probability that "0" is received.
(b) Find the probability that "1" is received.
(c) If "0" was received, what is the probability that "0" was sent?
(d) If "1" was received, what is the probability that "1" was sent?

Answer The sample space has two elements: s_1 = "0" and s_2 = "1."
We denote the event "the symbol at the channel input is s_i", using A_i, $i = 1, 2$:

$$\begin{aligned} A_1 &= \{\text{the symbol at the channel input is } s_1 = \text{``0''}\}, \\ A_2 &= \{\text{the symbol at the channel input is } s_2 = \text{``1''}\}. \end{aligned} \quad (1.117)$$

Similarly, we define the following events at the receiver:

$$\begin{aligned} B_1 &= \{\text{ the received symbol is } s_1 = \text{``0''}\}, \\ B_2 &= \{\text{ the received symbol is } s_2 = \text{``1''}\}. \end{aligned} \quad (1.118)$$

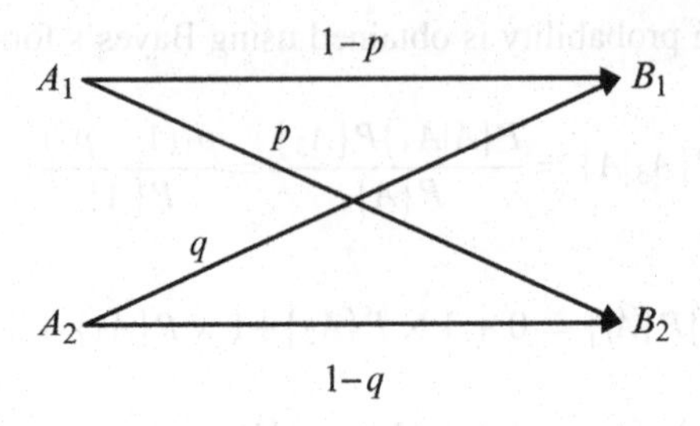

Fig. 1.12 Binary channel

The a priori probabilities are:

$$P\{A_1\} = P\{A_2\} = 0.5. \tag{1.119}$$

The transition probabilities (error probabilities) are (Fig. 1.12):

$$P\{B_1|A_2\} = q, \; P\{B_2|A_1\} = p. \tag{1.120}$$

From here we obtain:

$$P\{B_1|A_1\} = 1 - p, \; P\{B_2|A_2\} = 1 - q. \tag{1.121}$$

(a) A "0" is received, (event B_1), if either a "0" or a "1" was sent:

$$\begin{aligned} P\{B_1\} &= P\{B_1A_1\} + P\{B_1A_2\} = P\{B_1|A_1\}P\{A_1\} + P\{B_1|A_2\}P\{A_2\} \\ &= 0.5(1-p) + 0.5q. \end{aligned} \tag{1.122}$$

(b) A "1" is received, (event B_2), if either a "0" or a "1" was sent:

$$\begin{aligned} P\{B_2\} &= P\{B_2A_1\} + P\{B_2A_2\} = P\{B_2|A_1\}P\{A_1\} + P\{B_2|A_2\}P\{A_2\} \\ &= 0.5p + 0.5(1-q). \end{aligned} \tag{1.123}$$

(c) The probability $P\{A_1|B_1\}$ is obtained from Bayes' rule:

$$P\{A_1|B_1\} = \frac{P\{A_1B_1\}}{P\{B_1\}} = \frac{P\{B_1|A_1\}P\{A_1\}}{P\{B_1\}}. \tag{1.124}$$

Placing (1.121) and (1.122) into (1.124) we get:

$$P\{A_1|B_1\} = \frac{0.5(1-p)}{0.5(1-p) + 0.5q}. \tag{1.125}$$

(d) Similarly, we obtain:

$$P\{A_2|B_2\} = \frac{P\{A_2B_2\}}{P\{B_2\}} = \frac{P\{B_2|A_2\}P\{A_2\}}{P\{B_2\}} = \frac{0.5(1-q)}{0.5p + 0.5(1-q)}. \tag{1.126}$$

Exercise E.1.9 Find the probability of error P_e as well as the probability that the signal is transmitted correctly P_c, for the binary channel from Exercise E.1.8.

Answer An error occurs if either "0" or "1" is not transmitted correctly:

$$\begin{aligned} P_e &= P\{A_1B_2\} + P\{A_2B_1\} = P\{A_1\}P\{B_2|A_1\} + P\{A_2\}P\{B_1|A_2\} \\ &= 0.5p + 0.5q = 0.5(p+q). \end{aligned} \tag{1.127}$$

Similarly,

$$\begin{aligned} P_c &= P\{A_1B_1\} + P\{A_2B_2\} = P\{A_1\}P\{B_1|A_1\} + P\{A_2\}P\{B_2|A_2\} \\ &= 0.5(1-p) + 0.5(1-q) = 1 - 0.5(p+q). \end{aligned} \tag{1.128}$$

Note that

$$P_c = 1 - P_e. \tag{1.129}$$

1.9 Questions

Q.1.1. There are many situations in which we cannot determine exactly what will happen, but we can predict what will happen. What are some of the reasons for our inability to determine exactly what will happen?

Q.1.2. Is the sample space unique for a given experiment?

Q.1.3. Must the elements of a sample space always be mutually exclusive?

Q.1.4. Why do we need events if we already know the outcomes?

Q.1.5. An impossible event has zero probability. Is every event with zero probability impossible?

Q.1.6. How is a real experiment described mathematically?

Q.1.7. In probabilistic models, the probabilities are always considered to be known. However, in one experiment we may observe the outcomes but not the corresponding probabilities. What is the connection between the observed outcomes and assigned probabilities, and can the probability be measured exactly?

Q.1.8. What happens if probabilities are not assigned correctly?

Q.1.9. What is the shortcoming of the probability concept defined in (1.34) compared to the relative frequency concept?

Q.1.10. Are mutually exclusive events independent?

Q.1.11. If two events (A and B) are statistically independent, does it follows that A and the complement of B are also independent?

Q.1.12. For three events A_1, A_2, A_3 we know that the following is satisfied:

$$\begin{aligned} P\{A_1A_2\} &= P\{A_1\}P\{A_2\}. \\ P\{A_1A_3\} &= P\{A_1\}P\{A_3\}. \\ P\{A_2A_3\} &= P\{A_2\}P\{A_3\}. \end{aligned} \tag{1.130}$$

Is it correct to say that the events A_1, A_2, A_3 are mutually independent?

Q.1.13. Can the union of two events A and B be equal to their intersection?

Q.1.14. What is the usefulness of Bayes' rule?

1.10 Answers

A.1.1. The reasons may be the following [DAV87, p. 8]:

- We do not have enough data.
- We do not know all the causal forces at work.
- The forces are so complicated that the computation of an exact outcome will be so complicated and, thus, not useful.
- There is some basic indeterminacy in the physical word.

A.1.2. A sample space is not unique and the choice of a particular sample space depends on the desired result we would like to obtain by performing an experiment [HAD06, p. 21], [MIL04, p. 11]. For example, in rolling a die we may be interested only in the even number of dots. In that case, the outcomes are: 2, 4, and 6, and the sample space is $S = \{2, 4, 6\}$.

A.1.3. Elements in a sample space must always be mutually exclusive because the particular outcome in an experiment excludes the occurrence of another.

A.1.4. In practical experiments, the characteristics of outcomes may be of interest. Events are sets of outcomes that meet common characteristics.

A.1.5. Not every event with a probability of zero is impossible. For example, in the die rolling experiment we are interested only in an even number of dots. Thus, the sample space is $S = \{s_1, s_2, s_3\}$, where $s_1 = 2$, $s_2 = 4$, and $s_3 = 6$.

Let us now define the following events in the sample space S:

$$A = \{1 < s_i < 6\},$$
$$B = \{s_i < 2\},$$
$$C = \{s_i > 6\}. \tag{1.131}$$

Event A is: $A = \{s_1, s_2,\}$.

Event B is a null event, $B = \{\}$ (i.e., it does not contain any element in S). Therefore, its probability is zero, $P\{B\} = 0$. However, it is not an impossible event because in the die rolling experiment an outcome 1 (i.e., $s_i < 2$) is not an impossible outcome.

Conversely, event C is an impossible null event (an outcome greater than 6 cannot occur).

A.1.6. A practical experiment is described mathematically by the following steps [PEE93, p. 12]:

- Assignment of a sample space.
- Definition of events of interest.
- Assignment of probability to each event in such a way that the axioms of probability are satisfied.

A.1.7. "The answer is a strong intuitive belief that the probability of any outcome is roughly proportional to the number of times it comes up in a long independent repetition of the experiment. Therefore probability can never be measured exactly but only approximated" [BRE69, p. 3].

A.1.8. In that case, the mathematical model will fit poorly with the corresponding physical context.

A.1.9. All outcomes must be equally probable. Consequently, this concept depends on a priori analysis of the random experiment. However, the relative frequency concept is free of this shortcoming, since it depends on the observation of the experiment's results. However, the frequency approach requires that the experiment is repeated under the same conditions.

A.1.10. Since mutually exclusive events cannot occur together, then if one event occurs the other has zero probability of occurring. Therefore, if we know that one of them occurred, then we know that the other could not have occurred. As a consequence, mutually exclusive events are dependent [HAD06, p. 26].

A.1.11. From Fig. 1.6 we can observe that the events $A \cap \overline{B}$ and $A \cap B$ are mutually exclusive resulting in:

$$P\{A\} = P\{A \cap \overline{B}\} + P\{A \cap B\} = P\{A\overline{B}\} + P\{AB\}. \tag{1.132}$$

From here, we have:

$$\begin{aligned} P\{A\overline{B}\} &= P\{A\} - P\{AB\} = P\{A\} - P\{A\}P\{B\} \\ &= P\{A\}[1 - P\{B\}] = P\{A\}P\{\overline{B}\}. \end{aligned} \tag{1.133}$$

Therefore, if the events A and B are independent, then event A and the complement of event B are also independent.

A.1.12. The answer is NO, because it is necessary that (1.76) is satisfied, i.e., any one of them is independent of any other event formed by unions, intersections, or complements. We can easily see that:

$$\begin{aligned} P\{A_1A_2\} &= P\{A_1\}P\{A_2\}, \\ P\{A_1A_3\} &= P\{A_1\}P\{A_3\}, \\ P\{A_2A_3\} &= P\{A_2\}P\{A_3\}. \end{aligned} \tag{1.134}$$

do not ensure, for example, that events $\{A_1A_2\}$ and $\{A_3\}$ are independent. The former is ensured by adding

$$P\{A_1A_2A_3\} = P\{A_1\}P\{A_2\}P\{A_3\} = P\{A_1, A_2\}P\{A_3\} \\ = P\{(A_1, A_2), A_3\}. \tag{1.135}$$

A.1.13. It can, if the events are equivalent, $A = B$. Then,

$$(A + B) = (A \cup B) = A = B, \quad (AB) = (A \cap B) = A = B. \tag{1.136}$$

For example, [WEN86, p. 33], if, in a communication system, a message can only be distorted by a noise, occupying the same time interval, then,

$$A = \{\text{``the message is distorted''}\}. \tag{1.137}$$

$$B = \{\text{``there is a noise occupying the same time interval as the message''}\}, \tag{1.138}$$

are equivalent and (1.136) holds.

A.1.14. Bayes's rule is useful for calculating conditional probabilities $P\{A|B\}$, since in many problems it may be difficult to directly calculate $P\{A|B\}$. However, calculating $P\{B|A\}$ may be straightforward [MIL04, p. 23]. Additionally, in some cases $P\{B|A\}$ may be known as a characteristic of the system.

Chapter 2
Random Variable

2.1 What Is a Random Variable?

2.1.1 *Introduction*

The concept of random variable is one of the most important in probabilistic theory. In order to introduce this concept, we will relate random variables to the random experiment described in Chap. 1. As mentioned there, the outcomes s_i of a random experiment are random (i.e., it is not possible to predict them).

However, from a mathematical point of view, it would be useful to assign a certain nonrandom value to a random outcome. This idea is developed below.

To make this abstract idea more clear, let us consider two of the most exploited simple random experiments: rolling dice and coin tossing.

In a die rolling experiment, the outcomes are numbers of dots (1–6). In a coin tossing experiment, the outcomes are "heads" and "tails". In both cases, the outcomes are word-type outcomes. We can easily note that such descriptions are not convenient from a mathematical point of view.

Generally, it is more convenient to assign numerical values to the outcomes of random experiments so that mathematical expressions can be obtained as a function of numerical values. This idea leads us to the concept of the *random variable*.

2.1.2 *Definition of a Random Variable*

A real random variable X is a function which assigns a real number $X(s_i)$ to every random outcome s_i, according to certain rules (e.g., a function which maps the space S to the real axis x). In this text, we will consider only real random variables and thus real attribute can be omitted leaving only random variables. We will also use the abbreviation for random variable, *r.v.*

G.J. Dolecek, *Random Signals and Processes Primer with MATLAB*,
DOI 10.1007/978-1-4614-2386-7_2, © Springer Science+Business Media New York 2013

Every point from the space S must map to only one point on the real axis. This means that it is not possible to map the particular outcome s_i of the space S to two or more points on a real axis.

However, it is possible to map different outcomes s_i to the same point at x line.

Therefore, the mapping can be *one-to-one*, in which every outcome s_i is mapped onto only one point or *many-to-one*, in which different outcomes are mapped onto the same point, as shown in Fig. 2.1.

The numerical values x on the real axis are the *range* of the random variable. We will denote the random variable with capital letters, such as X, Y, Z, and the corresponding range with lowercase letters, such as x, y, and z. If the range of the random variable is discrete, then the random variable is *discrete*, as shown in Fig. 2.1. Otherwise it is a *continuous random variable* (see Fig. 2.2). Let us recall that a discrete set is a countable set of points that can also be infinite. A continuous set, on the other hand, is always uncountable and infinite.

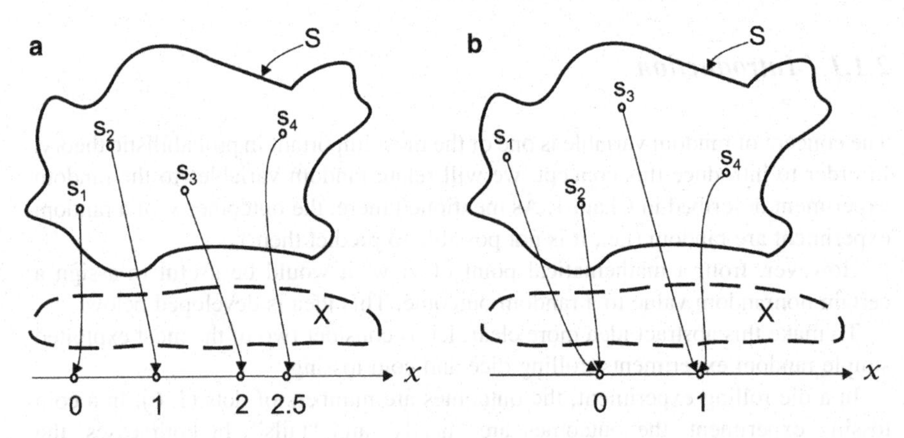

Fig. 2.1 Concept of a discrete random variable. (**a**) one-to-one mapping (**b**) many-to-one mapping

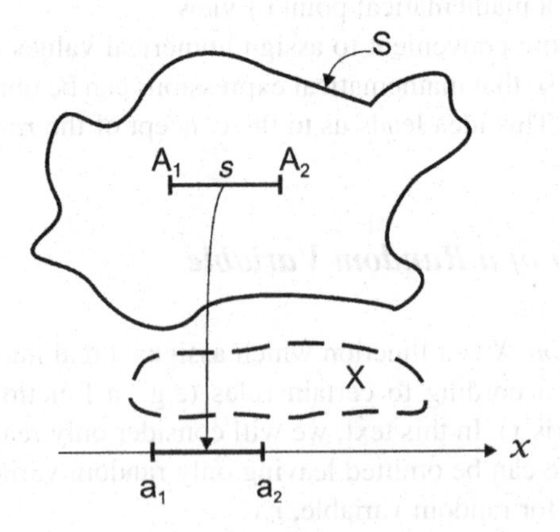

Fig. 2.2 Concept of a continuous random variable

In the case of discrete random variables, the discrete values can be integers or nonintegers, as shown in Fig. 2.1a. However, values between discrete points are not allowable.

If the range is continuous but also has some discrete points, the random variable is a *mixed random variable*.

This concept is illustrated in the following examples.

Example 2.1.1 We consider tossing a coin where the outcomes s_1 and s_2 denote "heads" and "tails", respectively. The space S is:

$$S = \{s_1, s_2\}. \tag{2.1}$$

The random variable X can be defined using the following rule (Fig. 2.3a):

$$\begin{aligned} X(s_1) &= 49, \\ X(s_2) &= 35. \end{aligned} \tag{2.2}$$

Note that this choice does not mean that the outcome s_1 is higher or more important than s_2. We can also denote (Fig. 2.3b):

$$\begin{aligned} X(s_2) &= 49, \\ X(s_1) &= 35. \end{aligned} \tag{2.3}$$

However, it is customary to assign the numerical values "0" and "1" where we have only two outcomes. It is irrelevant which outcome is assigned "0" or "1". Therefore, the customary mapping is shown in Fig. 2.3c, as:

$$\begin{matrix} X(s_1) = 1 & & X(s_1) = 0, \\ & \text{or} & \\ X(s_2) = 0 & & X(s_2) = 1. \end{matrix} \tag{2.4}$$

Note that the values of random variable X are numerically discrete values: $x_1 = 0$, $x_2 = 1$. Therefore, the random variable is discrete.

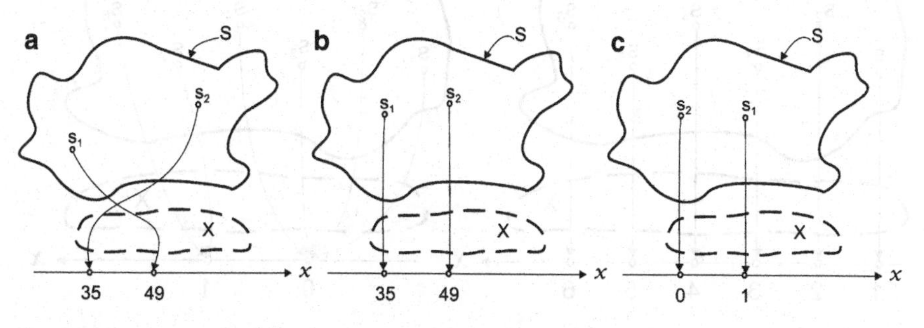

Fig. 2.3 The random variables for the coin tossing experiment

Example 2.1.2 Let us consider the die rolling experiment. Here, the random outcomes s_i, $i = 1, \ldots, 6$, are the number of dots (1, 2, 3, 4, 5, and 6). In this case, the most convenient approach is to map the number of dots to the corresponding numerical value as shown in Fig. 2.4a.

$$\begin{aligned} X(s_1) &= 1, \\ X(s_2) &= 2, \\ &\ldots \\ X(s_6) &= 6. \end{aligned} \tag{2.5}$$

This mapping is one-to-one. Note that outcomes s_i are not numerical values, but the number of dots.

Let us now define the random variable in a different way. The random variable $X(s_i)$ is defined depending on whether the number of dots is less than or equal to 4, or more than 4. Therefore, we have only two possibilities, and it is easiest to assign them numerical values of 0 and 1,

$$\begin{aligned} X(s_1, s_2, s_3, s_4) &= 0, \\ X(s_5, s_6) &= 1. \end{aligned} \tag{2.6}$$

Again, the random variable X has two discrete values: $x_1 = 0, x_2 = 1$. These are shown in Fig. 2.4b. This mapping is many-to-one, because four points in the sample space are mapped onto 0, and two points in the sample space are mapped onto 1.

Note that we can obtain the same random variables from different sample spaces (see Figs. 2.3c and 2.4b). We can also obtain different random variables from the same sample space, as shown in Fig. 2.4.

In the following example, we illustrate the concept of a continuous random variable.

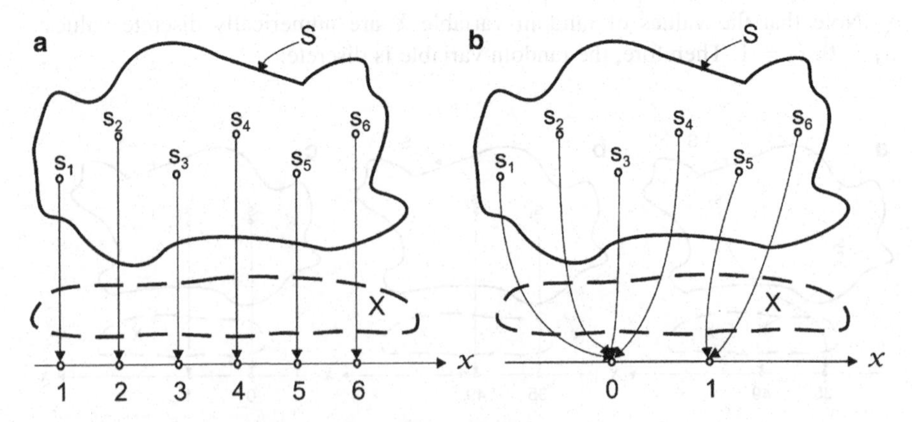

Fig. 2.4 The random variables for the die rolling experiment

Example 2.1.3 This experiment is a measurement of voltage and the outcomes are continuous values:

$$S = \{s \mid V_2 \leq s \leq V_1\}. \tag{2.7}$$

The random variable $X(s)$ is obtained by mapping the continuous space onto the continuous range on the x-axis, as shown in Fig. 2.5.

Note the following:

- The sample space for the continuous random variable must be continuous. Otherwise, some values in the continuous range x would not have the corresponding count-pairs in the sample space S.
- The continuous space does not always result in a continuous random variable. Generally, from the continuous space, we can obtain the continuous random variable, discrete random variable, and also mixed random variable as shown in Examples 2.1.4 and 2.1.5.

Example 2.1.4 Consider the continuous space S from Example 2.1.3. In this example, we are only interested in whether the voltage is positive or negative. That is we have only two possibilities and as mentioned before, it is easiest to assign the numerical values 0 and 1, as shown in Fig. 2.6. Therefore, we can define the random variable $X(s)$ in the following way,

$$\begin{aligned} X(s \geq 0) &= 0, \\ X(s < 0) &= 1. \end{aligned} \tag{2.8}$$

Note that the random variable is discrete and the sample space is continuous. Note also that as opposed to Example 2.1.3, where the numerical values in the x-axis correspond to the values of the voltage, in this example the numerical values 0 and 1 do not have anything to do with the physical values of the voltage in the sample space S.

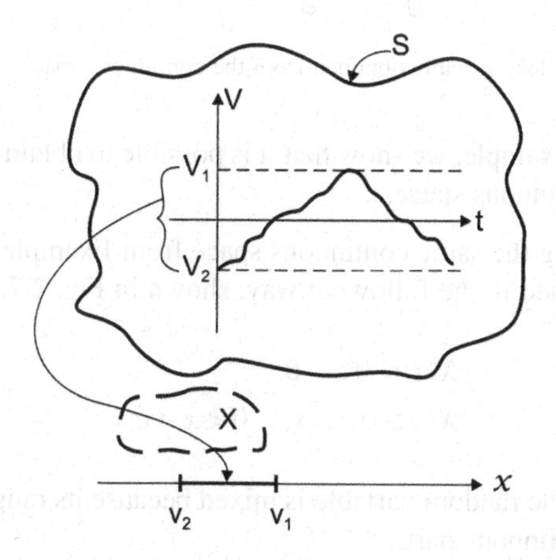

Fig. 2.5 The continuous random variable in Example 2.1.3

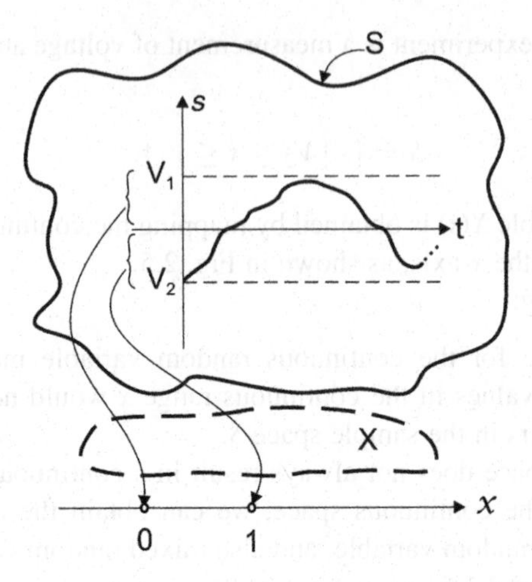

Fig. 2.6 The discrete random variable obtained from the continuous space

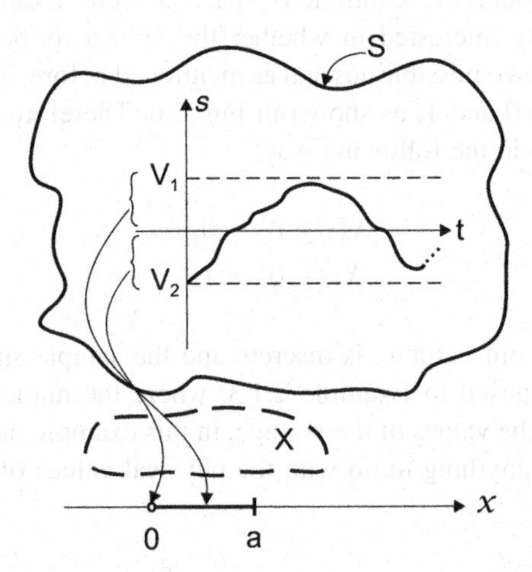

Fig. 2.7 The mixed random variable obtained from the continuous space

In the following example, we show that it is possible to obtain the mixed random variable from a continuous space.

Example 2.1.5 Using the same continuous space from Example 2.1.3, the random variable $X(s)$ is defined in the following way, shown in Fig. 2.7:

$$\begin{aligned} X(s \leq 0) &= 0, \\ X(s > 0) &= x, \quad 0 < x < a. \end{aligned} \tag{2.9}$$

In this example, the random variable is mixed because its range has one discrete point and also a continuous part.

2.2 Distribution Function

2.2.1 Definition

In order to completely describe a random variable, it is necessary to know not only all of the possible values it takes but also how often it takes these values. In other words, it is necessary to know the corresponding probabilities.

The probability that random variable X has values less than or equal to x is called *distribution function* or *distribution*,

$$F_X(x) = P\{X \leq x\}, \quad -\infty < x < \infty \tag{2.10}$$

Some authors also use the name *cumulative distribution function* (*CDF*).

This term includes both the discrete and continuous variables.

In the following examples, the calculation of the distribution functions is demonstrated. As defined in (2.10), the distribution is defined for all values of x. First, we divide all the x-axis into corresponding subintervals as defined by the corresponding values of the random variable. Next, we find the corresponding probability for each subinterval.

The following example illustrates the calculation of the distribution function for a discrete random variable.

Example 2.2.1 Consider a discrete random variable with only two values, 0 and 1 (Fig. 2.8a). The corresponding probabilities are:

$$\begin{aligned} P\{X = 1\} &= p = 3/4, \\ P\{X = 0\} &= q = 1/4. \end{aligned} \tag{2.11}$$

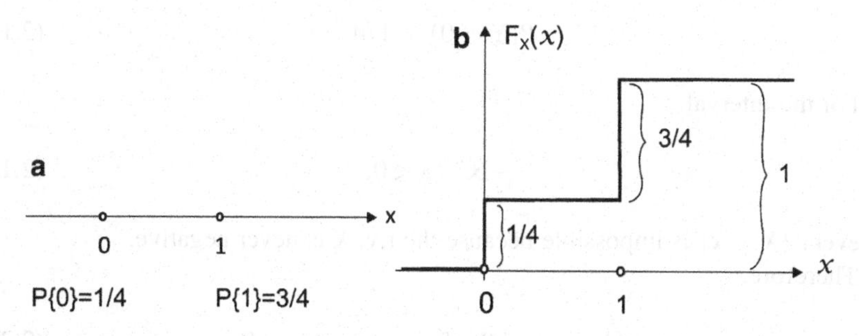

Fig. 2.8 Illustration of the distribution in Example 2.2.1

Solution Let us observe all values of x keeping in mind the following intervals:

(a) $1 < x < \infty$.
(b) $x = 1$.
(c) $0 < x < 1$.
(d) $x = 0$.
(e) $-\infty < x < 0$.

(a) For the values of X in the interval:

$$1 < x < \infty, \tag{2.12}$$

the event

$$\{X \leq x\} \tag{2.13}$$

is a certain event because the values of the random variable are 0 and 1, and thus always less than x–defined in (2.12)–yielding:

$$P\{X \leq x\} = 1 \quad \text{for} \quad 1 < x < \infty. \tag{2.14}$$

(b) For $x = 1$ (from (2.11)),

$$P(X = 1) = p = 3/4. \tag{2.15}$$

(c) For the values of x in the interval

$$0 < x < 1, \tag{2.16}$$

the probability of event $\{X \leq x\}$ corresponds to the probability that the r.v. takes the value 0,

$$P\{X \leq x\} = P\{X = 0\} = 1/4 \quad \text{for} \quad 0 < x < 1. \tag{2.17}$$

(d) For $x = 0$ (from (2.11)),

$$P(X = 0) = 1/4. \tag{2.18}$$

(e) For the interval

$$-\infty < x < 0, \tag{2.19}$$

the event $\{X \leq x\}$ is impossible because the r.v. X is never negative.
Therefore,

$$P\{X \leq x\} = 0 \quad \text{for} \quad -\infty \leq x < 0. \tag{2.20}$$

From (2.13) to (2.20), we arrive at the following expression for the distribution function:

$$F_X(x) = \begin{cases} 1 & \text{for} \quad x \geq 1 \\ 1/4 & \text{for} \quad 0 \leq x < 1 \\ 0 & \text{for} \quad x < 0 \end{cases} \tag{2.21}$$

This distribution is shown in Fig. 2.8b.
Note the following:

- The distribution function has a stair-step form.
- The distribution has jumps at the discrete values of the r.v. X (0 and 1). Each jump is equal to the probability that the r.v. takes the corresponding discrete value. Otherwise, it is constant.

Denoting the unit step function $u(x)$ as:

$$u(x) = \begin{cases} 1 & \text{for} \quad x \geq 0, \\ 0 & \text{for} \quad x < 0, \end{cases} \tag{2.22}$$

the distribution (2.21) can be expressed as,

$$F_X(x) = \sum_{k=1}^{2} P(x_k)u(x - x_k) = 1/4u(x) + 3/4u(x - 1). \tag{2.23}$$

In a general sense, the discrete random variable with the particular discrete values x_k, and the corresponding probabilities $P(x_k)$, has the distribution function,

$$F_X(x) = \sum_{k=-\infty}^{\infty} P(x_k)u(x - x_k). \tag{2.24}$$

The array of probabilities $P(x_k) = P\{X = x_k\}$, $k = 1, 2, 3, \ldots$, is called the *probability mass function*.

Example 2.2.2 Calculate the distribution function of the random variable which has only the discrete values 1, …, 6; assuming that all values have the same probability of occurrence,

$$P(x_k) = 1/6, \quad k = 1, \ldots, 6. \tag{2.25}$$

Solution From (2.24), we have:

$$F_X(x) = \sum_{k=1}^{6} P(x_k)u(x - x_k) = 1/6[u(x - 1) + \cdots + u(x - 6)]. \tag{2.26}$$

This distribution is plotted in Fig. 2.9.

In the following example, we can see the distribution of a continuous random variable.

Example 2.2.3 Consider the continuous random variable X defined in the continuous interval $[a, b]$, as shown in Fig. 2.10a for $a = 2$ and $b = 6$.

Solution For the interval of $6 \leq x < \infty$, we have:

$$P\{X \leq x\} = 1 \quad \text{for} \quad 6 \leq x < \infty. \tag{2.27}$$

For the interval $2 \leq x < 6$, the probability $P\{X \leq x\}$ depends on the ratio of the $(x - 2)$ and the total width $(6 - 2)$,

$$P\{X \leq x\} = \frac{x - 2}{4} \quad \text{for} \quad 2 \leq x < 6. \tag{2.28}$$

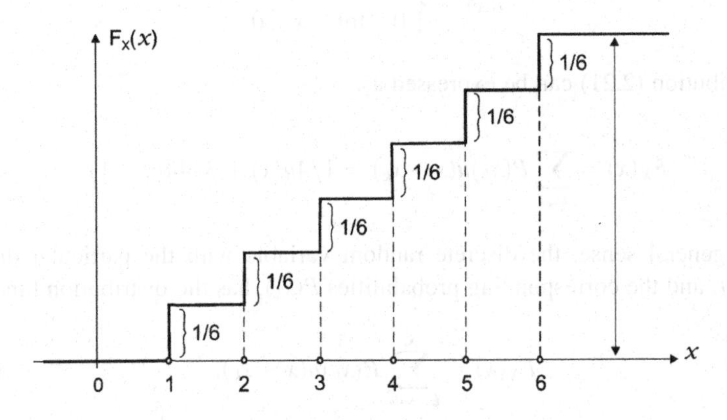

Fig. 2.9 Illustration of the distribution in Example 2.2.2

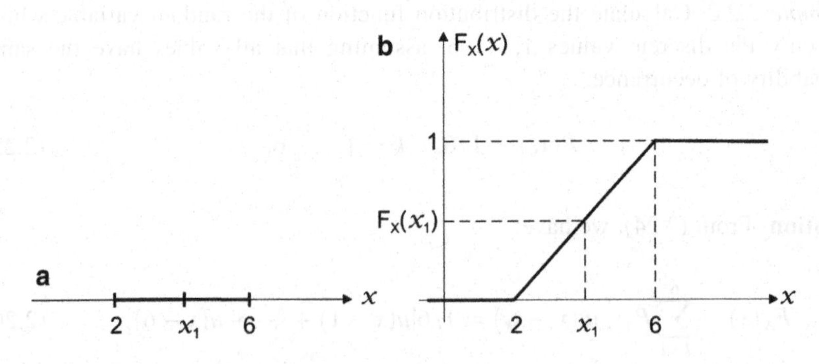

Fig. 2.10 Illustration of the distribution in Example 2.2.3

For the interval $-\infty < x < a$, we have:

$$P\{X \leq x\} = 0 \quad \text{for} \quad -\infty < x < a. \tag{2.29}$$

From (2.27) to (2.29), we arrive at:

$$F_X(x) = \begin{cases} 0 & \text{for} \quad -\infty < x < 2, \\ \dfrac{x-2}{4} & \text{for} \quad 2 \leq x \leq 6, \\ 1 & \text{for} \quad 6 < x < \infty. \end{cases} \tag{2.30}$$

This distribution is shown in Fig. 2.10b. Note that this distribution function does not have jumps like the distribution of a discrete variable (2.24).

The distribution of a mixed random variable has one or more jumps and the continuous part. To this end, it is possible to write the distribution of a mixed random variable [THO71, pp. 50–51] as:

$$F_X(x) = \alpha F_c(x) + F_d(x), \tag{2.31}$$

where F_c is a distribution of the continuous variable and F_d is a distribution of the discrete random variable, and

$$0 \leq \alpha \leq 1, \tag{2.32}$$

$$\alpha = 1 - \sum_i P\{X = x_i\}. \tag{2.33}$$

The proof can be found in [THO71, pp. 50–51].

A distribution where $\alpha = 0$ corresponds to a discrete r.v., while a distribution where $\alpha = 1$ corresponds to a continuous r.v.

Example 2.2.4 In this example, we will illustrate the distribution of a mixed random variable X; which has one discrete value in $x = 2$ with the probability $P\{2\} = 1/3$ and the continuous interval $2 < x < 6$ (as shown in Fig. 2.11a).

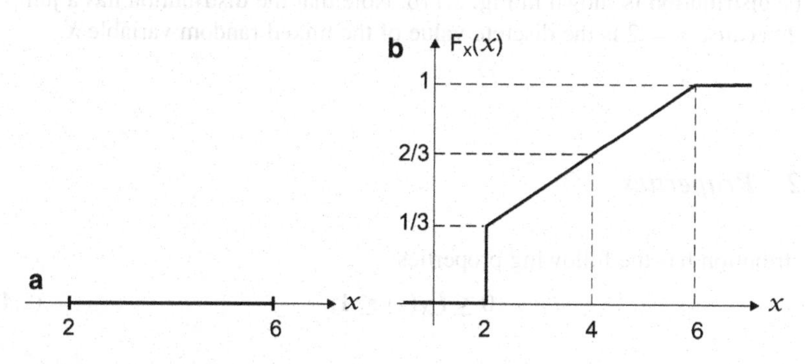

Fig. 2.11 Illustration of the distribution in Example 2.2.4

Solution From (2.32), we have:

$$\alpha = 1 - P\{2\} = 1 - 1/3 = 2/3. \tag{2.34}$$

Using (2.31) and (2.34), it follows that:

$$F_X(x) = \alpha F_c(x) + F_d(x) = \frac{2}{3}F_c(x) + F_d(x). \tag{2.35}$$

For the interval $6 < x < \infty$, the probability

$$P\{X \leq x\} = 1 \quad \text{for} \quad 6 < x < \infty. \tag{2.36}$$

For the interval $2 < x < 6$, the variable has a continuous range and the probability $P\{X \leq x\}$ depends on the ratio of $(x - 2)$ and the total width $(6 - 2) = 4$:

$$F_c(x) = \frac{2}{3}P\{X \leq x\} = \frac{2}{3}\frac{x-2}{4} \quad \text{for} \quad 2 < x < 6. \tag{2.37}$$

For $x = 2$, the variable has a discrete value

$$F_d(x) = P\{X = 2\} = 1/3. \tag{2.38}$$

For the interval $-\infty < x < 2$, the probability is:

$$P\{X \leq x\} = 0 \quad \text{for} \quad -\infty < x < a. \tag{2.39}$$

From (2.34) to (2.39), we arrive at:

$$F_X(x) = \begin{cases} 0 & \text{for} \quad x < 2, \\ \frac{2}{3}\frac{x-2}{4} + \frac{1}{3} & \text{for} \quad 2 \leq x \leq 6, \\ 1 & \text{for} \quad x > 6. \end{cases} \tag{2.40}$$

The distribution is shown in Fig. 2.11b. Note that the distribution has a jump at $x = 2$ because $x = 2$ is the discrete value of the mixed random variable X.

2.2.2 Properties

A distribution has the following properties:

P.1

$$0 \leq F_X(x) \leq 1. \tag{2.41}$$

This property is easy to justify knowing that according to (2.10) the distribution is the probability and thus must have the values between 0 and 1.

P.2

$$F_X(-\infty) = 0,$$
$$F_X(\infty) = 1. \tag{2.42}$$

The first equation follows from the fact that $P\{x \leq -\infty\}$ is the probability of the empty set and hence equal to 0. The second equation follows from the probability of a certain event which in turns is equal to 1.

P.3

$$F_X(x_1) \leq F_X(x_2) \quad \text{for} \quad x_1 < x_2. \tag{2.43}$$

This property states that distribution is a nondecreasing function. For $x_1 < x_2$, the event $X \leq x_1$ is included in the event $X \leq x_2$ and consequently $P\{X \leq x_1\} \leq P\{X \leq x_2\}$.

P.4

$$F_X(x^+) = F_X(x), \tag{2.44}$$

where the denotation x^+ implies $x + \varepsilon$, $\varepsilon > 0$, which is infinitesimally small as $\varepsilon \to 0$. This property states that distribution is a *continuous from the right* function. As we approach x from the right, the limiting value of the distribution should be the value of the distribution in this point, i.e., $F_X(x)$.

The properties (2.41)–(2.44), altogether, may be used to test if a given function could be a valid distribution function.

Next, we relate the probability that the variable X is in a given interval, with its distribution function,

$$P\{x_1 < X \leq x_2\} = P\{X \leq x_2\} - P\{X \leq x_1\}. \tag{2.45}$$

Using the definition (2.10) from (2.45), it follows:

$$P\{x_1 < X \leq x_2\} = F_X(x_2) - F_X(x_1). \tag{2.46}$$

Example 2.2.5 Using (2.45), find the following probabilities for the random variable X from Example 2.2.1:

(a) $P\{1.1 < X \leq 1.8\}$. (2.47)

(b) $P\{0.5 < X \leq 1.5\}$. (2.48)

(c) $P\{0.4 < X \leq 1\}$. (2.49)

Solution

(a) From (2.21), we have:

$$F_X(1.8) = 1; \quad F_X(1.1) = 1. \tag{2.50}$$

Therefore,

$$P\{1.1 < X \leq 1.8\} = F_X(1.8) - F_X(1.1) = 0. \tag{2.51}$$

This is the expected result, because the corresponding event is a null event; also, r.v. X takes only 0 and 1, as values, and cannot be in the interval defined in (2.47).

(b) From (2.21), we have:

$$F_X(1.5) = 1; \quad F_X(0.5) = 1/4. \tag{2.52}$$

Therefore,

$$P\{0.5 < X \leq 1.5\} = F_X(1.5) - F_X(0.5) = 3/4. \tag{2.53}$$

This is also the expected result, being as the desired probability (2.48) corresponds to the probability that $X = 1$, which, in turn, is equal to 3/4.

(c) From (2.21) and the property of (2.41), we have:

$$F_X(1) = 1; \quad F_X(0.4) = 1/4, \tag{2.54}$$

resulting in:

$$P\{0.4 < X \leq 1\} = F_X(1) - F_X(0.4) = 3/4. \tag{2.55}$$

Example 2.2.6 For the continuous random variable X from Example 2.2.3, find the following probabilities:

(a) $P\{3 < X \leq 4\}$. (2.56)

(b) $P\{X = 4\}$. (2.57)

Solution

(a) The probability

$$P\{3 < X \leq 4\} = F_X(4) - F_X(3) = 1/2 - 1/4 = 1/4. \tag{2.58}$$

(b) In this case, there is no interval but only a single point 4, thus the probability of the null event $\{4 < X \leq 4\}$ is zero.
Therefore,

$$P\{4 < X \leq 4\} = P\{X = 4\} = 0. \tag{2.59}$$

This is the expected result because the probability that the continuous variable takes the particular value is zero.

2.3 Probability Density Function

2.3.1 *Definition*

The distribution function introduced in Sect. 2.2 is a mathematical tool used to describe a random variable. However, it is often cumbersome to work with distribution especially, when dealing with the continuous random variable [MIL04, p. 53]. The easiest alternative is to use the probability density function (PDF).

A PDF $f_X(x)$, or shortly *density*, is the derivative of the distribution function $F_X(x)$,

$$f_X(x) = \frac{\mathrm{d}F_X(x)}{\mathrm{d}x}. \tag{2.60}$$

From here,

$$F_X(x) = \int_{-\infty}^{x} f_X(v)\mathrm{d}v. \tag{2.61}$$

Therefore, the distribution in one point x, corresponds to the area below the density function from $-\infty$ to the point x, as shown in Fig. 2.12.

Using the definition of the derivative, from (2.60), we can write:

$$\frac{\mathrm{d}F_X(x)}{\mathrm{d}x} = \lim_{\Delta x \to 0} \frac{F_X(x+\Delta x) - F_X(x)}{\Delta x}. \tag{2.62}$$

Using

$$F_X(x+\Delta x) - F_X(x) = P\{x < X \leq x + \Delta x\}, \tag{2.63}$$

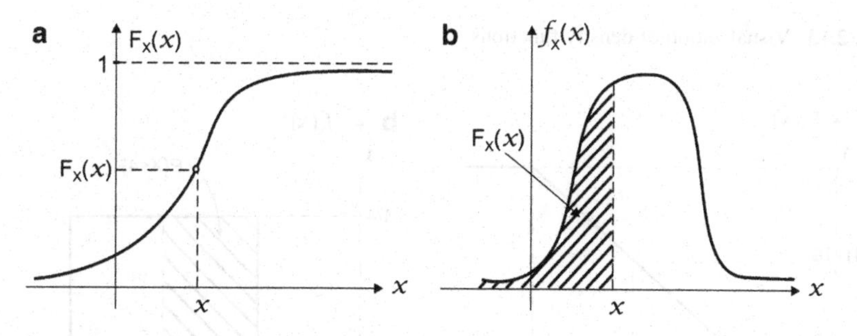

Fig. 2.12 (**a**) Distribution and (**b**) density

we have:

$$f_X(x) = \lim_{\Delta x \to 0} \frac{P\{x < X \leq x + \Delta x\}}{\Delta x} = \frac{P\{x < X \leq x + \mathrm{d}x\}}{\mathrm{d}x}. \tag{2.64}$$

This expression represents the probability that a random variable X lies in an infinitesimal interval around the point x, normalized by the length of the interval. Hence, the name density comes.

A visualization of this expression is given in Fig. 2.13, assuming $f_X(x) = f_X(x + \Delta x)$.

Example 2.3.1 Find the density function of the continuous variable from Example 2.2.3, assuming that $a = 2$ and $b = 6$. Also find the probability that the random variable is less than 4.

Solution For convenience, the distribution (2.30) is presented again in Fig. 2.14a for the given values a and b.

Using (2.60), we have:

$$f_X(x) = \begin{cases} 0 & \text{for} \quad x < 2, \\ \frac{\mathrm{d}}{\mathrm{d}x}\left(\frac{x-2}{4}\right) = \frac{1}{4} & \text{for} \quad 2 \leq x \leq 6, \\ 0 & \text{for} \quad x > 6. \end{cases} \tag{2.65}$$

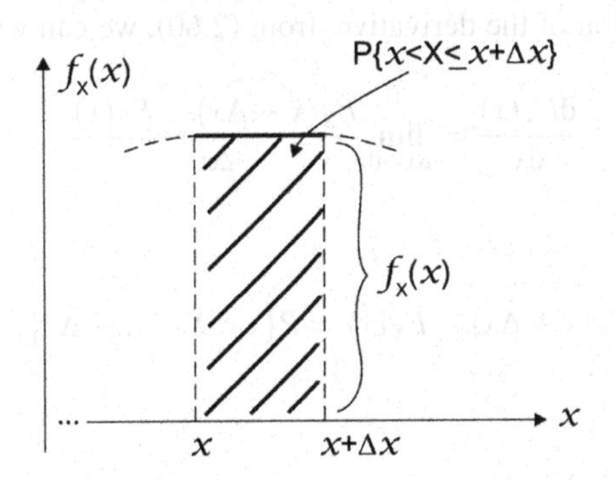

Fig. 2.13 Visualization of density functions

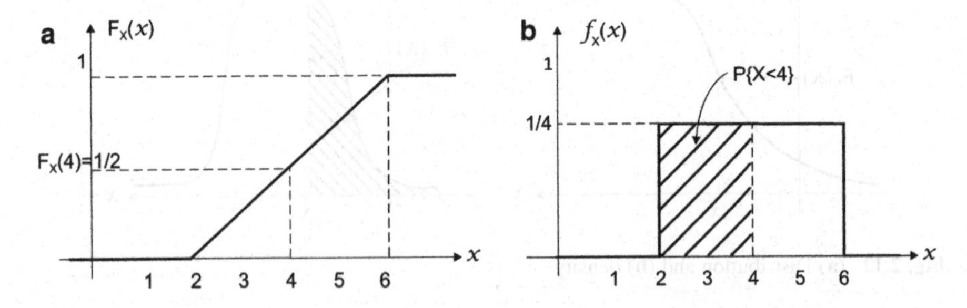

Fig. 2.14 Illustration of Example 2.3.1. (**a**) Distribution. (**b**) PDF

This density is shown in Fig. 2.14b.

The probability that the variable is less than 4 corresponds to the following integral of the PDF

$$P\{X \leq 4\} = \int_{2}^{4} f_X(x)\mathrm{d}x = \int_{2}^{4} \frac{1}{4}\mathrm{d}x = 1/2. \tag{2.66}$$

Note that this value corresponds to the area under the density, as shown in Fig. 2.14b, and alternately to the only one point $F_X(4) = 1/2$ at the distribution function in Fig. 2.14a.

2.3.2 *Delta Function*

As mentioned in Sect. 2.1, the distribution of a discrete random variable is in a stair-step form. In order to circumvent the problem in defining the derivative, we use the *unit-impulse function* $\delta(x)$, also called the *delta function*,

$$\delta(x) = \frac{\mathrm{d}u(x)}{\mathrm{d}x}. \tag{2.67}$$

From here

$$u(x) = \int_{-\infty}^{x} \delta(v)\mathrm{d}v. \tag{2.68}$$

A delta function $\delta(x)$ is a continuous function of x which is defined in the range $[-\infty, \infty]$. It can be interpreted as a function with infinite amplitude at $x = 0$ and zero duration with unitary area,

$$\delta(x) = \begin{cases} \infty & \text{for } x = 0 \\ 0 & \text{otherwise} \end{cases}; \quad \int_{-\infty}^{\infty} \delta(x)\mathrm{d}x = 1 \tag{2.69}$$

We usually present the delta function as an arrow occurring at $x = 0$ and with unity magnitude as shown in Fig. 2.15a. The interpretation of the delta function is shown in Fig. 2.15b, demonstrating that it has zero duration and infinite amplitude; but an area equal to 1, assuming $\Delta x \to 0$.

A general expression is obtained if we shift the delta function $\delta(x)$ by x_0,

$$\delta(x - x_0) = \begin{cases} \infty & \text{for } x = x_0, \\ 0 & \text{otherwise} \end{cases}; \quad \int_{-\infty}^{\infty} \delta(x - x_0)\mathrm{d}x = 1 \tag{2.70}$$

This function is presented in Fig. 2.15c. Taking the delayed delta function we can write:

$$\delta(x - x_0) = \frac{\mathrm{d}u(x - x_0)}{\mathrm{d}x}. \tag{2.71}$$

For any function $g(x)$, continuous at the point x_0, we have:

$$\int_{-\infty}^{\infty} g(x)\delta(x - x_0)\mathrm{d}x = g(x_0). \tag{2.72}$$

The interpretation of (2.72) is given in Fig. 2.15d.

2.3.3 Densities of Discrete and Mixed Random Variables

From (2.60) and the distribution of the discrete random variable (2.24), we arrive at the density of a discrete random variable in the form:

$$f_X(x) = \frac{\mathrm{d}}{\mathrm{d}x}\left(\sum_{i=-\infty}^{\infty} P\{x_i\}u(x - x_i)\right) = \sum_{i=-\infty}^{\infty} P\{x_i\}\frac{\mathrm{d}}{\mathrm{d}x}(u(x - x_i)). \tag{2.73}$$

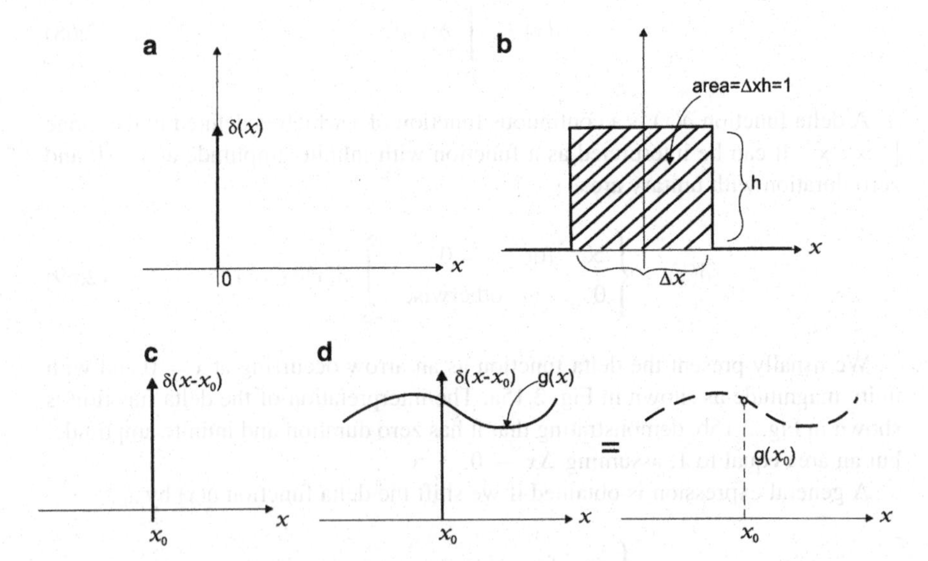

Fig. 2.15 Delta function: definition and characteristics

Using (2.71), we arrive at:

$$f_X(x) = \sum_{i=-\infty}^{\infty} P\{x_i)\delta(x - x_i). \tag{2.74}$$

Example 2.3.2 In this example, we are looking for the density function of the discrete variable from Example 2.2.2, where the distribution was $F_X(x) = 1/6u(x) + \cdots + 1/6u(x - 6)$. Find the probability that the random variable is also less than 4.

Solution Using (2.74), we have the density

$$f_X(x) = 1/6\sum_{k=1}^{6} \delta(x - k). \tag{2.75}$$

This density is shown in Fig. 2.16. The desired probability (not including point 4) is:

$$P\{X < 4\} = \frac{1}{6}\int_{-\infty}^{3}\sum_{k=1}^{6}\delta(x-k)\mathrm{d}x = \frac{1}{6}\sum_{k=1}^{6}\int_{-\infty}^{3}\delta(x-k)\mathrm{d}x = \frac{1}{6}(1+1+1) = \frac{1}{2}. \tag{2.76}$$

If we include point 4, we have:

$$\begin{aligned} P\{X \leq 4\} &= \frac{1}{6}\int_{-\infty}^{4}\sum_{k=1}^{6}\delta(x-k)\mathrm{d}x = \frac{1}{6}\sum_{k=1}^{6}\int_{-\infty}^{4}\delta(x-k)\mathrm{d}x \\ &= \frac{1}{6}(1+1+1+1) = \frac{2}{3}. \end{aligned} \tag{2.77}$$

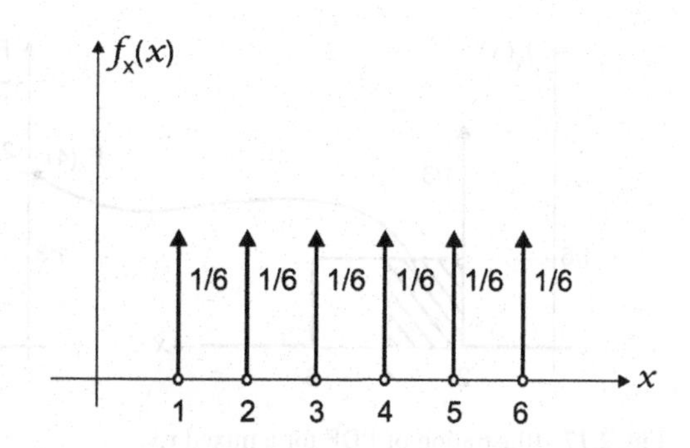

Fig. 2.16 Illustration of PDF for a discrete r.v.

A mixed random variable has a continuous density within range where the values are continuous and the delta functions in discrete points of x. From (2.31), we write the PDF of a mixed random variable:

$$f_X(x) = \alpha f_c(x) + \sum_i P\{X = x_i\}\delta(x - x_i), \tag{2.78}$$

where f_c is the density in the continuous range of the mixed r.v. and

$$\alpha = 1 - \sum_i P\{X = x_i\}. \tag{2.79}$$

The next example illustrates the density of the mixed random variable.

Example 2.3.3 Consider the mixed random variable from Example 2.2.4. Its distribution is again plotted in Fig. 2.17. Find its density as well as the probability that the r.v. is less than 4.

Solution From (2.40), we find the density to be:

$$f_X(x) = \begin{cases} 0 & \text{for} \quad x < 2, \\ \frac{1}{3}\delta(x-2) + \frac{2}{3} \times \frac{1}{4} & \text{for} \quad 2 \leq x \leq 6, \\ 0 & \text{for} \quad x > 6. \end{cases} \tag{2.80}$$

The density is shown in Fig. 2.17.

The desired probability is

$$P\{X < 4\} = \int_2^4 f_X(x)\mathrm{d}x = \int_2^4 \left(\frac{1}{3}\delta(x-2) + \frac{1}{6}\right)\mathrm{d}x = \frac{1}{3} + \frac{4-2}{6} = \frac{2}{3}. \tag{2.81}$$

This probability is illustrated in Fig. 2.17, demonstrating that the probability is the shaded area plus the area of the delta function. The probability is also illustrated in the distribution plot as the value of the distribution in the point $x = 4$.

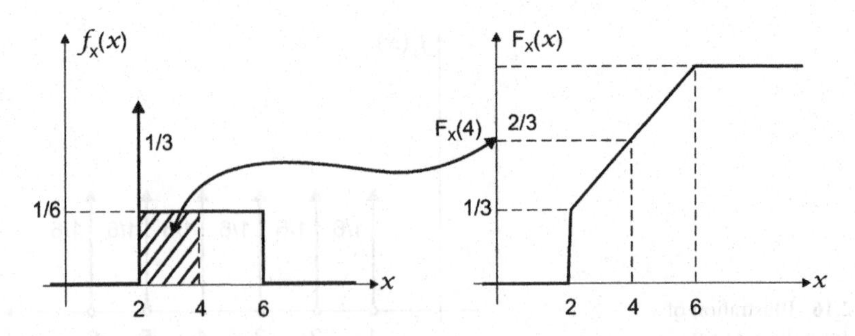

Fig. 2.17 Illustration of PDF for a mixed r.v.

2.3.4 Properties

The PDF has the following properties:

P.1

$$f_X(x) \geq 0 \quad \text{for any}\, x. \tag{2.82}$$

This property that the PDF is a nonnegative function easily follows from the third property of a distribution and its definition (2.60). As a result, the derivative of a nondecreasing function cannot be negative.

P.2

$$\int_{-\infty}^{\infty} f_X(x)\mathrm{d}x = 1. \tag{2.83}$$

This property that the PDF must integrate to one follows from (2.61). Replacing x with ∞, and using (2.42), we have:

$$F_X(\infty) = 1 = P\{X \leq \infty\} = \int_{-\infty}^{\infty} f_X(x)\mathrm{d}x. \tag{2.84}$$

P.3

$$P\{x_1 < X \leq x_2\} = \int_{x_1}^{x_2} f_X(x)\mathrm{d}x. \tag{2.85}$$

Using the definition of a distribution of (2.10) and (2.61), we have (see Fig. 2.18b):

$$\begin{aligned} P\{x_1 < X \leq x_2\} &= P\{X \leq x_2\} - P\{X \leq x_1\}, \\ &= F_X(x_2) - F_X(x_1), \\ &= \int_{-\infty}^{x_2} f_X(x)\mathrm{d}x - \int_{-\infty}^{x_1} f_X(x)\mathrm{d}x, \\ &= \int_{x_1}^{x_2} f_X(x)\mathrm{d}x. \end{aligned} \tag{2.86}$$

The probability that the random variable is in the interval $[x_1, x_2]$ is equal to the area under the density function in the interval, as demonstrated in Fig. 2.18.

The properties (2.82) and (2.83) serve as a test to determine whether a given function is a PDF. Both properties must be satisfied in order to determine that a function is a PDF.

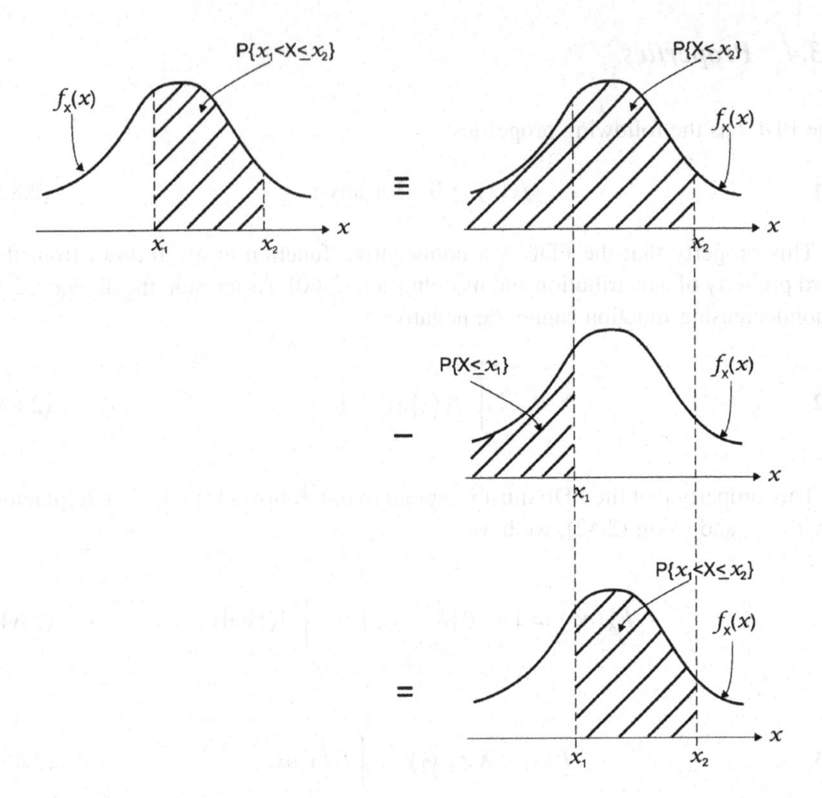

Fig. 2.18 PDF and probability

Example 2.3.4 The PDF of a variable X is given as

$$f_X(x) = a\mathrm{e}^{-b|x|}, \quad -\infty < x < \infty, \tag{2.87}$$

where both a and b are constant. Find the relation between a and b. Find the corresponding distribution as well as the probability that the variable is negative, if $a = 1$. Relate the probability to both the PDF and distribution. Do the same for the probability that the variable is in the interval [1, 2].

Solution The PDF must satisfy the property (2.83):

$$\int_{-\infty}^{\infty} f_X(x)\mathrm{d}x = 1 = \int_{-\infty}^{\infty} a\mathrm{e}^{-b|x|}\mathrm{d}x = \int_{-\infty}^{0} a\mathrm{e}^{bx}\mathrm{d}x + \int_{0}^{\infty} a\mathrm{e}^{-bx}\mathrm{d}x = \frac{a}{b} + \frac{a}{b} = \frac{2a}{b}. \tag{2.88}$$

From (2.88), we have:

$$b = 2a. \tag{2.89}$$

Taking $a = 1$, and using (2.89), the density (2.87) becomes:

$$f_X(x) = \mathrm{e}^{-2|x|}, \quad -\infty < x < \infty. \tag{2.90}$$

The distribution function is obtained from the PDF using (2.61).
For $x \leq 0$

$$F_X(x) = \int_{-\infty}^{x} f_X(x)\mathrm{d}x = \int_{-\infty}^{x} \mathrm{e}^{2x}\mathrm{d}x = \frac{\mathrm{e}^{2x}}{2}. \tag{2.91}$$

For $x \geq 0$

$$F_X(x) = \int_{-\infty}^{x} f_X(x)\mathrm{d}x = \int_{-\infty}^{0} \mathrm{e}^{2x}\mathrm{d}x + \int_{0}^{x} \mathrm{e}^{-2x}\mathrm{d}x = \frac{1}{2} + \frac{1-\mathrm{e}^{-2x}}{2} = 1 - \frac{\mathrm{e}^{-2x}}{2}. \tag{2.92}$$

The PDF and distribution are given in Fig. 2.19.
The probability that the X is negative is related to the PDF as:

$$P\{X \leq 0\} = \int_{-\infty}^{0} \mathrm{e}^{2x}\mathrm{d}x = \frac{1}{2}. \tag{2.93}$$

Similarly, the probability and the distribution are related as:

$$P\{X \leq 0\} = F_X(0). \tag{2.94}$$

Taking either (2.91) or (2.92), we have:

$$P\{X \leq 0\} = F_X(0) = 1 - \frac{\mathrm{e}^0}{2} = \frac{1}{2}. \tag{2.95}$$

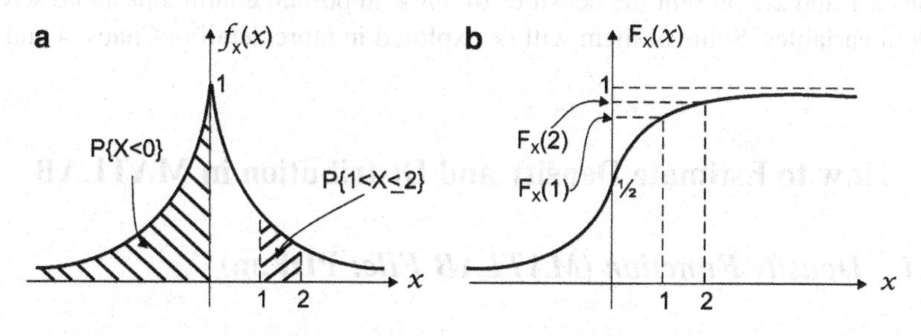

Fig. 2.19 Illustration of PDF, distribution, and probabilities from Example 2.3.4. **(a)** PDF. **(b)** Distribution

The probability is presented as a shaded area in the PDF, and as a point in the distribution.

Let us now find the probability that the random variable is in the interval [1, 2]. From (2.85), we write:

$$P\{1 < X \leq 2\} = \int_{1}^{2} e^{-2x} dx = \frac{e^{-2x}}{2}\bigg|_{2}^{1} = \frac{e^{-2} - e^{-4}}{2} = 0.0585. \tag{2.96}$$

The probability is presented in Fig. 2.19a as a shaded area below the PDF in the interval [1, 2].

From (2.86), we have:

$$P\{1 < X \leq 2\} = F_X(2) - F_X(1). \tag{2.97}$$

From (2.97) and (2.92), it follows:

$$\begin{aligned} P\{1 < X \leq 2\} = F_X(2) - F_X(1) &= \left(1 - \frac{e^{-4}}{2}\right) - \left(1 - \frac{e^{-2}}{2}\right), \\ &= \left(1 - \frac{0.0183}{2}\right) - \left(1 - \frac{0.1353}{2}\right), \\ &= 0.9909 - 0.9324, \\ &= 0.0585. \end{aligned} \tag{2.98}$$

The probability is presented as the difference of the values of the distributions in points 2 and 1, as demonstrated in Fig. 2.19b.

2.3.5 *Examples of Density Functions*

Tables 2.1 and 2.2 present the densities of a few important continuous and discrete random variables. Some of them will be explored in more detail in Chaps. 4 and 5.

2.4 How to Estimate Density and Distribution in MATLAB

2.4.1 *Density Function (MATLAB File:* PDF.m*)*

Here, we consider the uniform variable in the interval [0, 1]. The PDF and distribution are shown in Fig. 2.20.

Table 2.1 Continuous variables and densities

Variable	Density
Uniform in the interval $[x_1, x_2]$	$f_X(x) = \begin{cases} \dfrac{1}{x_2 - x_1} & \text{for } x_1 \le x \le x_2, \\ 0 & \text{otherwise.} \end{cases}$
Normal (Gaussian)	$f_X(x) = \dfrac{1}{\sqrt{2\pi}\sigma} e^{-(x-m)^2/2\sigma^2}$ for $-\infty < x < \infty$, m and σ^2 are parameters
Exponential	$f_X(x) = \begin{cases} \lambda e^{-\lambda x} & \text{for } x \ge 0, \\ 0 & \text{otherwise.} \end{cases}$, λ is a parameter
Laplacian	$f_X(x) = \dfrac{\lambda}{2} e^{-\lambda \lvert x \rvert}$ for $-\infty < x < \infty$, λ is a parameter
Gamma	$f_X(x) = \begin{cases} \dfrac{(x/b)^{k-1} e^{-x/b}}{b\Gamma(c)} & \text{for } x \ge 0, \\ 0 & \text{otherwise.} \end{cases}$, $\Gamma(c)$ is a Gamma function. b and c are parameters, $b > 0$, $c > 0$
Rayleigh	$f_X(x) = \begin{cases} \dfrac{x}{\sigma^2} e^{-x^2/2\sigma^2} & \text{for } x \ge 0, \\ 0 & \text{otherwise.} \end{cases}$, for any $\sigma > 0$
Weibull	$f_X(x) = \begin{cases} Kx^m & \text{for } x \ge 0, \\ 0 & \text{otherwise.} \end{cases}$, K and m are parameters
Cauchy	$f_x(x) = \dfrac{b/\pi}{b^2 + (x-a)^2}$ for $-\infty < x < \infty$, for any a and $b > 0$
Chi-Squared	$f_X(x) = \begin{cases} \dfrac{x^{c-1} e^{-x/2}}{2^c \Gamma(c)} & \text{for } x \ge 0, \\ 0 & \text{otherwise.} \end{cases}$

Table 2.2 Discrete variables and densities

Variable	Density
Uniform	$f_X(x) = \begin{cases} \dfrac{1}{N} & \text{for } x = x_1, \ldots x_N, \\ 0 & \text{otherwise.} \end{cases}$
Binomial	X is the number of successes in n trials $P\{X = k; n\} = \binom{n}{k} p^k q^{n-k}; p + q = 1,$ $f_X(x) = \sum_{k=0}^{n} P\{X = k; n\}\delta(x-k).$
Poisson	X is the number of arrivals in a given time t when the arrival rate is λ $P\{X = k\} = \dfrac{(\lambda t)^k}{k!} e^{-\lambda t}, \quad \lambda > 0, \quad k = 0, 1, 2, \ldots,$ $f_X(x) = \sum_{k=0}^{\infty} P\{X = k\}\delta(x-k).$

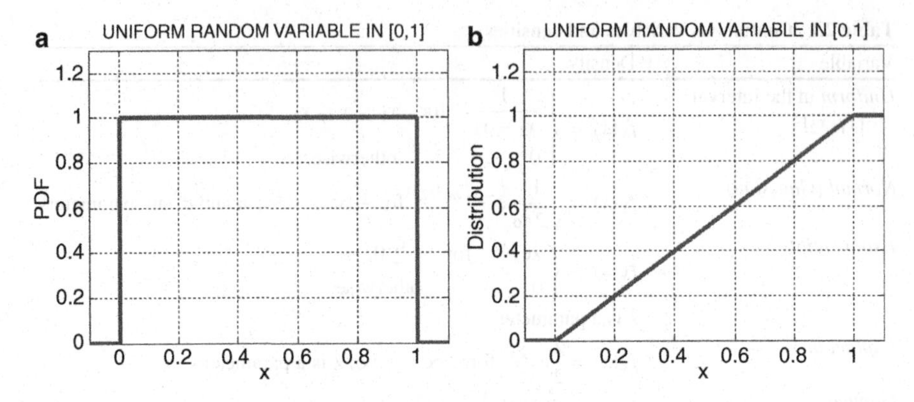

Fig. 2.20 (**a**) PDF and (**b**) distribution of the uniform random variable in the interval [0, 1]

The uniform variable in the interval [0, 1] can be generated in MATLAB using the file *rand.m*. Figure 2.21a shows the generated uniform variable X with $N = 10{,}000$ values, while Fig. 2.21b shows only the first 1,000 samples. From Fig. 2.21, it is evident why the variable is called "uniform" (Its values uniformly occupy all of the range).

In order to estimate the PDF of the random variable X, we divide the range of the variable [0, 1] into M equidistant cells Δx.

The *histogram*, NN = hist(x, M) shows the values Ni, $i = 1, \ldots, M$ (i.e., how many values of the random variable X are in each cell), where M is the number of cells. If we use $M = 10$, we get for example, the following result:

$$\mathrm{NN} = [952, 994, 993, 1{,}012, 985, 1{,}034, 1{,}006, 998, 1{,}047, 979]. \tag{2.99}$$

The result (2.99) shows that from $N = 10{,}000$ values of the random variable, 952 are in the first cell, 994 in the second cell, and so on. We can also note that the number of values in the cells do not vary significantly.

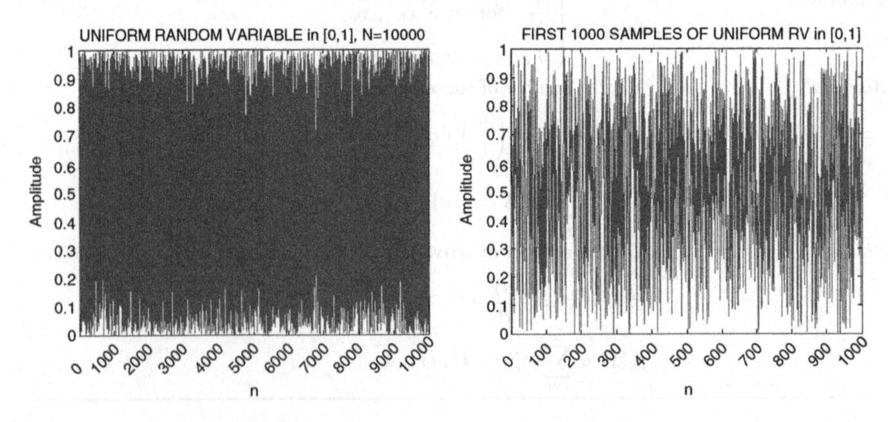

Fig. 2.21 Uniform variable generated in MATLAB

The plot of a histogram can be obtained by typing hist(X, 10) in the MATLAB prompt. Figure 2.22 shows the histogram.

Let N_i be the number of values of the random variable X in the ith cell. Then, the probability that the random variable belongs to the ith cell is approximated by a quantity Ni/N, called the *frequency ratio*,

$$P\{(i-1)\Delta x < X \leq i\Delta x\} \approx N_i/N = \text{hist}(X, M)/N. \tag{2.100}$$

The frequency ratio in MATLAB is obtained by dividing a histogram by N. The plot is shown in Fig. 2.23.

Next, from (2.99) and (2.100), we can estimate the PDF in ith cell as:

$$f_X(i\Delta x) \approx \frac{P\{(i-1)\Delta x < X \leq i\Delta x\}}{\Delta x} = \frac{N_i}{N}\frac{1}{\Delta x}. \tag{2.101}$$

According to (2.101), an estimation of the PDF within a given cell is obtained by dividing the probability that the random variable belongs to the cell (frequency ratio) by the length of cell Δx, which here is equal to $1/10 = 0.1$.

From (2.100) and (2.101), we get:

$$\text{PDF} \approx P\{X \text{ belongs to the cell } i,\ i = 1, \ldots, M\}/\Delta x = \frac{\text{hist}(x, M)}{N\Delta x}. \tag{2.102}$$

The PDF estimation is given in Fig. 2.24.

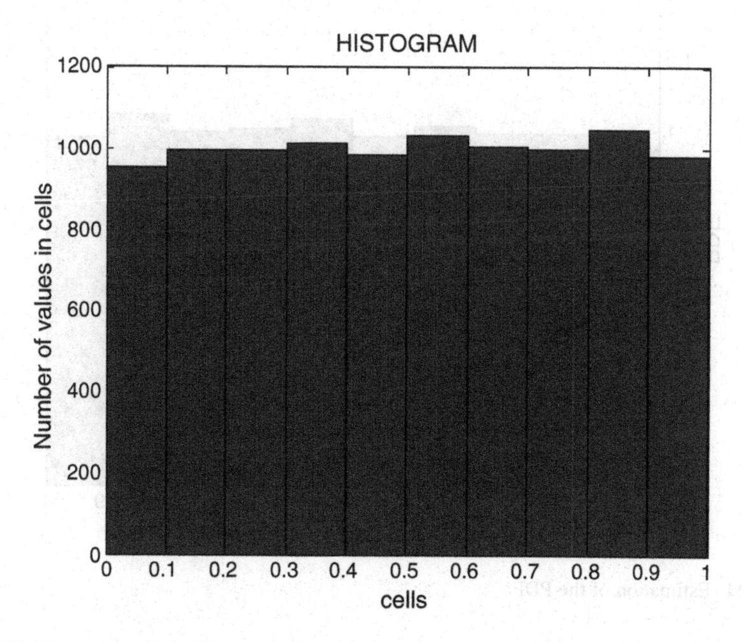

Fig. 2.22 Histogram, $M = 10$

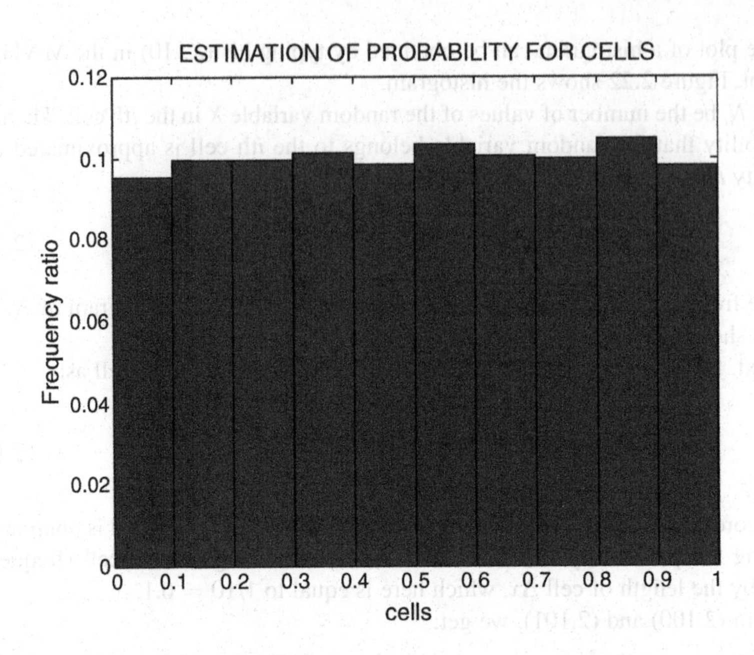

Fig. 2.23 Estimation of the probabilities

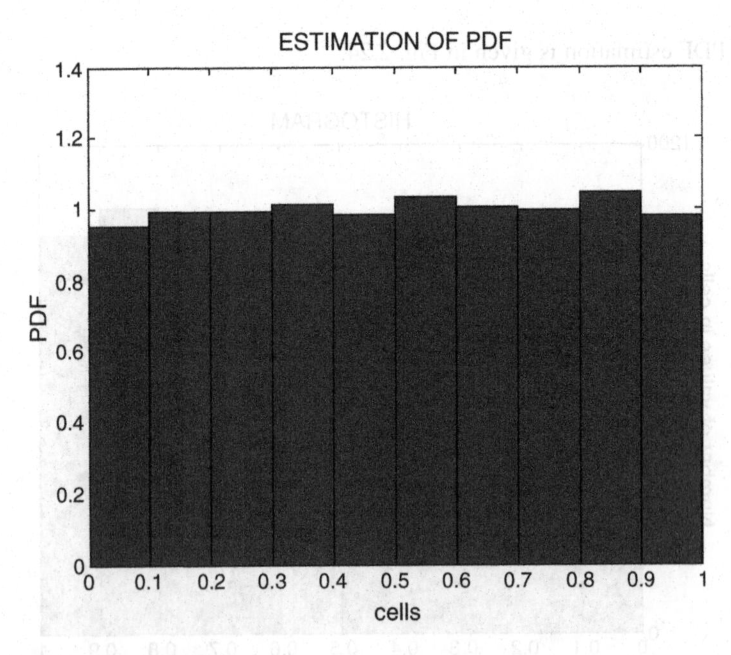

Fig. 2.24 Estimation of the PDF

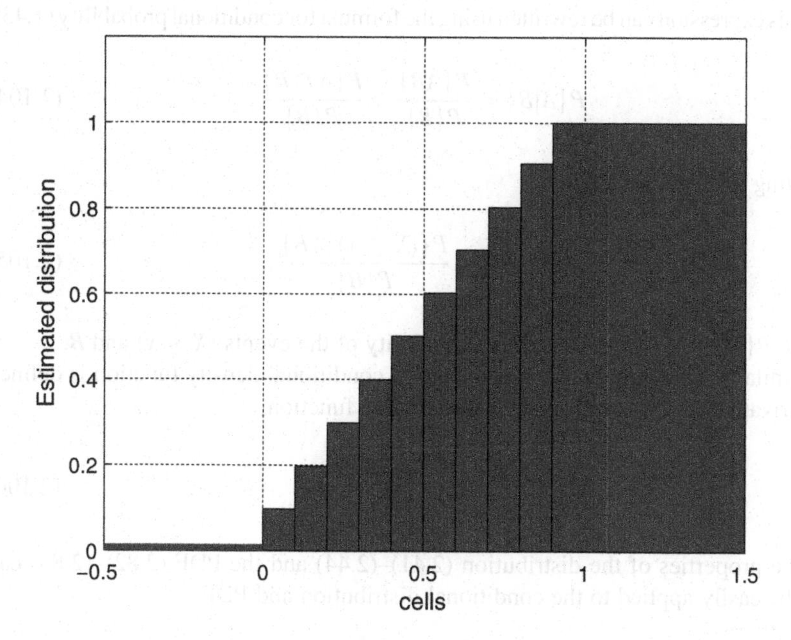

Fig. 2.25 Estimation of distribution function for the uniform r.v.

Note that all of the plots in Figs. 2.22–2.24 have the same form and that the only difference is the amplitude scaling. That is why some authors call the histogram plot the PDF estimation plot. However, we consider it is important to clarify the difference between histogram, probabilities, and PDF, because only the PDF estimation from Fig. 2.24 corresponds to the mathematical PDF from Fig. 2.20a, and the mathematical definition, introduced in (2.64).

2.4.2 *Distribution Function (MATLAB File:* Distribution.m*)*

Next, we estimate distribution from the estimated PDF using (2.61). The MATLAB function *cumsum.m* performs this estimation. The estimated distribution is given in Fig. 2.25.

2.5 Conditional Distribution and PDF

2.5.1 *Definition of Conditional Distribution and PDF*

A *conditional distribution* of a random variable X, given an event B, is defined as:

$$F_X(x|B) = P\{X \leq x|B\}. \tag{2.103}$$

This expression can be rewritten using the formula for conditional probability (1.43),

$$P\{A|B\} = \frac{P\{AB\}}{P\{B\}} = \frac{P\{A \cap B\}}{P\{B\}}, \tag{2.104}$$

resulting in:

$$F_X(x|B) = \frac{P\{(X \leq x) \cap B\}}{P\{B\}}, \tag{2.105}$$

where $P\{(X \leq x) \cap B\}$ is the joint probability of the events $(X \leq x)$ and B.

Similarly, as in a nonconditioned case, a conditional density function is defined as derivative of the corresponding distribution function,

$$f_X(x|B) = \frac{\mathrm{d}F_X(x|B)}{\mathrm{d}x}. \tag{2.106}$$

The properties of the distribution (2.41)–(2.44) and the PDF (2.82)–(2.85) can also be easily applied to the conditional distribution and PDF.

2.5.2 Definition of a Conditioned Event

There are a number of ways to define a conditioned event. In the following, we define the event B in terms of the random variable X.

We will consider two cases.

Case 1 Event B is defined as:

$$B = (X \leq b). \tag{2.107}$$

The corresponding probability is:

$$P\{B\} = P\{X \leq b\}, \tag{2.108}$$

where b is any real number.

According to (2.104) and (2.108), we have:

$$F_X(x|X \leq b) = \frac{P\{(X \leq x) \cap (X \leq b)\}}{P\{X \leq b\}}. \tag{2.109}$$

For $x \geq b$, event $(X \leq b)$ is the subset of event $(X \leq x)$, resulting in:

$$P\{(X \leq x) \cap (X \leq b)\} = P\{X \leq b\}. \tag{2.110}$$

Using (2.110), the conditional distribution (2.109) becomes

$$F_X(x|X \leq b) = \frac{P\{(X \leq x) \cap (X \leq b)\}}{P\{X \leq b\}} = \frac{P\{X \leq b\}}{P\{X \leq b\}} = 1. \tag{2.111}$$

For $x < b$, event $(X \leq x)$ is the subset of event $(X \leq b)$, resulting in:

$$P\{(X \leq x) \cap (X \leq b)\} = P\{X \leq x\} \tag{2.112}$$

and

$$F_X(x|X \leq b) = \frac{P\{(X \leq x) \cap (X \leq b)\}}{P\{X \leq b\}} = \frac{P\{X \leq x\}}{P\{X \leq b\}}. \tag{2.113}$$

The equations for (2.111) and (2.113) can be rewritten, using (2.10), as:

$$F_X(x|X \leq b) = \begin{cases} \dfrac{F_X(x)}{F_X(b)} & \text{for} \quad x < b, \\ 1 & \text{for} \quad x \geq b. \end{cases} \tag{2.114}$$

From here, we can use (2.106) and (2.61) to get the corresponding expression for the conditioned PDF:

$$f_X(x|X \leq b) = \begin{cases} \dfrac{f_X(x)}{F_X(b)} = \dfrac{f_X(x)}{\int_{-\infty}^{b} f_X(x)\mathrm{d}x} & \text{for} \quad x < b, \\ 0 & \text{for} \quad x \geq b. \end{cases} \tag{2.115}$$

Note that from (2.115), it follows that the total area under the conditioned density is equal to 1.

Example 2.5.1 Consider the uniform random variable in the interval $[-1, 2]$. Find the conditional distribution and the PDF of the random variable X given the event $B = \{X < 1\}$.

Solid lines in Fig. 2.26 denote the uniform PDF and the distribution function.

Solution Using (2.115), we have:

$$f_X(x|X \leq 1) = \begin{cases} \dfrac{f_X(x)}{F_X(1)} = \dfrac{f_X(x)}{\int_{-1}^{1} f_X(x)\mathrm{d}x} = \dfrac{1/3}{2/3} = \dfrac{1}{2} & \text{for} \quad -1 \leq x \leq 1, \\ 0 & \text{for} \quad x > 1. \end{cases} \tag{2.116}$$

The conditional density is shown in Fig. 2.26a, with the dashdot line.

From (2.114), we have:

$$F_X(x|X \leq 1) = \begin{cases} \dfrac{F_X(x)}{F_X(1)} = \dfrac{1/3}{2/3}(x+1) = \dfrac{1}{2}(x+1) & \text{for } -1 \leq x \leq 1, \\ 1 & \text{for } \quad x \geq 1. \end{cases} \tag{2.117}$$

The conditional distribution is shown in Fig. 2.26b with the dashdot line.

Case 2 The event B is defined as:

$$B = (a < X \leq b). \tag{2.118}$$

The corresponding probability is:

$$P\{B\} = P\{a < X \leq b\}. \tag{2.119}$$

The conditional distribution obtained from (2.105) and (2.119) is:

$$F_X(x|a < X \leq b) = \frac{P\{(X \leq x) \cap (a < X \leq b)\}}{P\{a < X \leq b\}}. \tag{2.120}$$

This expression depends on the relation between x, a, and b.
Let us consider first the case $x \leq a$. It follows that:

$$P\{(X \leq x) \cap (a < X \leq b)\} = 0 \tag{2.121}$$

and consequently,

$$F_X(x|a < X \leq b) = 0. \text{ for } x \leq a \tag{2.122}$$

For $(a < x \leq b)$, the probability is:

$$P\{(X \leq x) \cap (a < X \leq b)\} = P\{(a < X \leq x) \cap (a < X \leq b)\} = P\{a < X \leq x\} \tag{2.123}$$

resulting in (see (2.46)):

$$F_X(x|a < X \leq b) = \frac{P\{a < X \leq x\}}{P\{a < X \leq b\}} = \frac{F_X(x) - F_X(a)}{F_X(b) - F_X(a)}. \tag{2.124}$$

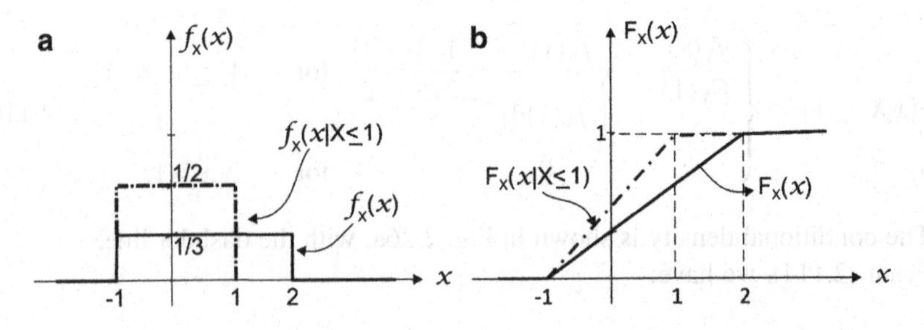

Fig. 2.26 Conditional PDF and distribution in Example 2.5.1. (**a**) PDFs. (**b**) Distributions

Similarly, for $x > b$, the probability is equal to:

$$P\{(X \le x) \cap (a < X \le b)\} = P\{(a < X \le b) \cap (a < X \le b)\} = P\{a < X \le b\}, \quad (2.125)$$

resulting in:

$$F_X(x | a < X \le b) = \frac{P\{a < X \le b\}}{P\{a < X \le b\}} = 1. \quad (2.126)$$

Finally, from (2.122), (2.124), and (2.126), we have:

$$F_X(x | a < X \le b) = \begin{cases} 1 & \text{for} \quad x > b, \\ \dfrac{F_X(x) - F_X(a)}{F_X(b) - F_X(a)} & \text{for} \quad a < x \le b, \\ 0 & \text{for} \quad x \le a. \end{cases} \quad (2.127)$$

Using (2.106), from (2.127), we obtain the corresponding density function as:

$$f_X(x | a < X \le b) = \begin{cases} 0 & \text{for} \quad x > b, \\ \dfrac{f_X(x) - F_X(a)}{F_X(b) - F_X(a)} & \text{for} \quad a < x \le b, \\ 0 & \text{for} \quad x \le a. \end{cases} \quad (2.128)$$

Example 2.5.2 Consider the uniform random variable X in the interval $[0, 4]$. Let the conditional event B be defined as $P\{0 < X \le 1\}$ (i.e., $a = 0$ and $b = 1$ in (2.118)).
From (2.128), we get:

$$f_X(x | 0 < X \le 1) = \begin{cases} 0 & \text{for} \quad x > 1, \\ \dfrac{f_X(x)}{F_X(1) - F_X(0)} = \dfrac{1/4}{1/4 - 0} = 1 & \text{for} \quad 0 < x \le 1, \\ 0 & \text{for} \quad x \le 0. \end{cases} \quad (2.129)$$

The PDF of the random variable X (solid line) and the conditional PDF (2.129) (dashdot line) are shown in Fig. 2.27a.

From (2.127), the conditional distribution is:

$$F_X(x | 0 < X \le 1) = \begin{cases} 1 & \text{for} \quad x > 1, \\ \dfrac{(1/4)x}{1/4 - 0} = x & \text{for} \quad 0 < x \le 1, \\ 0 & \text{for} \quad x \le 0. \end{cases} \quad (2.130)$$

The conditional distribution (dashdot line), along with the distribution of the random variable X, are shown in Figs. 2.27a,b. Observe that the conditioned distribution has all of the characteristics of the distribution itself.

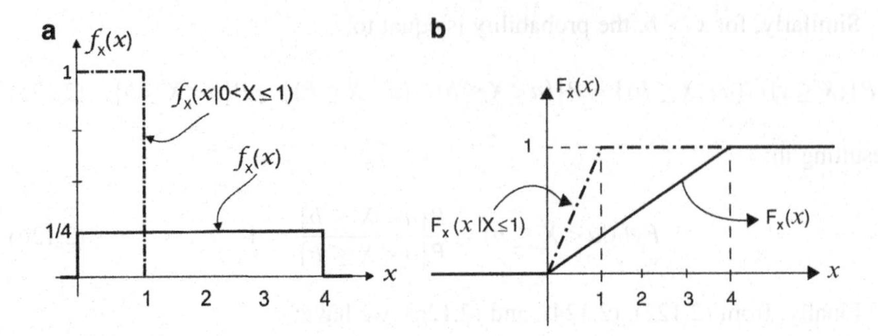

Fig. 2.27 Conditional PDF and distribution from Example 2.5.2. (**a**) Densities. (**b**) Distributions

2.6 Transformation of a Random Variable

2.6.1 Introduction

Random signals often pass through devices which transform their characteristics. A typical example of this is noise in telecommunications systems. The problem arises when trying finding the characteristics of the random variable after its transformation, if the characteristics of the random variable before transformation are known. In denoting random variables before and after transformation (as X and Y, respectively), we have:

$$Y = g(X), \tag{2.131}$$

where g denotes the functional relation between X and Y (Fig. 2.28). The same relation holds for the corresponding ranges

$$y = g(x). \tag{2.132}$$

The result depends on the type of the transformation. Here, we consider a *unique* transformation in which the same input value always corresponds to the same output value. The unique transformation can be a *monotone* and a *nonmonotone*. In a monotone transformation there is only one output value for each input value, as shown in Fig. 2.29a. However, in a nonmonotone transformation, two or more input values correspond to only one output value, as illustrated in Fig. 2.29b.

2.6.2 Monotone Transformation

Consider the monotone transformation of the input random variable X, $Y = g(X)$. If the input random variable X is in the interval

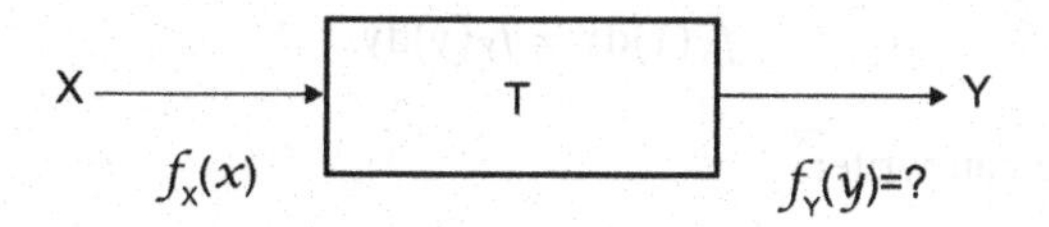

Fig. 2.28 Transformation of the input random variable

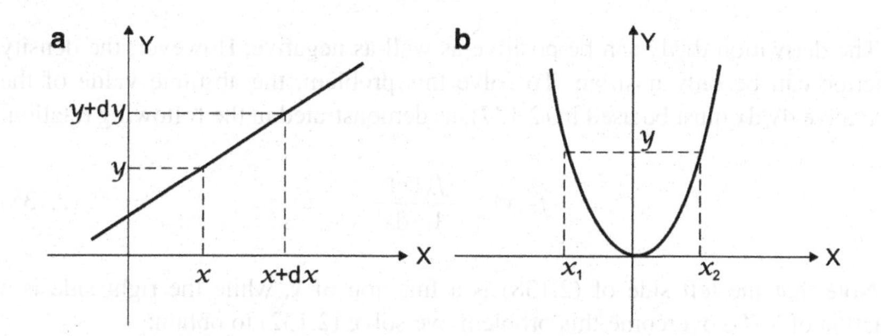

Fig. 2.29 (**a**) Monotone and (**b**) nonmonotone transformation

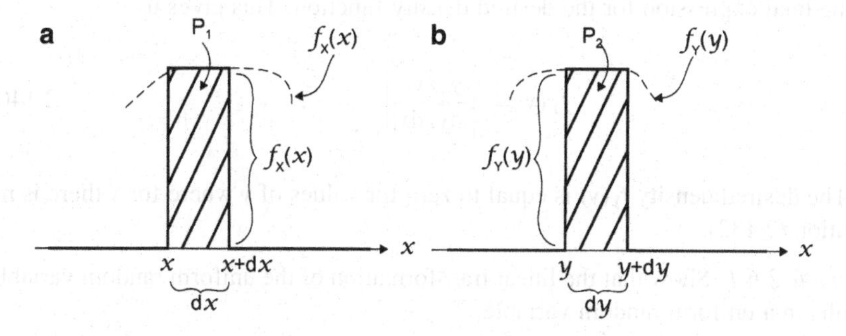

Fig. 2.30 Probabilities and densities of the input and output random variables

$$x < X \leq x + \mathrm{d}x, \tag{2.133}$$

then the output random variable will be in the interval

$$y < Y \leq y + \mathrm{d}y. \tag{2.134}$$

It follows that the corresponding probabilities of events (2.133) and (2.134) must be equal to:

$$P_1 = P\{x < X \leq x + \mathrm{d}x\} = P_2 = P\{y < Y \leq y + \mathrm{d}y\}. \tag{2.135}$$

This relation can be expressed using the corresponding densities, as shown in Fig. 2.30.

$$f_X(x)\mathrm{d}x = f_Y(y)\mathrm{d}y. \tag{2.136}$$

From here, we can write:

$$f_Y(y) = \frac{f_X(x)}{\mathrm{d}y/\mathrm{d}x}. \tag{2.137}$$

The derivation dy/dx can be positive as well as negative. However, the density function can be only positive. To solve this problem, the absolute value of the derivative dy/dx must be used in (2.137), as demonstrated in the following relation:

$$f_Y(y) = \frac{f_X(x)}{|\mathrm{d}y/\mathrm{d}x|}. \tag{2.138}$$

Note that the left side of (2.138) is a function of y, while the right side is a function of x. To overcome this problem, we solve (2.132) to obtain:

$$x = g^{-1}(y), \tag{2.139}$$

in the final expression for the desired density function. This gives us:

$$f_Y(y) = \left| \frac{f_X(x)}{|\mathrm{d}y/\mathrm{d}x|} \right|_{x=g^{-1}(y)}. \tag{2.140}$$

The desired density $f_Y(y)$ is equal to zero for values of y where for y there is no solution (2.132).

Example 2.6.1 Show that the linear transformation of the uniform random variable results in a uniform random variable.

Solution Let the input variable X be in the interval $[x_1, x_2]$ with the corresponding density, as shown in Fig. 2.31.

$$f_X(x) = \begin{cases} \dfrac{1}{x_2 - x_1} & \text{for} \quad x_1 \le x \le x_2, \\ 0 & \text{otherwise} \end{cases} \tag{2.141}$$

A linear transformation is given as:

$$y = ax + b, \tag{2.142}$$

where a and b are constants.

From (2.142), we obtain:

$$x = (y - b)/a \tag{2.143}$$

and

$$\left|\frac{\mathrm{d}y}{\mathrm{d}x}\right| = |a|. \tag{2.144}$$

From (2.140) and (2.143)–(2.144), we have:

$$f_Y(y) = \left.\frac{f_X(x)}{|\mathrm{d}y/\mathrm{d}x|}\right|_{x=(y-b)/a} = \left.\frac{f_X(x)}{|a|}\right|_{x=(y-b)/a} = \frac{1}{|a|}\frac{1}{\dfrac{y_2-b}{a}-\dfrac{y_1-b}{a}}$$

$$= \frac{a}{|a|}\frac{1}{y_2-y_1}, \tag{2.145}$$

where

$$\frac{a}{|a|} = \begin{cases} 1 & \text{for} \quad a > 0, \\ -1 & \text{for} \quad a < 0. \end{cases} \tag{2.146}$$

Using (2.146), we can rewrite (2.145) in the following forms:
For $a > 0$:

$$f_Y(y) = \begin{cases} \dfrac{1}{y_2-y_1} & \text{for} \quad y_1 \le y \le y_2, \\ 0 & \quad \text{otherwise.} \end{cases} \tag{2.147}$$

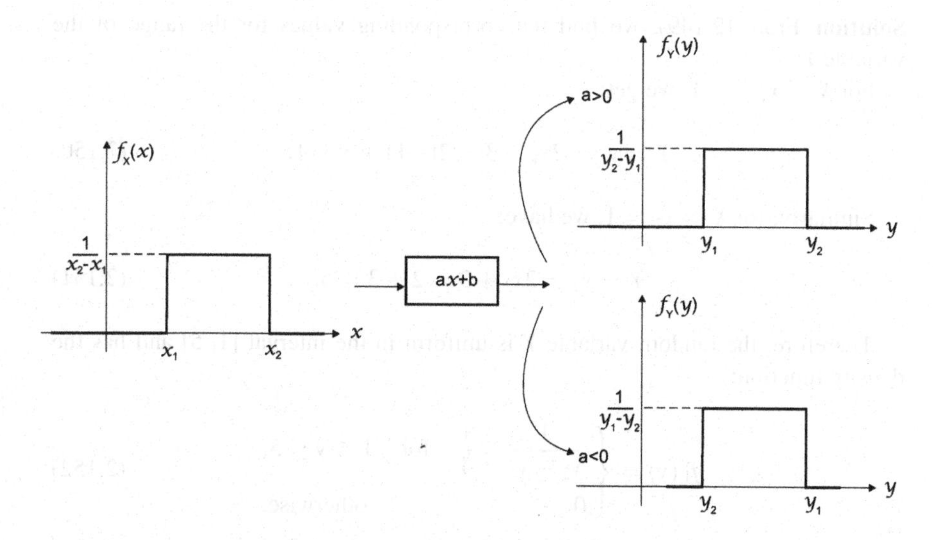

Fig. 2.31 Linear transformation of a uniform random variable

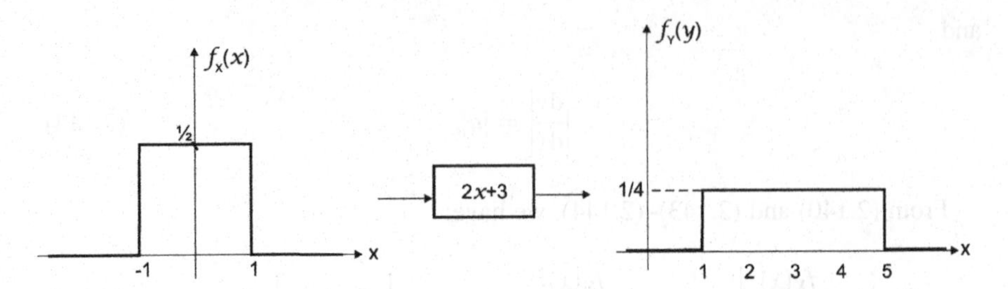

Fig. 2.32 Illustration of the linear transformation of the uniform random variable in Example 2.6.2

For $a < 0$:

$$f_Y(y) = \begin{cases} \dfrac{1}{y_1 - y_2} & \text{for} \quad y_2 \le y \le y_1, \\ 0 & \text{otherwise.} \end{cases} \tag{2.148}$$

The results obtained in (2.147) and (2.148) show that the linear transformation of the uniform random variable in a given interval results in a uniform variable in a different interval, as shown in Fig. 2.31.

Example 2.6.2 Using the result from Example 2.6.1, find the PDF for the random variable

$$Y = 2X + 3, \tag{2.149}$$

where X is a uniform random variable in $[-1, 1]$.

Solution From (2.149), we find the corresponding values for the range of the variable Y:

For $X = x_1 = -1$, we get:

$$Y = y_1 = 2x_1 + 3 = 2(-1) + 3 = 1. \tag{2.150}$$

Similarly, for $X = x_2 = 1$, we have:

$$Y = y_2 = 2x_2 + 3 = 2 + 3 = 5. \tag{2.151}$$

Therefore, the random variable Y is uniform in the interval [1, 5] and has the density function:

$$f_Y(y) = \begin{cases} \dfrac{1}{y_2 - y_1} = \dfrac{1}{4} & \text{for} \quad 1 \le y \le 5, \\ 0 & \text{otherwise.} \end{cases} \tag{2.152}$$

The PDF (2.152) is shown in Fig. 2.32 along with the input PDF.

Example 2.6.3 Using the result from Example 2.6.1, find the PDF for the random variable

$$Y = -0.5X + 2, \tag{2.153}$$

if X is a uniform random variable in $[-1, 1]$.

Solution From (2.153), we have:

$$y_1 = -0.5(-1) + 2 = 2.5, \tag{2.154}$$

$$y_2 = -0.5 + 2 = 1.5. \tag{2.155}$$

Using (2.154), (2.155), and (2.148), we arrive at:

$$f_Y(y) = \begin{cases} \dfrac{1}{y_1 - y_2} = 1 & \text{for} \quad 1.5 \leq y \leq 2.5, \\ 0 & \text{otherwise.} \end{cases} \tag{2.156}$$

The input and output PDFs are shown in Fig. 2.33.

Example 2.6.4 The random variable X is uniform in the interval $[0, 1]$. Find the transformation which will result in an uniform random variable in the interval $[y_1, y_2]$.

Solution From Example 2.6.1, it follows that the transformation is a linear transformation

$$Y = aX + b, \tag{2.157}$$

where X is the given uniform random variable in the interval $[0, 1]$, and the variable Y is uniform in the interval $[y_1, y_2]$.

From (2.157), using $x_1 = 0$ and $x_2 = 1$, we get:

$$\begin{aligned} y_1 &= ax_1 + b = b, \\ y_2 &= ax_2 + b = a + b. \end{aligned} \tag{2.158}$$

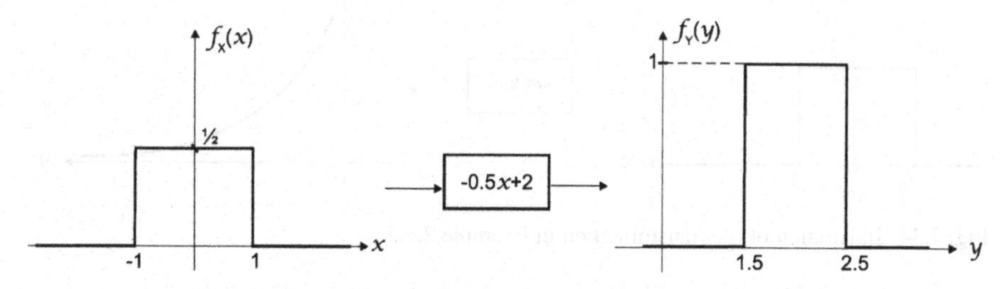

Fig. 2.33 Illustration of the linear transformation of the uniform random variable in Example 2.6.3

From (2.158), we have:

$$b = y_1, \quad a = y_2 - b = y_2 - y_1. \tag{2.159}$$

Placing (2.159) into (2.157), we get:

$$Y = (y_2 - y_1)X + y_1. \tag{2.160}$$

Therefore, the linear transformation (2.160) of the uniform variable in the interval [0, 1] results in a uniform random variable in the interval $[y_1, y_2]$.

Example 2.6.5 The random variable is uniform in the interval [0, 1]. Find the PDF for the random variable

$$Y = -\ln X. \tag{2.161}$$

Solution The transformation (2.161) is a monotone transformation (as shown in Fig. 2.34a).

From (2.161), we have:

$$\begin{aligned} y &= -\ln x \\ \left|\frac{dy}{dx}\right| &= e^y. \end{aligned} \tag{2.162}$$

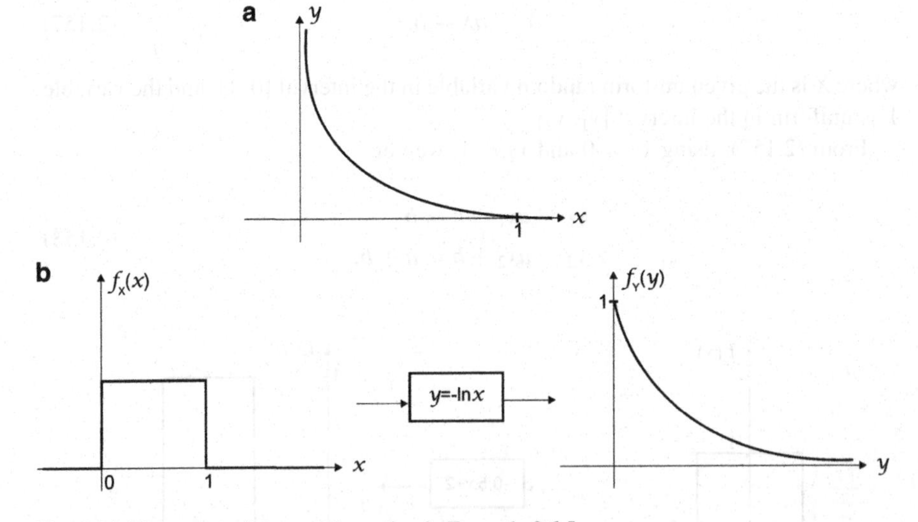

Fig. 2.34 Illustration of the transformation in Example 2.6.5

Finally, from (2.162) and (2.140), we obtain the desired PDF:

$$f_Y(y) = \begin{cases} \mathrm{e}^{-y} & \text{for} \quad y \geq 0, \\ 0 & \quad\text{otherwise}. \end{cases} \tag{2.163}$$

which is shown in Fig. 2.34b.

2.6.3 Nonmonotone Transformation

In the case of the nonmonotone transformation,

$$Y = g(X), \tag{2.164}$$

more than one value of the input random variable corresponds to only one value of the output random variable. Consider the simplest case, in which two input values correspond to only one output value. In the case of continuous random variables, which means that the output random variable Y is in the interval

$$(y, y + dy), \tag{2.165}$$

if the input random variable is either in the interval ($\mathrm{d}x_1 = \mathrm{d}x_2 = \mathrm{d}x$):

$$(x_1, x_1 + \mathrm{d}x) \text{ or } (x_2 - \mathrm{d}x, x_2), \tag{2.166}$$

as shown in Fig. 2.35.

Since the events (2.166) are exclusive, according to Axiom III (1.13), we can write:

$$P\{y < Y \leq y + \mathrm{d}y\} = P\{x_1 < X \leq x_1 + \mathrm{d}x\} + P\{x_2 - \mathrm{d}x < X \leq x_2\}. \tag{2.167}$$

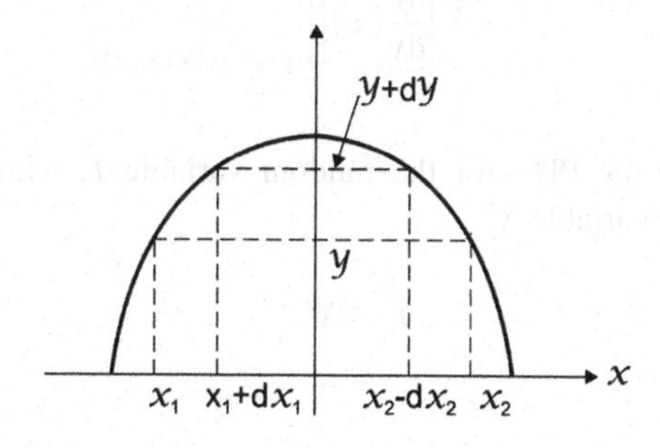

Fig. 2.35 Nonmonotone transformation

The relation presented in (2.167) can be expressed in terms of the corresponding density functions (see (2.64)), knowing that dx is an infinitesimal value,

$$\begin{aligned} P\{x_1 < X \leq x_1 + \mathrm{d}x\} &= f_X(x_1)\mathrm{d}x, \\ P\{x_2 - \mathrm{d}x < X \leq x_2\} = P\{x_2 < X \leq x_2 + \mathrm{d}x\} &= f_X(x_2)\mathrm{d}x. \end{aligned} \tag{2.168}$$

From (2.167) and (2.168), we have:

$$f_Y(y)\mathrm{d}y = f_X(x_1)\mathrm{d}x + f_X(x_2)\mathrm{d}x. \tag{2.169}$$

From here, we arrive at:

$$f_Y(y) = \frac{f_X(x_1)}{\left|\frac{\mathrm{d}y}{\mathrm{d}x}(x_1)\right|} + \frac{f_X(x_2)}{\left|\frac{\mathrm{d}y}{\mathrm{d}x}(x_2)\right|}. \tag{2.170}$$

Next, the right side of (2.170) is expressed in terms of y, resulting in:

$$f_Y(y) = \left. \frac{f_X(x_1)}{\left|\frac{\mathrm{d}y}{\mathrm{d}x}(x_1)\right|} + \frac{f_X(x_2)}{\left|\frac{\mathrm{d}y}{\mathrm{d}x}(x_2)\right|} \right|_{\substack{x_1 = g^{-1}(y) \\ x_2 = g^{-1}(y)}}, \tag{2.171}$$

where x_1 and x_2 are the solutions of (2.164).

For the values of y, for which (2.164) does not have the real solutions, it follows:

$$f_Y(y) = 0. \tag{2.172}$$

The results of (2.171) and (2.172) can be generalized to include a case in which (2.164) has N real solutions

$$f_Y(y) = \left. \sum_{i=1}^{N} \frac{f_X(x_i)}{\left|\frac{\mathrm{d}y}{\mathrm{d}x}(x_i)\right|} \right|_{x_i = g^{-1}(y), i=1,\ldots,N}. \tag{2.173}$$

Example 2.6.6 Find the PDF for the random variable Y, which is obtained by squaring the random variable X,

$$Y = X^2. \tag{2.174}$$

Solution The relations between the corresponding ranges,

$$y = x^2, \tag{2.175}$$

does not have any real solution for $y < 0$. Therefore,

$$f_Y(y) = 0 \quad \text{for} \quad y < 0. \tag{2.176}$$

For $y \geq 0$, (2.175) has two solutions:

$$\begin{aligned} x_1 &= \sqrt{y}, \\ x_2 &= -\sqrt{y}. \end{aligned} \tag{2.177}$$

From (2.175) and (2.177), we have:

$$\begin{aligned} \frac{\mathrm{d}y}{\mathrm{d}x}(x_1) &= 2x_1 = 2\sqrt{y}, \\ \frac{\mathrm{d}y}{\mathrm{d}x}(x_2) &= 2x_2 = -2\sqrt{y}. \end{aligned} \tag{2.178}$$

Finally, from (2.170) and (2.178), we obtain:

$$f_Y(y) = \begin{cases} \dfrac{f_X(\sqrt{y}) + f_X(-\sqrt{y})}{2\sqrt{y}} & \text{for} \quad y \geq 0, \\ 0 & \text{for} \quad y < 0. \end{cases} \tag{2.179}$$

If the density function $f_x(x)$ is even,

$$f_X(x) = f_X(-x), \tag{2.180}$$

from (2.179), we obtain:

$$f_Y(y) = \begin{cases} \dfrac{f_X(\sqrt{y})}{\sqrt{y}} & \text{for} \quad y \geq 0, \\ 0 & \text{for} \quad y < 0. \end{cases} \tag{2.181}$$

Example 2.6.7 Consider the uniform random variable X in the interval $[-1, 5]$ (as shown in Fig. 2.36a). The random variable Y is the absolute value of random variable X, as shown in Fig. 2.36b,

$$Y = |X|. \tag{2.182}$$

Note that for the values of X in the interval $[-1, 1]$, the transformation is nonmonotone because two input values correspond to only one output value. However, for $1 < X < 5$, the transformation is monotone,

$$y = \begin{cases} |x| & \text{for } -1 < x < 1, \\ x & \text{for } \quad 1 < x < 5. \end{cases} \tag{2.183}$$

Find the density of the variable Y.

Solution The density of the variable X is given as:

$$f_X(x) = \begin{cases} 1/6 & \text{for } \quad -1 \leq x \leq 5, \\ 0 & \quad \text{otherwise.} \end{cases} \tag{2.184}$$

For $-1 < x < 1$, it follows: $0 < y < 1$. Using (2.170), we have:

$$f_Y(y) = 2f_X(x) = 1/3 \quad \text{for} \quad 0 \leq y \leq 1. \tag{2.185}$$

For $1 < x < 5$, it follows: $1 < y < 5$. Using (2.140), we have:

$$f_Y(y) = f_X(x) = 1/6 \quad \text{for} \quad 1 \leq y \leq 5. \tag{2.186}$$

From (2.182), there is no solution for $y < 0$ and

$$f_Y(y) = 0 \quad \text{for} \quad y < 0. \tag{2.187}$$

The PDF of the r.v. Y is shown in Fig. 2.36c.

2.6.4 *Transformation of Discrete Random Variables*

The monotone and nonmonotone transformations considered for the continuous random variables in Sects. 2.6.2 and 2.6.3 may be easily applied to the discrete random variables, taking into account that the PDF and distribution for a discrete random variable are given as:

$$\begin{aligned} f_X(x) &= \sum_i P\{x_i\}\delta(x - x_i), \\ F_X(x) &= \sum_i P\{x_i\}u(x - x_i). \end{aligned} \tag{2.188}$$

If the transformation

$$Y = g(X) \tag{2.189}$$

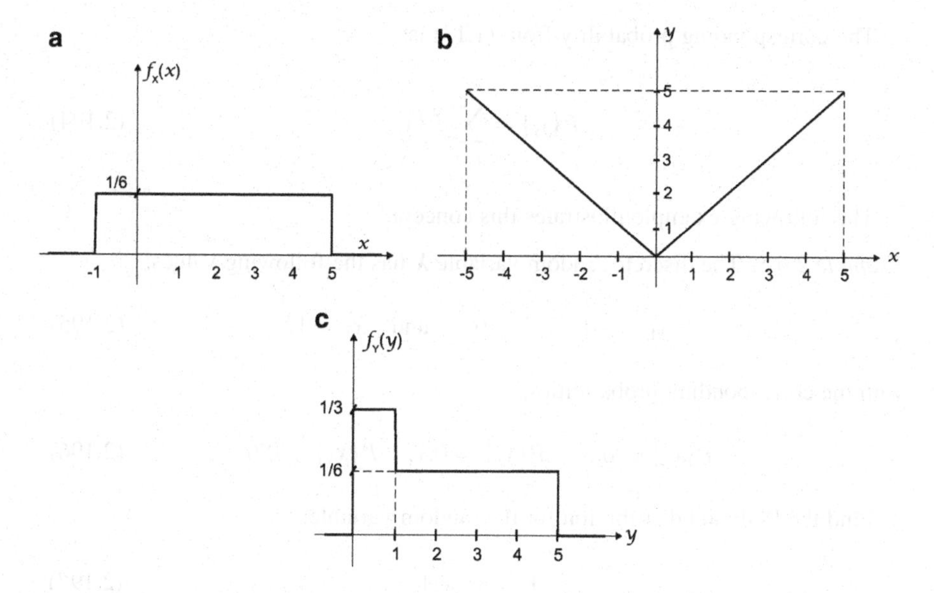

Fig. 2.36 Densities and transformation in Example 2.6.7. (**a**) Input PDF. (**b**) Transformation. (**c**) Output PDF

is monotone, there is a unique correspondence between X and Y, and each value of X corresponds to only one value of Y, resulting in:

$$P\{y_i\} = P\{x_i\} \tag{2.190}$$

and

$$f_Y(y) = \sum_i P\{y_i\}\delta(y - y_i),$$
$$F_Y(y) = \sum_i P\{y_i\}u(y - y_i), \tag{2.191}$$

where

$$y_i = g(x_i). \tag{2.192}$$

For the nonmonotone transformation, N values of the random variable X generally correspond to only one value of the random variable Y, where N is an integer,

$$Y = y_i' \quad \text{if} \quad X = x_1, \text{ or } X = x_2, \text{ or} \ldots X = x_N. \tag{2.193}$$

The corresponding probability from (1.13) is:

$$P\{y_i'\} = \sum_{j=1}^{N} P\{x_j\}. \tag{2.194}$$

The following example illustrates this concept.

Example 2.6.8 The discrete random variable X has the following values:

$$x_1 = -1, \quad x_2 = 0, \quad \text{and} \quad x_3 = 1, \tag{2.195}$$

with the corresponding probabilities:

$$P\{x_1\} = 0.3, \quad P\{x_2\} = 0.1, \quad P\{x_3\} = 0.6. \tag{2.196}$$

Find the PDF and distribution for the random variable:

$$Y = X^2 + 1. \tag{2.197}$$

Solution From (2.195) and (2.197), it follows:

$$\begin{aligned} y_1 &= x_1^2 + 1 = 2, \\ y_2 &= x_2^2 + 1 = 1, \\ y_3 &= x_3^2 + 1 = 2. \end{aligned} \tag{2.198}$$

From (2.198), we can see that the random variable Y takes only the value $y_1{}' = 2$, if X is equal to either x_1 or x_3, and the value $y_2{}' = 1$ if X is equal to x_2. Consequently, we have:

$$y_1{}' = y_1 = y_3 = 2 \tag{2.199}$$

and

$$P\{y_1{}'\} = P\{y_3\} = P\{y_1\} = P\{2\} = P\{x_1\} + P\{x_3\} = 0.9. \tag{2.200}$$

Similarly, from (2.198), we have:

$$y_2{}' = y_2 = 1 \tag{2.201}$$

and

$$P\{y_2{}'\} = P\{y_2\} = P\{1\} = P\{x_2\} = 0.1. \tag{2.202}$$

From (2.200) and (2.202), we have:

$$
\begin{aligned}
f_Y(y) &= \sum_i P\{y_i'\}\delta(y - y_i') = 0.9\delta(y-2) + 0.1\delta(y-1), \\
F_Y(y) &= \sum_i P\{y_i'\}u(y - y_i') = 0.9u(y-2) + u(y-1).
\end{aligned}
\tag{2.203}
$$

2.7 Mean Value

The probability density and distribution functions provide all of the information about the particular random variable being considered. However, there are many applications in which we do not need that much information but where we need some parameters that are representative of a particular distribution without finding the entire density or distribution functions [HAD06, p. 110], [NGU09, p. 86]. The mean value of a random variable is one of the single most important parameters associated with a random variable [PAP65, p. 138], [BRE69, p. 56]. The mean value plays an important role in the characterization of the random variable when a partial description is either needed or only possible.

2.7.1 What Is a Mean Value?

We are all familiar with the term "average" or "mean value" when referring to a finite known set of values. For example, an average grade of an exam, average salary, an average monthly temperature in a town, etc. As an example, denote the temperature measured in ith day as t_i, then the average temperature t_{av} is obtained by adding all t_i and dividing by the number of the days

$$
t_{av} = \frac{t_1 + t_2 + \cdots + t_{30}}{30}. \tag{2.204}
$$

The temperature t_{av} can be viewed as the "most likely" or "expected" temperature in a month, yet may never happen. For example, if the average temperature calculated in (2.204) is 16.5°C, it may never occur that the measured temperature was 16.5°C. However, generally speaking we have enough information about the weather during this month (i.e., we do not expect snow, but we are also not expecting very hot weather). Similarly, an average salary in a country tells us about the standard of the life in this country; the average grade on an exam tells us about the general success of students on this exam; an average speed of 110 km/h in a car tells that the driver drives very fast, etc.

Generally, when a large collection of numbers is assembled, we are not interested in the individual numbers, but rather in the average value. The average temperature, for example, gives us more information than would a single temperature on any particular day.

If we have a known set of values x_i, $i = 1, \ldots, N$, then the *average value* m_{av} is just the sum of values S divided by the total number of items, N:

$$m_{av} = \frac{S}{N} = \frac{x_1 + x_2 + \cdots + x_N}{N}. \tag{2.205}$$

Note that we always average numbers and not things [BRO97, p. 22]. There is no such thing as the average of sea, for example. However, there is an average temperature of the sea, which is a numerical value.

In order to interpret the meaning of the value m_{av}, Fig. 2.37 presents the set of the values x_i. In this example, we are looking for point y, which is in the smallest squared distance ("squared" is used to take equally both positive and negative distances into account), of all points x_i. The sum of squared distances from y to all x_i is:

$$D = \sum_{i=1}^{N} (y - x_i)^2. \tag{2.206}$$

The minimum distance occurs for

$$\frac{dD}{dy} = 0. \tag{2.207}$$

From (2.206) and (2.207), we have:

$$2\sum_{i=1}^{N} (y - x_i) = 0 \tag{2.208}$$

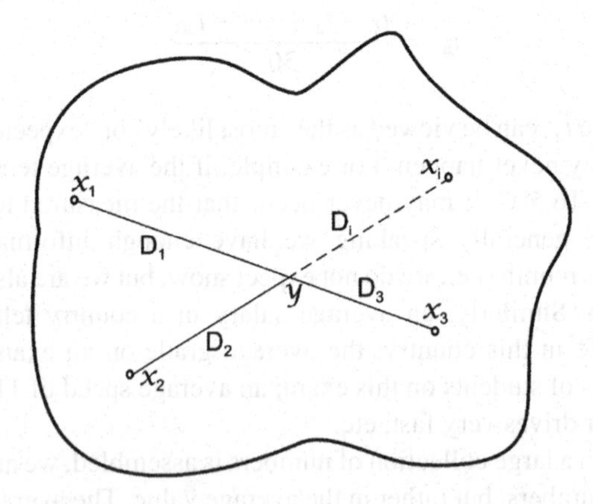

Fig. 2.37 Interpretation of the mean value

or

$$2\sum_{i=1}^{N} y - 2\sum_{i=1}^{N} x_i = 2\,Ny - 2\sum_{i=1}^{N} x_i = 0. \qquad (2.209)$$

From (2.209), we arrive at:

$$y = \frac{x_1 + x_2 + \cdots + x_N}{N}. \qquad (2.210)$$

Note that this is the same expression as we saw in (2.205). Therefore, the average m_{av} of the set of numbers x_i, $i = 1, \ldots, N$, can be viewed as the number, which is simultaneously closest to all the numbers in the set or the *center of gravity* of the set.

The next question is: "can we apply the same concept to a random variable?"

2.7.2 Concept of a Mean Value of a Discrete Random Variable

Consider the discrete random variable X with the values $x_1, \ldots, x_N$. We cannot directly apply the formula (2.205), because in (2.205) each value x_i occurs only once, and the values x_i of the random variable X occur with a certain probability. In order to apply (2.205), we must know the exact number of occurrences for each value x_i. The required values can be obtained in one experiment.

Let the experiment be performed M times. We will measure how many times each value x_i occurs in the experiment:

$$\begin{array}{ll} X = x_1 & N_1 \text{ times}, \\ X = x_2 & N_2 \text{ times}, \\ \cdots & \cdots \\ X = x_N & N_N \text{times}, \end{array} \qquad (2.211)$$

where

$$M = N_1 + N_2 + \ldots + N_N. \qquad (2.212)$$

Now we have the finite set of values: N_1 values of x_1, N_2 values of x_2, etc.

$$S = N_1 x_1 + N_2 x_2 + \ldots + N_N x_N \qquad (2.213)$$

and we can find the mean value of the set (2.211) using (2.205):

$$m_{\text{emp}} = \frac{S}{M} = \frac{N_1 x_1 + \cdots + N_N x_N}{M} = \sum_{i=1}^{N} \frac{N_i}{M} x_i. \qquad (2.214)$$

The subscript "emp" indicates that the mean value is obtained in the experiment. This mean value is called the *empirical mean value*.

If we repeat the experiment again, we will obtain another empirical mean value. All of these mean values depend on the corresponding experiment (i.e., the average value of the outcomes is not predictable).

Example 2.7.1 A die rolling experiment is performed three times. In each experiment, the die is rolled 100 times and the number of dots facing up is counted and noted. The results are shown in Table 2.3.

The mean value obtained in the first experiment is:

$$m_{\mathrm{emp}_1} = \frac{1 \times 18 + 2 \times 20 + 3 \times 17 + 4 \times 15 + 5 \times 14 + 6 \times 16}{100} = 3.35. \tag{2.215}$$

Similarly, in the second and third experiments, we have:

$$m_{\mathrm{emp}_2} = \frac{1 \times 13 + 2 \times 22 + 3 \times 18 + 4 \times 15 + 5 \times 17 + 6 \times 15}{100} = 3.46. \tag{2.216}$$

$$m_{\mathrm{emp}_3} = \frac{1 \times 12 + 2 \times 16 + 3 \times 19 + 4 \times 14 + 5 \times 16 + 6 \times 18}{100} = 3.45. \tag{2.217}$$

The obtained mean values are different. Similarly, in each experiment, we will obtain a different empirical mean value.

However, we are interested in a mean value which does not depend on the experiment. Intuitively, we can say that the empirical mean values will vary a little if the number of repetitions of each experiment M is very high ($M \gg 1$).

To this end, note that the ratio N_i/M in (2.214) represents the relative frequency which approaches the probability of the event x_i, if $M \gg 1$ (see (1.38)).

Table 2.3 The results obtained in three experiments

Experiment 1		Experiment 2		Experiment 3	
Number of dots	Occurrence	Number of dots	Occurrence	Number of dots	Occurrence
1	18	1	13	1	12
2	20	2	22	2	16
3	17	3	18	3	19
4	15	4	15	4	14
5	14	5	17	5	16
6	16	6	15	6	18

$$\frac{N_i}{M} \to P\{X = x_i\}, \text{ as } M \gg 1 \tag{2.218}$$

In this way, (2.214) becomes independent of the experiment and becomes the characteristic of the random variable X, called the *mean*, or *expected*, *value* of the random variable, $E\{X\}$.

$$m_{\text{emp}} \to E\{X\}, \text{as } M \gg 1 \tag{2.219}$$

From (2.214), (2.218), and (2.219), we have:

$$E\{X\} = \sum_{i=1}^{N} x_i P\{X = x_i\}. \tag{2.220}$$

Other denotations used in literature are: m, $\overline{X}$, and $\langle X \rangle$. These denotations will also be used throughout this text. When one would like to emphasize that m is the mean value of the random variable X, one write m_X rather than m.

The next examples show that $E\{X\}$ should not be interpreted as a value that we would "expect" X to take.

Example 2.7.2 Find the mean value of the random variable X using only two values: $x_1 = 1$ and $x_2 = 0$ where $P\{x_1\} = P\{x_2\} = 0.5$.

Solution From (2.220), we have:

$$E\{X\} = 1 \times 0.5 + 0 \times 0.5 = 0.5. \tag{2.221}$$

If X can only take the values 0 or 1, then we can never expect X to take the value 0.5 (see Fig. 2.38).

Example 2.7.3 Consider a random variable with six values: 1, 2, 3, 4, 5, and 6, where all six values of the random variable have the same probability (i.e., $P\{x_i\} = 1/6$, $i = 1, \ldots, 6$). The mean value is:

$$E\{X\} = 1 \times 1/6 + 2 \times 1/6 + 3 \times 1/6 + 4 \times 1/6 + 5 \times 1/6 + 6 \times 1/6 \\ = 3.5. \tag{2.222}$$

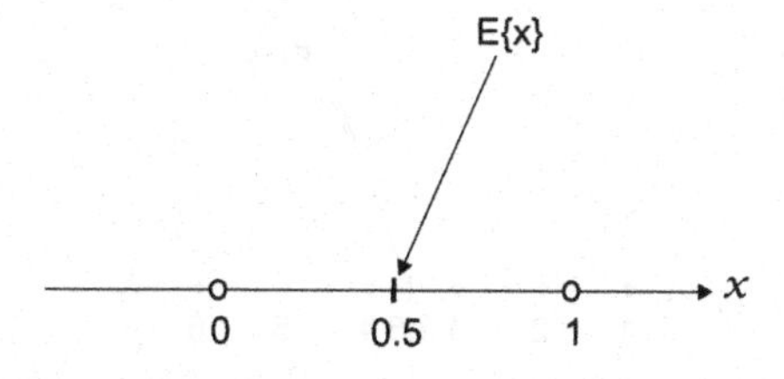

Fig. 2.38 Illustration of Example 2.7.2

The mean value also never occurs as a value of the random variable, as shown in Fig. 2.39.

2.7.2.1 General Expressions

More general expression of (2.220) can be given as:

$$E\{X\} = \sum_{i=-\infty}^{\infty} x_i P\{X = x_i\}. \tag{2.223}$$

However, the infinite sum (2.223) may not converge and in that case, a mean value does not exist. The condition for the existence of a mean value is the absolute convergence of the sum (2.223), as given in (2.224):

$$\sum_{i=-\infty}^{\infty} |x_i| P\{X = x_i\} < \infty. \tag{2.224}$$

Example 2.7.4 Consider a random variable X with the values $x_k = k, k = 0,1, \ldots, \infty$. The probability that random variable takes the value k is given as:

$$P\{X = k\} = \frac{4/\pi^2}{k^2}, \tag{2.225}$$

where

$$\sum_{k=0}^{\infty} \frac{4/\pi^2}{k^2} = 1. \tag{2.226}$$

The mean value of the random variable X is:

$$E\{X\} = \sum_{k=0}^{\infty} kP\{X = k\} = \sum_{k=0}^{\infty} k\frac{4/\pi^2}{k^2} = \frac{4}{\pi^2}\sum_{k=0}^{\infty} \frac{1}{k}. \tag{2.227}$$

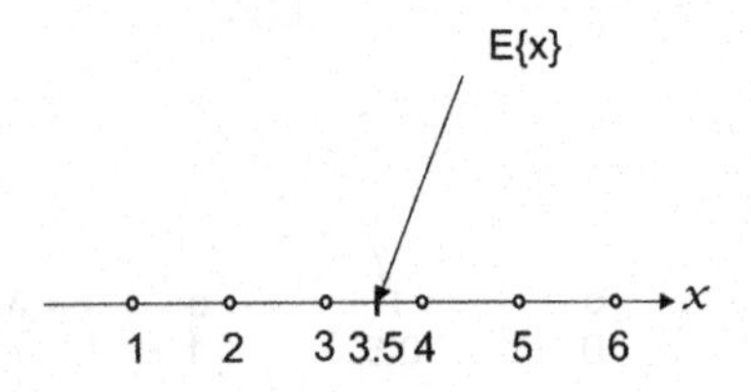

Fig. 2.39 Illustration of Example 2.7.3

However, the sum of (2.227) does not converge

$$\sum_{k=0}^{\infty} \frac{1}{k} \to \infty, \tag{2.228}$$

and consequently, the mean value of the random variable X does not exist.

If we consider an arbitrary function of x_i, $g(x_i)$, instead of x_i, the expression (2.214) becomes:

$$m_{\text{empg}} = \frac{N_1 g(x_1) + \cdots + N_N g(x_n)}{M} = \sum_{i=1}^{N} \frac{N_i}{M} g(x_i). \tag{2.229}$$

For a high M value ($M \gg 1$), the expression (2.229) becomes the expected value of $g(X)$:

$$E\{g(X)\} = \sum_{i=1}^{N} g(x_i) P\{X = x_i\}. \tag{2.230}$$

If we consider that the random variable $g\{X\}$ is a new variable Y, and $y_i = g(x_i)$, then, according to (2.220), we have:

$$E\{g(x)\} = E\{Y\} = \sum_i y_i P\{Y = y_i\}. \tag{2.231}$$

Generally, the number of values of the random variables X and Y may be different. The expression in (2.231) is an alternative relation used to find the expected value of the function of the variable X, $g(X)$. Note that in (2.231), we have to find the values y_i and the probabilities $P\{Y = y_i\}$.

Example 2.7.5 Consider the random variable X with values $x_1 = -1$, $x_2 = 1$, $x_3 = -2$, $x_4 = 2$. The corresponding probabilities are: $P\{x_i\} = 1/4$, for $i = 1, \ldots, 4$. Find the expected value of X^2.

Solution Using (2.230), we have:

$$\begin{aligned} E\{g(X)\} = E\{X^2\} &= \sum_{i=1}^{4} g(x_i) P\{X = x_i\}, \\ &= (-1)^2 \times 1/4 + 1^2 \times 1/4 + (-2)^2 \times 1/4 + 2^2 \times 1/4 = 5/2. \end{aligned} \tag{2.232}$$

In order to use (2.231), we define the random variable $Y = X^2$, and find the values x_i^2, $i = 1, \ldots, 4$:

$$x_1^2 = x_2^2 = 1; \quad x_3^2 = x_4^2 = 4. \tag{2.233}$$

The random variable Y has only two values:

$$y_1 = 1 \quad \text{and} \quad y_2 = 4. \tag{2.234}$$

The value y_1 occurs if $X = x_1$ or $X = x_2$. Therefore, according to (1.13), the corresponding probability is:

$$P\{y_1\} = P\{x_1\} + P\{x_2\} = 1/4 + 1/4 = 1/2. \tag{2.235}$$

Similarly, we find:

$$P\{y_2\} = P\{x_3\} + P\{x_4\} = 1/4 + 1/4 = 1/2. \tag{2.236}$$

From (2.231) and (2.234)–(2.236), we have:

$$E\{X^2\} = E\{Y\} = \sum_{i=1}^{2} y_i P\{Y = y_i\} = 1 \times 1/2 + 4 \times 1/2 = 5/2. \tag{2.237}$$

This is the same result as was obtained in (2.232).

A more general expression (2.230) is obtained considering the infinite value for N:

$$E\{g(X)\} = \sum_{i=-\infty}^{\infty} g(x_i) P\{X = x_i\}. \tag{2.238}$$

The following condition must be satisfied in order to show that the expected value (2.238) exists

$$\sum_{i=-\infty}^{\infty} |g(x_i)| P\{X = x_i\} < \infty. \tag{2.239}$$

2.7.2.2 Properties

We now review some properties of the mean value for discrete random variable:

P.1 The mean value of the constant is the constant itself.

Consider the expected value of the constant b. In this case, there is no random variable but only one discrete value b with the probability $P\{b\} = 1$. From (2.220), it follows:

$$E\{b\} = b \times 1 = b; \quad b = \text{const}. \tag{2.240}$$

P.2 The mean value of the mean value is the mean value itself.

Being $E\{X\} = \text{const}$, from (2.240), it easily follows:

$$E\{E\{X\}\} = E\{X\}. \tag{2.241}$$

P.3 The expected value of a random variable X multiplied by the constant a is equal to the product of the constant a and the expected value of the normal random variable: $E\{aX\} = aE\{X\}$, $a = \text{const}$.

From (2.230), the expected value of aX, where a is a constant, is:

$$E\{aX\} = \sum_{i=1}^{N} ax_i P\{X = x_i\} = a \sum_{i=1}^{N} x_i P\{X = x_i\} = aE\{X\}. \tag{2.242}$$

(Because a is constant it can be moved in front of the sum, which itself represents $E\{X\}$).

P.4 This is a combination of P.1 and P.2: $E\{aX + b\} = aE\{X\} + b$

Combining (2.240) and (2.242), we have:

$$\begin{aligned} E\{aX + b\} &= \sum_{i=1}^{N} (ax_i + b) P\{X = x_i\} \\ &= a \sum_{i=1}^{N} x_i P\{X = x_i\} + b \sum_{i=1}^{N} P\{X = x_i\}. \end{aligned} \tag{2.243}$$

The first sum in (2.243) is $E\{X\}$ while the second sum is equal to 1, yielding:

$$E\{aX + b\} = aE\{X\} + b. \tag{2.244}$$

P.5 The linear transformation $g(X)$ of the random variable X and its expectation are commutative operations.

From (2.244), it follows that if the function of the random variable X is a linear function

$$g(X) = aX + b, \tag{2.245}$$

then the linear transformation of the random variable and its expectation are commutative operations:

$$E\{g(X)\} = aE\{X\} + b = g(E\{X\}). \tag{2.246}$$

However, in a general case in which the transformation $g(X)$ is not linear

$$E\{g(X)\} \neq g(E\{X\}). \tag{2.247}$$

Example 2.7.6 Consider the random variable X from Example 2.7.5, where we found

$$E\{X\} = 0; \quad E\{X^2\} = 5/2. \tag{2.248}$$

The transformation of the variable X was a quadratic function (i.e., a nonlinear function)

$$g(X) = X^2. \tag{2.249}$$

The corresponding mean value is:

$$E\{g(X)\} = E\{X^2\} = 5/2. \tag{2.250}$$

However, the quadratic transformation of the mean value is:

$$g(E\{X\}) = (E\{X\})^2 = 0^2 = 0. \tag{2.251}$$

Therefore,

$$E\{g(X)\} \neq g(E\{X\}). \tag{2.252}$$

2.7.3 *Mean Value of a Continuous Random Variable*

We can apply a similar approach to that used to derive the mean value of a discrete random variable, in order to develop the expression for the mean value of a continuous random variable. Consider the continuous random variable X with the continuous range x from 0 to A (Fig. 2.40). We divide the range x into k small intervals Δx such that, if the random variable belongs to the ith interval $[(i - 1)\,\Delta x, i\Delta x]$, we can consider that it is equal to the end of the interval (i.e., equal to $i\Delta x$).

In other words, if $X\varepsilon[(i-1)\Delta x, i\Delta x]$ it means that $X = i\,\Delta x$, $i = 1, \ldots, k$.

Imagine that we perform an experiment N times and that we obtain the number of times that the variable X was in each of the intervals $i\Delta x$, $i = 1, \ldots, k$:

$$\begin{aligned}
X &= \Delta x && \cdots && N_1 \text{ times},\\
X &= 2\,\Delta x && \cdots && N_2 \text{ times},\\
& && \cdots && \cdots\\
X &= i\,\Delta x && \cdots && N_i \text{ times},\\
& && \cdots && \cdots\\
X &= k\,\Delta x && \cdots && N_k \text{ times}.
\end{aligned} \tag{2.253}$$

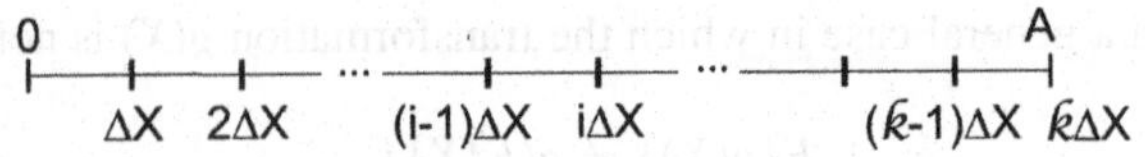

0 ΔX 2ΔX ... (i-1)ΔX iΔX ... (k-1)ΔX kΔX A

Fig. 2.40 Continuous range of the variable X

where

$$M = N_1 + N_2 + \cdots + N_i + \cdots + N_k. \quad (2.254)$$

The average value of X in this experiment is obtained by adding all of the values (2.253) and dividing by M:

$$m_{\text{emp}} = \frac{N_1\Delta x + \cdots + N_i i\Delta x + \cdots + N_k k\Delta x}{M} = \sum_{i=1}^{k} \frac{N_i}{M} i\Delta x. \quad (2.255)$$

We rewrite (2.255) as:

$$m_{\text{emp}} = \sum_{i=1}^{K} \frac{N_i}{M} i\Delta x = \sum_{i=1}^{K} \left(\frac{N_i}{M}\frac{1}{\Delta x}\right)(i\Delta x)\Delta x. \quad (2.256)$$

Considering that M is a high enough value ($M> > 1$), the frequency ratio N_i/M approaches the probability

$$\frac{N_i}{M} \to P\{(i-1)\Delta x < X \leq i\Delta x\}, \text{ as } M \gg 1, \quad (2.257)$$

Taking into account that Δx is small enough to become the infinitesimal interval dx, the expression in parenthesis in (2.256) becomes the density function:

$$\left(\frac{N_i}{M}\frac{1}{\Delta x}\right) \to f_X(x). \quad (2.258)$$

Similarly, the sum in (2.256) becomes the integral and $(i\Delta x)$ becomes x, yielding:

$$\underset{\Delta x \to \mathrm{d}x}{m_{\text{emp}}} \to E\{X\} = \int_0^A x f_X(x)\mathrm{d}x, \text{ as } M \gg 1, \ \Delta x \to \mathrm{d}x \quad (2.259)$$

The expression on the right side of (2.259) is the average, or expected, value m for the continuous variable X

$$E\{X\} = m = \int_0^A x f_X(x)\mathrm{d}x. \quad (2.260)$$

The result (2.260) may be generalized to include any continuous range of the random variable:

$$m = E\{X\} = \int_{-\infty}^{\infty} x f_X(x) \mathrm{d}x. \tag{2.261}$$

Example 2.7.7 Find the mean values of the random variables X_1 and X_2 with the corresponding density functions shown in Fig. 2.41:

$$\begin{aligned} f_{X_1}(x) &= \begin{cases} 1/2 & \text{for} \quad -1 \le x \le 1, \\ 0 & \text{otherwise.} \end{cases} \\ f_{X_2}(x) &= \begin{cases} 1/2 & \text{for} \quad 0 \le x \le 2, \\ 0 & \text{otherwise.} \end{cases} \end{aligned} \tag{2.262}$$

Solution From (2.261), we have:

$$\begin{aligned} m_{X_1} &= \int_{-1}^{1} x \frac{1}{2} \mathrm{d}x = \frac{1}{4} - \frac{1}{4} = 0, \\ m_{X_2} &= \int_{0}^{2} x \frac{1}{2} \mathrm{d}x = \frac{2^2}{4} - 0 = 1. \end{aligned} \tag{2.263}$$

Example 2.7.8 We want to find the mean value of the random variable X with the density function, shown in Fig. 2.42.

The density function is described as:

$$f_X(x) = \begin{cases} \frac{1}{4}(x+2) & \text{for} \quad -2 \le x \le 0, \\ -\frac{1}{4}(x-2) & \text{for} \quad 0 \le x \le 2. \end{cases} \tag{2.264}$$

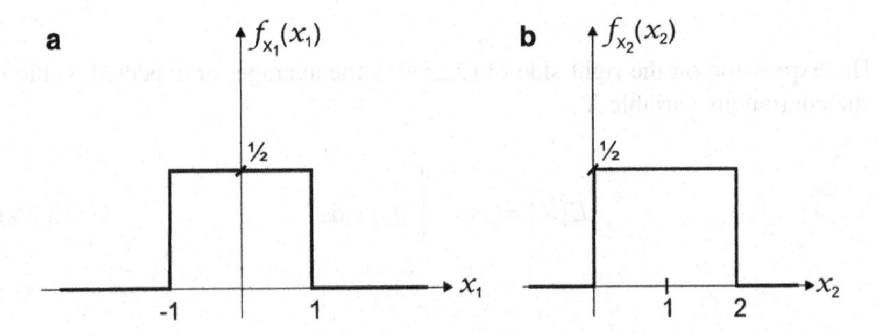

Fig. 2.41 Density functions in Example 2.7.7

The mean value is:

$$m_X = \int_{-2}^{0} x\frac{1}{4}(x+2)\mathrm{d}x - \int_{0}^{2} x\frac{1}{4}(x-2)\mathrm{d}x$$
$$= -\frac{(-2)^3}{12} - \frac{(-2)^2}{4} - \left(\frac{2^3}{12} - \frac{2^2}{4}\right) = 0. \quad (2.265)$$

Example 2.7.9 Find the average value of the random variable with the density function:

$$f_X(x) = \begin{cases} 0.5\mathrm{e}^{-0.5x} & \text{for} \quad 0 \le x < \infty, \\ 0 & \text{otherwise}, \end{cases} \quad (2.266)$$

as shown in Fig. 2.43.

Solution From (2.261), we have:

$$m = \int_{0}^{\infty} 0.5x\mathrm{e}^{-0.5x}\mathrm{d}x. \quad (2.267)$$

Using the integral:

$$\int x\mathrm{e}^{ax}\mathrm{d}x = \frac{\mathrm{e}^{ax}}{a^2}(ax-1), \quad (2.268)$$

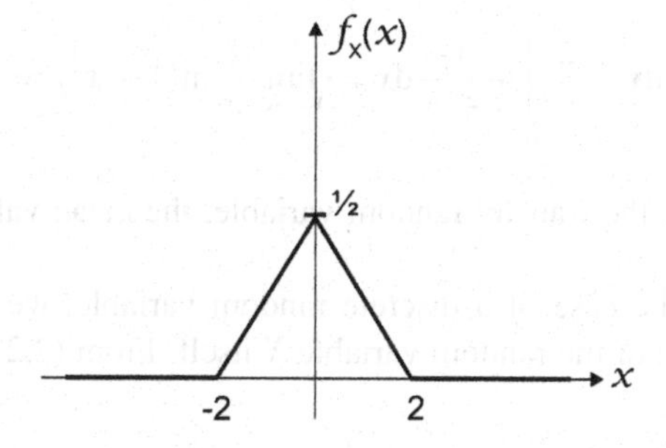

Fig. 2.42 Density function in Example 2.7.8

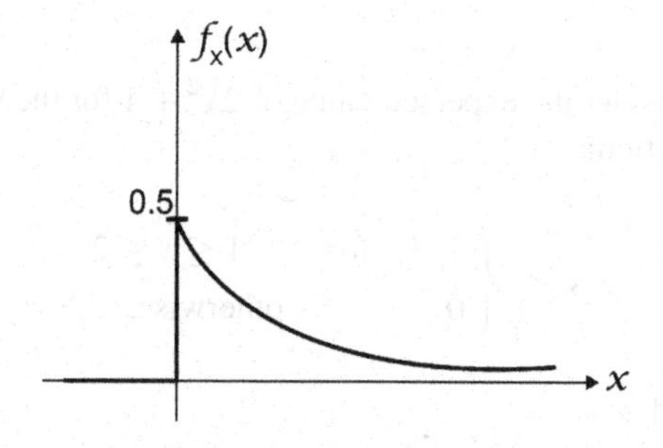

Fig. 2.43 Density function in Example 2.7.9

we arrive at:

$$m = 0.5 \int_0^\infty x e^{-0.5x} \mathrm{d}x = 0.5 \left[\frac{e^{-0.5x}}{(-0.5)^2} (-0.5x - 1) \right] \Bigg|_0^\infty = \frac{0.5}{0.5^2} = 2. \quad (2.269)$$

The following condition must be satisfied in order for a mean value to exist (the integral at the right side of (2.261) must converge absolutely)

$$\int_{-\infty}^{\infty} |x| f_X(x) \mathrm{d}x < \infty. \quad (2.270)$$

If the condition in (2.270) is not satisfied, the mean value of the random variable does not exist, as shown in the next example.

Example 2.7.10 Let X be a continuous random variable in the range $-\infty < x < \infty$ with the density function:

$$f_X(x) = \frac{1}{\pi(1 + x^2)}, \quad -\infty < x < \infty. \quad (2.271)$$

A random variable that has this density function is called a *Cauchy random variable*. However, in Cauchy random variables, the condition in (2.270) is not satisfied (see Integral 15, Appendix B):

$$\frac{1}{\pi} \int_{-\infty}^{\infty} \frac{|x|}{1 + x^2} \mathrm{d}x = \frac{2}{\pi} \int_0^\infty \frac{x}{1 + x^2} \mathrm{d}x = \lim_{X \to \infty} \frac{1}{\pi} \ln(1 + x^2) = \infty. \quad (2.272)$$

Consequently, for the Cauchy random variable, the mean value $E\{X\}$ does not exist.

Similarly, as in the case of a discrete random variable, we may consider the function $g(X)$ instead of the random variable X itself. From (2.270), we obtain:

$$E\{g(X)\} = \int_{-\infty}^{\infty} g(x) f_x(x) \mathrm{d}x. \quad (2.273)$$

Example 2.7.11 Consider the expected value of $2X^2 + 1$ for the random variable X with the density function:

$$f_X(x) = \begin{cases} 1/3 & \text{for} \quad -1 \leq x \leq 2, \\ 0 & \text{otherwise}, \end{cases} \quad (2.274)$$

as shown in Fig. 2.44.

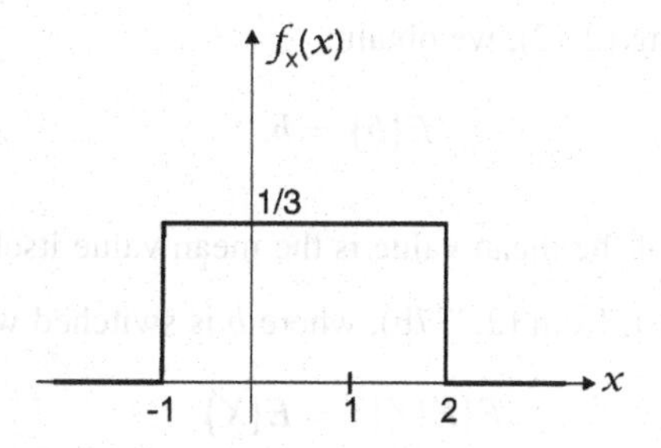

Fig. 2.44 Density function in Example 2.7.11

Solution From (2.273), we have:

$$E\{g(X)\} = E\{2X^2 + 1\} = \int_{-1}^{2} g(x) f_X(x) \mathrm{d}x = \int_{-1}^{2} (2x^2 + 1) \frac{1}{3} \mathrm{d}x = 3. \tag{2.275}$$

2.7.3.1 Properties

We can easily verify that the properties considered for the discrete random variable also stand for a continuous variable.

P.1 The mean value of the constant is the constant itself.

Consider the expected value of the constant b. In this case, there is no random variable but only one discrete value b with the probability $P\{b\} = 1$. The corresponding density function is a delta function $\delta(x - b)$ (Fig. 2.45)

$$f_X(x) = \delta(x - b). \tag{2.276}$$

From (2.261), we have:

$$E\{b\} = \int_{-\infty}^{\infty} x f_X(x) \mathrm{d}x = \int_{-\infty}^{\infty} x \delta(x - b) \mathrm{d}x. \tag{2.277a}$$

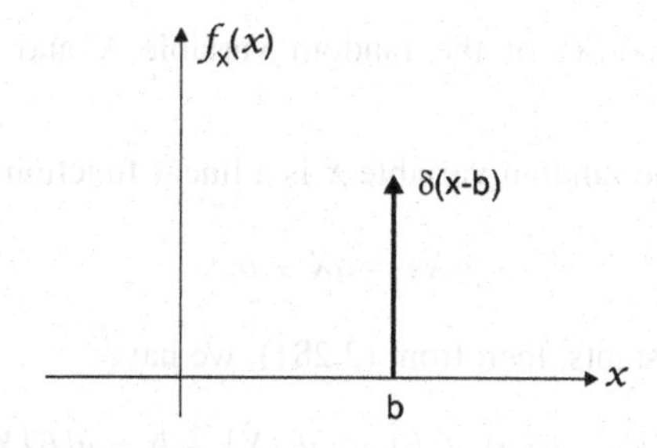

Fig. 2.45 Density function of the constant b

From (2.277a), using (2.72), we obtain:

$$E\{b\} = b. \tag{2.277b}$$

P.2 The mean value of the mean value is the mean value itself.

Being $E\{X\}$ = const, from (2.277b), where b is switched with $E\{X\}$, we have:

$$E\{E\{X\}\} = E\{X\}. \tag{2.278}$$

P.3 $E\{aX\} = aE\{X\}$, a = const.

From (2.273), the expected value of aX, where a is the constant, is:

$$E\{aX\} = \int_{-\infty}^{\infty} axf_X(x)\mathrm{d}x = a\int_{-\infty}^{\infty} xf_X(x)\mathrm{d}x = aE\{X\}. \tag{2.279}$$

(Since a is constant, it can be moved in front of the integral, which itself represents $E\{X\}$).

P.4 $E\{aX + b\} = aE\{X\} + b$.

Combining (2.277b) and (2.279), we have:

$$E\{aX+b\} = \int_{-\infty}^{\infty} (ax+b)f_X(x)\mathrm{d}x = a\int_{-\infty}^{\infty} xf_X(x)\mathrm{d}x + b \times \int_{-\infty}^{\infty} f_X(x)\mathrm{d}x = aE\{X\} + b. \tag{2.280}$$

Note that the first integral on the right side of (2.280) represents the expected value of the random variable X, while the second integral is equal to 1 (see the characteristic of the density function (2.83)). This results in:

$$E\{aX+b\} = aE\{X\} + b. \tag{2.281}$$

P.5 A linear function $g(X)$ of the random variable X and its expectation are commutative operations.

If the function of the random variable X is a linear function,

$$g(X) = aX + b, \tag{2.282}$$

where a and b are constants, then from (2.281), we have:

$$E\{g(X)\} = E\{aX+b\} = aE\{X\} + b = g(E\{X\}). \tag{2.283}$$

From here, we conclude that the linear transformation of the continuous random variable and its expectation are commutative operations.

However, in general case, when $g(X)$ is not a linear function, it stands that:

$$E\{g(X)\} \neq g(E\{X\}). \tag{2.284}$$

The next example illustrates (2.284).

Example 2.7.12 Find the expected value of

$$g(X) = 2X^2 + 1, \tag{2.285}$$

where X is the random variable considered in Example 2.7.11.

Using the result from Example 2.7.11 (2.275) the expected value of the transformation of the random variable X is:

$$E\{g(X)\} = E\{2X^2 + 1\} = 3. \tag{2.286}$$

Using (2.274), the mean value of r.v. X is:

$$E\{X\} = \int_{-1}^{2} x\frac{1}{3}\mathrm{d}x = \frac{1}{2}. \tag{2.287}$$

The transformation (2.285) of the expected value $E\{X\}$ is:

$$g(E\{X\}) = 2(E\{X\})^2 + 1 = 2\left(\frac{1}{2}\right)^2 + 1 = \frac{3}{2}, \tag{2.288}$$

thus confirming that in this example, the relation (2.284) holds because the transformation was not a linear transformation.

2.7.4 General Expression of Mean Value for Discrete, Continuous, and Mixed Random Variables

The expression (2.261) is more general than (2.223) because we can apply (2.261) to find the mean value of a discrete random variable. However, (2.223) cannot be used to find the mean value of a continuous random variable because the probability that a continuous random variable takes any particular value in its range is zero, as explained in Sect. 2.3.

Knowing that the density function of a discrete random variable is:

$$f_X(x) = \sum_i P\{X = x_i\}\delta(x - x_i), \tag{2.289}$$

placing (2.289) into (2.261), we have:

$$m = \int_{-\infty}^{\infty} x \sum_i P\{X = x_i\}\delta(x - x_i)\mathrm{d}x. \tag{2.290}$$

Interchanging the sum and the integral in (2.290), and knowing that

$$\int_{-\infty}^{\infty} x\delta(x - x_i)\mathrm{d}x = \begin{cases} x_i & \text{for} \quad x = x_i, \\ 0 & \text{otherwise}. \end{cases} \tag{2.291}$$

we arrive at:

$$m = \sum_i P\{X = x_i\} \int_{-\infty}^{\infty} x\delta(x - x_i)\mathrm{d}x = \sum_i x_i P\{X = x_i\}. \tag{2.292}$$

This is the same result as in (2.223).

To this end, we will use the expression (2.261) for the mean value in order to explain different characteristics of random variables considered in the rest of the book.

Example 2.7.13 We want to find the mean value of the discrete variable X from Example 2.7.2 using (2.261).

Solution The density function is:

$$f_X(x) = 0.5\delta(x) + 0.5\delta(x - 1). \tag{2.293}$$

From (2.261), we have:

$$\begin{aligned} m_X &= \int_{-\infty}^{\infty} x f_X(x)\mathrm{d}x = \int_{-\infty}^{\infty} x[0.5\delta(x) + 0.5\delta(x - 1)]\mathrm{d}x, \\ &= \int_{-\infty}^{\infty} 0.5x\delta(x)\mathrm{d}x + \int_{-\infty}^{\infty} 0.5x\delta(x - 1)\mathrm{d}x. \end{aligned} \tag{2.294}$$

The first integral on the right side of (2.294) is:

$$\int_{-\infty}^{\infty} x\delta(x)\mathrm{d}x = 0 \tag{2.295}$$

Similarly, the second integral in (2.294) is:

$$\int_{-\infty}^{\infty} x\delta(x-1)\mathrm{d}x = 1 \tag{2.296}$$

From (2.294) to (2.296), we have:

$$m_X = 0.5 \times 0 + 0.5 \times 1 = 0.5. \tag{2.297}$$

Note that this is the same result as in Example 2.7.2

The next example illustrates that the general expression (2.261) can also be used in the calculation of the mean value for a mixed random variable.

Example 2.7.14 Consider a mixed random variable with the density function:

$$f_X(x) = \begin{cases} \frac{1}{3}\delta(x) & \text{for} \quad x = 0, \\ \frac{2}{3}x & \text{for} \quad 0 < x \leq 1, \\ -\frac{2}{3}(x-2) & \text{for} \quad 1 \leq x \leq 2, \\ 0 & \text{otherwise.} \end{cases} \tag{2.298}$$

From (2.261) and (2.298), we have:

$$E\{X\} = \int_{-\infty}^{\infty} x f_X(x)\mathrm{d}x = \int_{-\infty}^{\infty} \frac{x}{3}\delta(x)\mathrm{d}x + \int_{0}^{1} \frac{2x^2}{3}\mathrm{d}x + \int_{1}^{2} -\frac{2x(x-2)}{3}\mathrm{d}x. \tag{2.299}$$

Knowing that

$$\int_{-\infty}^{\infty} x\delta(x)\mathrm{d}x = 0 \tag{2.300}$$

the first integral in the right side of (2.299) is equal to zero, that is:

$$\int_{-\infty}^{\infty} \frac{x}{3}\delta(x)\mathrm{d}x = 0 \tag{2.301}$$

The second integral on the right side of (2.299) is:

$$\int_{0}^{1} \frac{2x^2}{3}\mathrm{d}x = \left.\frac{2x^3}{9}\right|_{0}^{1} = \frac{2}{9}. \tag{2.302}$$

The third integral in (2.299) is:

$$\int_{1}^{2} -\frac{2x(x-2)}{3}\mathrm{d}x = \frac{2x^3}{9}\Big|_2^1 + \frac{2x^2}{3}\Big|_1^2 = \frac{4}{9}. \quad (2.303)$$

From (2.299), and using (2.301)–(2.303), we arrive at:

$$E\{X\} = 2/9 + 4/9 = 2/3. \quad (2.304)$$

2.7.5 Conditional Mean Values

There are some applications where we need to compute mean value conditioned on certain events.

The mean value of the random variable X, conditioned on a given event A, can be calculated using the general expression (2.261) by replacing the PDF with conditional PDF as shown in (2.305):

$$E\{X|A\} = \int_{-\infty}^{\infty} x f_X(x|A)\mathrm{d}x. \quad (2.305)$$

Similarly, for the function $g(X)$ of the random variable X, conditioned on event A, the corresponding mean value is given as

$$E\{g(X)|A\} = \int_{-\infty}^{\infty} g(x) f_X(x|A)\mathrm{d}x. \quad (2.306)$$

The next issue is how to define event A.

One way to define event A is to let it be dependent on the random variable X, for example,

$$A = \{X \le a\}. \quad (2.307)$$

Then the corresponding conditional density is (see (2.115)):

$$f_X(x|A) = f_X(x|X \le a) = \frac{f_X(x)}{P\{X \le A\}} = \frac{f_X(x)}{\int_{-\infty}^{a} f_X(x)\mathrm{d}x}. \quad (2.308)$$

Placing (2.308) into (2.305), we arrive at:

$$E\{X|A\} = \frac{\int_{-\infty}^{a} x f_X(x)\mathrm{d}x}{\int_{-\infty}^{a} f_X(x)\mathrm{d}x}. \tag{2.309}$$

Although the general expression for a continuous random variable can also be used for a discrete variable (as mentioned in Sect. 2.7.4), for the sake of convenience, we also give the corresponding expressions for the discrete random variables:

$$\begin{aligned} E\{X|A\} &= \sum_k x_k P_X(x_k|A), \\ E\{g(X)|A\} &= \sum_k g(x_k) P_X(x_k|A). \end{aligned} \tag{2.310}$$

Example 2.7.15 Consider a uniform random variable in the interval [−2, 5], as shown in Fig. 2.46.

Find the conditional mean value for two cases:

(a) $A_1 = \{X < 0\}$.
(b) $A_2 = \{X > 1\}$.

Solution
(a) The probability of the event A_1 (see Fig. 2.46) is:

$$P\{A_1\} = P\{X < 0\} = 2/7. \tag{2.311}$$

From (2.309), using (2.311), we get:

$$E\{X|A_1\} = \frac{\int_{-2}^{0} x f_X(x)\mathrm{d}x}{P\{A_1\}} = \frac{1/7 \int_{-2}^{0} x\mathrm{d}x}{2/7} = -1. \tag{2.312}$$

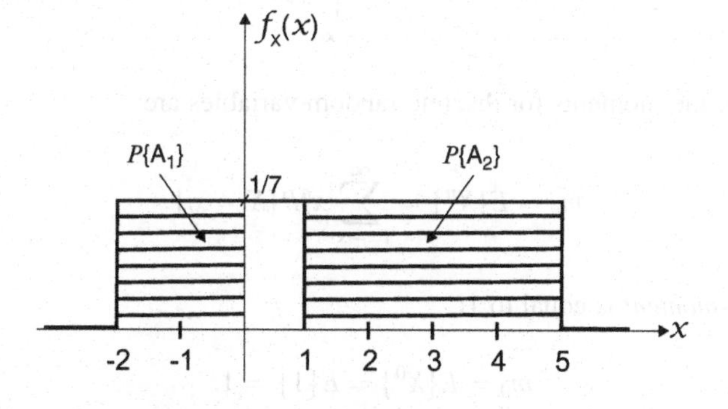

Fig. 2.46 Uniform random variable and probabilities of events A_1 and A_2

(b) The probability of event A_2 (see Fig. 2.46) is:

$$P\{A_2\} = P\{X>1\} = 4/7. \tag{2.313}$$

From (2.309), using (2.313), we arrive at:

$$E\{X|A_2\} = \frac{\int_1^5 x f_X(x)\mathrm{d}x}{P\{A_2\}} = \frac{1/7\int_1^5 x\mathrm{d}x}{4/7} = 3. \tag{2.314}$$

2.8 Moments

The mean value gives little information about a random variable; that is, it represents a sort of central value around which a random variable takes its values. However, it does not tell us, for example, about the variation, or spread of these values. In this section, we introduce different parameters based on the mean value operation, which describe the random variable more completely.

2.8.1 Moments Around the Origin

One simple type of functions whose mean value might be of interest is the power function of the random variable X

$$g(X) = X^n \quad \text{for} \quad n = 1, 2, 3, \ldots. \tag{2.315}$$

The mean value of this function is called nth *moment* and it is denoted as m_n:

$$m_n = E\{X^n\} = \int_{-\infty}^{\infty} x^n f_x(x)\mathrm{d}x. \tag{2.316}$$

Similarly, the moments for discrete random variables are:

$$m_n = E\{X^n\} = \sum_{i=-\infty}^{\infty} x_i^n P\{X = x_i\}. \tag{2.317}$$

The *zero-moment* is equal to 1:

$$m_0 = E\{X^0\} = E\{1\} = 1. \tag{2.318}$$

The most important moments are the first and the second.

The *first moment* is the mean value of the r.v. X,

$$m_1 = E\{X^1\} = E\{X\}. \tag{2.319}$$

The *second moment* is called the *mean squared value* and is a measure of the strength of the random variable

$$m_2 = E\{X^2\}. \tag{2.320}$$

The square root of the mean squared value $\sqrt{E\{X^2\}}$ is called the *rms (root mean squared)* value of X

$$rms(X) = \sqrt{E\{X^2\}}. \tag{2.321}$$

The name "moments" comes from mechanics. If we think of $f_X(x)$ as a mass distributed along the x-axis [THO71, p. 83], [CHI97, p. 94], then:

- The first moment calculates the *center of gravity of the mass.*
- The second moment is the central moment of inertia of the distribution of mass around the center of gravity. (2.322a)

Next, we interpret moments of X which represents a random voltage across a 1 Ω resistor. Then:

- $m_1 = E\{X\}$ represents the *DC (direct-current) component,*
- $m_2 = E\{X^2\}$ represents *the total power* (including DC). (2.322b)

Example 2.8.1 We want to find the first three moments of the discrete random variable X with values 0 and 1 and the corresponding probabilities $P\{0\} = P\{1\} = 0.5$.

Solution The random variable X is a discrete variable and its moments can be calculated using (2.317) as shown in the following.

The first moment:

$$m_1 = 0 \times 0.5 + 1 \times 0.5 = 0.5. \tag{2.323}$$

The second moment:

$$m_2 = 0 \times 0.5 + 1^2 \times 0.5 = 0.5. \tag{2.324}$$

Third moment:

$$m_3 = 0 \times 0.5 + 1^3 \times 0.5 = 0.5 \tag{2.325}$$

Note that all three moments are equal. The same is true for all other moments

$$m_n = 0 \times 0.5 + 1^n \times 0.5 = 0.5. \tag{2.326}$$

Example 2.8.2 In this example, we want to calculate the first three moments of the uniform random variable in the interval [0, 1].

The density function is:

$$f_X(x) = \begin{cases} 1 & \text{for} \quad 0 \leq x \leq 1, \\ 0 & \text{otherwise.} \end{cases} \tag{2.327}$$

Solution The random variable X is a continuous variable and its moments are calculated using (2.316):

$$m_1 = \int_0^1 x\mathrm{d}x = \frac{1}{2},$$

$$m_2 = \int_0^1 x^2\mathrm{d}x = \frac{1}{3},$$

$$m_3 = \int_0^1 x^3\mathrm{d}x = \frac{1}{4}. \tag{2.328}$$

2.8.2 Central Moments

The moments described in the previous section depend on the mean value. If, for example, the mean value is very high, then the deterministic component of the variable is dominant, and the variable nearly loses its "randomness".

To overcome this situation, another class of moments, taken around the mean value, are useful. The mean value is subtracted from the signal, and thus the corresponding mean value does not depend on the mean value of the random variable.

The moments around the mean value are called *central moments* and are defined as expected value of the function

$$g(X) = (X - E\{X\})^n. \tag{2.329}$$

Central moments are denoted by μ_n

$$\mu_n = E\{(X - E\{X\})^n\}. \tag{2.330}$$

From (2.330), it follows that the zero central moment is equal to 1, while the first central moment is equal to zero

$$\begin{aligned} \mu_0 &= 1, \\ \mu_1 &= 0. \end{aligned} \tag{2.331}$$

We want to represent the average spread of the random variable in positive as well in negative directions from the mean value. One possible way to measure this variation would be to consider the quantity $E\{|X - E\{X\}|\}$. However, it turns out to be useless to deal with this quantity, as shown below,

$$E\{|X - E\{X\}|\} = \begin{cases} E\{X - E\{X\}\} & \text{for} \quad X \geq E\{X\}, \\ E\{-(X - E\{X\})\} & \text{for} \quad X \leq E\{X\}. \end{cases} \tag{2.332}$$

Note that both expressions on the right side of (2.332) lead to a zero value. As a consequence, an easier way to describe the variations of the random variable around its expected value is a variance, presented below.

2.8.2.1 Variance

The most important central moment is the *second central moment*, μ_2, because it presents the average spread of the random variable in positive as well in negative directions from its mean value. It has its own name (*variance*) and denotation, σ^2 or VAR(*X*). When one would like to emphasize that σ^2 is the variance of the random variable *X*, one writes σ_X^2 rather than σ^2.

$$\mu_2 = \sigma_X^2 = \mathrm{VAR}(X) = E\left\{(X - E\{X\})^2\right\}. \tag{2.333}$$

From (2.273), we get the following expression for the variance:

$$\sigma_X{}^2 = \int_{-\infty}^{\infty} (x - E\{X\})^2 f_X(x)\mathrm{d}x. \tag{2.334}$$

This expression can be used for both continuous and discrete random variables as explained in Sect. 2.7.4. However, we can also give the corresponding expression for the discrete variable *X*:

$$\sigma_X{}^2 = \sum_{i=-\infty}^{\infty} (x_i - E\{X\})^2 P(X = x_i). \tag{2.335}$$

From (2.334), we have an alternative expression for the variance:

$$
\begin{aligned}
{\sigma_X}^2 &= \int_{-\infty}^{\infty} (x - E\{X\})^2 f_X(x)\mathrm{d}x \\
&= \int_{-\infty}^{\infty} (x^2 - 2xE\{X\} + (E\{X\})^2) f_X(x)\mathrm{d}x \\
&= \int_{-\infty}^{\infty} x^2 f_X(x)\mathrm{d}x - 2E\{X\} \int_{-\infty}^{\infty} x f_X(x)\mathrm{d}x + (E\{X\})^2 \int_{-\infty}^{\infty} f_X(x)\mathrm{d}x.
\end{aligned}
\tag{2.336}
$$

Note that $E\{X\}$, being a constant, is moved outside of the second and third integrals on the right side of (2.336). According to (2.316), the first, second, and third integrals on the right side of (2.336) are equal to the second, first, and zero moments, respectively.

Therefore, from (2.336), we have:

$$
{\sigma_X}^2 = m_2 - 2{m_1}^2 + {m_1}^2 m_0 = m_2 - {m_1}^2 \tag{2.337}
$$

or equivalently

$$
{\sigma_X}^2 = E\{X^2\} - (E\{X\})^2. \tag{2.338}
$$

From (2.334), it follows that

$$
{\sigma_X}^2 \geq 0. \tag{2.339}
$$

If the variance from (2.333) is zero, there is no variation about the mean value, and consequently the variable X becomes a single constant $X = E\{X\}$. As a result, the random variable is no longer a "random variable". In this way, the variance is a measure of the "randomness" of the random variable.

The positive square root of the variance is called the *standard deviation*

$$
\sigma_X = \sqrt{{\sigma_X}^2}, \tag{2.340}
$$

and it is also a measure of the spread of the random variable around the mean value.

Example 2.8.3 Using both expressions for variance (2.335) and (2.338), find the variance of the discrete r.v. from Example 2.8.1.

Solution From (2.335), we have:

$$\sigma_X{}^2 = \sum_{i=-\infty}^{\infty} (x_i - E\{X\})^2 P(X = x_i) = \left(0 - \frac{1}{2}\right)^2 \frac{1}{2} + \left(1 - \frac{1}{2}\right)^2 \frac{1}{2} = \frac{1}{4}. \quad (2.341)$$

Using (2.338), we obtain the same result as shown in the following equation:

$$\sigma_X{}^2 = E\{X^2\} - (E\{X\})^2 = \frac{1}{2} - \left(\frac{1}{2}\right)^2 = \frac{1}{4}. \quad (2.342)$$

Example 2.8.4 In this example, we calculate the variance of the continuous variable from Example 2.8.2, using (2.334) and (2.338).

Solution Using the definition of the variance (2.334), we have:

$$\sigma_X{}^2 = \int_{x=-\infty}^{\infty} (x - E\{X\})^2 f_X(x)\mathrm{d}x = \int_{x=0}^{1} (x - 0.5)^2 \mathrm{d}x = \frac{1}{12}. \quad (2.343)$$

From (2.338), we get:

$$\sigma_X{}^2 = E\{X^2\} - (E\{X\})^2 = \frac{1}{3} - \left(\frac{1}{2}\right)^2 = \frac{1}{12}. \quad (2.344)$$

Example 2.8.5 We want to calculate the variance of uniform random variables with the same width, but which are differently positioned on the x-axis (i.e., uniform variables with equal density functions but different ranges), as shown in Fig. 2.47.

The first two moments of the r.v. X_1 are:

$$m_1 = \int_{-0.5}^{0.5} x\mathrm{d}x = 0, \quad m_2 = \int_{-0.5}^{0.5} x^2 \mathrm{d}x = \frac{1}{3}\left[\frac{1}{8} + \frac{1}{8}\right] = \frac{1}{12} \quad (2.345)$$

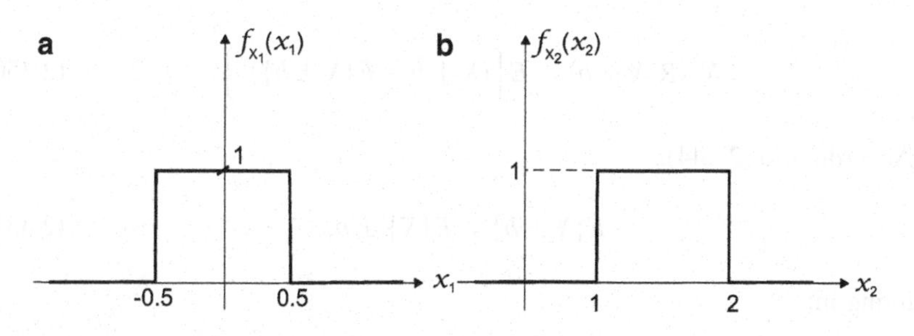

Fig. 2.47 Uniform densities in Example 2.8.5

yielding

$$\sigma_{X_1}^2 = \frac{1}{12} - 0 = \frac{1}{12}. \tag{2.346}$$

Similarly, for the r.v. X_2, we get:

$$m_1 = \int_1^2 x\mathrm{d}x = \frac{3}{2}, \quad m_2 = \int_1^2 x^2\mathrm{d}x = \frac{1}{3}[8 - 1] = \frac{7}{3}, \tag{2.347}$$

resulting in:

$$\sigma_{X_2}^2 = \frac{7}{3} - \frac{9}{4} = \frac{1}{12}. \tag{2.348}$$

Note that the random variables X_1 and X_2 have equal variance given that both variables have the same uniform range.

Generally speaking, the variance of the uniform random variable with a range width Δ can be presented as:

$$\sigma^2 = \frac{\Delta^2}{12}. \tag{2.349}$$

2.8.2.2 Properties of Variance

The variance of a random variable X has the useful properties described in the continuation.

P.1 Adding a constant to the random variable does not affect its variance: $\mathrm{VAR}(X \pm b) = \mathrm{VAR}(X)$, for any constant b.

From (2.333), we write:

$$\mathrm{VAR}(X \pm b) = E\left\{(X \pm b - E\{X \pm b\})^2\right\}. \tag{2.350}$$

According to (2.244),

$$E\{X \pm b\} = E\{X\} \pm b, \tag{2.351}$$

resulting in:

$$\mathrm{VAR}(X \pm b) = E\{(X - E\{X\})^2\} = \mathrm{VAR}(X). \tag{2.352}$$

P.2 If a random variable is multiplied by any constant a, then its variance is multiplied by the squared value of a: $\mathrm{VAR}\{aX\} = a^2\mathrm{VAR}\{X\}$.

From the definition of the variance (2.279), we have:

$$\mathrm{VAR}(aX) = E\left\{(aX - E\{aX\})^2\right\}. \tag{2.353}$$

Keeping in mind that $E\{aX\} = aE\{X\}$ (see (2.279)), from (2.353) it follows:

$$\begin{aligned}\mathrm{VAR}(aX) &= E\left\{(aX - aE\{X\})^2\right\},\\ &= E\left\{a^2(X - E\{X\})^2\right\} = a^2E\left\{(X - E\{X\})^2\right\} = a^2\mathrm{VAR}\{X\}.\end{aligned} \tag{2.354}$$

P.3 This is a combination of P.1 and P.2: $\mathrm{VAR}\{aX \pm b\} = \mathrm{VAR}\{aX\} = a^2\mathrm{VAR}\{X\}$.

From (2.352) and (2.353), the variance of the linear transformation of the r.v. X, where a and b are constants, is:

$$\mathrm{VAR}\{aX + b\} = \mathrm{VAR}\{aX\} = a^2\mathrm{VAR}\{X\}. \tag{2.355}$$

From (2.355), we note that

$$\mathrm{VAR}\{aX + b\} \neq a\mathrm{VAR}\{X\} + b. \tag{2.356}$$

Therefore, as opposed to the mean value, in which the mean of linear transformation of the random variable results in the same linear transformation of its mean, the relation (2.356) indicates that the variance of a linearly transformed r. v. X does not result in a linear transformation of the variance of X. The reason for this is because the variance is not a linear function, as evident from (2.353).

2.8.2.3 Standarized Random Variable

Properties (2.352) and (2.354) can be used to normalize the variable (i.e., to obtain the corresponding random variable with zero mean and the variance of one). Such random variable is called the *standardized random variable*. The use of standardized random variables frequently simplifies discussion. The standardization of random variables eliminates the effects of origin and scale.

Consider a random variable X with mean value m_X and variance σ^2; then the random variable

$$Y = X - m_X \tag{2.357}$$

has zero mean:

$$E\{Y\} = E\{X - m_X\} = E\{X\} - m_X = 0. \tag{2.358}$$

The standardized random variable Y, derived from the r.v. X, using

$$Y = \frac{X - m_X}{\sigma} \tag{2.359}$$

has zero mean and variance equal to 1.

The mean value is:

$$E\{Y\} = E\left\{\frac{X - m_X}{\sigma}\right\} = \frac{1}{\sigma}E\{X - m_X\} = \frac{1}{\sigma}[E\{X\} - m_X] = \frac{1}{\sigma}[m_X - m_X]$$
$$= 0. \tag{2.360}$$

The variance is:

$$\mathrm{VAR}\{Y\} = E\left\{\left[\frac{X - m_X}{\sigma} - E\left\{\frac{X - m_X}{\sigma}\right\}\right]^2\right\}. \tag{2.361}$$

According to (2.360), the second term on the right side of (2.361) is zero, resulting in:

$$\mathrm{VAR}\{Y\} = E\left\{\left[\frac{X - m_X}{\sigma}\right]^2\right\} = \frac{1}{\sigma^2}E\{(X - m_X)^2\} = \frac{\sigma^2}{\sigma^2} = 1. \tag{2.362}$$

2.8.3 Moments and PDF

As mentioned before, moments are often used as statistical descriptors of certain characteristics of the random variable, which in turn is completely described using probability density. Thus, moments may be considered as parameters of probability density, as discussed below.

Consider the mean squared deviation of a random variable X around the constant a:

$$s = \int_{-\infty}^{\infty} (x - a)^2 f_X(x)\mathrm{d}x. \tag{2.363}$$

The value s will vary depending on the choice of a. Let us find the value of a, for which s will be minimum

$$\frac{\mathrm{d}s}{\mathrm{d}a} = 0. \tag{2.364}$$

From (2.363) and (2.364), we have:

$$-2\int_{-\infty}^{\infty}(x-a)f_X(x)\mathrm{d}x = -2\int_{-\infty}^{\infty}xf_X(x)\mathrm{d}x + 2a\int_{-\infty}^{\infty}f_X(x)\mathrm{d}x = 0. \tag{2.365}$$

In (2.365), the second integral is 1, resulting in:

$$a = \int_{-\infty}^{\infty}xf_X(x)\mathrm{d}x = m. \tag{2.366}$$

This result shows us that the first moment (mean value) is the point at which the mean squared deviation s is minimal; consequently, this minimal mean squared deviation s is equal to the variance.

Next we relate the shape of the density function with the moments.

2.8.3.1 The Shape of Density Function and Moments

Density and First Moment (Mean Value)

If the density function is symmetric, then the point of symmetry corresponds to the mean value. To this end, consider that the density function is symmetrical regarding point a, as shown in Fig. 2.48.

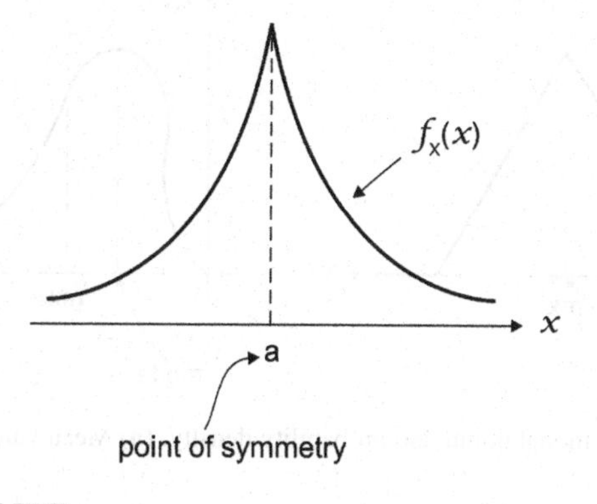

Fig. 2.48 Symmetrical PDF

From Fig. 2.48, we have:

$$f_X(a-x) = f_X(a+x). \tag{2.367}$$

The mean value can be expressed as:

$$m_X = \int_{-\infty}^{\infty} x f_X(x)\mathrm{d}x = \int_{-\infty}^{\infty} (x-a) f_X(x)\mathrm{d}x + \int_{-\infty}^{\infty} a f_X(x)\mathrm{d}x. \tag{2.368}$$

Keeping in mind that the second integral in (2.368) is equal to a, we arrive at:

$$m_X = -\int_{a}^{\infty} x f_X(a-x)\mathrm{d}x + \int_{a}^{\infty} x f_X(a+x)\mathrm{d}x + a = a. \tag{2.369}$$

Using (2.367) to (2.369), it follows that $m_X = a$, i.e., the mean value is equal to the symmetry point a.

The value of r.v. for which the PDF has maximum, is called the *mode*. Generally, the mean value is not equal to the mode. If the density is symmetrical, then the mode is the point of symmetry, and thus corresponds to the mean value (see Fig. 2.49a). Otherwise, the mode and the mean value are different, as shown in Fig. 2.49b.

Density and Third Central Moment

The third central moment, μ_3, is a measure of the symmetry of the random variable and is also called a *skew* of the density function. If the density has symmetry around

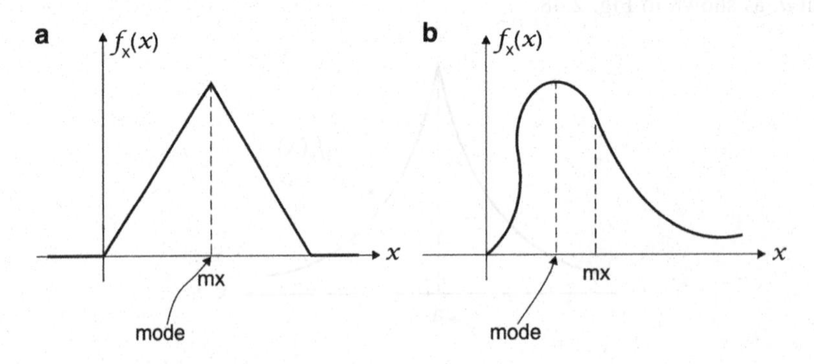

Fig. 2.49 Mean value, modal point, and probability density. (**a**) Mean value. (**b**) Modal point

the mean then $\mu_3 = 0$ and also all odd central moments μ_n, $n = 3, 5, 7, \ldots$, are equal to zero, as shown below.

$$\mu_n = \int_{-\infty}^{\infty} (x-m)^n f_X(x)\mathrm{d}x = \int_{-\infty}^{m} (x-m)^n f_X(x)\mathrm{d}x + \int_{m}^{\infty} (x-m)^n f_X(x)\mathrm{d}x. \quad (2.370)$$

After brief calculations, we arrive at:

$$\mu_n = \int_{0}^{\infty} [(-x)^n + x^n] f_X(m-x)\mathrm{d}x = \int_{0}^{\infty} [(-x)^n + x^n]\, f_X(m+x)\mathrm{d}x = 0 \quad (2.371)$$

for n is odd.

The normalized third central moment is called the *coefficient of skewness*

$$c_s = \frac{\mu_3}{\sigma^3} = \frac{E\{(X-m)^3\}}{\sigma^3}. \quad (2.372)$$

The word "skew" means to pull in one direction [AGR07, p. 41]. A density is skewed to the left if the left tail is longer than the right tail (see Fig. 2.50a). Similarly, a density is skewed to the right if the right tail is longer than the left tail (see Fig. 2.50b).

The coefficient c_s (2.372) is positive if the density is skewed to the right and negative if it is skewed to the left.

Fourth Central Moment and PDF

The *fourth central moment* is a measure of the "peakedness" of density near the mean value.

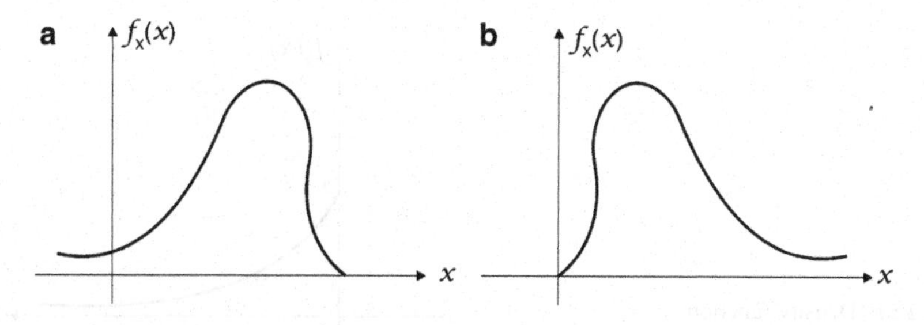

Fig. 2.50 PDF skewed to the (**a**) left and (**b**) right

The normalized fourth central moment is called the *coefficient of kurtosis*, and is written as:

$$c_k = \frac{\mu_4}{\sigma^4} = \frac{E\{(X-m)^4\}}{\sigma^4}. \tag{2.373}$$

The larger the coefficient of kurtosis c_k, the larger the peak of the density near the mean is.

Example 2.8.6 Find the mean value, mode, coefficient of skewness, and coefficient of kurtosis for the random variable X with density $f_X(x) = \lambda e^{-\lambda x}$, shown in Fig. 2.51.

Solution The density is asymmetrical and, therefore, its mode at $x = 0$ is not a mean value.

Using the integral

$$\int_0^\infty x^n e^{-ax} dx = \frac{\Gamma(n+1)}{a^{n+1}}, \tag{2.374}$$

where $\Gamma(x)$ is the Gamma function for x and the Gamma function for integer n is given as follows:

$$\Gamma(n) = (n-1)!, \tag{2.375}$$

where ! stands for the factorial of n

$$n! = 1 \times 2 \times 3 \times \ldots \times n, \tag{2.376}$$

we calculate the first moment (mean value):

$$m_1 = \int_0^\infty x\lambda e^{-\lambda x} dx = \lambda\frac{\Gamma(2)}{\lambda^2} = \frac{1!}{\lambda} = \frac{1}{\lambda}. \tag{2.377}$$

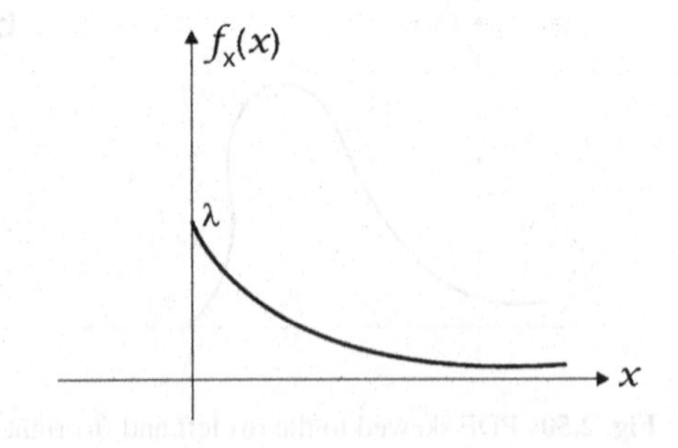

Fig. 2.51 Density function in Example 2.8.6

The second moment is equal to:

$$m_2 = \int_0^{\infty} x^2 \lambda e^{-\lambda x} dx = \lambda \frac{\Gamma(3)}{\lambda^3} = \frac{2!}{\lambda^2} = \frac{2}{\lambda^2}. \tag{2.378}$$

From (2.377) and (2.378), using (2.337), the variance is:

$$\sigma^2 = m_2 - m_1^2 = \frac{2}{\lambda^2} - \frac{1}{\lambda^2} = \frac{1}{\lambda^2}. \tag{2.379}$$

The third central moment is:

$$\mu_3 = \int_0^{\infty} (x - 1/\lambda)^3 \lambda e^{-\lambda x} dx. \tag{2.380}$$

After brief calculations, we arrive at:

$$\mu_3 = 2/\lambda^3. \tag{2.381}$$

The nonzero third central moment is the result of asymmetry in the PDF. Similarly, we get the fourth central moment:

$$\mu_4 = \int_0^{\infty} (x - 1/\lambda)^4 \lambda e^{-\lambda x} dx = 9/\lambda^4. \tag{2.382}$$

From (2.372) and (2.379)–(2.381), the coefficient of skewness is given by:

$$c_s = \frac{\mu_3}{\sigma^3} = \frac{E\{(X - m)^3\}}{\sigma^3} = 2. \tag{2.383}$$

The positive value (2.383) shows that the density function is skewed to the right, as it is shown in Fig. 2.51.

Similarly, from (2.373), (2.379), and (2.382), the coefficients of kurtosis is given as:

$$c_k = \frac{\mu_4}{\sigma^4} = \frac{E\{(X - m)^4\}}{\sigma^4} = 9. \tag{2.384}$$

The large coefficient of Kurtosis indicates that the density has a sharp peak near its mean. This is confirmed in Fig. 2.51.

Higher values of the moments do not have a similar physical interpretation, as do the first four moments.

2.8.4 *Functions Which Give Moments*

It is well known that the Fourier and Laplace transforms are crucial in the description of properties of deterministic signals and systems. For similar reasons, we can apply Fourier and Laplace transforms to the PDF, because the PDF is a deterministic function. Here, we consider the characteristic function and the moment generating function (MGF). Both functions are very useful in determining the moments of the random variable and particularly in simplifying the description of the convolution of probability densities, as we will see later in Chap. 3.

2.8.4.1 Characteristic Function

The *characteristic function* of the random variable X denoted as $\phi_X(\omega)$ is defined as the expected value of the complex function:

$$e^{j\omega X} = \cos(\omega X) + j\sin(\omega X), \tag{2.385}$$

where $j = \sqrt{-1}$.

Using the definition of the expected value of the function of a random variable (2.273), we have:

$$\phi_X(\omega) = E\{e^{j\omega X}\} = \int_{-\infty}^{\infty} e^{j\omega x} f_X(x)\mathrm{d}x. \tag{2.386}$$

Note that the integral in the right side of (2.386) is the Fourier transform of density function $f_X(x)$, (with a reversal in the sign of the exponent). As a consequence, the importance of (2.386) comes in knowing that the Fourier transform has many useful properties. Additionally, using the inverse Fourier transformation, we can find the density function from its characteristic function,

$$f_X(x) = \frac{1}{2\pi}\int_{-\infty}^{\infty} \phi_X(\omega)e^{-j\omega x}\mathrm{d}\omega. \tag{2.387}$$

From (2.385), we can easily see that, for $\omega = 0$, the characteristic function is equal to 1.

Similarly,

$$|\phi_X(\omega)| \le \int_{-\infty}^{\infty} |e^{j\omega x}| f_X(x)\mathrm{d}x = \int_{-\infty}^{\infty} f_X(x)\mathrm{d}x = 1. \tag{2.388}$$

Let us summarize the previously mentioned characteristics as:

$$\begin{aligned} \phi_X(0) &= 1, \\ |\phi_X(\omega)| &\leq 1. \end{aligned} \tag{2.389}$$

Example 2.8.7 We want to find the characteristic function of the exponential random variable with the density function:

$$f_X(x) = \begin{cases} \lambda \mathrm{e}^{-\lambda x} & \text{for} \quad x \geq 0. \\ 0 & \quad\quad\;\; \text{otherwise,} \end{cases} \tag{2.390}$$

Using (2.385), the characteristic function is obtained as:

$$\phi_X(\omega) = \int_0^\infty \mathrm{e}^{j\omega x} f_X(x)\mathrm{d}x = \int_0^\infty \mathrm{e}^{j\omega x} \lambda \mathrm{e}^{-\lambda x}\mathrm{d}x = \lambda \int_0^\infty \mathrm{e}^{-x(\lambda - j\omega)}\mathrm{d}x = \frac{\lambda}{\lambda - j\omega}. \tag{2.391}$$

Note that the characteristic function is a complex function. Next, we verify that the properties in (2.389) are satisfied:

$$\phi_X(0) = \frac{\lambda}{\lambda - j0} = 1. \tag{2.392}$$

$$|\phi_X(\omega)| = \frac{\lambda}{\sqrt{\lambda^2 + \omega^2}} \leq 1. \tag{2.393}$$

Generally speaking, the characteristic functions are complex. However, if the density function is even, the characteristic function is real.

$$\phi_X(\omega) = \int_{-\infty}^\infty \cos(\omega x) f_X(x)\mathrm{d}x, \tag{2.394}$$

From here, the density function is:

$$f_X(x) = \frac{1}{2\pi} \int_{-\infty}^\infty \phi(\omega)\cos(\omega x)\mathrm{d}\omega. \tag{2.395}$$

Example 2.8.8 Find the characteristic function for the uniform variable in the interval $[-1, 1]$.

Solution Using (2.385), we get:

$$\phi_X(\omega) = E\{\mathrm{e}^{j\omega X}\} = \int_{-1}^1 \mathrm{e}^{j\omega x}\frac{1}{2}\mathrm{d}x = \frac{1}{2j\omega}(\mathrm{e}^{j\omega} - \mathrm{e}^{-j\omega}) = \frac{\sin(\omega)}{\omega}. \tag{2.396}$$

Note that the characteristic function (2.396) is real because the density function is even. We can obtain the same result using (2.394):

$$\phi_X(\omega) = \int_{-1}^{1} \cos(\omega x)\frac{1}{2}(x)\mathrm{d}x = \frac{\sin(\omega)}{\omega}. \tag{2.397}$$

Let us now find the characteristic function of the transformed random variable

$$Y = g(X). \tag{2.398}$$

Using (2.398) and the definition in (2.386), we get:

$$\phi_Y(\omega) = E\{\mathrm{e}^{j\omega Y}\} = \int_{-\infty}^{\infty} \mathrm{e}^{j\omega y} f_Y(y)\mathrm{d}y = E\{\mathrm{e}^{j\omega g(X)}\} = \int_{-\infty}^{\infty} \mathrm{e}^{j\omega g(x)} f_X(x)\mathrm{d}x. \tag{2.399}$$

Example 2.8.9 Find the characteristic function of the random variable for the density function that is given in Fig. 2.52.

From (2.385), we have:

$$\phi_X(\omega) = E\{\mathrm{e}^{j\omega X}\} = \int_{a-b}^{a+b} \mathrm{e}^{j\omega x}\frac{1}{2b}\mathrm{d}x = \left.\frac{\mathrm{e}^{j\omega x}}{2bj\omega}\right|_{a-b}^{a+b} = \frac{\mathrm{e}^{j\omega a}}{b\omega}\sin(b\omega). \tag{2.400}$$

Note that the characteristic function is complex because the density, although it is symmetric, is not even.

We can obtain the same result using (2.399), where the density function from Fig. 2.52 is obtained as the linear transformation

$$X = bY + a, \tag{2.401}$$

of the random variable Y uniform in the interval $[-1, 1]$.

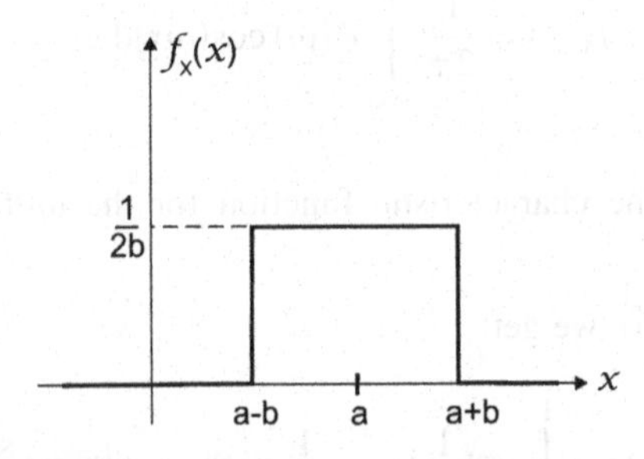

Fig. 2.52 Density function in Example 2.8.9

$$\phi_X(\omega) = \int_{-1}^{1} e^{j\omega(by+a)} \frac{1}{2} dy = \frac{e^{j\omega a}}{\omega b} \frac{e^{j\omega b} - e^{-j\omega b}}{2j} = \frac{e^{j\omega a}}{\omega b} \sin(b\omega). \tag{2.402}$$

Now we will demonstrate how the characteristic function is also useful to express the density function of the transformed random variable $Y = g(X)$ in terms of the PDF of the variable X.

From

$$y = g(x), \tag{2.403}$$

we get:

$$x = g^{-1}(y). \tag{2.404}$$

Considering (2.404), we can write:

$$\begin{aligned} f_X(x)|_{x=g^{-1}(y)} &= f_X(y), \\ dx &= g_1(y)dy, \end{aligned} \tag{2.405}$$

Placing (2.405) into (2.386), we arrive at:

$$\phi_Y(\omega) = E\{e^{j\omega Y}\} = \int_{-\infty}^{\infty} e^{j\omega g(x)} f_X(y) g_1(y) dy = \int_{-\infty}^{\infty} e^{j\omega y} h(y) dy. \tag{2.406}$$

Comparing the two integrals in (2.406), we can express density function as:

$$f_Y(y) = h(y) = f_x(y) g_1(y). \tag{2.407}$$

The difficulty arises when the transform is not unique, because it is not possible to simply express the transformed PDF in the form (2.407).

Example 2.8.10 Find the density function from Fig. 2.52, using (2.401).

Solution

$$\begin{aligned} \phi_Y(\omega) = E\{e^{j\omega Y}\} &= \int_{-1}^{1} e^{j\omega g(x)} f_X(x) dx = \int_{a-b}^{a+b} e^{j\omega y} f_X\left(\frac{y-a}{b}\right) \frac{1}{b} dy \\ &= \int_{a-b}^{a+b} e^{j\omega y} \frac{1}{2} \frac{1}{b} dy = \int_{a-b}^{a+b} e^{j\omega y} h(y) dy. \end{aligned} \tag{2.408}$$

From here

$$h(y)=\begin{cases}1/2b & a-b\le y\le a+b,\\ 0 & \text{otherwise.}\end{cases} \tag{2.409}$$

2.8.4.2 Moment Theorem

One alternative expression for the characteristic function is obtained by developing $e^{j\omega x}$:

$$e^{j\omega x}=1+j\omega x+\frac{(j\omega x)^2}{2!}+\cdots+\frac{(j\omega x)^n}{n!}+\cdots. \tag{2.410}$$

Placing (2.410) into (2.386), we obtain:

$$\phi_X(\omega)=\int_{-\infty}^{\infty}f_X(x)\mathrm{d}x+j\omega\int_{-\infty}^{\infty}xf_X(x)\mathrm{d}x+\frac{(j\omega)^2}{2!}\int_{-\infty}^{\infty}x^2f_X(x)\mathrm{d}x+\cdots$$
$$+\frac{(j\omega)^n}{n!}\int_{-\infty}^{\infty}x^nf_X(x)\mathrm{d}x+\cdots. \tag{2.411}$$

Note that the integrals in (2.411) may not exist. However, if they do exist they are moments m_i, as shown in (2.412):

$$m_i=\int_{-\infty}^{\infty}x^if_X(x)\mathrm{d}x. \tag{2.412}$$

Consequently, the characteristic function can be represented by its moments,

$$\phi_X(\omega)=1+j\omega m_1+\frac{(j\omega)^2}{2!}m_2+\cdots+\frac{(j\omega)^n}{n!}m_n+\cdots. \tag{2.413}$$

Finding successive derivatives of (2.413), $k=1,\ldots,n$, evaluated at $\omega=0$, we can find nth moments, as shown in the following equation:

$$\left.\frac{\mathrm{d}\phi(\omega)}{d\omega}\right|_{\omega=0}=jm_1,$$

$$\ldots$$

$$\left.\frac{\mathrm{d}^n\phi(\omega)}{d\omega^n}\right|_{\omega=0}=j^nm_n. \tag{2.414}$$

From here, we can find moments from the corresponding characteristic functions:

$$m_1 = \frac{1}{j}\frac{\mathrm{d}\phi(\omega)}{\mathrm{d}\omega}\bigg|_{\omega=0}$$

$$\ldots$$

$$m_n = \frac{1}{j^n}\frac{\mathrm{d}^n\phi(\omega)}{\mathrm{d}\omega^n}\bigg|_{\omega=0} \tag{2.415}$$

The expressions (2.415) are also known as the *moment theorem*.

Example 2.8.11 Find the mean value and variance of the exponential variable from Example 2.8.7, using the moment theorem.

Solution Based on the result (2.391), the characteristic function is:

$$\phi_X(\omega) = \frac{\lambda}{\lambda - j\omega}. \tag{2.416}$$

From (2.415) and (2.416), we have:

$$m_1 = \frac{1}{j}\frac{\mathrm{d}\phi(\omega)}{\mathrm{d}\omega}\bigg|_{\omega=0} = \frac{1}{\lambda}$$

$$m_2 = \frac{1}{j^2}\frac{\mathrm{d}^2\phi(\omega)}{\mathrm{d}\omega^2}\bigg|_{\omega=0} = \frac{2}{\lambda^2}. \tag{2.417}$$

This is the same result as (2.377) and (2.378).
The variance is:

$$\sigma^2 = m_2 - m_1^2 = \frac{2}{\lambda^2} - \frac{1}{\lambda^2} = \frac{1}{\lambda^2}, \tag{2.418}$$

which is the same result as in (2.379).

The corresponding equation for the characteristic function of the discrete random variable is:

$$\phi_X(\omega) = \sum_i \mathrm{e}^{j\omega x_i} P(X = x_i). \tag{2.419}$$

The use of (2.419) is illustrated in the following example.

Example 2.8.12 Find the characteristic function of a discrete variable X which takes with the equal probabilities each of its possible values: C and $-C$.

$$P\{X = C\} = P\{X = -C\} = 1/2. \tag{2.420}$$

From (2.419) and (2.420), we have:

$$\phi_X(\omega) = \sum_{i=1}^{2} e^{j\omega x_i} P(X = x_i) = e^{j\omega C}\frac{1}{2} + e^{-j\omega C}\frac{1}{2} = \frac{e^{j\omega C} + e^{-j\omega C}}{2}$$
$$= \cos(\omega C). \tag{2.421}$$

Note that the characteristic function is real because the corresponding PDF is an even function.

2.8.4.3 Moment Generating Function

A *moment generating function* (MGF) of a random variable X, denoted as $M_X(s)$, is defined as the expected value of the complex function e^{Xs}, where s is a complex variable.

Using the definition of the expected value of the function random variable (2.273), we have:

$$M_X(s) = E\{e^{sX}\} = \int_{-\infty}^{\infty} e^{sx} f_X(x) dx. \tag{2.422}$$

The values of s for which the integral (2.422) converges, define a region in complex plane, called *ROC* (*region of convergence*). Note that except for a minus sign difference in the exponent, the expression (2.422) is the *Laplace transform* of the density function $f_X(x)$.

Expanding e^{sX} in the ROC, we obtain:

$$e^{sX} = 1 + sX + \frac{1}{2!}(Xs)^2 + \frac{1}{3!}(Xs)^3 + \cdots. \tag{2.423}$$

The expectation of (2.423) is equal to:

$$E\{e^{sX}\} = \int_{-\infty}^{\infty} e^{sx} f_X(x) dx = \int_{-\infty}^{\infty} \left(1 + sx + \frac{1}{2!}x^2 s^2 + \frac{1}{3!}x^3 s^3 + \cdots\right) f_X(x) dx$$
$$= 1 + sE\{X\} + \frac{1}{2!}s^2 E\{X^2\} + \frac{1}{3!}s^3 E\{X^3\} + \cdots$$
$$= 1 + sm_1 + \frac{1}{2!}s^2 m_2 + \frac{1}{3!}s^3 m_3 + \cdots. \tag{2.424}$$

Given MGF, the moments are derived, by taking derivatives with respect to s, and evaluating the result at $s = 0$,

$$\begin{aligned} m_1 &= \left.\frac{\mathrm{d}M_X(s)}{\mathrm{d}s}\right|_{s=0}, \\ m_2 &= \left.\frac{\mathrm{d}^2M_X(s)}{\mathrm{d}s^2}\right|_{s=0}, \\ m_3 &= \left.\frac{\mathrm{d}^3M_X(s)}{\mathrm{d}s^3}\right|_{s=0}, \\ &\cdots \\ m_n &= \left.\frac{\mathrm{d}^nM_X(s)}{\mathrm{d}s^n}\right|_{s=0}. \end{aligned} \tag{2.425}$$

When the MGF is evaluated at $s = j\omega$, the result is the Fourier transform, i.e., MGF function becomes the characteristic function. As a difference to MGF, the characteristic function always converges, and this is the reason why it is often used instead of MGF, [PEE93, p. 82].

Example 2.8.13 Find the MGF function of the exponential random variable and find its variance using (2.425).

Solution

$$M_X(s) = E\{\mathrm{e}^{sX}\} = \int_{-\infty}^{\infty} \mathrm{e}^{sx} f_X(x)\mathrm{d}x = \int_0^{\infty} \mathrm{e}^{sx}\lambda\mathrm{e}^{-\lambda x}\mathrm{d}x = \lambda\int_0^{\infty} \mathrm{e}^{-(\lambda-s)x}\mathrm{d}x. \tag{2.426}$$

This integral converges for Re{s} < λ, where Re{s} means the real value of s. From (2.426), we easily found:

$$M_X(s) = \frac{\lambda}{\lambda - s}. \tag{2.427}$$

The first and second moments are obtained taking the first and second derivatives of (2.427), as indicated in (2.428).

$$\begin{aligned} m_1 &= \left.\frac{\mathrm{d}M_X(s)}{\mathrm{d}s}\right|_{s=0} = \left.\frac{\mathrm{d}}{\mathrm{d}s}\left(\frac{\lambda}{\lambda - s}\right)\right|_{s=0} = \left.\frac{\lambda}{(\lambda - s)^2}\right|_{s=0} = \frac{1}{\lambda}, \\ m_2 &= \left.\frac{\mathrm{d}^2M_X(s)}{\mathrm{d}s^2}\right|_{s=0} = \left.\frac{\mathrm{d}}{\mathrm{d}s}\left(\frac{\lambda}{(\lambda - s)^2}\right)\right|_{s=0} = \left.\frac{2\lambda}{(\lambda - s)^3}\right|_{s=0} = \frac{2}{\lambda^2}. \end{aligned} \tag{2.428}$$

From here, we can easily find the variance as:

$$\sigma^2 = m_2 - m_1^2 = \frac{2}{\lambda^2} - \frac{1}{\lambda^2} = \frac{1}{\lambda^2}. \tag{2.429}$$

This is the same result as that obtained in (2.379).

2.8.5 Chebyshev Inequality

Sometimes we need to estimate a bound on the probability of how much a random variable can deviate from its mean value. A mathematical description of this statement is provided by *Chebyshev inequality*.

The Chebyshev inequality states this bounding in two different forms for r.v. X in terms of the expected value m_X and the variance σ_X^2.

The Chebyshev inequality is very crude, but is useful in situations in which we have no knowledge of a given random variable, other than its mean value and variance. However, if we know a density function, then the precise bounds can be found simply calculating the probability of a desired deviation from the mean value.

2.8.5.1 First Form of Inequality

The probability that the absolute deviation of the random variable X from its expected value m_X is more than ε is less than the variance σ_X^2 divided by ε^2,

$$P\{|X - m_X| \geq \varepsilon\} \leq \sigma_X^2/\varepsilon^2. \tag{2.430}$$

In the continuation, we prove the formula (2.430).

Using the definition of the variance, we can write:

$$\begin{aligned}\sigma_X^2 &= \int_{-\infty}^{\infty} (x - m_X)^2 f_X(x)\mathrm{d}x \\ &= \int_{-\infty}^{m_X-\varepsilon} (x - m_X)^2 f_X(x)\mathrm{d}x + \int_{m_X-\varepsilon}^{m_X+\varepsilon} (x - m_X)^2 f_X(x)\mathrm{d}x + \int_{m_X+\varepsilon}^{\infty} (x - m_X)^2 f_X(x)\mathrm{d}x.\end{aligned} \tag{2.431}$$

Omitting the middle integral (see Fig. 2.53a), we can write:

$$\sigma_X^2 \geq \int_{-\infty}^{m_X-\varepsilon} (x - m_X)^2 f_X(x)\mathrm{d}x + \int_{m_X+\varepsilon}^{\infty} (x - m_X)^2 f_X(x)\mathrm{d}x. \tag{2.432}$$

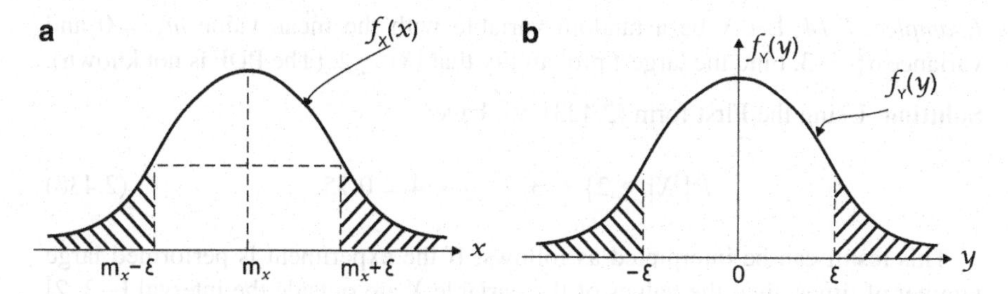

Fig. 2.53 Illustration of Chebyshev theorem (first form)

We introduce the zero mean variable shown in Fig. 2.53b (see Sect. 2.8.2.3)

$$Y = X - m_X, \tag{2.433}$$

resulting in:

$$\sigma_X^2 \geq \int_{-\infty}^{-\varepsilon} y^2 f_Y(y) \mathrm{d}y + \int_{\varepsilon}^{\infty} y^2 f_Y(y) \mathrm{d}y. \tag{2.434}$$

Replacing y by value $(-\varepsilon)$ in (2.434) makes this inequality even stronger:

$$\begin{aligned} \sigma_X^2 &\geq \int_{-\infty}^{-\varepsilon} \varepsilon^2 f_Y(y) \mathrm{d}y + \int_{\varepsilon}^{\infty} \varepsilon^2 f_Y(y) \mathrm{d}y, = \varepsilon^2 \left[\int_{-\infty}^{-\varepsilon} f_Y(y) \mathrm{d}y + \int_{\varepsilon}^{\infty} f_Y(y) \mathrm{d}y \right], \\ &= \varepsilon^2 P\{|Y| \geq \varepsilon\} = \varepsilon^2 P\{|X - m_X| \geq \varepsilon\}. \end{aligned} \tag{2.435}$$

Finally, from (2.435), we have:

$$P\{|X - m_X| \geq \varepsilon\} \leq \sigma_X^2 / \varepsilon^2. \tag{2.436}$$

2.8.5.2 Second Form of Inequality

Replacing $\varepsilon = k\sigma_X$ in (2.430), we can obtain the alternative form of Chebyshev inequality

$$P\{|X - m_X| \geq k\sigma_X\} \leq 1/k^2. \tag{2.437}$$

The inequality (2.437) imposes a limit of $1/k^2$, where k is a constant, to the probability at the left side of (2.437).

Example 2.8.14 Let X be a random variable with the mean value $m_X = 0$ and variance $\sigma_X^2 = 3$. Find the largest probability that $|X| \geq 2$. (The PDF is not known).

Solution Using the First form (2.423), we have:

$$P\{|X| \geq 2\} \leq 3/2^2 = 3/4 = 0.75. \tag{2.438}$$

This result can be interpreted as follows: If the experiment is performed large number of times, then the values of the variable X are outside the interval $[-2, 2]$ approximately less than 75% of the time.

The same result is obtained using the second form (2.434)

$$P\left\{|X| \geq k\sqrt{3}\right\} \leq 1/k^2. \tag{2.439}$$

Denoting

$$k\sqrt{3} = 2, \tag{2.440}$$

we get:

$$1/k^2 = 3/4 = 0.75, \tag{2.441}$$

which is the same result as (2.438).

Let us now consider the random variable Y with known PDF, and with the same mean value $m_Y = 0$ and variance $\sigma_Y^2 = 3$. Consider that the random variable Y is the uniform r.v. in the interval $[-3, 3]$, as shown in Fig. 2.54. In this case, we can find the exact probability that $|Y| \geq 2$, because we know the PDF. We can easily verify that the mean value is equal to 0 and the variance is equal to 3.

$$\begin{aligned} m_Y &= 0, \\ \sigma_Y^2 &= 3. \end{aligned} \tag{2.442}$$

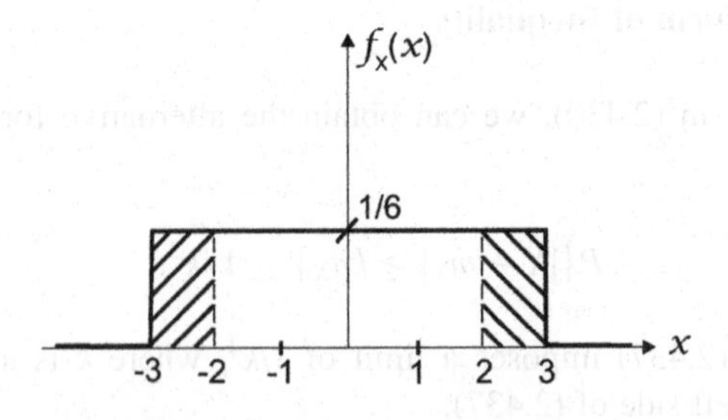

Fig. 2.54 Illustration of Chebyshev inequality for uniform variable

The desired probability corresponds to the shaded area under the density function as shown in Fig. 2.54,

$$P\{|Y| \geq 2\} = \int_{-3}^{-2} f_Y(y)\mathrm{d}y + \int_{2}^{3} f_Y(y)\mathrm{d}y = \int_{-3}^{-2} \frac{1}{6}\mathrm{d}y + \int_{2}^{3} \frac{1}{6}(y)\mathrm{d}y = \frac{1}{3} < \frac{3}{4}. \quad (2.443)$$

Comparing results (2.438) and (2.443), we note that in the absence of more knowledge of the random variable, the result provided by Chebyshev inequality is quite crude.

In this example knowing only mean value and the variance, we found that the desired probability is less than 3/4. Knowing that the variable has a uniform density we can find the exact value of probability, which is 1/3.

Example 2.8.15 Consider the case that the variance of the random variable X is $\sigma_X^2 = 0$. We know that in this case there is no random variable any more, because all values of random variables are equal to its mean value. We will use the Chebyshev inequality to prove this statement.

From (2.430), we have:

$$P\{|X - m_X| \geq \varepsilon\} \leq 0. \quad (2.444)$$

However, the probability cannot be less than zero, and the equality stands in (2.444), i.e.,

$$P\{|X - m_X| \geq \varepsilon\} = 0. \quad (2.445)$$

In other words, the probability that the random variable takes the value different than its mean value m_X is zero, i.e., all values of the random variable are equal to its mean value.

2.9 Numerical Exercises

Exercise E.2.1 A telephone call can come in random during a time interval from 0 P.M. until 8 P.M. Find the PDF and the distribution of the random variable X, defined in the following way,

$$X = \begin{cases} 1 & \text{call in the interval } [2, 5], \\ 0 & \text{otherwise} \end{cases} \quad (2.446)$$

if the probability that the call appears in the interval [2, 5] is:

$$p = (5 - 2)/8 = 3/8. \quad (2.447)$$

Answer The random variable X is the discrete random variable and it has only two values: 0 and 1, with the probabilities

$$P\{0\} = 5/8, \quad P\{1\} = 3/8. \tag{2.448}$$

The corresponding PDF has two delta functions in the discrete points 0 and 1, as shown in Fig. 2.55a,

$$f_X(x) = P\{0\}\delta(x) + P\{1\}\delta(x-1) = 5/8\delta(x) + 3/8\delta(x-1). \tag{2.449}$$

The distribution function has steps at the discrete points 0, and 1, as shown in Fig. 2.55b.

$$F_X(x) = P\{0\}u(x) + P\{1\}u(x-1) = 5/8u(x) + 3/8u(x-1). \tag{2.450}$$

Exercise E.2.2 Solve the previous problem but define the random variable X as the random instant in time in which the call is received.

Answer This example illustrates that one can define different variables for the same random experiment. In this case, the random variable X is the continual variable in the interval [0, 8].

If $x > 8$ (the call has been received), then

$$F_X(x) = P\{X \leq x\} = 1. \tag{2.451}$$

For $0 \leq x \leq 8$,

$$F_X(x) = P\{X \leq x\} = \frac{x}{8-0} = \frac{x}{8}. \tag{2.452}$$

For $x < 0$ (the call cannot be received), resulting in:

$$F_X(x) = P\{X \leq x\} = 0. \tag{2.453}$$

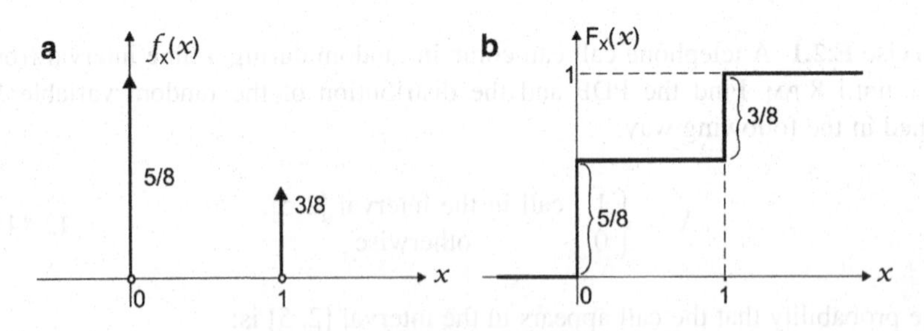

Fig. 2.55 (**a**) PDF and (**b**) distribution in Exercise E.2.1

From (2.451) to (2.453), the distribution is:

$$F_X(x) = \begin{cases} 1 & \text{for} \quad x \geq 8, \\ x/8 & \text{for} \quad 0 \leq x \leq 8, \\ 0 & \text{for} \quad x \leq 1. \end{cases} \tag{2.454}$$

The PDF is obtained by using the derivation of the distribution

$$f_X(x) = \begin{cases} 1/8 & \text{for} \quad 0 \leq x \leq 8, \\ 0 & \qquad \text{otherwise.} \end{cases} \tag{2.455}$$

The distribution and the PDF are shown in Fig. 2.56.

Exercise E.2.3 The distribution of the random variable is given in Fig. 2.57a. Determine whether the random variable is discrete or continuous and find the PDF; also find the probability that the random variable is less than 2.5.

Answer The random variable is continuous because its distribution is not in the form of step functions.

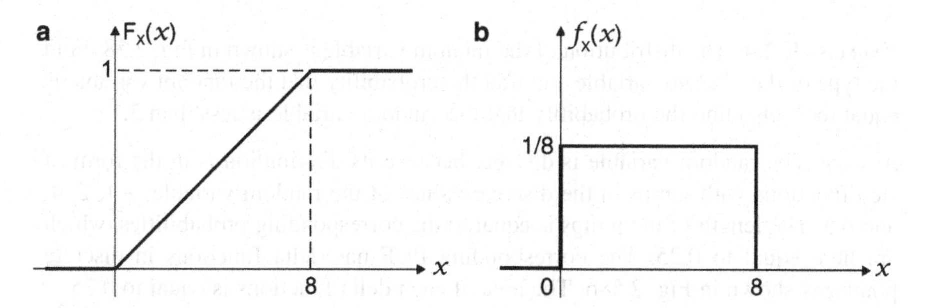

Fig. 2.56 Distribution (**a**) and density (**b**)

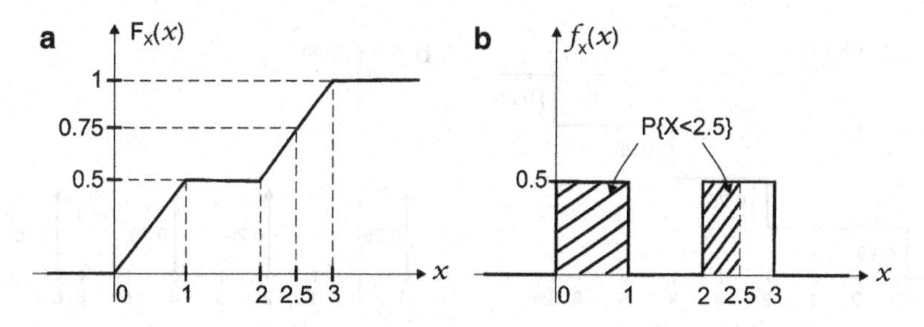

Fig. 2.57 Distribution (**a**), PDF (**b**), and probability

From Fig. 2.57a, we have:

$$F_X(x) = \begin{cases} 0 & \text{for} \quad x \leq 0, \\ 0.5x & \text{for} \quad 0 < x \leq 1, \\ 0.5 & \text{for} \quad 1 < x \leq 2, \\ 0.5(x-1) & \text{for} \quad 2 < x \leq 3, \\ 1 & \text{for} \quad x > 3. \end{cases} \tag{2.456}$$

The PDF is obtained as the derivative of the distribution (2.456)

$$f_X(x) = \begin{cases} 0 & \text{for} \quad x \leq 0, \\ 0.5 & \text{for} \quad 0 < x \leq 1, \\ 0 & \text{for} \quad 1 < x \leq 2, \\ 0.5 & \text{for} \quad 2 < x \leq 3, \\ 0 & \text{for} \quad x > 3. \end{cases} \tag{2.457}$$

The PDF is shown in Fig. 2.57b.

The probability that X is less than 2.5 is equal to the shaded area in Fig. 2.57b:

$$P\{X < 2.5\} = \int_0^{2.5} f_X(x)\mathrm{d}x = 0.5 + 0.5 \times 0.5 = 0.75. \tag{2.458}$$

Exercise E.2.4 The distribution of the random variable is shown in Fig. 2.58. Find the type of the random variable and find the probability that the random variable is equal to 3; also find the probability that the random variable is less than 3.

Answer The random variable is discrete because its distribution is in the form of step functions with jumps in the discrete values of the random variable: -1, 2, 4, and 6.5. The length of the jumps is equal to the corresponding probabilities, which are here equal to 0.25. The corresponding PDF has delta functions in discrete points, as shown in Fig. 2.58b. The area of each delta functions is equal to 0.25.

The r.v. does not take the value 3 and consequently

$$P\{X = 3\} = 0. \tag{2.459}$$

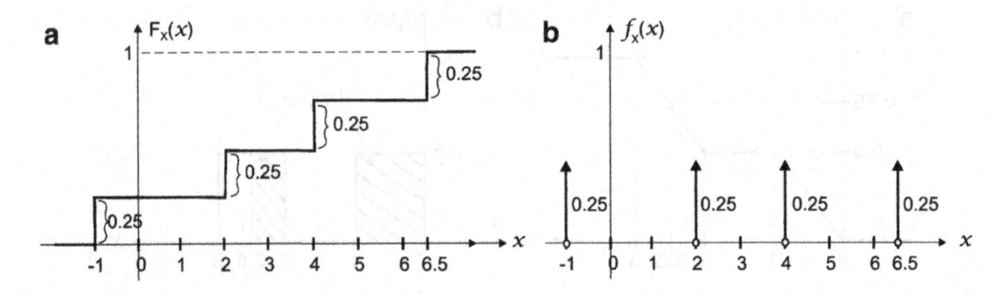

Fig. 2.58 Distribution (**a**) and PDF (**b**)

However, the r.v. X is less than 3 if it is equal to either -1 or 2, resulting in:

$$P\{X<3\} = 0.25 + 0.25 = 0.5. \tag{2.460}$$

This probability corresponds to the value of the distribution in the distribution plot for $x = 3$ and to the area below delta functions from $-\infty$ until 3 in the PDF plot, i.e., again equal to 0.5.

Exercise E.2.5 The PDF of the random variable X is given in Fig. 2.59a. Find the type of the random variable and plot the corresponding distribution function. Find the probability that the random variable is equal to 1 and to 2, as well. Additionally, find the probability that the random variable is greater than 1.

Answer The random variable is mixed (i.e., it is continuous in the interval [0, 2] and has one discrete value $X = 2$, indicated by the delta function in point 2 of the PDF plot). The corresponding distribution function is the integral of the PDF, as demonstrated in Fig. 2.59b.

The r.v is continuous in the interval [0, 2] and thus it does not take any particular value in this interval, resulting in:

$$P\{X = 1\} = 0. \tag{2.461}$$

However, the random variable takes the discrete value 2 with the probability equal to 1/4

$$P\{X = 2\} = 1/4. \tag{2.462}$$

Similarly, we have:

$$P\{X>1\} = \int_1^2 f_X(x)\mathrm{d}x = \frac{3}{8} + \frac{1}{4} = \frac{5}{8}. \tag{2.463}$$

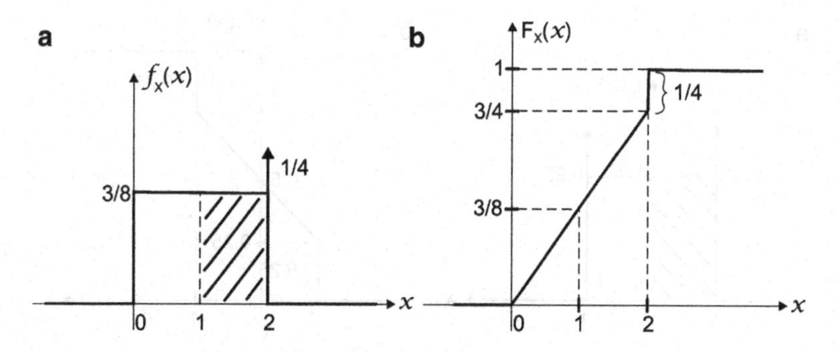

Fig. 2.59 (**a**) PDF and (**b**) distribution of the mixed random variable

The probability (2.463) corresponds to the shaded area plus the area of the delta function in Fig. 2.59b.

The probability (2.463) can be rewritten as:

$$P\{X>1\} = 1 - P\{X \leq 1\} = 1 - \int_{0}^{1} f_X(x)\mathrm{d}x = 1 - \frac{3}{8} = \frac{5}{8} = 1 - F_X(1). \quad (2.464)$$

Exercise E.2.6 The PDF of the random variable X is given as:

$$f_X(x) = a\mathrm{e}^{-b|x|}, \quad -\infty < x < \infty, \quad (2.465)$$

where a and b are constants. Find the relation between a and b.

Answer From the PDF property (2.83), we have:

$$\int_{-\infty}^{\infty} f_X(x)\mathrm{d}x = 1 = \int_{-\infty}^{\infty} a\mathrm{e}^{-b|x|}\mathrm{d}x = a\left[\int_{-\infty}^{0} \mathrm{e}^{bx}\mathrm{d}x + \int_{0}^{\infty} \mathrm{e}^{-bx}\mathrm{d}x\right] = \frac{2a}{b}. \quad (2.466)$$

From (2.465), it follows $b = 2a$ and

$$f_X(x) = a\mathrm{e}^{-2a|x|}, \quad -\infty < x < \infty. \quad (2.467)$$

Exercise E.2.7 The random variable X is uniform in the interval $[-1, 1]$ and with the probability equal to 0.25, it takes both values at the interval ends (i.e. 1 and -1). Plot the PDF and distribution and find the probability that X is negative. Show this result in the PDF and distribution plots.

Answer The random variable X is the mixed variable: continuous in the interval $[-1, 1]$ and discrete in the points -1 and 1. It takes the discrete values -1 and 1, both with a probability of 0.25. As a consequence, the PDF (Fig. 2.60a) has two

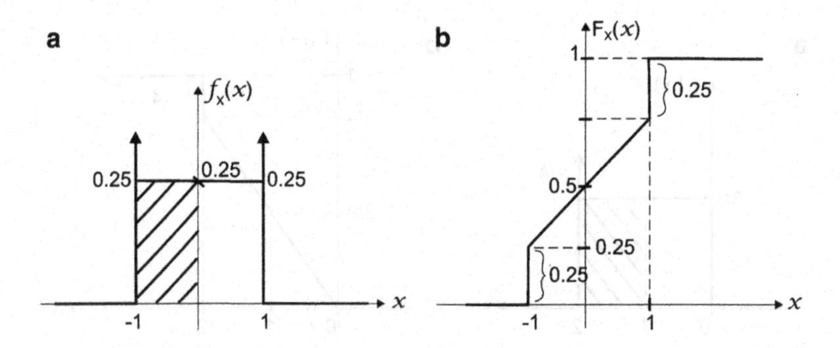

Fig. 2.60 (**a**) PDF and (**b**) distribution of the mixed variable

delta functions at points -1 and 1, each with an area of 0.25. The density function in the interval $[-1, 1]$ is equal to 0.25 so that the area below the PDF is equal to 1.

The distribution has jumps at the discrete points -1 and 1, as shown in Fig. 2.60b.

The probability that X is negative corresponds to the area below the PDF (the shaded area and the area of the delta function in Fig. 2.60a):

$$P\{X < 0\} = 0.25 + 0.25 = 0.5. \tag{2.468}$$

The same result follows from a distribution function:

$$P\{X < 0\} = F_X(0) = 0.5. \tag{2.469}$$

Exercise E.2.8 The PDF of the random variable X is given as (Fig. 2.61):

$$f_X(x) = 0.5\mathrm{e}^{-|x|}, \quad -\infty < x < \infty. \tag{2.470}$$

Find the third central moment and show that the Chebyshev inequality (in the form (2.437)) is satisfied for $k = 2$.

Answer The PDF is symmetrical and the mean value is equal to 0.

$$m_X = 0. \tag{2.471}$$

The third central moment is zero because the PDF is symmetrical, which is confirmed in:

$$\begin{aligned}\mu_3 &= \overline{(X - m_X)^3} = \overline{X^3} = 0.5\int_{-\infty}^{\infty} x^3\mathrm{e}^{-|x|}\mathrm{d}x, = 0.5\left[\int_{-\infty}^{0} x^3\mathrm{e}^{x}\mathrm{d}x + \int_{0}^{\infty} x^3\mathrm{e}^{-x}\mathrm{d}x\right],\\ &= 0.5\left[-\int_{0}^{\infty} x^3\mathrm{e}^{-x}\mathrm{d}x + \int_{0}^{\infty} x^3\mathrm{e}^{-x}\mathrm{d}x\right] = 0.\end{aligned} \tag{2.472}$$

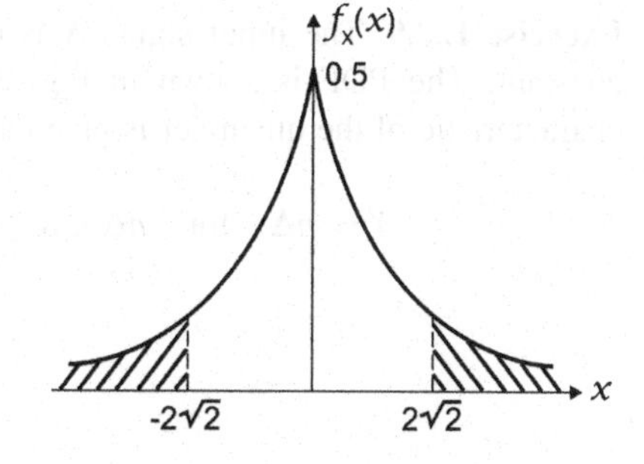

Fig. 2.61 PDF and demonstration of the Chebyshev inequality in Exercise E.2.8

The variance is the second central moment:

$$\sigma_X{}^2 = \mu_2 = \overline{(X - m_X)^2} = \overline{X^2} = 0.5\int_{-\infty}^{\infty} x^2 \mathrm{e}^{-|x|}\mathrm{d}x = 0.5\left[\int_{-\infty}^{0} x^2\mathrm{e}^{x}\mathrm{d}x + \int_{0}^{\infty} x^2\mathrm{e}^{-x}\mathrm{d}x\right]$$

$$= 0.5\left[\int_{0}^{\infty} x^2\mathrm{e}^{-x}\mathrm{d}x + \int_{0}^{\infty} x^2\mathrm{e}^{-x}\mathrm{d}x\right] = \int_{0}^{\infty} x^2\mathrm{e}^{-x}\mathrm{d}x. \tag{2.473}$$

Using integral 1 from Appendix A and the value of the Gamma function from Appendix D, we arrive at:

$$\sigma_X{}^2 = \Gamma(3) = 2! = 2. \tag{2.474}$$

The standard deviation is:

$$\sigma_X = \sqrt{2}. \tag{2.475}$$

From (2.437), given $k = 2$ and using (2.471) and (2.475), we have:

$$P\{|X - m_X| \geq k\sigma_X\} = P\left\{|X| \geq 2\sqrt{2}\right\} \leq 1/4. \tag{2.476}$$

This probability is the shaded area below the PDF in Fig. 2.61.
Next, we find the exact probability using the symmetry property of the PDF:

$$P\left\{|X| \geq 2\sqrt{2}\right\} = 2\int_{2\sqrt{2}}^{\infty} 0.5\mathrm{e}^{-x}\mathrm{d}x = \mathrm{e}^{-2\sqrt{2}} = 0.0591. \tag{2.477}$$

This probability is less than the probability (0.25) of (2.476), thus confirming the Chebyshev inequality.

Exercise E.2.9 The input signal X is uniform in the interval $[0, a]$, where a is constant. The PDF is shown in Fig. 2.62a. The signal is quantized where the characteristic of the quantizer is given as (Fig. 2.62b):

$$Y = n\Delta \quad \text{for} \quad n\Delta < X \leq (n+1)\,\Delta, \quad n = 1, \ldots, 4 \tag{2.478}$$

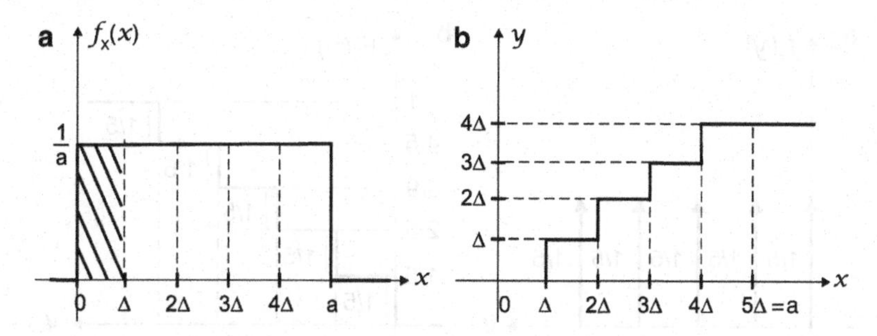

Fig. 2.62 PDF of the input variable and the characteristic of the quantizer. (**a**) PDF. (**b**) Quantizer

and

$$\Delta = a/5. \tag{2.479}$$

Find the mean value of the random variable Y at the output of the quantizer and plot both the PDF and the distribution.

Answer The output random variable Y is the discrete random variable, with the discrete values:

$$y_0 = 0, \quad y_1 = \Delta, \quad y_2 = 2\Delta, \quad y_3 = 3\Delta, \quad y_4 = 4\Delta \tag{2.480}$$

and the corresponding probabilities

$$\begin{aligned} P\{Y = 0\} &= P\{0 < X \le \Delta\} = 1/5, \\ P\{Y = \Delta\} &= P\{\Delta < X \le 2\Delta\} = 1/5, \\ P\{Y = 2\Delta\} &= P\{2\Delta < X \le 3\Delta\} = 1/5, \\ P\{\mathrm{Y} = 3\Delta\} &= P\{3\Delta < X \le 4\Delta\} = 1/5, \\ P\{Y = 4\Delta\} &= P\{4\Delta < X \le 5\Delta\} = 1/5. \end{aligned} \tag{2.481}$$

The PDF and the distribution are shown in Fig. 2.63.
The mean value is:

$$m_Y = \sum_{i=0}^{4} y_i P\{y_i\} = \frac{1}{5}[\Delta + 2\Delta + 3\Delta + 4\Delta] = 2\Delta. \tag{2.482}$$

Note that the PDF of Y has the point of symmetry at 2Δ, which is equal to the mean value.

Exercise E.2.10 The PDF of the variable X is shown in Fig. 2.64a.

$$f_X(x) = 0.5\mathrm{e}^{-|x|}, \quad -\infty < x < \infty. \tag{2.483}$$

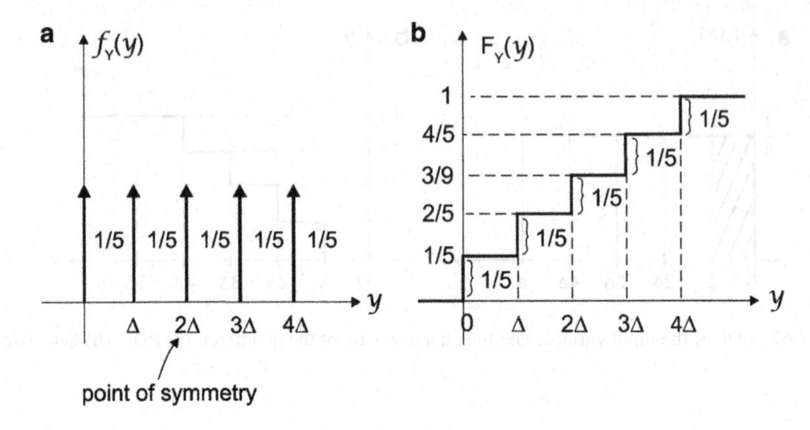

Fig. 2.63 (**a**) PDF and (**b**) distribution of the variable Y

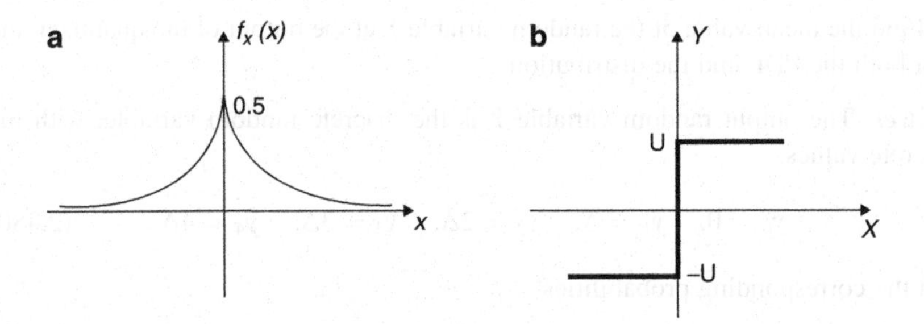

Fig. 2.64 Input random variable and its transformation. (**a**) PDF. (**b**) Transformation

Find the PDF, distribution, and variance of the random variable Y, when the random variables X and Y are related, as shown in Fig. 2.64b.

Answer The output random variable Y is discrete, taking the value U when the variable X is positive and the value $-U$ when the variable X is negative.

Therefore,

$$P\{Y = U\} = P\{X>0\} = 0.5; P\{Y = -U\} = P\{X<0\} = 0.5. \qquad (2.484)$$

The corresponding PDF will have delta functions at the discrete points $-U$ and U, each with the area equal to 1/2 (Fig. 2.65a). The distribution will have two jumps, each equal to 0.5 at the discrete points $-U$ and U, as shown in Fig. 2.65b.

Note that the PDF is symmetrical around $y = 0$, resulting in $E\{Y\} = 0$. The corresponding variance is:

$$\begin{aligned}\sigma_Y^2 &= E\{Y\}^2 - (E\{Y\})^2 \\ &= 0.5U^2 + 0.5(-U)^2 = U^2. \end{aligned} \qquad (2.485)$$

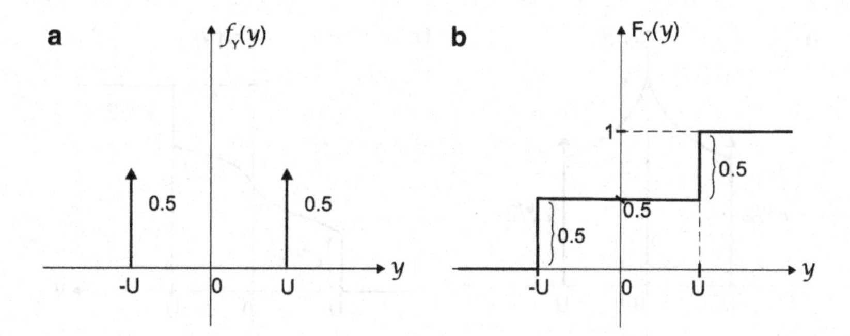

Fig. 2.65 Output (**a**) PDF and (**b**) distribution

Exercise E.2.11 For convenience, the PDF of the variable X, which is the same as in Exercise E.2.10, is shown again in Fig. 2.66a. The transformation of the random variable X is shown in Fig. 2.66b. Find the PDF, distribution, and mean value of the output variable Y.

Answer The output variable Y is a mixed variable with two discrete values, $-U$ and U, and continuous interval $[-U, U]$.

The variable Y has a discrete value of $-U$, when the input variable is less than $-U$, resulting in:

$$P\{Y = -U\} = P\{X \leq -U\} = 0.5 \int_{-\infty}^{-U} \mathrm{e}^{x} \mathrm{d}x = 0.5\mathrm{e}^{-U}. \tag{2.486}$$

Similarly,

$$P\{Y = U\} = P\{X \geq U\} = 0.5 \int_{U}^{\infty} \mathrm{e}^{-x} \mathrm{d}x = 0.5\mathrm{e}^{-U} = P\{Y = -U\}. \tag{2.487}$$

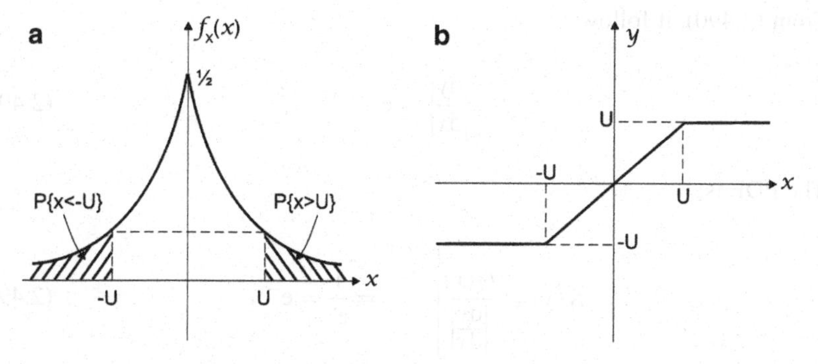

Fig. 2.66 Input (**a**) PDF and (**b**) limiter

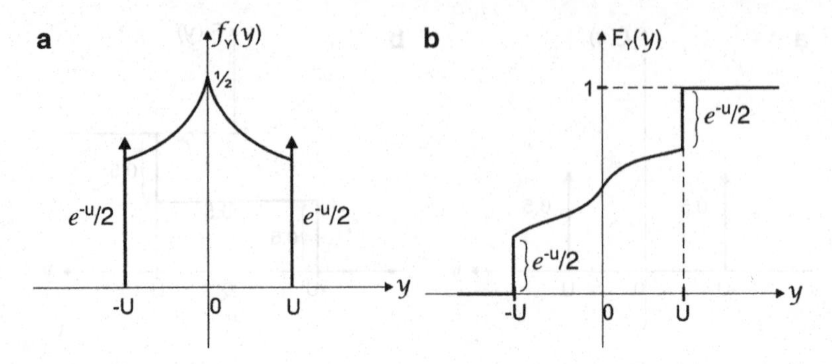

Fig. 2.67 Output (**a**) PDF and (**b**) distribution

For $-U < X < U$, the random variable $Y = X$; in this interval, the corresponding PDFs are equal, as shown in Fig. 2.67a. The output distribution is given in Fig. 2.67b. The output variable is symmetrical around $y = 0$. As a consequence, its mean value is equal to 0.

Exercise E.2.12 The random variable X is uniform in the interval [0, 1]. Find the PDF of the random variable Y if

$$Y = -\ln X. \tag{2.488}$$

Answer The transformation (2.488) is monotone and the input value always corresponds to only one output value. From (2.488), we have:

$$x = \mathrm{e}^{-y} \tag{2.489}$$

and

$$\frac{\mathrm{d}x}{\mathrm{d}y} = -\mathrm{e}^{-y}. \tag{2.490}$$

From (2.490), it follows:

$$\left|\frac{\mathrm{d}y}{\mathrm{d}x}\right| = \mathrm{e}^{y}. \tag{2.491}$$

The PDF is:

$$f_Y(y) = \left.\frac{f_X(x)}{\left|\frac{\mathrm{d}y}{\mathrm{d}x}\right|}\right|_{x=\mathrm{e}^{-y}} = \frac{1}{\mathrm{e}^{y}} = \mathrm{e}^{-y}. \tag{2.492}$$

If $X \to 0$, $Y \to \infty$, while if $X \to 1$, $Y \to 0$. Therefore, the output variable is defined only in the interval $[0, \infty]$,

$$f_Y(y) = \mathrm{e}^{-y} u(y), \tag{2.493}$$

where $u(y)$ is a step function

$$u(y) = \begin{cases} 1 & \text{for} \quad y \geq 0, \\ 0 & \qquad \text{otherwise.} \end{cases} \tag{2.494}$$

Exercise E.2.13 The random variable X is uniform in the interval [0, 1]. Find which monotone transformation of the variable X will result in a random variable Y with the PDF

$$f_Y(y) = \lambda \mathrm{e}^{-\lambda y} u(y). \tag{2.495}$$

Answer The PDF of the variable X is:

$$f_X(x) = \begin{cases} 1 & 0 \leq x \leq 1 \\ 0 & \text{otherwise.} \end{cases} \tag{2.496}$$

For the monotone transformation, we can write:

$$\int_0^x f_X(x) \mathrm{d}x = \int_0^y f_Y(y) \mathrm{d}y. \tag{2.497}$$

Placing (2.495) and (2.496) into (2.497), we get:

$$\int_0^x \mathrm{d}x = x = \int_0^y \lambda \mathrm{e}^{-\lambda y} \mathrm{d}y = 1 - \mathrm{e}^{-\lambda y}. \tag{2.498}$$

From (2.498), we arrive at:

$$y = -\frac{1}{\lambda} \ln(1 - x) \tag{2.499}$$

or

$$Y = -\frac{1}{\lambda} \ln(1 - X). \tag{2.500}$$

Exercise E.2.14 The random variable X has the following PDF:

$$f_X(x) = \begin{cases} 0.5(x+1) & -1 \leq x \leq 1, \\ 0 & \text{otherwise.} \end{cases} \tag{2.501}$$

as shown in Fig. 2.68. Find the PDF of the random variable Y where

$$Y = 1 - X^2. \tag{2.502}$$

Answer From (2.502), it follows:

$$x_{1,2} = \pm\sqrt{1-y} \tag{2.503}$$

and

$$\left|\frac{\mathrm{d}y}{\mathrm{d}x}\right| = |2x| = 2\sqrt{1-y}. \tag{2.504}$$

The PDF is obtained using (2.170), (2.501), (2.503), and (2.504):

$$\begin{aligned} f_Y(y) &= \left.\frac{f_X(x)}{\left|\frac{\mathrm{d}y}{\mathrm{d}x}\right|}\right|_{x_1=\sqrt{1-y}} + \left.\frac{f_X(x)}{\left|\frac{\mathrm{d}y}{\mathrm{d}x}\right|}\right|_{x_2=-\sqrt{1-y}} \\ &= \frac{0.5(\sqrt{1-y}+1)}{2\sqrt{1-y}} + \frac{0.5(-\sqrt{1-y}+1)}{2\sqrt{1-y}}. \end{aligned} \tag{2.505}$$

From here, we have:

$$f_Y(y) = \begin{cases} \dfrac{1}{2\sqrt{1-y}} & 0 \leq y \leq 1, \\ 0 & \text{otherwise.} \end{cases} \tag{2.506}$$

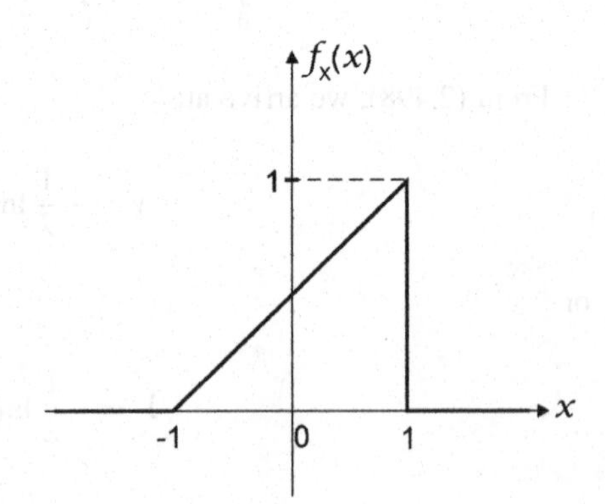

Fig. 2.68 PDF of the input variable X

Exercise E.2.15 The random variable X is uniform in the interval $[-\pi, \pi]$. Find the PDF of the variable Y if

$$Y = \sin X. \tag{2.507}$$

Answer The PDF $f_X(x)$ is shown in Fig. 2.69a.

The random variable Y is a "sin function" and, consequently, takes only the values in the interval $[-1, 1]$, resulting in:

$$f_Y(y) = 0 \quad \text{for} \quad |Y|>1. \tag{2.508}$$

For $|y| < 1$, there are two solutions x_1 and x_2 for

$$y = \sin x, \tag{2.509}$$

as shown in Fig. 2.68b. From (2.509), it follows:

$$\left|\frac{dy}{dx}\right| = |\cos x| = \sqrt{1 - \sin^2 x} = \sqrt{1 - y^2}. \tag{2.510}$$

Finally, from (2.170) and (2.510), the PDF is given as

$$\begin{aligned} f_Y(y) &= \frac{f_X(x)}{\left|\frac{dy}{dx}\right|}\Bigg|_{x=x_1} + \frac{f_X(x)}{\left|\frac{dy}{dx}\right|}\Bigg|_{x=x_2} = 2\frac{1}{2\pi\sqrt{1-y^2}} \\ &= \frac{1}{\pi\sqrt{1-y^2}} \quad \text{for} \quad -1 \le y \le 1. \end{aligned} \tag{2.511}$$

For other values of y, the PDF is zero.

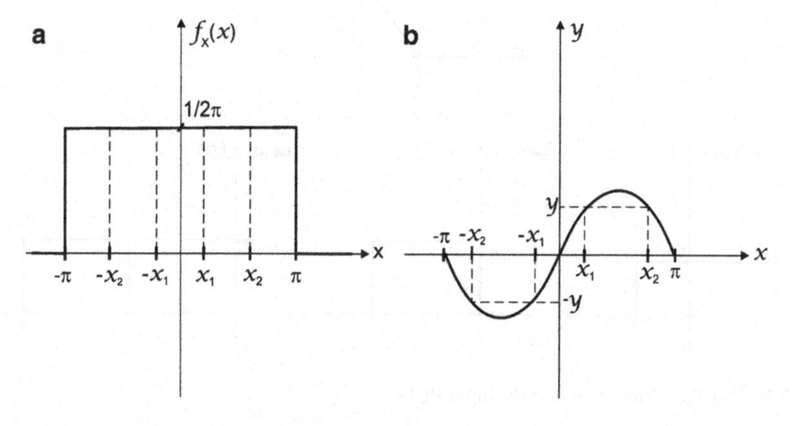

Fig. 2.69 Input (**a**) PDF and the (**b**) transformation

Exercise E.2.16 Find the PDF of the random variable Y

$$Y = A\sin(\omega_0 t + X), \tag{2.512}$$

where X is uniform in the interval $[0, 2\pi]$, and A is constant.

Answer As in Exercise E.2.14, the nonmonotone transformation results in two values of the input variable corresponding to only one value of Y.

From (2.512), it follows:

$$\left|\frac{\mathrm{d}y}{\mathrm{d}x}\right| = |A\cos(\omega_0 t + x)| = A\sqrt{1 - \frac{y^2}{A^2}} = \sqrt{A^2 - y^2} \tag{2.513}$$

and

$$f_Y(y) = \begin{cases} \dfrac{1}{\pi\sqrt{A^2 - y^2}} & -A \le y \le A \\ 0 & \text{otherwise.} \end{cases} \tag{2.514}$$

Exercise E.2.17 Find the PDF of the random variable Y (Fig. 2.70a)

$$Y = \begin{cases} X & x>0, \\ 0 & \text{otherwise.} \end{cases} \tag{2.515}$$

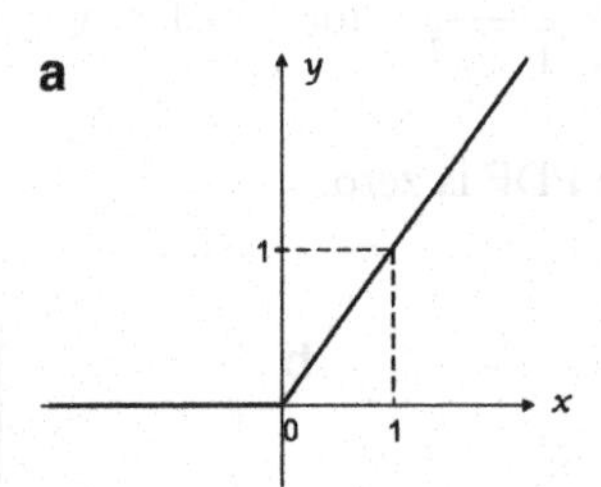

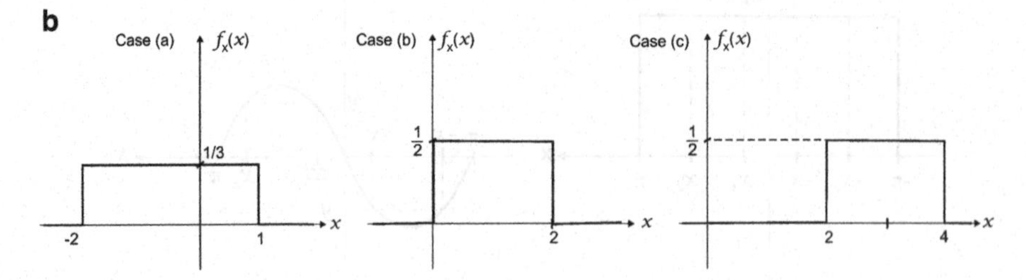

Fig. 2.70 (**a**) Transformation and (**b**) input PDFs

for three cases (Fig. 2.70b):

(a) X is uniform in the interval [−2, 1]
(b) X is uniform in the interval [0, 2]
(c) X is uniform in the interval [2, 4],

Answer

(a) The random variable Y is a mixed variable that has the discrete value $y = 0$, and it is continuous in the range [0, 1].

$$P\{Y = 0\} = P\{X < 0\} = 2/3. \tag{2.516}$$

For $0 < x < 1$, the output random variable Y is continuous and equal to X, resulting in:

$$f_Y(y) = f_X(x) = 1/3. \tag{2.517}$$

From (2.516) and (2.517), we arrive at:

$$f_Y(y) = \begin{cases} 2/3\delta(y) + 1/3 & 0 \leq y \leq 1, \\ 0 & \text{otherwise}. \end{cases} \tag{2.518}$$

(b) The random variables X and Y are both positive and, from (2.515), the random variable is equal to X in the interval [0, 2].

$$f_Y(y) = \begin{cases} 1/2 & 0 \leq y \leq 2, \\ 0 & \text{otherwise}. \end{cases} \tag{2.519}$$

(c) The random variable Y is equal to X in the interval [2, 4]

$$Y = X \quad \text{for} \quad 2 \leq x \leq 4 \tag{2.520}$$

and

$$f_Y(y) = \begin{cases} 1/2 & 2 \leq y \leq 4, \\ 0 & \text{otherwise}. \end{cases} \tag{2.521}$$

Exercise E.2.18 The discrete random variable X has discrete values 1, 2, 3, 6, and 8, with the corresponding probabilities:

$$\begin{aligned} P\{X = 1\} = P\{X = 8\} &= 0.1, \\ P\{X = 2\} = P\{X = 3\} &= 0.3, \\ P\{X = 6\} &= 0.2. \end{aligned} \tag{2.522}$$

Find the PDF and distribution of the random variable Y

$$Y = \min\{X, 4\}. \tag{2.523}$$

Answer According to (2.523), the variable Y is:

$$Y = \begin{cases} X & x < 4, \\ 4 & x \geq 4. \end{cases} \tag{2.524}$$

From (2.524), it follows that for values of X equal to 1, 2, and 3, $Y = X$; thus, Y takes the same values 1, 2, and 3 as does the variable X, with the corresponding probabilities

$$\begin{aligned} P\{Y = 1\} &= P\{X = 1\} = 0.1, \\ P\{Y = 2\} &= P\{X = 2\} = 0.3, \\ P\{Y = 3\} &= P\{X = 3\} = 0.3. \end{aligned} \tag{2.525}$$

Similarly, from (2.524), for the values of X equal to 6 and 8, the random variable Y will have only one value equal to 4, with the following probability:

$$P\{Y = 4\} = P\{X = 6\} + P\{X = 8\}. \tag{2.526}$$

Finally, from (2.525) and (2.526), we get the distribution:

$$F_Y(y) = 0.1u(y-1) + 0.3u(y-2) + 0.3u(y-3) + 0.3u(y-4) \tag{2.527}$$

and the PDF

$$f_Y(y) = 0.1\delta(y-1) + 0.3\delta(y-2) + 0.3\delta(y-3) + 0.3\delta(y-4), \tag{2.528}$$

where u and δ are the step and delta functions, respectively.

Exercise E.2.19 A message consisting of two pulses is transmitted over a channel. Due to noise on the channel, each or both of the pulses can be destroyed independently with the probability p. The random variable X represents the number of the destroyed pulses. Find the characteristic function of the variable X.

Answer The variable X is a discrete variable which takes values 0, 1, and 2 with the probabilities,

$$\begin{aligned} &P\{X = 0\} = (1-p)^2, \\ &P\{X = 1\} = p(1-p) + (1-p)p = 2p(1-p), \\ &P\{X = 2\} = p^2. \end{aligned} \tag{2.529}$$

From (2.419) and (2.529), the characteristic function is:

$$\phi_X(\omega) = \sum_{i=0}^{2} e^{j\omega x_i} P(X = x_i),$$
$$= e^{j\omega 0}(1-p)^2 + e^{j\omega}2p(1-p) + e^{j\omega 2}p^2 = \left[p - (pe^{j\omega} + 1)\right]^2. \quad (2.530)$$

Exercise E.2.20 Find the characteristic function of the uniform variable in the interval [0, 1].

Answer Using the result from Example 2.8.9, for $a = b = 0.5$, we have:

$$\phi_X(\omega) = \frac{e^{0.5j\omega}}{0.5\omega} \sin 0.5\omega. \quad (2.531)$$

Exercise E.2.21 Using the moment theorem, find the variance of the random variable X with the corresponding characteristic function

$$\phi_X(\omega) = ae^{j\omega} + b, \quad (2.532)$$

where a and b are constants

$$a + b = 1. \quad (2.533)$$

Answer The variance determined by the two first moments is:

$$\sigma_X^2 = m_2 - m_1^2. \quad (2.534)$$

From (2.415), we have:

$$m_1 = \frac{\phi^{(1)}(\omega)\Big|_{\omega=0}}{j} = a,$$
$$m_2 = \frac{\phi^{(2)}(\omega)\Big|_{\omega=0}}{j^2} = a. \quad (2.535)$$

From (2.533) to (2.535), we have:

$$\sigma_X^2 = m_2 - m_1^2 = a - a^2 = a(1-a) = ab. \quad (2.536)$$

Exercise E.2.22 A random variable X is uniform in the interval $[-\pi/2, \pi/2]$. Find the PDF of the variable Y

$$Y = \mathrm{sin}X, \tag{2.537}$$

using the characteristic function.

Answer The characteristic function of the random variable

$$Y = g(X) = \mathrm{sin}X \tag{2.538}$$

is found to be

$$\phi_{g(X)} = \int_{-\pi/2}^{\pi/2} \mathrm{e}^{j\omega g(x)} f_X(x)\mathrm{d}x = \int_{-\pi/2}^{\pi/2} \mathrm{e}^{j\omega \sin x} \frac{1}{\pi}\mathrm{d}x. \tag{2.539}$$

From (2.538), we have:

$$y = \mathrm{sin}x \tag{2.540}$$

and

$$\mathrm{d}y = \cos x\mathrm{d}x = \sqrt{1-y^2}\mathrm{d}x. \tag{2.541}$$

Placing (2.540) and (2.541) into (2.539), we get:

$$\phi_Y(\omega) = \int_{-1}^{1} \frac{\mathrm{e}^{j\omega y}}{\pi\sqrt{1-y^2}}\mathrm{d}y. \tag{2.542}$$

Using the definition of the characteristic function of the variable Y and (2.542), we can write:

$$\phi_Y(\omega) = \int_{-1}^{1} \frac{\mathrm{e}^{j\omega y}}{\pi\sqrt{1-y^2}}\mathrm{d}y = \int_{-1}^{1} \mathrm{e}^{j\omega y} f_Y(y)\mathrm{d}y. \tag{2.543}$$

From (2.543) we can easily find:

$$f_Y(y) = \begin{cases} \dfrac{1}{\pi\sqrt{1-y^2}} & -1 \le y \le 1, \\ 0 & \text{otherwise.} \end{cases} \tag{2.544}$$

Exercise E.2.23 The random variable X has the Simpson PDF as shown in Fig. 2.71. Suppose that one has observed that the values of X are only in the interval

$$1 < X \leq 1.5 \qquad (2.545)$$

Find the conditional PDF and distribution, given the condition (2.545).

Answer The probability of the event (2.545) corresponds to the shaded area in Fig. 2.71, it is equal to 1/8:

$$P\{1 < X \leq 1.5\} = F_X(1.5) - F_X(1) = (0.5 \times 0.5)/2 = 1/8. \qquad (2.546)$$

From (2.127) and (2.546), we have:

$$F_X(x|1 < X \leq 1.5) = \begin{cases} 1 & \text{for} \quad x > 1.5, \\ \dfrac{F_X(x) - F_X(1)}{F_X(1.5) - F_X(1)} = \dfrac{(x-1)^2}{2}\dfrac{1}{1/8} = 4(x-1)^2 & \text{for} \quad 1 < x \leq 1.5, \\ 0 & \text{for} \quad x \leq 1. \end{cases} \qquad (2.547)$$

From (2.128), the conditional PDF is:

$$f_X(x|1 < X \leq 1.5) = \begin{cases} 0 & \text{for} \quad x > 1.5, \\ \dfrac{f_X(x) - F_X(1)}{F_X(1.5) - F_X(1)} = 8(x-1) & \text{for} \quad 1 < x \leq 1.5, \\ 0 & \text{for} \quad x \leq 1. \end{cases} \qquad (2.548)$$

The conditional PDF is also shown in Fig. 2.71 with a dotted line.

Exercise E.2.24 Find the mean value for the conditional PDF in Exercise E.2.23.

Answer From (2.548), we have:

$$m = \int_1^{1.5} x f_X(x|1 < X \leq 1.5)\mathrm{d}x = 8\int_1^{1.5} x(x-1)\mathrm{d}x = 4/3. \qquad (2.549)$$

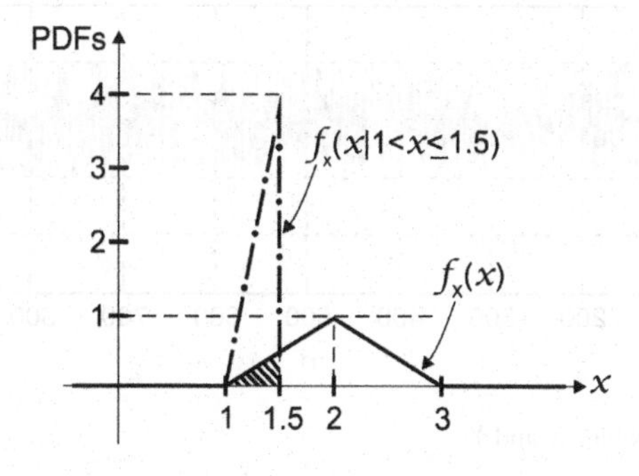

Fig. 2.71 Original and conditional PDFs

2.10 MATLAB Exercises

Exercise M.2.1 (MATLAB file: *exercise_M_2_1.m*). Show that the linear transformation of the uniform random variable results in a uniform variable, using as an example, the uniform random variable X in the interval [0, 1] and the transformation:

$$Y = -2X + 3. \tag{2.550}$$

Solution The random variables X and Y are shown in Fig. 2.72.

We can easily note that the variable Y is also uniform but in a different interval (i.e., in the interval [1, 3]) as presented in the numerical Example 2.6.1.

From (2.550), we have:

$$X = x_1 = 0, \quad Y = y_1 = 3, \tag{2.551}$$

$$X = x_2 = 1; \quad Y = y_2 = 1. \tag{2.552}$$

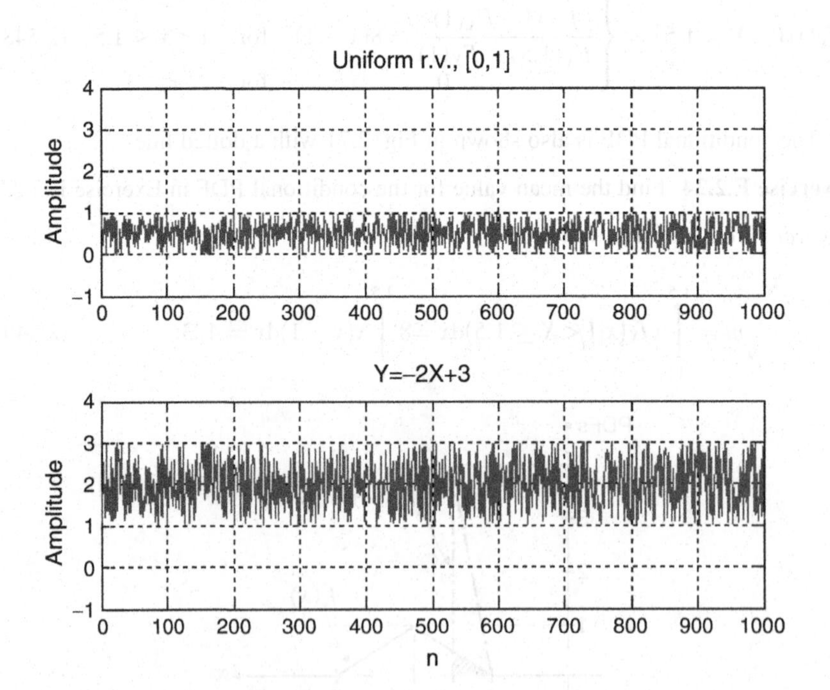

Fig. 2.72 Random variables X and Y

The PDF of the variable Y is

$$f_Y(y) = \begin{cases} \dfrac{1}{y_1 - y_2} = \dfrac{1}{3-1} = \dfrac{1}{2} & \text{for} \quad 1 < y < 3, \\ 0 & \text{otherwise}. \end{cases} \tag{2.553}$$

The PDF of the random variable Y is estimated and plotted using the file *density.m* as shown in Fig. 2.73.

Exercise M.2.2 (MATLAB file: *exercise_M_2_2.m*). Find the linear transformation

$$Y = 2X + 2 \tag{2.554}$$

of the uniform random variable X in the interval [0, 1], estimate and plot the PDF of the random variable Y.

Solution From (2.554) and using $x_1 = 0$ and $x_2 = 2$, it follows:

$$y_1 = 2x_1 + 2 = 2, \quad y_2 = 2x_2 + 2 = 4. \tag{2.555}$$

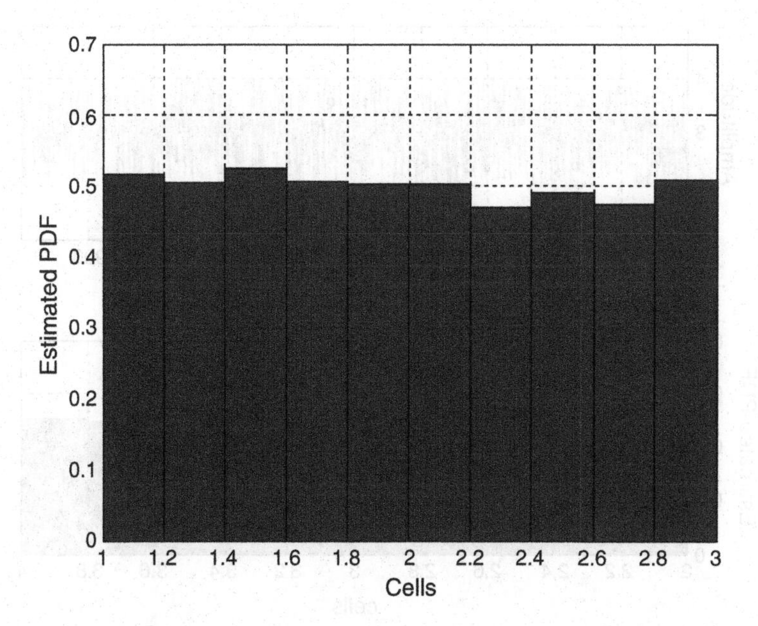

Fig. 2.73 Estimated density of the random variable Y

From Example 2.6.1 and (2.555), we have:

$$f_Y(y) = \begin{cases} \frac{1}{4-2} = \frac{1}{2} & \text{for} \quad 2 < y < 4, \\ 0 & \text{otherwise.} \end{cases} \tag{2.556}$$

The random variable and its estimated PDF are shown in Fig. 2.74.

Exercise M.2.3 (MATLAB file: *exercise_M_2_3.m*). Using the MATLAB file *rand.m* which generates the uniform random variable X in the interval [0, 1], write the MATLAB program to generate the uniform variable in the desired interval $[R_1, R_2]$. Consider, for example, $R_1 = -2$ and $R_2 = 4$.

Solution Using the result of Example 2.6.4, we have:

$$Y = (R_2 - R_1)X + R_1. \tag{2.557}$$

The corresponding MATLAB file is *unif.m*. The transformed variable Y and its estimated PDF are shown in Fig. 2.75.

Exercise M.2.4 (MATLAB file: *exercise_M_2_4.m*). Estimate and plot the PDF and distribution of the uniform random variable X in the interval [0.5, 3.5].

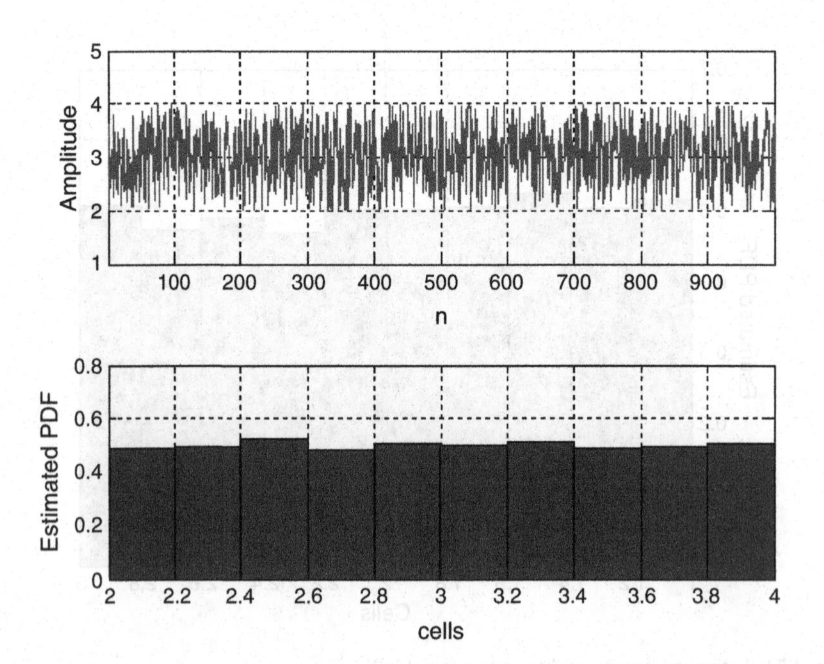

Fig. 2.74 Transformed uniform variable and its estimated PDF

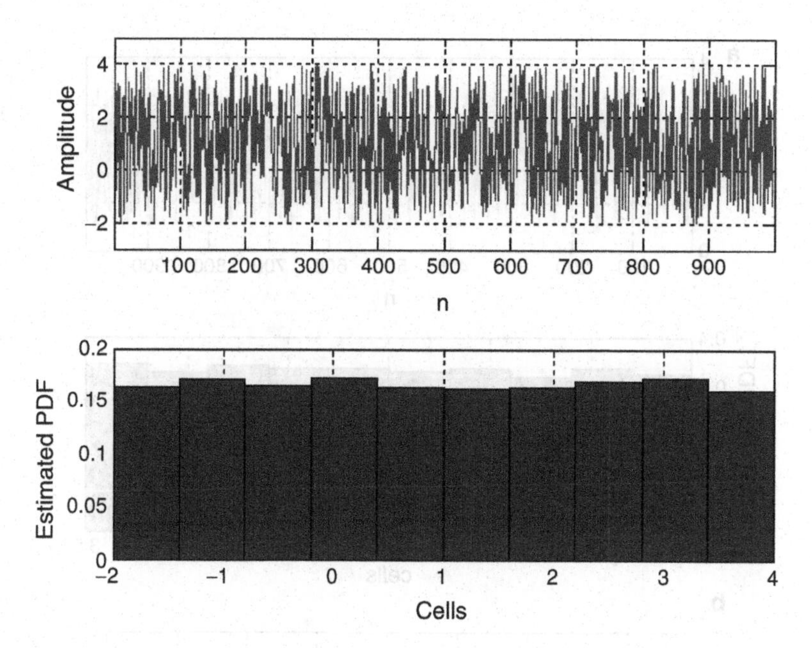

Fig. 2.75 Random variable and PDF from Exercise M.2.3

Solution The random variable X is generated using *unif.m* for $N = 10{,}000$ and $R_1 = 0.5, R_2 = 3.5$.

The PDF is estimated using *density.m* for NN = 10. The random variable Y and its estimated distribution are shown in Fig. 2.76a. The distribution is estimated using the file *distrib.m* and is shown in Fig. 2.76b.

Exercise M.2.5 (MATLAB file: *exercise_M_2_5.m*). Generate the uniform random variable in the interval [3, 7] and estimate its PDF, mean value, and variance.

Solution The random variable is generated using *unif.m* for $N = 10{,}000$. The PDF is estimated using *density.m* for 10 cells, as shown in Fig. 2.77.

The point of symmetry of the PDF is 5 and corresponds to the mean value. The estimated mean value is obtained using *mean.m* and is equal to 4.9937. The variance is calculated using (2.349), resulting in $\sigma^2 = 4^2/12 = 1.3333$. The variance is estimated using *var.m* and the obtained result is 1.341. Note that there is error in the estimations.

Consider now the estimation using $N = 100{,}000$. The estimated values of the mean and variance are 5.0026 and 1.3377, respectively. Increasing N to 500,000, the estimated mean value and variance are 4.9995 and 1.3318, respectively.

Exercise M.2.6 (MATLAB file: *exercise_M_2_6.m*). Generate the uniform random variables X_1 and X_2, having the same variance $\sigma^2 = 1/3$ and mean values of 4 and 2, respectively. Estimate the corresponding PDFs.

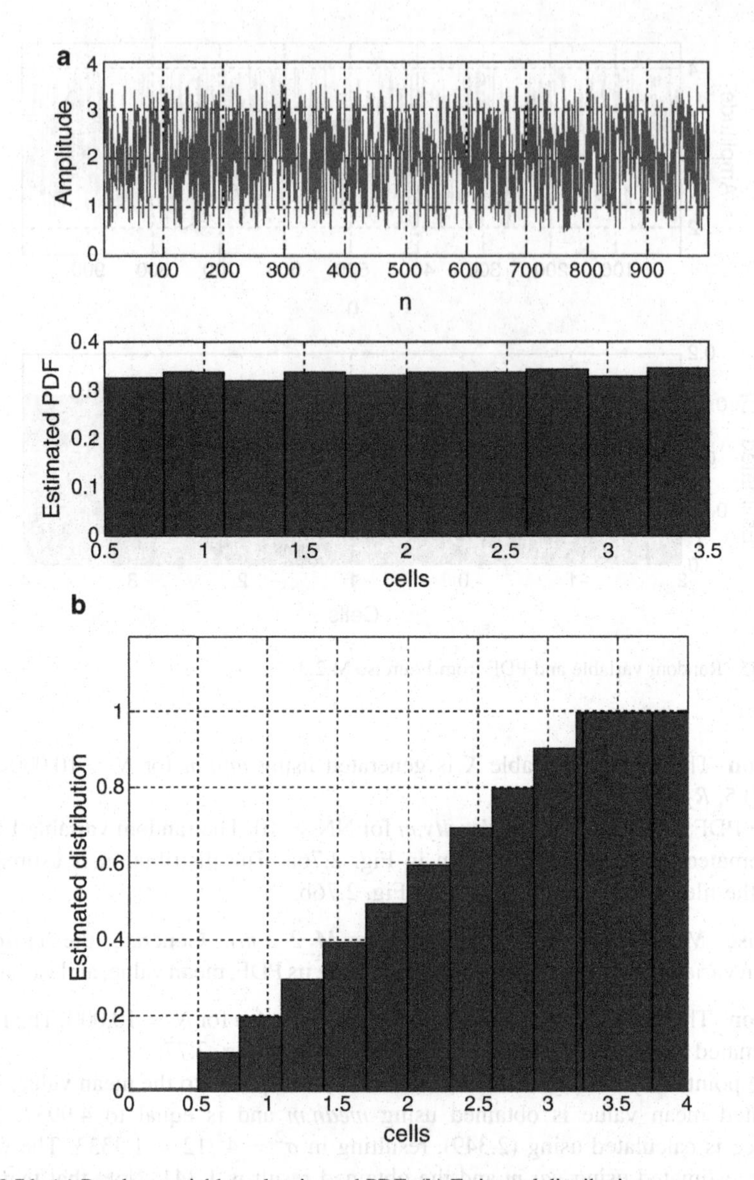

Fig. 2.76 (**a**) Random variable and estimated PDF. (**b**) Estimated distribution

Solution From the expression for the variance of uniform variable (2.349) it follows:

$$\Delta^2/12 = 1/3, \tag{2.558}$$

and both random variables have the same width of the range $\Delta = 2$.

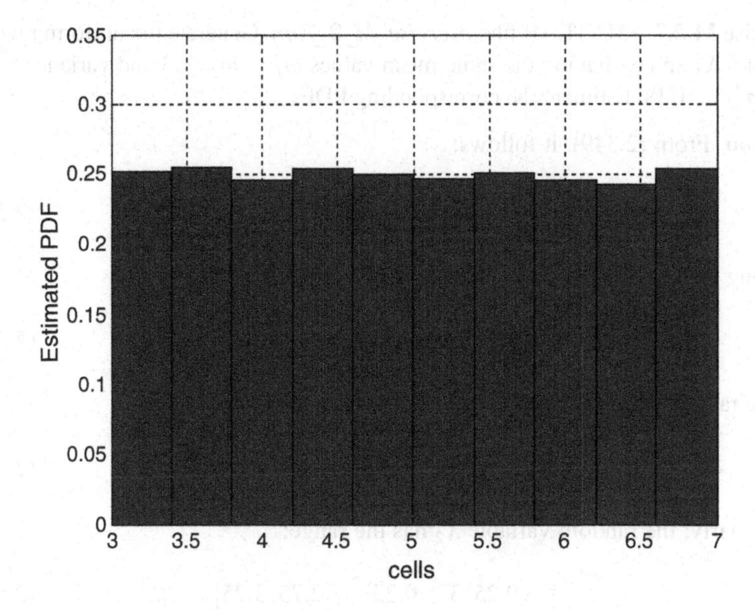

Fig. 2.77 Uniform variable and its PDF

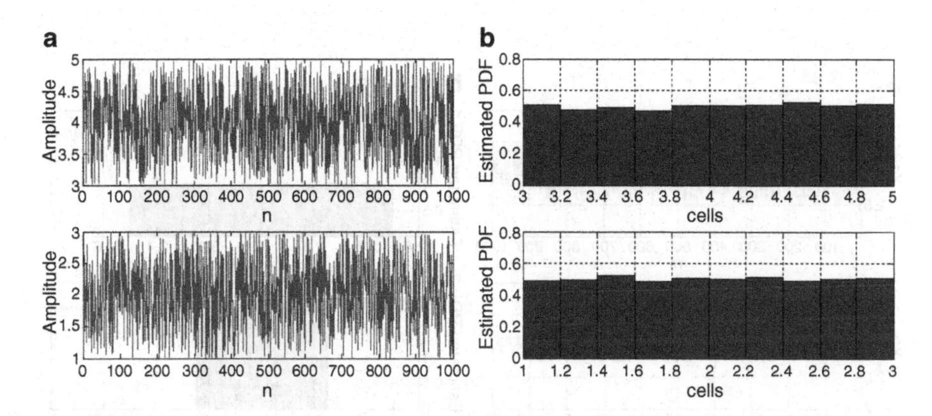

Fig. 2.78 Example of uniform variables with the same variance. (**a**) Variables. (**b**) Estimated PDFs

Consequently, both of the PDFs are equal, but in different intervals. Knowing that the mean value is in the point of symmetry of Δ, it follows that the random variable X_1 is in the interval [3, 5], while the random variable X_2 is in the interval [1, 3]. The variables are generated using *unif.m* and are shown in Fig. 2.78a. The estimated PDFs obtained from *density.m* are shown in Fig. 2.78b.

Exercise M.2.7 (MATLAB file: *exercise_M_2_7.m*). Generate the uniform random variables X_1 and X_2 having the same mean values $m_1 = m_2 = 3$ and variances $\sigma^2{}_1 = 1/12$, $\sigma^2{}_2 = 1/48$. Estimate the corresponding PDFs.

Solution From (2.349), it follows:

$$\Delta_1^2/12 = 4\Delta_2^2/12 \tag{2.559}$$

resulting in:

$$\Delta_1 = 2\Delta_2 = 1, \quad \Delta_2 = 0.5. \tag{2.560}$$

The range of the variable X_1 is:

$$[3 - 0.5, 3 + 0.5] = [2.5,\ 3.5]. \tag{2.561}$$

Similarly, the random variable X_2 has the range:

$$[3 - 0.25, 3 + 0.25] = [2.75, 3.25]. \tag{2.562}$$

The generated variables and estimated PDFs are shown in Fig. 2.79.

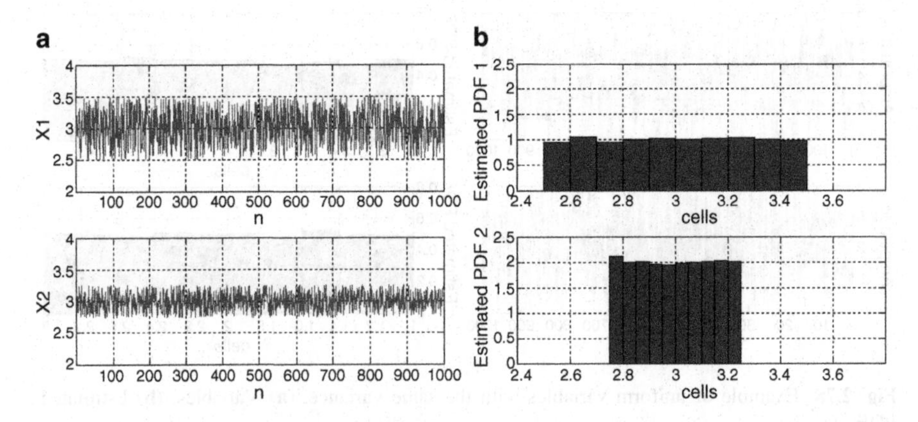

Fig. 2.79 Example of uniform variables with the same mean value and $\sigma_1{}^2 = 1/12$ and $\sigma_2{}^2 = 1/48$. **(a)** Signals X_1 and X_2. **(b)** Estimated PDFs

Exercise M.2.8 (MATLAB file: *exercise_M_2_8.m*). Generate the random variable:

$$Y = -\ln X, \tag{2.563}$$

where X is the uniform random variable in the range [0, 1]. Estimate PDF of variable Y.

Solution The variable and its estimated density are shown in Fig. 2.80.

Exercise M.2.9 (MATLAB file: *exercise_M_2_9.m*). Generate the random variable Y and estimate its PDF if

$$Y = \sqrt{-2\ln X}, \tag{2.564}$$

where X is the uniform random variable in the range [0, 1].

Solution The variable Y and its estimated density are shown in Fig. 2.81.

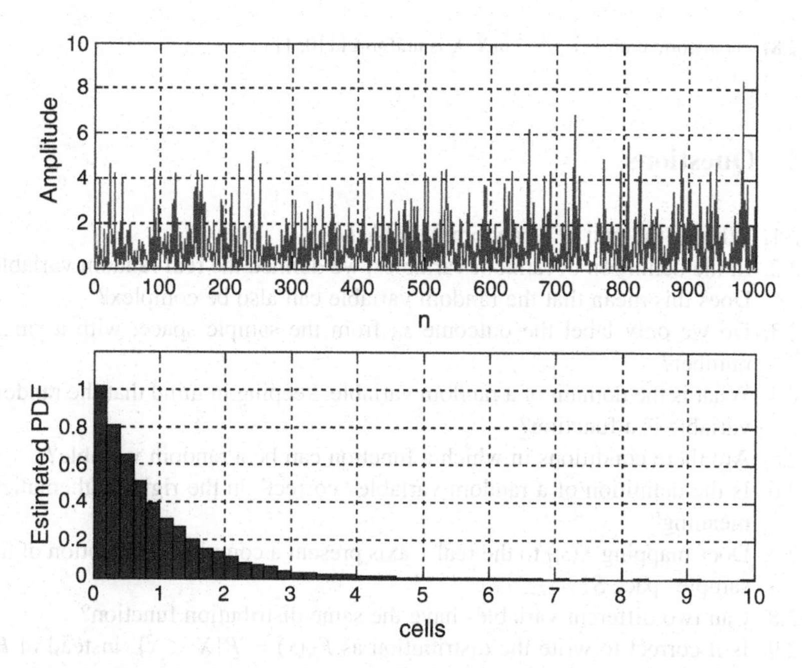

Fig. 2.80 Transformation $Y = -\ln X$, X is uniform in [0, 1]

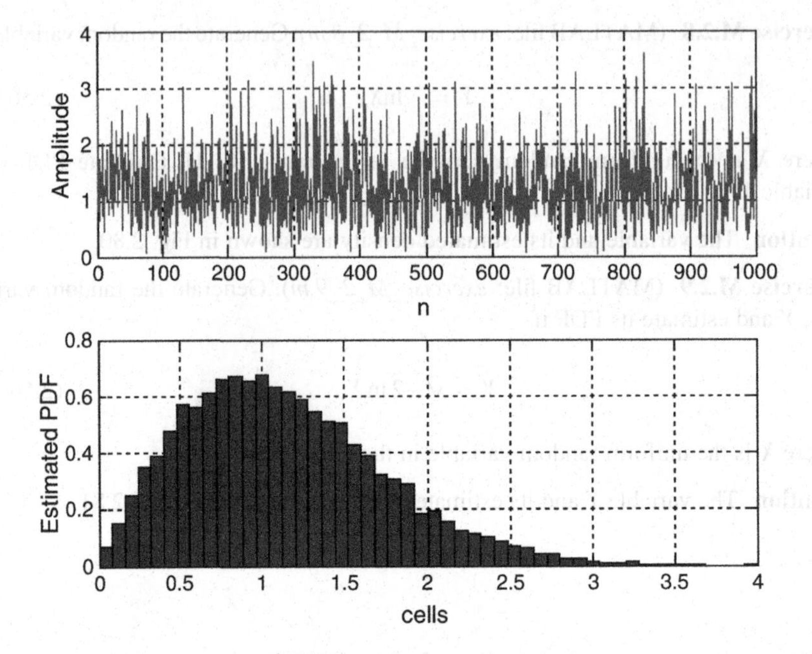

Fig. 2.81 Transformation $Y = \sqrt{-2\ln X}$, X is uniform in [0, 1]

2.11 Questions

Q.2.1. What is the unit of a random variable?

Q.2.2. In the definition of random variables, we defined the real random variable. Does this mean that the random variable can also be complex?

Q.2.3. Do we only label the outcome s_i, from the sample space, with a single number?

Q.2.4. What is the domain of a random variable, keeping in mind that the random variable is a function?

Q.2.5. Are there conditions in which a function can be a random variable?

Q.2.6. Is the definition of a random variable "correct" in the rigid mathematical meaning?

Q.2.7. Does mapping $X(s_i)$ to the real x-axis present a complete description of the sample space S?

Q.2.8. Can two different variables have the same distribution function?

Q.2.9. Is it correct to write the distribution as $F_X(x) = P\{X < x\}$, instead of $F_X(x) = P\{X \leq x\}$?

Q.2.10. Is the PDF of a discrete random variable a discrete function in itself?

Q.2.11. The distribution function for the discrete, continuous, and mixed random variables is given as $F_X(x)$. Does this mean that the distribution is a continuous function in all cases?

Q.2.12. A discrete random variable is only defined in discrete points. Does this mean that its distribution function only exists in discrete points?

Q.2.13. Why must the PDF of a discrete random variable only have delta functions in the discrete values of a random variable?

Q.2.14. Imagine a continuous random variable X in the interval $[a, b]$ with a finite probability of $0 < P \leq 1$. What is the probability that the random variable X takes a particular value in the interval $[a, b]$?

Q.2.15. Does variance exist for all random variables?

Q.2.16. Is standard deviation always a measure of the width of the PDF?

2.12 Answers

A.2.1. The random variable is dimensionless. The outcomes of the experiment can be different physical values and things. However, the values of random variables are only numerical values without dimension (units). However, when random signals are concerned and numerical values of a random variable correspond to the numerical values of a physical signal, some authors consider a random variable as a physical quantity; for example, random voltage in electrical engineering.

A.2.2. YES. The random variable can also be defined as complex in terms of real random variables X and Y:

$$Z = X + jY, \tag{2.565}$$

where $j = \sqrt{-1}$.
However, complex random variables are beyond the scope of this book and we will only considering real random variables.

A.2.3. NO. We can also label s_i with a couple of numbers which lead to the two-dimensional r.v. or n-tuple numbers leading to the n-dimensional r.v. Additionally, we can also sometimes assign varying functions to each outcome, leading us to the random process.

A.2.4. The domain of the random variable is the sample space S. Therefore, it would be more correct to write $X(S)$. However, since we understand that r.v is a function of the sample space, we do not always show the dependence on S. Instead, in most cases, we just denote random variable such as X, Y, and Z.

A.2.5. Two following conditions must be satisfied, [PAP65, p. 88]:

- The set $(X \leq x)$ must be an event for any real number x.
- The probabilities of the events $(X = \infty)$ and $(X = -\infty)$ are zero,

$$P\{X = \infty\} = 0, \quad P\{X = -\infty\} = 0.$$

(These events are not generally empty (i.e., we allow that the r.v. be equal to ∞ or $-\infty$ for some outcomes, [PAP65, p. 87])

A.2.6. NO. A random variable X represents the function or rule which assigns a real number to each outcome s_i in a sample space S. However, there is nothing random in $X(s_i)$ because this function is fixed and deterministic, and is usually chosen by us [LEO94, p. 85], [KAY06, p. 107].

What is truly random is the input argument s_i and consequently the random output of the function. However, traditionally, this terminology is widely accepted.

A.2.7. NO. In order to describe the r.v. completely, we must also know how often the values of x are taken. That is we must include probabilistic measure.

A.2.8. Every random variable uniquely defines its distribution. Each random variable can have only one distribution. However, there is an arbitrary number of different random variables which have the same distribution. Let, for example, X be a discrete r.v. which takes only two values: -1 and 1, each with a probability of 1/2. Consider the variable $Y = -X$. It is obvious that X and Y are different variables. However, both of these variables have the same distribution [GNE82, p. 130]:

$$F_X(x) = \begin{cases} 0 & \text{for} \quad x < -1, \\ 1/2 & \text{for} \quad -1 \leq x \leq 1, \\ 1 & \text{for} \quad x > 1. \end{cases}$$
$$F_Y(y) = \begin{cases} 0 & \text{for} \quad y < -1, \\ 1/2 & \text{for} \quad -1 \leq y \leq 1, \\ 1 & \text{for} \quad y > 1. \end{cases} \tag{2.566}$$

A.2.9. The definition of the distribution function $F_X(x) = P\{X \leq x\}$ includes both cases: the discrete and continuous random variables. However, for continuous random variables, the probability $P\{X = x\}$ is zero, and we may write $F_X(x) = P\{X < x\}$.

A.2.10. The PDF of the discrete random variables has only delta functions which are, by definition, continuous functions. Therefore, the PDF of the discrete random variable is a continuous function.

A.2.11. The distribution function of a discrete random variable X is a discontinuous step function, with jumps at the discrete values of x which are equal to the corresponding probabilities. However, between the jumps, the function $F_X(x)$ remains constant, and the distribution function is "continuous from the right."

For continuous random variables, the distribution function is continuous for any x and in addition, it has a derivative everywhere except possibly at certain points.

If a random variable is mixed, then its distribution function is continuous in certain intervals and has discontinuities (jumps) at particular points [WEN86, p. 112].

A.2.12. By definition, the distribution function exists for any random variable, whether discrete, continuous, or mixed, for all values of x.

A.2.13. For discrete variable, the probability that the variable takes a particular x_i is the finite number $P(x_i)$. However, in the PDF, the probability is defined as an area below the PDF in a given interval. Then, the interval is approaching zero for the specific point x_i. In order for the probability (the area) to remain the same, the height must approach infinity.

A.2.14. Zero. The probability that the continuous random variable takes a particular value in its range is zero because the continuous range is enumerable. However, there is a nonzero probability that the random variable takes a value in the infinitesimal interval $\mathrm{d}x$ around the particular value x, $P\{x < X \leq x + \mathrm{d}x\} = f_X(x)\mathrm{d}x$, where $f_X(x)$ is the value of the PDF of the r.v. X for x. However, as $f_X(x)$ is a finite value for a continual r.v. and $\mathrm{d}x = 0$ in the discrete point, $P\{X = x\}$ must be also zero.

A.2.15. Variance does not exist for all random variables as, for example, in the case of a Cauchy random variable. For a more detailed discussion, see [KAY06, p. 357].

A.2.16. This is true for a PDF which has only one peak (an unimodal PDF). However, if a PDF has two peaks (a bimodal PDF), the standard deviation measures the range of the most expected values and not the widths of any of the peaks [HEL91, pp. 113–114].

Chapter 3
Multidimensional Random Variables

3.1 What Is a Multidimensional Random Variable?

We have already explained that one random variable is obtained by mapping a sample space S to the real x-axis. However, there are many problems in which the outcomes of the random experiment need to be mapped from the space S to a two-dimensional space (x_1, x_2), leading to a two-dimensional random variable (X_1, X_2). Since the variables X_1 and X_2 are the result of the same experiment, then it is necessary to make a joint characterization of these two random variables.

In general, the mapping of the outcomes from S to a N-dimensional space leads to a N-dimensional random variable.

Fortunately, many problems in engineering can be solved by considering only two random variables [PEE93, p. 101]. This is why we emphasize the two random variables and the generalization of a two-variable case will lead to a N-dimensional case.

3.1.1 *Two-Dimensional Random Variable*

Consider the sample space S, as shown in Fig. 3.1, and a two-dimensional space where x_1 and x_2 are real axes, $-\infty < x_1 < \infty$, $-\infty < x_2 < \infty$. Mapping of the outcome s_i to a point (x_1, x_2) in a two-dimensional space, leads to a two-dimensional random variable (X_1, X_2).

If the sample space is discrete, the resulting two-dimensional random variable will also be discrete. However, as in the case of a one variable, the continuous sample space may result in continuous, discrete, or mixed two-dimensional random variables. The range of a discrete two-dimensional random variable is made of points in a two-dimensional space, while the range of a continuous two-dimensional variable is a continuous area in a two-dimensional space. Similarly, a mixed two-dimensional random variable has, aside from continuous area, additional

G.J. Dolecek, *Random Signals and Processes Primer with MATLAB*, DOI 10.1007/978-1-4614-2386-7_3,

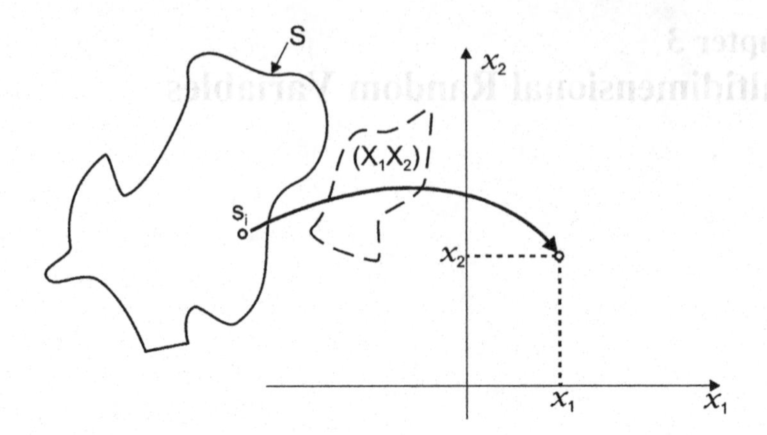

Fig. 3.1 Obtaining a two-dimensional random variable

discrete points. Figure 3.2 shows examples of discrete, continuous, and mixed ranges.

The following example illustrates this concept.

Example 3.1.1 Two messages have been sent through a communication system. Each of them, independently of another, can be transmitted with or without error. Define the two-dimensional variable for this case.

Solution The sample space has four outcomes, shown in Fig. 3.3.

We used the following denotation:

$$
\begin{aligned}
s_1 &= \{\text{The first and second messages are transmitted correctly}\},\\
s_2 &= \{\text{The first message is transmitted correctly, while the second one incurs an error}\},\\
s_3 &= \{\text{The second message is transmitted correctly, while the first message incurs an error}\},\\
s_4 &= \{\text{Both messages incur an error in transmission}\}.
\end{aligned}
\tag{3.1}
$$

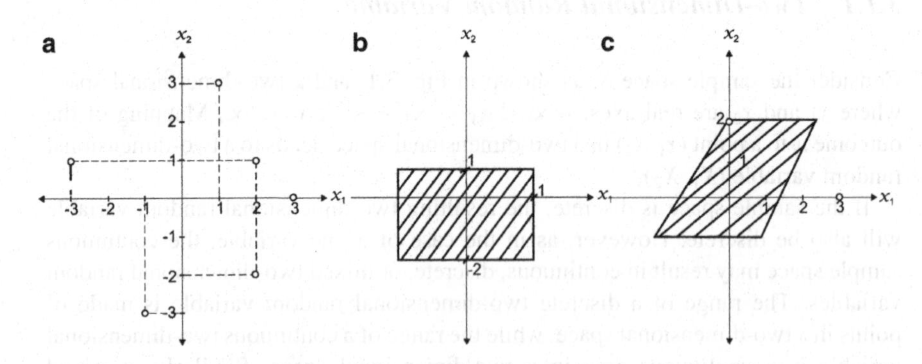

Fig. 3.2 Examples of discrete, continuous and mixed ranges. **(a)** Discrete. **(b)** Continuous. **(c)** Mixed

We can see here that, in order to mathematically model this experiment, it is more convenient to use the numerical assignations to the outcomes s_i, instead of words in (3.1). Similarly, as in the case of one variable, it is convenient to assign the numbers 0 and 1, for binary outcomes (the message is either correct or incorrect).

Let us assign the number 1 if the message is correct. Otherwise, 0 is assigned. The first message is mapped to the x_1-axis, while the second message is mapped to the x_2-axis, as shown in Fig. 3.3. The range of two-dimensional variable (X_1, X_2) is the following points in the (x_1, x_2) space:

$$(x_1 = 1, x_2 = 1), (x_1 = 1, x_2 = 0), (x_1 = 0, x_2 = 1), (x_1 = 0, x_2 = 0). \quad (3.2)$$

Example 3.1.2 Consider the experiment of tossing two coins. There are four outcomes in the sample space S:

$$\begin{aligned} s_1 &= \{\text{First coin comes up heads; Second coin comes up heads}\}, \\ s_2 &= \{\text{First coin comes up heads; Second coin comes up tails}\}, \\ s_3 &= \{\text{First coin comes up tails; Second coin comes up heads}\}, \\ s_4 &= \{\text{First coin comes up tails; Second coin comes up tails}\}. \end{aligned} \quad (3.3)$$

Adopting the denotation from Example 1.1.1, we rewrite the outcomes (3.3) in the following form:

$$\begin{aligned} s_1 &= \{\text{H}, \text{H}\}, \\ s_2 &= \{\text{H}, \text{T}\}, \\ s_3 &= \{\text{T}, \text{H}\}, \\ s_4 &= \{\text{T}, \text{T}\}. \end{aligned} \quad (3.4)$$

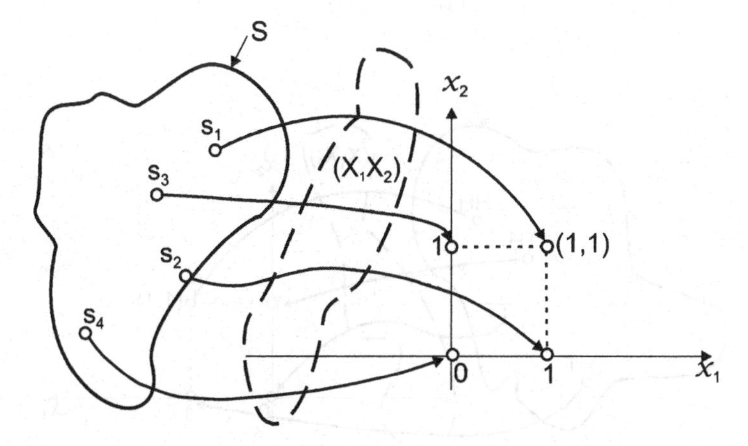

Fig. 3.3 Sample space and two-dimensional r.v. in Example 3.1.1

Let X_1 and X_2 indicate the occurrence of heads and tails for the first and second coins, respectively.

The values of the random variable X_1 for the first coin tossing are $x_1 = 1$ if heads occurs, and $x_1 = 0$, if tails occurs.

Similarly, for the variable X_2, the value $x_2 = 1$ indicates the occurrence of heads, and $x_2 = 0$ indicates tails in the second coin toss.

The mapping from the sample space S to the (x_1, x_2) space is shown in Fig. 3.4. Note that the values of the random variables are the same as in Example 3.1.1. Therefore, one can obtain the same two-dimensional random variable, from different random experiments.

The next example illustrates that, like in the one variable example, one can obtain different two-dimensional random variables from the same experiment.

Example 3.1.3 Consider the same outcomes as in Example 3.1.1 for the following mapping:

$$X_1 \text{ indicates at least one head occurs.}$$
$$X_2 \text{ indicates at least one tail occurs.} \tag{3.5}$$

Therefore, the values of the random variable X_1 are $x_1 = 1$, if at least one heads occurs and $x_1 = 0$ if no heads occurs. Similarly, the values of X_2 are $x_2 = 1$ if at least one tails occurs and $x_2 = 0$ if no tails occurs. This mapping is shown in Fig. 3.5.

In a similar way, one can define a N-dimensional random variable by mapping the outcomes of the space S to N-dimensional space, thus obtaining the N-dimensional random variable,

$$(X_1, X_2, \ldots, X_N) \tag{3.6}$$

with the range:

$$(x_1, x_2, \ldots, x_N). \tag{3.7}$$

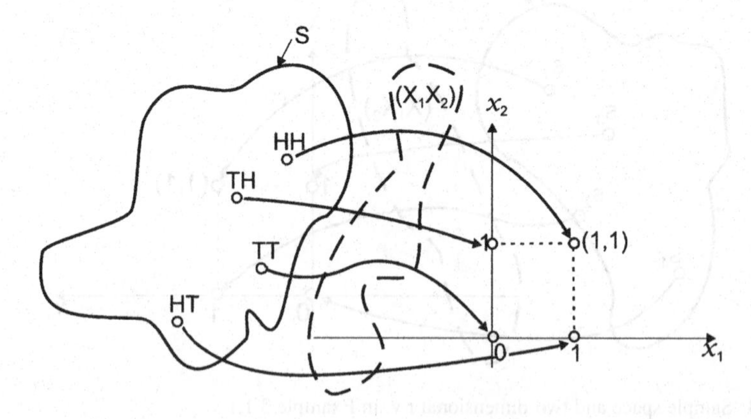

Fig. 3.4 Mapping the sample space S to (x_1, x_2) space in Example 3.1.2

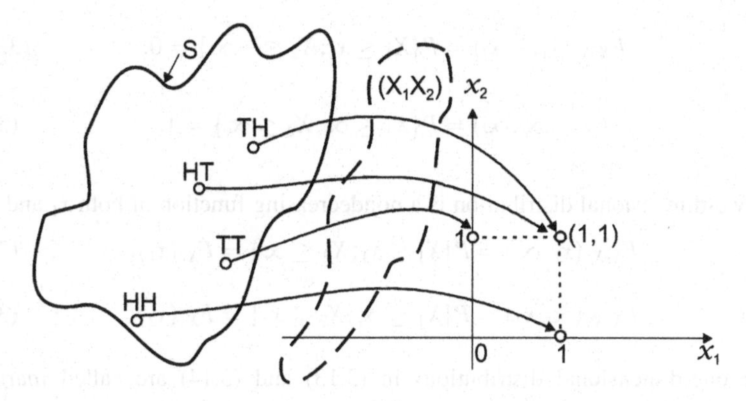

Fig. 3.5 Mapping from the space S to the space (x_1, x_2) in Example 3.1.3

3.2 Joint Distribution and Density

3.2.1 Joint Distribution

The *cumulative distribution function, joint distribution function* or, shortly, *joint distribution* of a pair of random variables X_1 and X_2 is defined as the probability of joint events:

$$\{X_1 \leq x_1; X_2 \leq x_2\}, \tag{3.8}$$

and is denoted as $F_{X_1X_2}(x_1, x_2)$.

Therefore, we have:

$$F_{X_1X_2}(x_1,x_2) = P\{X_1 \leq x_1; X_2 \leq x_2\}, \quad -\infty < x_1 < \infty, \\ -\infty < x_2 < \infty, \tag{3.9}$$

where x_1 and x_2 are values within the two-dimensional space as shown in (3.9). The joint distribution is also called *two-dimensional distribution*, or *second distribution*. In this context, distribution of one random variable is also called *one-dimensional distribution*, or *first distribution*.

From the properties of a one-dimensional distribution, we easily find the following properties for a two-dimensional distribution:

P.1
$$0 \leq F_{X_1X_2}(x_1,x_2) \leq 1. \tag{3.10}$$

P.2
$$F_{X_1X_2}(-\infty,-\infty) = P\{X_1 \leq -\infty; X_2 \leq -\infty\} = 0. \tag{3.11a}$$

$$F_{X_1X_2}(-\infty,x_2) = P\{X_1 \leq -\infty; X_2 \leq x_2\} = 0. \tag{3.11b}$$

$$F_{X_1X_2}(x_1, -\infty) = P\{X_1 \leq x_1; X_2 \leq -\infty\} = 0. \tag{3.11c}$$

P.3

$$F_{X_1X_2}(\infty, \infty) = P\{X_1 \leq \infty; X_2 \leq \infty\} = 1. \tag{3.12}$$

P.4 Two-dimensional distribution is a nondecreasing function of both x_1 and x_2.

P.5

$$F_{X_1X_2}(x_1, \infty) = P\{X_1 \leq x_1; X_2 \leq \infty\} = F_{X_1}(x_1). \tag{3.13}$$

$$F_{X_1X_2}(\infty, x_2) = P\{X_1 \leq \infty; X_2 \leq x_2\} = F_{X_2}(x_2). \tag{3.14}$$

The one-dimensional distributions in (3.13) and (3.14) are called *marginal distributions.*

Next, we express the probability that the pair of variables X_1 and X_2 is in a given space, in term of its two-dimensional distribution function:

$$\begin{aligned}
&P\{x_{11} < X_1 \leq x_{12}, x_{21} < X_2 \leq x_{22}\} = P\{x_{11} < X_1 \leq x_{12}, X_2 \leq x_{22}\} \\
&\quad -P\{x_{11} < X_1 \leq x_{12}, X_2 < x_{21}\} = P\{X_1 \leq x_{12}, X_2 \leq x_{22}\} - P\{X_1 < x_{11}, X_2 \leq x_{22}\} \\
&\quad - [P\{X_1 \leq x_{12}, X_2 < x_{21}\} - P\{X_1 < x_{11}, X_2 < x_{21}\}] \\
&\quad = F_{X_1X_2}(x_{12}, x_{22}) - F_{X_1X_2}(x_{11}, x_{22}) - F_{X_1X_2}(x_{12}, x_{21}) + F_{X_1X_2}(x_{11}, x_{21}).
\end{aligned} \tag{3.15}$$

If a two-dimensional variable is discrete, then its distribution has two-dimensional unit step functions $u(x_{1i}, x_{2j}) = u(x_{1i})u(x_{2j})$ in the corresponding discrete points (x_{1i}, x_{2j}), as shown in the following equation (see [PEE93, pp. 358–359]):

$$F_{X_1X_2}(x_1, x_2) = \sum_i \sum_j P\{X_1 = x_{1i}, X_2 = x_{2j}\} u(x_1 - x_{1i}) u(x_2 - x_{2j}). \tag{3.16}$$

Example 3.2.1 Find the joint distribution function from Example 3.1.1 considering that the probability that the first and second messages are correct (denoted as p_1 and p_2, respectively), and that they are independent.

$$\begin{aligned}
P\{x_1 = 1, x_2 = 1\} &= P\{x_1 = 1\}P\{x_2 = 1\} = p_1p_2, \\
P\{x_1 = 1, x_2 = 0\} &= P\{x_1 = 1\}P\{x_2 = 0\} = p_1(1 - p_2), \\
P\{x_1 = 0, x_2 = 1\} &= P\{x_1 = 0\}P\{x_2 = 1\} = p_2(1 - p_1), \\
P\{x_1 = 0, x_2 = 0\} &= P\{x_1 = 0\}P\{x_2 = 0\} = (1 - p_1)(1 - p_2).
\end{aligned} \tag{3.17}$$

The joint distribution is:

$$\begin{aligned}
F_{X_1X_2}(x_1, x_2) = &+p_1p_2u(x_1 - 1)u(x_2 - 1) + p_1(1 - p_2)u(x_1 - 1)u(x_2) \\
&+ (1 - p_1)p_2u(x_1)u(x_2 - 1) + (1 - p_1)(1 - p_2)u(x_1)u(x_2).
\end{aligned} \tag{3.18}$$

Example 3.2.2 Find the joint distribution in Example 3.1.2.

Solution All outcomes have the same probability of occurrence, resulting in:

$$\begin{aligned} P\{X_1 = 1, X_2 = 1\} &= P\{X_1 = 0, X_2 = 0\} = P\{X_1 = 1, X_2 = 0\} \\ &= P\{X_1 = 0, X_2 = 1\} = 1/4. \end{aligned} \tag{3.19}$$

From (3.19), we have:

$$\begin{aligned} F_{X_1X_2}(x_1, x_2) = &+ 1/4[u(x_1 - 1)u(x_2 - 1) + u(x_1 - 1)u(x_2) + u(x_1)u(x_2 - 1) \\ &+ u(x_1)u(x_2)]. \end{aligned} \tag{3.20}$$

Example 3.2.3 Find the joint distribution in Example 3.1.3.

Solution All outcomes have the same probability of occurrence. However, the probabilities of the values of the random variables are:

$$\begin{aligned} &P\{x_1 = 1, x_2 = 1\} = P\{\text{H, T}\} + P\{\text{T, H}\} = 1/4 + 1/4 = 1/2, \\ &P\{x_1 = 1, x_2 = 0\} = P\{\text{H, H}\} = 1/4, \\ &P\{x_1 = 0, x_2 = 1\} = P\{\text{T, T}\} = 1/4. \end{aligned} \tag{3.21}$$

The joint distribution is:

$$F_{X_1X_2}(x_1, x_2) = 1/2u(x_1 - 1)u(x_2 - 1) + 1/4u(x_1 - 1)u(x_2) + 1/4u(x_1)u(x_2 - 1). \tag{3.22}$$

Example 3.2.4 Find the probability that a random point (x_1, x_2) will fall in the area A, as shown in Fig. 3.6.

Solution Area A is defined as:

$$\{x_1 \leq a, x_2 \leq c\} - \{x_1 \leq a, x_2 \leq b\}. \tag{3.23}$$

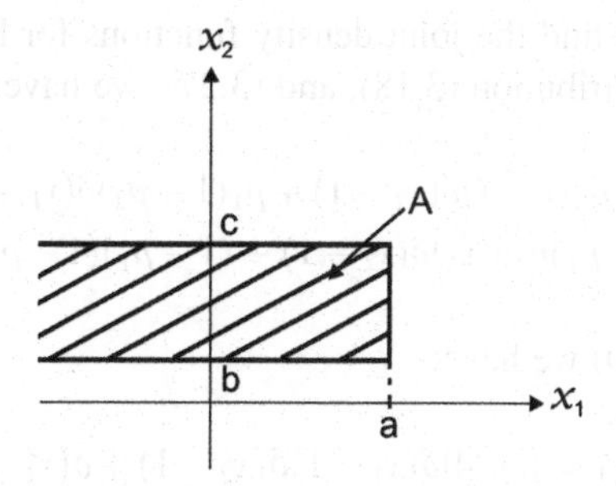

Fig. 3.6 Area A in Example 3.2.4

From here, we have:

$$P\{(X_1X_2) \in A\} = P\{x_1 \leq a, x_2 \leq c\} - P\{x_1 \leq a, x_2 \leq b\}. \tag{3.24}$$

Using (3.9), (3.24) can be rewritten as:

$$P\{(X_1,X_2) \in A\} = F_{X_1X_2}(a,c) - F_{X_1X_2}(a,b). \tag{3.25}$$

In general, for N random variables

$$(X_1,X_2,\ldots,X_N),$$

the joint distribution, denoted as $F_{X_1,\ldots,X_N}(x_1,\ldots,x_N)$, is then defined as:

$$F_{X_1,\ldots,\,X_N}(x_1,\ldots,x_N) = P\{X_1 \leq x_1,\ldots,X_1 \leq x_N\}. \tag{3.26}$$

3.2.2 *Joint Density Function*

The *joint probability density function (PDF)* or *joint density function* or *two-dimensional density function* or shortly, *joint PDF*, for pair of random variables X_1 and X_2, is denoted as, $f_{X_1X_2}(x_1,x_2)$, and defined as:

$$f_{X_1X_2}(x_1,x_2) = \frac{\partial^2 F_{X_1X_2}(x_1,x_2)}{\partial x_1 \partial x_2}. \tag{3.27}$$

For discrete variables the derivations are not defined in the step discontinuities, implying the introduction of delta functions at pairs of discrete points (x_{1i}, x_{2j}). Therefore, the joint PDF is equal to (see [PEE93, pp. 358–359], [HEL91, p. 147]):

$$f_{X_1X_2}(x_1,x_2) = \sum_i \sum_j P\{X_1 = x_{1i}, X_2 = x_{2j}\}\delta(x_1 - x_{1i})\delta(x_2 - x_{2j}). \tag{3.28}$$

Example 3.2.5 We can find the joint density functions for Examples 3.2.1, 3.2.2, and 3.2.3. Using the distribution (3.18), and (3.27), we have:

$$\begin{aligned} f_{X_1X_2}(x_1,x_2) = &+ p_1p_2\delta(x_1-1)\delta(x_2-1) + p_1(1-p_2)\delta(x_1-1)\delta(x_2) \\ &+ (1-p_1)p_2\delta(x_1)\delta(x_2-1) + (1-p_1)(1-p_2)\delta(x_1)\delta(x_2). \end{aligned} \tag{3.29}$$

Similarly, from (3.20) we have:

$$\begin{aligned} f_{X_1X_2}(x_1,x_2) = &+1/4[\delta(x_1-1)\delta(x_2-1) + \delta(x_1-1)\delta(x_2) \\ &+ \delta(x_1)\delta(x_2-1) + \delta(x_1)\delta(x_2)]. \end{aligned} \tag{3.30}$$

From (3.22), we get:

$$f_{X_1X_2}(x_1, x_2) = 1/2\delta(x_1 - 1)\delta(x_2 - 1) + 1/4\delta(x_1 - 1)\delta(x_2) + 1/4\delta(x_1)\delta(x_2 - 1). \quad (3.31)$$

Next, in Fig. 3.7 we compare one-dimensional and two-dimensional PDFs.

The shaded area in Fig. 3.7a represents the probability that the random variable X is in the infinitesimal interval $[x, x + \mathrm{d}x]$:

$$A = P\{x < X \leq x + \mathrm{d}x\}. \quad (3.32)$$

From here, considering that $\mathrm{d}x$ is an infinitesimal interval, the PDF in the interval $[x, x + \mathrm{d}x]$ is constant, resulting in:

$$f_X(x)\mathrm{d}x = A = P\{x < X \leq x + \mathrm{d}x\}. \quad (3.33)$$

Similarly, the volume in Fig. 3.7b presents the probability that the random variables X_1 and X_2 are in the intervals $[x_1, x_1 + \mathrm{d}x_1]$ and $[x_2, x_2 + \mathrm{d}x_2]$, respectively.

The equivalent probability

$$P\{x_1 < X_1 \leq x_1 + \mathrm{d}x_1, x_2 < X_2 \leq x_2 + \mathrm{d}x_2\} \quad (3.34)$$

corresponds to the elemental volume V, with a base of $(\mathrm{d}x_1\ \mathrm{d}x_2)$ and height of $f_{X_1X_2}(x_1, x_2)$:

$$f_{X_1X_2}(x_1, x_2)\mathrm{d}x_1\mathrm{d}x_2 = V = P\{x_1 < X_1 \leq x_1 + \mathrm{d}x_1, x_2 < X_2 \leq x_2 + \mathrm{d}x_2\}. \quad (3.35)$$

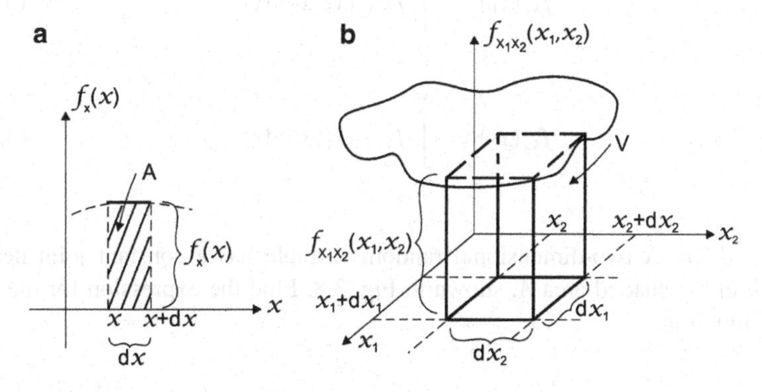

Fig. 3.7 One-dimensional and two-dimensional PDFs

Therefore, as opposed to one-dimensional random variable, where the probability represents the area below the density function, for two-dimensional variable, the probability is a volume below a joint density function.

The properties of one-dimensional PDF can be easily applied to the two-dimensional PDF:

P.1

$$f_{X_1X_2}(x_1, x_2) \geq 0. \tag{3.36}$$

P.2

$$\int_{-\infty}^{\infty}\int_{-\infty}^{\infty} f_{X_1X_2}(x_1, x_2)\mathrm{d}x_1\,\mathrm{d}x_2 = 1. \tag{3.37}$$

P.3

$$F_{X_1X_2}(x_1, x_2) = \int_{-\infty}^{x_1}\int_{-\infty}^{x_2} f_{X_1X_2}(x_1, x_2)\mathrm{d}x_1\,\mathrm{d}x_2. \tag{3.38}$$

P.4

$$F_{X_1}(x_1) = \int_{-\infty}^{x_1}\int_{-\infty}^{\infty} f_{X_1X_2}(x_1, x_2)\mathrm{d}x_1\,\mathrm{d}x_2. \tag{3.39}$$

$$F_{X_2}(x_2) = \int_{-\infty}^{x_2}\int_{-\infty}^{\infty} f_{X_1X_2}(x_1, x_2)\mathrm{d}x_1\,\mathrm{d}x_2. \tag{3.40}$$

P.5

$$P\{x_{11} < X_1 \leq x_{12}, x_{21} < X_2 \leq x_{22}\} = \int_{x_{21}}^{x_{22}}\int_{x_{11}}^{x_{12}} f_{X_1X_2}(x_1, x_2)\mathrm{d}x_1\,\mathrm{d}x_2. \tag{3.41}$$

P.6

$$f_{X_1}(x_1) = \int_{-\infty}^{\infty} f_{X_1X_2}(x_1, x_2)\mathrm{d}x_2. \tag{3.42a}$$

$$f_{X_2}(x_2) = \int_{-\infty}^{\infty} f_{X_1X_2}(x_1, x_2)\mathrm{d}x_1. \tag{3.42b}$$

Example 3.2.6 A two-dimensional random variable has a constant joint density function in the shaded area A, shown in Fig. 3.8. Find the expression for the joint density function.

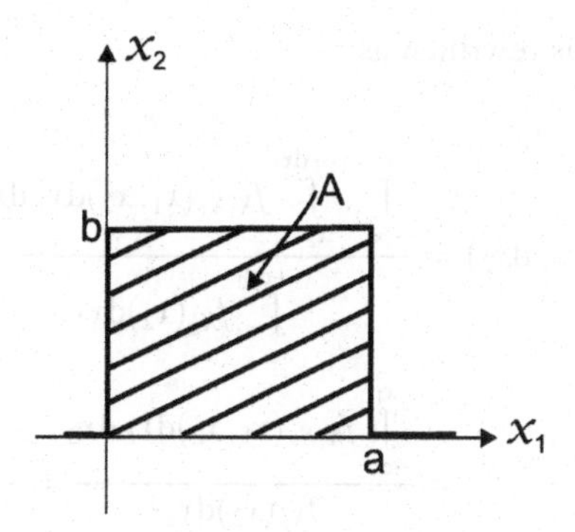

Fig. 3.8 The range of (X_1, X_2) in Example 3.2.6

Solution From (3.37), we have:

$$f_{X_1X_2}(x_1, x_2) = \begin{cases} 1/ab & \text{for} \quad (x_1, x_2) \in A \\ 0 & \quad \text{otherwise} \end{cases}. \tag{3.43}$$

For N random variables, N-dimensional density function is defined as,

$$f_{X_1X_2,\ldots,X_N}(x_1, x_2, \ldots, x_N) = \frac{\partial^N F_{X_1X_2,\ldots,X_N}(x_1, x_2, \ldots, x_N)}{\partial x_1 \partial x_2, \ldots, \partial x_N}. \tag{3.44}$$

3.2.3 Conditional Distribution and Density

In some practical problems, it is necessary to find the distribution or PDF of one variable given the specific value of another variable. The corresponding conditional distribution denoted as, $F_{X_1}(x_1|X_2 = x_2)$, can be expressed in terms of the joint density function.

We consider a general case in which X_1 and X_2 are both continuous variables and thus the conditional event B is defined as:

$$B = \{x_2 < X_2 \leq x_2 + dx_2\}. \tag{3.45}$$

From (1.73) and (3.45), we have:

$$F_{X_1}(x_1|B) = \frac{P\{X_1 \leq x_1, B\}}{P\{B\}} = \frac{P\{X_1 \leq x_1, x_2 < X_2 \leq x_2 + dx_2\}}{P\{x_2 < X_2 \leq x_2 + dx_2\}}. \tag{3.46}$$

Using (3.45), (3.46) is rewritten as:

$$F_{X_1}(x_1|x_2 < X_2 \leq x_2 + \mathrm{d}x_2) = \frac{\int\limits_{-\infty}^{x_1}\int\limits_{x_2}^{x_2+\mathrm{d}x_2} f_{X_1X_2}(x_1,x_2)\mathrm{d}x_1\,\mathrm{d}x_2}{\int\limits_{x_2}^{x_2+\mathrm{d}x_2} f_{X_2}(x_2)\mathrm{d}x_2}$$

$$= \frac{\int\limits_{-\infty}^{x_1} f_{X_1X_2}(x_1,x_2)\mathrm{d}x_1\,\mathrm{d}x_2}{f_{X_2}(x_2)\mathrm{d}x_2} = \frac{\int\limits_{-\infty}^{x_1} f_{X_1X_2}(x_1,x_2)\mathrm{d}x_1}{f_{X_2}(x_2)}. \tag{3.47}$$

If $\mathrm{d}x_2$ is approaching zero, and for each x_2 for which $f_{X_2}(x_2) \neq 0$, we finally obtain:

$$F_{X_1}(x_1|X_2 = x_2) = \frac{\int\limits_{-\infty}^{x_1} f_{X_1X_2}(x_1,x_2)\mathrm{d}x_1}{f_{X_2}(x_2)}. \tag{3.48}$$

Similarly, we have:

$$F_{X_2}(x_2|X_1 = x_1) = \frac{\int\limits_{-\infty}^{x_2} f_{X_1X_2}(x_1,x_2)\mathrm{d}x_2}{f_{X_1}(x_1)}. \tag{3.49}$$

From (3.48) and (3.49), using (3.27), we obtain the corresponding PDFs:

$$f_{X_1}(x_1|X_2 = x_2) = \frac{f_{X_1X_2}(x_1,x_2)}{f_{X_2}(x_2)}. \tag{3.50}$$

$$f_{X_2}(x_2|X_1 = x_1) = \frac{f_{X_1X_2}(x_1,x_2)}{f_{X_1}(x_1)}. \tag{3.51}$$

Consider now that the condition for event B is defined as an event in which the other variable X_2 lies in the given interval $[x_{21}, x_{22}]$, resulting in:

$$F_{X_1}(x_1|x_{21} < X_2 \leq x_{22}) = \frac{\int\limits_{-\infty}^{x_1}\int\limits_{x_{21}}^{x_{22}} f_{X_1X_2}(x_1,x_2)\mathrm{d}x_1\,\mathrm{d}x_2}{\int\limits_{x_{21}}^{x_{22}} f_{X_2}(x_2)\mathrm{d}x_2}$$

$$= \frac{F_{X_1X_2}(x_1,x_{22}) - F_{X_1X_2}(x_1,x_{21})}{F_{X_2}(x_{22}) - F_{X_2}(x_{21})}, \tag{3.52}$$

where the denominator in (3.52) is:

$$F_{X_2}(x_{22}) - F_{X_2}(x_{21}) = P\{x_{21} < X_2 \leq x_{22}\} \neq 0. \tag{3.53}$$

From (3.27) and (3.52), the corresponding conditional density is:

$$f_{X_1}(x_1|x_{21} < X_2 \leq x_{22}) = \frac{\int\limits_{x_{21}}^{x_{22}} f_{X_1X_2}(x_1, x_2)\mathrm{d}x_2}{\int\limits_{x_{21}}^{x_{22}} \int\limits_{-\infty}^{\infty} f_{X_1X_2}(x_1x_2)\mathrm{d}x_1\,\mathrm{d}x_2}. \tag{3.54}$$

Example 3.2.7 Consider a random variable X_1 with the density function:

$$f_{X_1}(x_1) = \begin{cases} \lambda\,\mathrm{e}^{-\lambda x_1} & \text{for} \quad x_1 > 0 \\ 0 & \text{otherwise} \end{cases} \tag{3.55}$$

and the conditional density

$$f_{X_2}(x_2|x_1) = \begin{cases} x_1\,\mathrm{e}^{-x_1 x_2} & \text{for} \quad x_1 > 0, x_2 > 0 \\ 0 & \text{otherwise} \end{cases}. \tag{3.56}$$

Find the conditional density $f_{X_1}(x_1|x_2)$.

Solution From (3.51), (3.55), and (3.56), we get:

$$f_{X_1X_2}(x_1, x_2) = f_{X_2}(x_2|x_1)f_{X_1}(x_1). \tag{3.57}$$

Placing (3.55) and (3.56) into (3.57), we arrive at:

$$f_{X_1X_2}(x_1, x_2) = \begin{cases} \lambda x_1\,\mathrm{e}^{-x_1(\lambda+x_2)} & \text{for} \quad x_1 > 0, x_2 > 0 \\ 0 & \text{otherwise} \end{cases}. \tag{3.58}$$

From here, using (3.42b), we find:

$$f_{X_2}(x_2) = \int\limits_0^{\infty} f_{X_1X_2}(x_1, x_2)\,\mathrm{d}x_1 = \int\limits_0^{\infty} \lambda x_1\,\mathrm{e}^{-x_1(\lambda+x_2)}\,\mathrm{d}x_1. \tag{3.59}$$

Using integral 1 from Appendix A, we obtain:

$$f_{X_2}(x_2) = \begin{cases} \frac{\lambda}{(\lambda+x_2)^2} & \text{for} \quad x_2 > 0 \\ 0 & \text{otherwise} \end{cases}. \tag{3.60}$$

The desired conditional density is obtained using (3.50):

$$f_{X_1}(x_1|x_2) = \frac{f_{X_1X_2}(x_1,x_2)}{f_{X_2}(x_2)}. \tag{3.61}$$

Finally, placing (3.58) and (3.60) into (3.61), we have:

$$f_{X_1}(x_1|x_2) = \begin{cases} x_1(\lambda + x_2)^2 \,\mathrm{e}^{-x_1(\lambda+x_2)} & \text{for} \quad x_1 > 0 \\ 0 & \text{otherwise} \end{cases}. \tag{3.62}$$

3.2.4 *Independent Random Variables*

The two random variables, X_1 and X_2, are *independent* if the events

$$\{X_1 \le x_1\} \quad \text{and} \quad \{X_2 \le x_1\}$$

are independent for any value of x_1 and x_2.

Using (1.105), we can write:

$$P\{X_1 \le x_1, X_2 \le x_1\} = P\{X_1 \le x_1\}P\{X_2 \le x_1\}. \tag{3.63}$$

From (3.9) and (2.10), the joint distribution is:

$$F_{X_1X_2}(x_1,x_2) = F_{X_1}(x_1)F_{X_1}(x_1). \tag{3.64}$$

Similarly, for the joint density we have:

$$f_{X_1X_2}(x_1,x_2) = f_{X_1}(x_1)f_{X_1}(x_1). \tag{3.65}$$

Therefore, if the random variables X_1 and X_2 are independent, then their joint distributions and joint PDFs are equal to the products of the marginal distributions and densities, respectively.

Example 3.2.8 Determine whether or not the random variables X_1 and X_2 are independent, if the joint density is given as:

$$f_{X_1X_2}(x_1,x_2) = \begin{cases} 1/2 & \text{for} \quad 0 < x_1 < 2; 0 < x_2 < 1 \\ 0 & \text{otherwise} \end{cases}. \tag{3.66}$$

Solution The joint density (3.66) can be rewritten as:

$$f_{X_1X_2}(x_1,x_2) = f_{X_1}(x_1)f_{X_2}(x_2), \tag{3.67}$$

where

$$f_{X_1}(x_1) = \begin{cases} 1/2 & \text{for} \quad 0 < x_2 < 2 \\ 0 & \text{otherwise} \end{cases}. \tag{3.68}$$

$$f_{X_2}(x_2) = \begin{cases} 1 & \text{for} \quad 0 < x_2 < 1 \\ 0 & \text{otherwise} \end{cases}. \tag{3.69}$$

From (3.65) and (3.67)–(3.69), we can conclude that the variables are independent.

The result (3.65) can be generalized to N jointly independent random variables:

$$F_{X_1,\ldots,X_N}(x_1,\ldots,x_N) = \prod_{i=1}^{N} F_{X_i}(x_i). \tag{3.70}$$

$$f_{X_1,\ldots,X_N}(x_1,\ldots,x_N) = \prod_{i=1}^{N} f_{X_i}(x_i). \tag{3.71}$$

3.3 Expected Values and Moments

3.3.1 Expected Value

In order to find the mean value of two joint random variables X_1 and X_2, we will apply the similar procedure to that which we used in the case of one random variable (see Sect. 2.7), starting with a random experiment.

Consider two discrete random variables X_1 and X_2 with the possible values x_{1i} and x_{2j}, respectively.

The experiment is performed N times under the same conditions,

$$N = \sum_{i=1}^{N_1} \sum_{j=1}^{N_2} N_{ij}, \tag{3.72}$$

and as a result the following values are obtained:

$$\begin{array}{llll} X_1 = x_{11}, & \text{and} \quad X_2 = x_{21}, & N_{11} \text{ times.} \\ \ldots & \ldots & \ldots \\ X_1 = x_{1i}, & \text{and} \quad X_2 = x_{2j}, & N_{ij} \text{ times.} \\ \ldots & \ldots & \ldots \\ X_1 = x_{1N_1}, & \text{and} \quad X_2 = x_{2N_2}, & N_{N_1N_2} \text{ times.} \end{array} \tag{3.73}$$

As indicated in Sect. 2.7.2, we now have a finite set of values x_{1i} and x_{2j}, and we can calculate the arithmetic mean value of the products, also called the *empirical mean value*, since it is obtained from the experiment,

$$\overline{X_1X_2}_{\text{emp}} = \frac{\sum_{i=1}^{N_1}\sum_{j=1}^{N_2} N_{ij}x_{1i}x_{2j}}{N}. \tag{3.74}$$

For a high enough value N, the ratio N_{ij}/N becomes a good approximation of the probability,

$$\frac{N_{ij}}{N} \to P\{X_1 = x_{1i}, X_2 = x_{2j}\}, \tag{3.75}$$

and the empirical mean value becomes independent on experiment and approaches the *mean value of joint random variables* X_1 *and* X_2,

$$\overline{X_1X_2} = \sum_i\sum_j x_{1i}x_{2j}P\{X_1 = x_{1i}, X_2 = x_{2j}\}, \tag{3.76}$$

or, in a general case,

$$\overline{X_1X_2} = \sum_{i=-\infty}^{\infty}\sum_{j=-\infty}^{\infty} x_{1i}x_{2j}P\{X_1 = x_{1i}, X_2 = x_{2j}\}. \tag{3.77}$$

This result can be generalized *for* N discrete random variables,

$$X_1, \ldots, X_N, \tag{3.78}$$

as given in the following equation:

$$\overline{X_1 \ldots X_N} = \sum_{i=-\infty}^{\infty} \ldots \sum_{j=-\infty}^{\infty} x_{1i} \ldots x_{Nj}P\{X_1 = x_{1i}, \ldots, X_N = x_{Nj}\}. \tag{3.79}$$

Similarly,

$$\overline{g(X_1 \ldots X_N)} = \sum_{i=-\infty}^{\infty} \ldots \sum_{j=-\infty}^{\infty} g(x_{1i} \ldots x_{Nj})P\{X_1 = x_{1i}, \ldots, X_N = x_{Nj}\}. \tag{3.80}$$

Example 3.3.1 The discrete random variables X_1 and X_2, both take the discrete values -1 and 1, with the following probabilities:

$$\begin{aligned}
&P\{X_1 = x_{11} = -1, X_2 = x_{21} = -1\} = 1/4,\\
&P\{X_1 = x_{12} = 1, X_2 = x_{22} = 1\} = 1/2,\\
&P\{X_1 = x_{11} = -1, X_2 = x_{22} = 1\} = 1/8,\\
&P\{X_1 = x_{12} = 1, X_2 = x_{21} = -1\} = 1/8.
\end{aligned} \tag{3.81}$$

Find the mean values:

(a) $E\{X_1X_2\}$. (3.82)

(b) $E\{X_1^2 + X_2\}$. (3.83)

Solution

(a) From (3.76), we have:

$$\overline{X_1X_2} = (-1)\times(-1)\times 1/4 + 1\times 1\times 1/2 + (-1)\times 1\times 1/8 + 1\times(-1)\times 1/8 = 1/2. \tag{3.84}$$

(b) Using (3.80), we get:

$$\begin{aligned}
\overline{X_1^2 + X_2} &= \sum_{i=1}^{2}\sum_{j=1}^{2}(x_{i1}^2 + x_{2j})P\{X_1 = x_{1i}, X_2 = x_{2j}\}\\
&= \left[(-1)^2 + (-1)\right]\times 1/4 + \left[1^2 + 1\right]\times 1/2\\
&\quad + \left[(-1)^2 + 1)\right]\times 1/8 + \left[1^2 + (-1)\right]\times 1/8 = 5/4.
\end{aligned} \tag{3.85}$$

Using a similar approach to that taken in Sect. 2.7.3, we can express the mean value of the joint continuous random variables X_1 and X_2, using the joint density function $f_{X_1X_2}(x_1, x_2)$:

$$\overline{X_1X_2} = \int_{-\infty}^{\infty}\int_{-\infty}^{\infty} x_1x_2 f_{X_1X_2}(x_1, x_2)\mathrm{d}x_1\,\mathrm{d}x_2. \tag{3.86}$$

Instead, we can consider the function $g(X_1, X_2)$ resulting in:

$$\overline{g(X_1X_2)} = \int_{-\infty}^{\infty}\int_{-\infty}^{\infty} g(x_1x_2) f_{X_1X_2}(x_1, x_2)\mathrm{d}x_1\,\mathrm{d}x_2. \tag{3.87}$$

The mean value for the two random variables can be generalized for N continuous random variables:

$$X_1, \ldots, X_N. \tag{3.88}$$

$$\overline{X_1,\ldots,X_N} = \int_{-\infty}^{\infty} \cdots \int_{-\infty}^{\infty} x_1,\ldots,x_N f_{X_1,\ldots,X_N}(x_1,\ldots,x_N)\mathrm{d}x_1,\ldots,\mathrm{d}x_N. \tag{3.89}$$

Similarly, the expression (3.87) can be generalized for N random variables (the mean value of a function of N variables),

$$g(X_1,\ldots,X_N) \tag{3.90}$$

$$\overline{g(X_1,\ldots,X_N)} = \int_{-\infty}^{\infty} \cdots \int_{-\infty}^{\infty} g(x_1,\ldots,x_N) f_{X_1,\ldots,X_N}(x_1,\ldots,x_N)\mathrm{d}x_1,\ldots,\mathrm{d}x_N. \tag{3.91}$$

3.3.1.1 Mean Value of the Sum of Random Variables

Consider the sum of two random variables X_1 and X_2. We can write:

$$g(X_1,X_2) = X_1 + X_2, \tag{3.92}$$

and use (3.87) to find the desired mean value.

$$\begin{aligned}
\overline{g(X_1X_2)} = \overline{X_1 + X_2} &= \int_{-\infty}^{\infty}\int_{-\infty}^{\infty} (x_1 + x_2) f_{X_1X_2}(x_1,x_2)\mathrm{d}x_1\,\mathrm{d}x_2 \\
&= \int_{-\infty}^{\infty}\int_{-\infty}^{\infty} x_1 f_{X_1X_2}(x_1,x_2)\mathrm{d}x_1\,\mathrm{d}x_2 + \int_{-\infty}^{\infty}\int_{-\infty}^{\infty} x_2 f_{X_1X_2}(x_1,x_2)\mathrm{d}x_1\,\mathrm{d}x_2 \\
&= \int_{-\infty}^{\infty} x_1\mathrm{d}x_1 \int_{-\infty}^{\infty} f_{X_1X_2}(x_1,x_2)\mathrm{d}x_2 + \int_{-\infty}^{\infty} x_2\mathrm{d}x_2 \int_{-\infty}^{\infty} f_{X_1X_2}(x_1,x_2)\mathrm{d}x_1.
\end{aligned} \tag{3.93}$$

Using (3.42a) and (3.42b), we have:

$$\int_{-\infty}^{\infty} f_{X_1X_2}(x_1,x_2)\mathrm{d}x_2 = f_{X_1}(x_1), \quad \int_{-\infty}^{\infty} f_{X_1X_2}(x_1,x_2)\mathrm{d}x_1 = f_{X_2}(x_2). \tag{3.94}$$

Placing (3.94) into (3.93) and using (2.261) we obtain:

$$\overline{X_1 + X_2} = \int_{-\infty}^{\infty} x_1 f_{X_1}(x_1)\mathrm{d}x_1 + \int_{-\infty}^{\infty} x_2 f_{X_2}(x_2)\mathrm{d}x_2 = \overline{X_1} + \overline{X_2}. \tag{3.95}$$

Equation (3.95) shows the following statement:

The mean value of the sum of two random variables is equal to the sum of the corresponding mean values.

Note that no conditions have been imposed to obtain the result (3.95). A more general result, which includes the N random variables, can be expressed as:

$$\overline{\sum_{k=1}^{N} X_k} = \sum_{k=1}^{N} \overline{X_k}. \tag{3.96}$$

Example 3.3.2 Verify the relation (3.95) for the discrete random variables from Example 3.3.1.

Solution From (3.80), we have:

$$\overline{X_1 + X_2} = \sum_{i=1}^{2}\sum_{j=1}^{2} (x_{1i} + x_{2j}) P\{X_1 = x_{1i}, X_2 = x_{2j}\} = 3/4 - 1/4 = 1/2. \tag{3.97}$$

To verify the result (3.97), we found the following probabilities for random variables X_1 and X_2, using (1.67):

$$\begin{aligned} P\{X_1 = x_{11} = -1\} = P\{X_1 = x_{11} = -1, X_2 = x_{21} = -1\} \\ + P\{X_1 = x_{11} = -1, X_2 = x_{22} = 1\} = 3/8. \end{aligned} \tag{3.98}$$

$$\begin{aligned} P\{X_1 = x_{12} = 1\} = P\{X_1 = x_{12} = 1, X_2 = x_{21} = -1\} \\ + P\{X_1 = x_{12} = 1, X_2 = x_{22} = 1\} = 5/8. \end{aligned} \tag{3.99}$$

$$\begin{aligned} P\{X_2 = x_{21} = -1\} = P\{X_2 = x_{21} = -1, X_1 = x_{11} = -1\} \\ + P\{X_2 = x_{21} = -1, X_1 = x_{12} = 1\} = 3/8. \end{aligned} \tag{3.100}$$

$$\begin{aligned} P\{X_2 = x_{21} = 1\} = P\{X_2 = x_{21} = 1, X_1 = x_{11} = -1\} \\ + P\{X_2 = x_{21} = 1, X_1 = x_{12} = 1\} = 5/8. \end{aligned} \tag{3.101}$$

From (2.220) and (3.98) and (3.99), we have:

$$\overline{X_1} = \sum_{i=1}^{2} x_{1i} P\{X_1 = x_{1i}\} = -1 \times 3/8 + 1 \times 5/8 = 1/4. \tag{3.102}$$

Similarly, from (3.100) and (3.101), it follows:

$$\overline{X_2} = \sum_{i=1}^{2} x_{2i} P\{X_2 = x_{2i}\} = -1 \times 3/8 + 1 \times 5/8 = 1/4. \tag{3.103}$$

From (3.95), (3.97), (3.102), and (3.103), we arrive at:

$$\overline{X_1 + X_2} = 1/2 = \overline{X_1} + \overline{X_2} = 1/4 + 1/4 = 1/2. \tag{3.104}$$

3.3.2 *Joint Moments Around the Origin*

The expected value of joint random variables X_1 and X_2,

$$E\{X_1^n X_2^k\}. \tag{3.105}$$

is called the *joint moment* m_r of the *order* r, where

$$r = n + k. \tag{3.106}$$

Equation (3.105) presents the expected value of the function $g(X_1, X_2)$ of the random variables X_1 and X_2, and thus can be obtained using (3.87):

$$m_r = E\{X_1^n X_2^k\} = \int_{-\infty}^{\infty}\int_{-\infty}^{\infty} x_1^n x_2^k f_{X_1X_2}(x_1, x_2)\mathrm{d}x_1\,\mathrm{d}x_2. \tag{3.107}$$

This result can be generalized for N random variables $X_1, \ldots, X_N$ in order to obtain the joint moment around the origin of the order

$$r = \sum_{i=1}^{N} n_i, \tag{3.108}$$

$$E\{X_1^{n_1}, \ldots, X_N^{n_N}\} = \int_{-\infty}^{\infty} \cdots \int_{-\infty}^{\infty} x_1^{n_1}, \ldots, x_N^{n_N} f_{X_1,\ldots,X_N}(x_1, \ldots, x_N)\mathrm{d}x_1, \ldots, \mathrm{d}x_N. \tag{3.109}$$

Special importance has been placed on the second moment, where $n = k = 1$, which is called the *correlation* and has its proper denotation $R_{X_1X_2}$,

$$R_{X_1X_2} = E\{X_1X_2\} = \int_{-\infty}^{\infty}\int_{-\infty}^{\infty} x_1x_2 f_{X_1X_2}(x_1, x_2)\mathrm{d}x_1\,\mathrm{d}x_2. \tag{3.110}$$

Some important relations between variables X_1 and X_2 can be expressed using the correlation.

For example, if the correlation can be written as a product of the expected values of X_1 and X_2,

$$E\{X_1, X_2\} = E\{X_1\}E\{X_2\}, \tag{3.111}$$

then it is said that the variables are *uncorrelated*.

If the correlation is zero, the variables are said to be *orthogonal*,

$$R_{X_1X_2} = E\{X_1, X_2\} = 0. \tag{3.112}$$

Note that if the variables are uncorrelated and one or both variables have a zero mean value, it follows that they are also orthogonal.

Example 3.3.3 The random variables X_1 and X_2 are related in the following form:

$$X_2 = -2X_1 + 5. \tag{3.113}$$

Determine whether or not the variables are correlated and orthogonal, if the random variable X_1 has the mean, and the squared mean values equal to 2 and 5, respectively.

Solution In order to determine if the variables are correlated and orthogonal, we first have to find the correlation:

$$\begin{aligned} R_{X_1X_2} &= E\{X_1X_2\} = E\{X_1(-2X_1 + 5)\} = E\{-2X_1^2 + 5X_1\} \\ &= -2E\{X_1^2\} + 5E\{X_1\} = -2 \times 5 + 5 \times 2 = 0. \end{aligned} \tag{3.114}$$

Therefore, according to (3.112) the variables are orthogonal.

Next, we have to verify that the condition (3.111) is satisfied. To this end, we have to find the mean value of X_2, being as the mean value of X_1 has been found to be equal to 2.

$$E\{X_2\} = -2E\{X_1\} + 5 = -4 + 5 = 1. \tag{3.115}$$

Since

$$E\{X_1X_2\} = R_{X_1X_2} = 0 \neq E\{X_1\}E\{X_2\} = 2 \times 1 = 2, \tag{3.116}$$

the variables are correlated.

3.3.3 Joint Central Moments

The *joint central moment of order r*, of two random variables X_1 and X_2, with corresponding mean values $E\{X_1\}$ and $E\{X_2\}$, is defined as:

$$\mu_{nk} = E\left\{(X_1 - E\{X_1\})^n (X_2 - E\{X_2\})^k\right\}, \tag{3.117}$$

where the order r is:

$$r = n + k. \tag{3.118}$$

Using the expression for the mean value of the function of two random variables (3.87), we arrive at:

$$\mu_{nk} = \int_{-\infty}^{\infty}\int_{-\infty}^{\infty} (x_1 - \overline{X_1})^n (x_2 - \overline{X_2})^k f_{X_1X_2}(x_1, x_2)\mathrm{d}x_1\,\mathrm{d}x_2. \tag{3.119}$$

This expression can be generalized for N random variables $X_1, \ldots, X_N$:

$$\mu_{n_1,\ldots,n_N} = \int_{-\infty}^{\infty}\cdots\int_{-\infty}^{\infty} (x_1 - \overline{X_1})^{n_1}, \ldots, (x_N - \overline{X_N})^{n_N} f_{X_1,\ldots,X_N}(x_1, \ldots, x_N)\mathrm{d}x_1, \ldots, \mathrm{d}x_N. \tag{3.120}$$

The second central moment μ_{11}, called *covariance* is especially important:

$$\begin{aligned} C_{X_1X_2} = \mu_{11} &= E\{(X_1 - \overline{X_1})(X_2 - \overline{X_2})\} \\ &= \int_{-\infty}^{\infty}\int_{-\infty}^{\infty} (x_1 - \overline{X_1})(x_2 - \overline{X_2}) f_{X_1X_2}(x_1, x_2)\mathrm{d}x_1\,\mathrm{d}x_2. \end{aligned} \tag{3.121}$$

Let us first relate the covariance to the *independent variables*, where the joint PDF is equal to the product of the marginal PDFs, resulting in the following covariance:

$$\begin{aligned} C_{X_1X_2} = \mu_{11} &= \int_{-\infty}^{\infty}\int_{-\infty}^{\infty} (x_1 - \overline{X_1})(x_2 - \overline{X_2}) f_{X_1}(x_1) f_{X_2}(x_2)\mathrm{d}x_1\,\mathrm{d}x_2 \\ &= \left[\int_{-\infty}^{\infty} x_1 f_{X_1}(x_1)\mathrm{d}x_1 - \overline{X_1}\int_{-\infty}^{\infty} f_{X_1}(x_1)\mathrm{d}x_1\right]\left[\int_{-\infty}^{\infty} x_2 f_{X_2}(x_2)\mathrm{d}x_2 - \overline{X_2}\int_{-\infty}^{\infty} f_{X_2}(x_2)\mathrm{d}x_2\right] \\ &= \left[\overline{X_1} - \overline{X_1}\right]\left[\overline{X_2} - \overline{X_2}\right] = 0. \end{aligned} \tag{3.122}$$

From (3.122) it follows that the covariance is equal to zero for the independent variables.

Using (3.95), (3.121) can be simplified as:

$$C_{X_1X_2} = \overline{X_1X_2} - \overline{X_1}\,\overline{X_2}. \tag{3.123}$$

Let us now relate covariance with the correlated and orthogonal random variables.

From (3.123) and (3.111), it follows that the covariance is equal to zero if the random variables are uncorrelated.

Therefore, from summing the above statements it follows that *the covariance equals to zero if the random variables are either independent or dependent but uncorrelated.*

Additionally, from (3.112) it follows that *if the random variables are orthogonal, then the covariance is equal to the negative product of their mean values.*

$$C_{X_1X_2} = -E\{X_1\}E\{X_2\}. \tag{3.124}$$

3.3.3.1 Variance of the Sum of Random Variables

Consider the sum of two random variables X_1 and X_2, which is itself a random variable X:

$$X = X_1 + X_2. \tag{3.125}$$

By applying the definition (2.333) of the variance of the random variable X and using (3.88) and (3.125), we get:

$$\begin{aligned}\sigma_X^2 = \overline{(X-\overline{X})^2} &= \overline{(X_1 + X_2 - \overline{X_1 + X_2})^2} \\ &= \overline{(X_1 - \overline{X_1})^2} + \overline{(X_2 - \overline{X_2})^2} + 2\overline{(X_1 - \overline{X_1})(X_2 - \overline{X_2})}.\end{aligned} \tag{3.126}$$

The first two terms in (3.126) are the corresponding variances of the variables X_1 and X_2, respectively, while the third averaged product is the covariance.

Therefore, (3.126) reduces to:

$$\sigma_X^2 = \sigma_{X_1}^2 + \sigma_{X_2}^2 + 2C_{X_1X_2}. \tag{3.127}$$

Equation (3.127) states that *the variance of the sum of the variables X_1 and X_2 is equal to the sum of the corresponding variances if their covariance is equal to zero (i.e., the variables are either independent or uncorrelated).*

Therefore, if the random variables X_1 and X_2 are either independent or uncorrelated ($C_{X_1X_2} = 0$), then

$$\sigma_{X_1+X_2}^2 = \sigma_{X_1}^2 + \sigma_{X_2}^2. \tag{3.128}$$

The result (3.128) can be generalized to the sum of N either independent or uncorrelated variables $X_1, \ldots, X_N$:

$$\sigma^2_{\sum_{i=1}^{N} X_i} = \sum_{i=1}^{N} \sigma_{X_i}^2. \tag{3.129}$$

3.3.4 *Independence and Correlation*

Random variables are independent if one variable does not have any influence over the values of another variable, and vice versa. For nonrandom variables, dependency means that if we know one variable, we can find the exact values of another variable.

However, dependence between random variables can have different degrees of dependency.

If we can express the relation between random variables X_1 and X_2 with an exact mathematical expression, then this dependency is called *functional dependency*, as seen for the example ($X_2 = 2X_1 + 2$). In the opposite case, we do not have the exact mathematical expression but the tendency, as for example, if X_1 is the height and X_2 is the weight of people in a population. In the majority of cases, higher values of X_1 correspond to higher values of X_2, but there is no mathematical relation to express this relation.

To this end, the dependence between variables is expressed using some characteristics, like covariance. However, the covariance contains the information, not only of dependency of random variables, but also the information about the dissipation of variables around their mean values. If, for example, the dissipations of random variables X_1 and X_2 around their mean values were very small, then the covariance $C_{X_1X_2}$ would be small for any degree of dependency in the variables.

This problem is solved by introducing the *correlation coefficient* ($\rho_{X_1X_2}$):

$$\rho_{X_1X_2} = \frac{C_{X_1X_2}}{\sigma_{X_1}\sigma_{X_2}} = \frac{\overline{(X_1 - \overline{X_1})(X_2 - \overline{X_2})}}{\sigma_{X_1}\sigma_{X_2}} = \frac{\overline{X_1X_2} - \overline{X_1}\,\overline{X_2}}{\sigma_{X_1}\sigma_{X_2}}. \tag{3.130}$$

What are the values that the correlation coefficient can take?

(a) *The variables are equal*
Consider a case in which the random variables are equal, $X_2 = X_1$, and therefore, there is a maximum degree of dependency between them. From (3.130), and using the definition of variance (2.333), we have:

$$\rho_{X_1X_2} = \frac{C_{X_1X_2}}{\sigma_{X_1}\sigma_{X_2}} = \frac{\overline{(X_1 - \overline{X_1})(X_1 - \overline{X_1})}}{\sigma_{X_1}\sigma_{X_1}} = \frac{\overline{(X_1 - \overline{X_1})^2}}{\sigma_{X_1}^2} = \frac{\sigma_{X_1}^2}{\sigma_{X_1}^2} = 1. \tag{3.131}$$

Therefore, for the maximum dependency of random variables, the correlation coefficient is equal to 1.

(b) *Linear dependency of random variables*
Consider the linear dependency for random variables,

$$X_2 = aX_1 + b, \tag{3.132}$$

where a and b are deterministic constants. Figure 3.9 shows this dependency for $a > 0$ and $a < 0$.

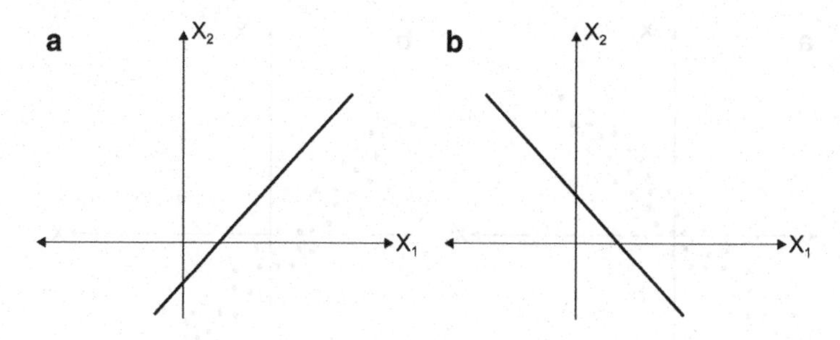

Fig. 3.9 Linear dependency

The covariance is equal to:

$$C_{X_1X_2} = \overline{(X_1 - \overline{X_1})(aX_1 + b - \overline{aX_1 + b})} = \overline{(X_1 - \overline{X_1})a(X_1 - \overline{X_1})}$$
$$= a\overline{(X_1 - \overline{X_1})^2} = a\sigma_{X_1}^2. \tag{3.133}$$

From (2.355), the variance of the random variable X_2 is

$$\sigma_{X_2}^2 = a^2\sigma_{X_1}^2. \tag{3.134}$$

Finally, using (3.133) and (3.134), we can obtain the correlation coefficient:

$$\rho_{X_1X_2} = \frac{C_{X_1X_2}}{\sigma_{X_1}\sigma_{X_2}} = \frac{a\sigma_{X_1}^2}{\sigma_{X_1}|a|\sigma_{X_1}} = \frac{a}{|a|} = \begin{cases} 1 & \text{for} \quad a > 0 \\ -1 & \text{for} \quad a < 0 \end{cases}. \tag{3.135}$$

Therefore, *the maximum absolute value of the correlation coefficient is equal to 1* (*if the variables are linearly dependent*), as given in (3.135).

If

$$\rho_{X_1X_2} > 0, \tag{3.136}$$

then there is a *positive correlation*, as shown in Fig. 3.10a, in contrast, to the case in which

$$\rho_{X_1X_2} < 0, \tag{3.137}$$

there is a *negative correlation.*

Note that in (3.136) and (3.137) there is no functional relation between variables, but only tendency: if one variable increases, in the majority of cases, the other variable either increases (positive correlation) or decreases (negative correlation).

Therefore, the coefficient of correlation presents the degree of linear dependency.

If there is no linear dependency between the variables (like in Fig. 3.11a) the coefficient of correlation is equal to zero. As a consequence, the random variables are dependent, but uncorrelated.

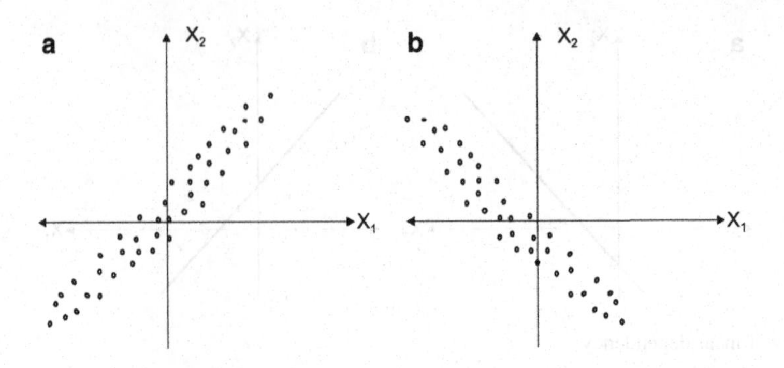

Fig. 3.10 Positive and negative correlations

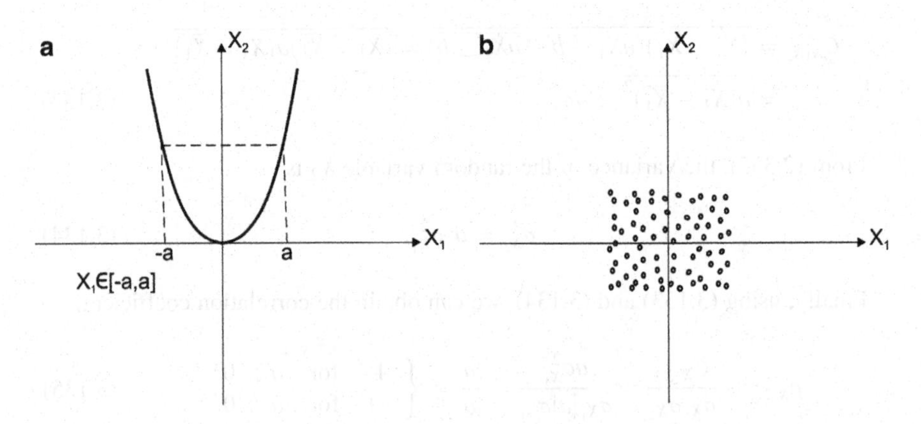

Fig. 3.11 Dependent uncorrelated and independent variables

If the random variables are independent, then obviously there is no relation between them (not even a linear relation), resulting in a zero value of the coefficient of correlation, as shown in Fig. 3.11b.

From (3.135) to (3.137), the values of the correlation coefficient are:

$$-1 \leq \rho_{X_1X_2} \leq 1. \tag{3.138}$$

Example 3.3.4 Consider dependent random variables X and Y, where

$$Y = X^2. \tag{3.139}$$

Determine whether or not the random variables X and Y are correlated and find the coefficient of correlation for the following two cases:

(a) The random variable X is uniform in the interval $[-1, 1]$.
(b) The random variable X is uniform in the interval $[0, 2]$.

Solution

(a) The expected values for the random variables X and Y are:

$$\overline{X} = 0; \overline{Y} = \overline{X^2} = \int_{-1}^{1} \frac{1}{2} x^2 \, dx = \frac{1}{3}. \tag{3.140}$$

From (3.121) and (3.140), the covariance is equal to:

$$C_{XY} = \overline{(X - \overline{X})(Y - \overline{Y})} = \overline{X(X^2 - 1/3)} = \overline{X^3} - 1/3\overline{X} = 0. \tag{3.141}$$

The obtained result shows that the covariance is equal to zero, and thus the coefficient of correlation (3.130) is also zero. As such, the random variables are uncorrelated.

Therefore, variables X and Y are dependent and uncorrelated.

(b) In this case, the random variables are also dependent.

The corresponding mean values are:

$$\overline{X} = 1, \qquad \overline{Y} = \overline{X^2} = \int_{0}^{2} \frac{1}{2} x^2 \, dx = \frac{4}{3}. \tag{3.142}$$

From (3.121) and (3.142), the covariance is equal to:

$$C_{XY} = \overline{(X - \overline{X})(Y - \overline{Y})} = \overline{(X - 1)(X^2 - 4/3)} = \overline{X^3} - \overline{X^2} - 4/3\overline{X} + 4/3. \tag{3.143}$$

To calculate (3.143), we first need the third moment of the random variable X:

$$\overline{X^3} = \int_{0}^{2} \frac{1}{2} x^3 \, dx = 2. \tag{3.144}$$

Placing (3.142) and (3.144) into (3.143), we get:

$$C_{XY} = 2 - 4/3 - 4/3 + 4/3 = 2/3. \tag{3.145}$$

This result indicates that the random variables are correlated. To find the coefficient of correlation we need the corresponding variances.

From (3.142), we have:

$$\sigma_X^2 = \overline{X^2} - \overline{X}^2 = 4/3 - 1 = 1/3, \tag{3.146}$$

$$\sigma_Y^2 = \overline{Y^2} - \overline{Y}^2 = \overline{X^4} - \overline{X^2}^2, \tag{3.147}$$

$$\overline{X^4} = \int_0^2 \frac{1}{2} x^4 \, dx = 16/5. \tag{3.148}$$

Placing (3.148) and (3.142) into (3.147), we can obtain the variance of the random variable Y:

$$\sigma_Y^2 = 16/5 - (4/3)^2 = 1.422. \tag{3.149}$$

The standard deviations are obtained from (3.146) and (3.149):

$$\sigma_X = \sqrt{\sigma_X^2} = \sqrt{1/3} = 0.5774, \tag{3.150}$$

$$\sigma_Y = \sqrt{\sigma_Y^2} = \sqrt{1.422} = 1.1925. \tag{3.151}$$

Using values for the covariance (3.145) and standard deviations (3.150) and (3.151), we calculate the coefficient of correlation using (3.130):

$$\rho_{XY} = \frac{C_{XY}}{\sigma_X \sigma_Y} = \frac{2/3}{0.5774 \times 1.1925} = 0.9682. \tag{3.152}$$

As opposed to case (a), in case (b), the variables are dependent and correlated. Based on the previous discussion, we can conclude that the dependence is a stronger condition than correlation. That is, *if the variables are independent, they are also uncorrelated.* However, *if the random variables are dependent they can be either correlated or uncorrelated*, as summarized in Fig. 3.12.

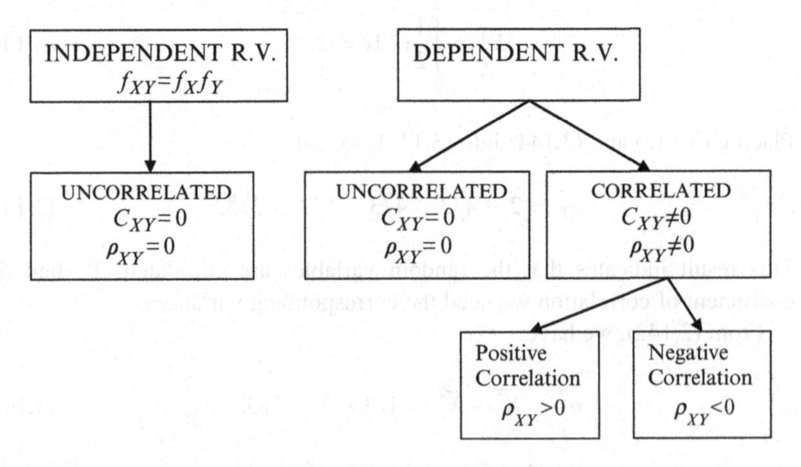

Fig. 3.12 Correlation and dependency between r.v.

Therefore, if random variables are uncorrelated, they can be either dependent or independent. However, there are some exceptions such as the case of uncorrelated normal random variables. If normal random variables are uncorrelated it follows that they are independent (see Chap. 4).

3.4 Transformation of Random Variables

3.4.1 *One-to-One Transformation*

Consider two random variables X_1 and X_2 with a known joint PDF $f_{X_1X_2}(x_1, x_2)$. As a result of the transformation

$$Y_1 = g_1(X_1, X_2)$$
$$Y_2 = g_2(X_1, X_2) \tag{3.153}$$

new random variables Y_1 and Y_2 are obtained, as shown in Fig. 3.13.

We are looking for the joint density of the transformed random variables (3.153). The result depends on the type of transformation. Here we consider a simple case in which the infinitesimal area in the (x_1, x_2) system has a one-to-one correspondence to the infinitesimal area in the (y_1, y_2) system, as shown in Fig. 3.14. In other words,

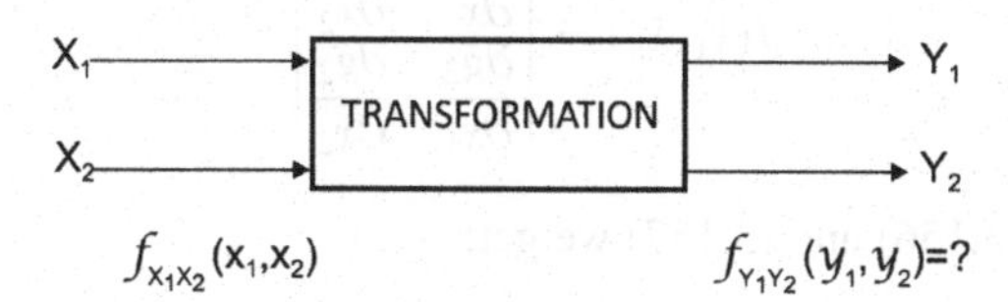

Fig. 3.13 Transformation of two random variables

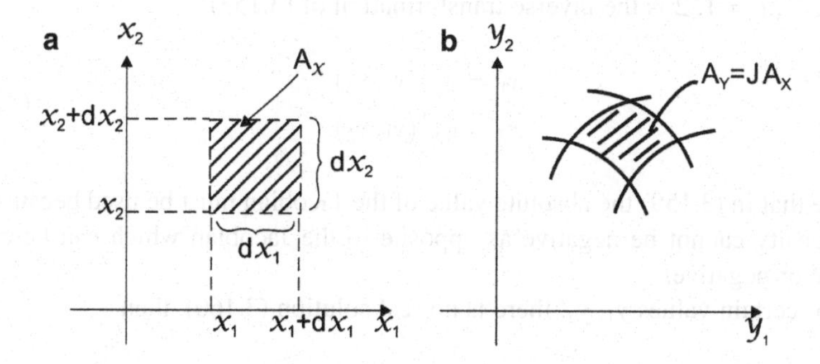

Fig. 3.14 Mapping from (x_1, x_2) space onto (y_1, y_2) space

the elementary area ($dx_1\,dx_2$) is mapped one-to-one onto a corresponding infinitesimal area ($dy_1\,dy_2$). As a result, the corresponding probabilities are equal:

$$P\{x_1 < X_1 \le x_1 + dx_1, x_2 < X_2 \le x_2 + dx_2\} = P\{y_1 < Y_1 \le y_1 + dy_1, y_2 < Y_2 \le y_2 + dy_2\}. \quad (3.154)$$

The probabilities in (3.154) can be expressed in terms of their corresponding joint densities:

$$\begin{aligned} P\{x_1 < X_1 \le x_1 + dx_1, x_2 < X_2 \le x_2 + dx_2\} &= f_{X_1X_2}(x_1, x_2)dx_1\,dx_2, \\ P\{y_1 < Y_1 \le y_1 + dy_1, y_2 < Y_2 \le y_2 + dy_2\} &= f_{Y_1Y_2}(y_1, y_2)dy_1\,dy_2. \end{aligned} \quad (3.155)$$

From (3.154) and (3.155), we have:

$$f_{Y_1Y_2}(y_1, y_2)dy_1\,dy_2 = f_{X_1X_2}(x_1, x_2)dx_1\,dx_2. \quad (3.156)$$

The infinitesimal areas ($dy_1\,dy_2$) and ($dx_1\,dx_2$) are related as,

$$dx_1\,dx_2 = \frac{dy_1\,dy_2}{J(x_1x_2)}, \quad (3.157)$$

where $J(x_1, x_2)$ is the *Jacobian* of the transformation (3.153) [PAP65, p. 201]:

$$J(x_1, x_2) = \begin{vmatrix} \dfrac{\partial g_1}{\partial x_1} & \dfrac{\partial g_1}{\partial x_2} \\ \dfrac{\partial g_2}{\partial x_1} & \dfrac{\partial g_2}{\partial x_2} \end{vmatrix}. \quad (3.158)$$

Finally, from (3.156) and (3.157) we get:

$$f_{Y_1Y_2}(y_1, y_2) = \left.\frac{f_{X_1X_2}(x_1, x_2)}{|J(x_1, x_2)|}\right|_{\substack{x_1 = g_1^{-1}(y_1, y_2) \\ x_2 = g_2^{-1}(y_1, y_2)}}, \quad (3.159)$$

where g_i^{-1}, $i = 1, 2$ is the inverse transformation of (3.153),

$$\begin{aligned} x_1 &= g_1^{-1}(y_1, y_2) \\ x_2 &= g_2^{-1}(y_1, y_2). \end{aligned} \quad (3.160)$$

Note that in (3.159) the absolute value of the Jacobian must be used because the joint density cannot be negative as opposite to the Jacobian which can be either positive or negative.

If for certain values y_1, y_2, there is no real solution (3.160), then

$$f_{Y_1Y_2}(y_1, y_2) = 0. \quad (3.161)$$

Example 3.4.1 The joint density function of random variables X_1 and X_2 is given as:

$$f_{X_1X_2}(x_1,x_2) = \frac{1}{2\pi\sigma^2}\mathrm{e}^{-\frac{x_1^2+x_2^2}{2\sigma^2}}; \quad -\infty < x_1 < \infty, \quad -\infty < x_2 < \infty. \tag{3.162}$$

Find the joint density function of the random variables Y_1 and Y_2 if,

$$\begin{aligned} X_1 &= Y_1 \cos Y_2, \\ X_2 &= Y_1 \sin Y_2. \end{aligned} \tag{3.163}$$

Solution The given transformation (3.163) can be rewritten as:

$$\begin{aligned} y_2 &= \tan^{-1}\left(\frac{x_2}{x_1}\right) = g_2(x_1,x_2) \\ y_1 &= \sqrt{x_1^2+x_2^2} = g_1(x_1,x_2). \end{aligned} \tag{3.164}$$

The Jacobian of the transformation (3.164), according to (3.158), is:

$$J = \begin{vmatrix} \dfrac{x_1}{\sqrt{x_1^2+x_2^2}} & \dfrac{x_2}{\sqrt{x_1^2+x_2^2}} \\ \dfrac{-x_2}{x_1^2+x_2^2} & \dfrac{x_1}{x_1^2+x_2^2} \end{vmatrix} = \frac{x_1^2+x_2^2}{(x_1^2+x_2^2)\sqrt{x_1^2+x_2^2}} = \frac{1}{\sqrt{x_1^2+x_2^2}} = \frac{1}{y_1}. \tag{3.165}$$

Using (3.159), we get:

$$f_{Y_1Y_2}(y_1,y_2) = \begin{cases} \dfrac{y_1}{2\pi\sigma^2}\mathrm{e}^{-\frac{y_1^2}{2\sigma^2}} & \text{for} \quad y_1 \geq 0; 0 \leq y_2 \leq 2\pi, \\ 0 & \text{otherwise.} \end{cases} \tag{3.166}$$

In the following, we demonstrate how the expression (3.159) can be used to obtain the PDF of a random variable which is the function of two random variables:

$$Y = g(X_1,X_2). \tag{3.167}$$

In order to apply the expression (3.159), we introduce the auxiliary variable ρ [PAP65, p. 204], as shown in Fig. 3.15.

The variable ρ will later be eliminated and therefore we chose it for convenience:

$$\rho = X_1 \quad \text{or} \quad \rho = X_2. \tag{3.168}$$

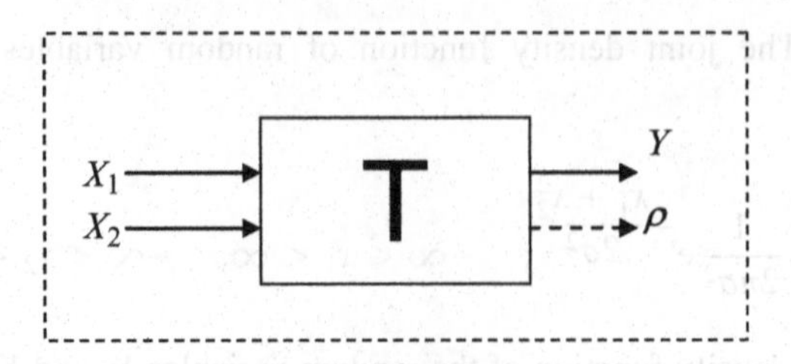

Fig. 3.15 Transformation of two input random variables

Now we have the transformation:

$$\begin{aligned} Y &= g(X_1, X_2), \\ \rho &= X_1, \end{aligned} \tag{3.169}$$

or

$$\begin{aligned} Y &= g(X_1, X_2). \\ \rho &= X_2. \end{aligned} \tag{3.170}$$

The joint PDF $f_{Y\rho}(y, z)$ is obtained from (3.159). Finally, the required PDF $f_Y(y)$ is obtained from $f_{Y\rho}(y, z)$, using (3.42a, 3.42b):

$$f_Y(y) = \int_{-\infty}^{\infty} f_{Y\rho}(y, z)\, \mathrm{d}z. \tag{3.171}$$

Example 3.4.2 Find the PDF of the variable Y

$$Y = X_1 X_2, \tag{3.172}$$

if it is known then the joint PDF is $f_{X_1 X_2}(x_1, x_2)$.

Solution The auxiliary variable

$$\rho = X_1 \tag{3.173}$$

is introduced, resulting in the following transformation:

$$\begin{aligned} y &= x_1 x_2. \\ z &= x_1. \end{aligned} \tag{3.174}$$

From here,

$$x_1 = z, \qquad x_2 = y/z. \tag{3.175}$$

Using (3.159), (3.158), and (3.175) we get:

$$J = \begin{vmatrix} x_2 & x_1 \\ 1 & 0 \end{vmatrix} = -x_1 = -z, \tag{3.176}$$

and

$$f_{Y\rho}(y,z) = \frac{f_{X_1X_2}(x_1,x_2)}{|z|} = \frac{f_{X_1X_2}(z,y/z)}{|z|}. \tag{3.177}$$

Finally, the desired PDF is obtained from (3.171) and (3.174):

$$f_Y(y) = \int_{-\infty}^{\infty} \frac{1}{|z|} f_{X_1X_2}(z,y/z)\mathrm{d}z. \tag{3.178}$$

3.4.2 Nonunique Transformation

In this case, the infinitesimal area ($\mathrm{d}y_1\ \mathrm{d}y_2$) corresponds to two or more infinitesimal areas ($\mathrm{d}x_1\ \mathrm{d}x_2$), resulting in:

$$f_{Y_1Y_2}(y_1,y_2) = \left. \frac{\sum_i f_{X_1X_2}(x_1^i,x_2^i)}{|J(x_1^i,x_2^i)|} \right|_{\substack{x_1^i = g_1^{-1}(y_1,y_2) \\ x_2^i = g_2^{-1}(y_1,y_2)}}. \tag{3.179}$$

Example 3.4.3 Find the joint PDF of the random variables Y_1 and Y_2,

$$Y_1 = \sqrt{X_1^2 + X_2^2}; \qquad Y_2 = \frac{X_1}{X_2}, \tag{3.180}$$

if the given joint PDF of the random variables X_1 and X_2 is given as:

$$f_{X_1X_2}(x_1,x_2) = \frac{1}{2\pi\sigma^2}\mathrm{e}^{-\frac{(x_1^2+x_2^2)}{2\sigma^2}}. \tag{3.181}$$

Solution From (3.180), we have:

$$y_1 = \sqrt{x_1^2 + x_2^2}; \qquad y_2 = \frac{x_1}{x_2}. \tag{3.182}$$

The Jacobian of the transformation (3.180) is:

$$J(x_1,x_2)=\begin{vmatrix}\dfrac{x_1}{\sqrt{x_1^2+x_2^2}} & \dfrac{x_2}{\sqrt{x_1^2+x_2^2}}\\ \dfrac{1}{x_2} & -\dfrac{x_1}{x_2^2}\end{vmatrix}=-\frac{\left(\frac{x_1}{x_2}\right)^2+1}{\sqrt{x_1^2+x_2^2}}. \tag{3.183}$$

Placing (3.182) into (3.183), we get:

$$J(y_1,y_2)=-\frac{y_2^2+1}{y_1}. \tag{3.184}$$

Equations (3.182) have two solutions for x_1 and x_2:

$$\begin{aligned} x_1^1&=\frac{y_1y_2}{\sqrt{1+y_2^2}}, & x_2^1&=\frac{y_1}{\sqrt{1+y_2^2}},\\ x_1^2&=-\frac{y_1y_2}{\sqrt{1+y_2^2}}, & x_2^2&=-\frac{y_1}{\sqrt{1+y_2^2}}. \end{aligned} \tag{3.185}$$

From (3.179), (3.181), and (3.185), it follows:

$$\begin{aligned} f_{Y_1Y_2}(y_1,y_2)&=\frac{f_{X_1X_2}(x_1^1,x_2^1)+f_{X_1X_2}(x_1^2,x_2^2)}{|J(y_1,y_2)|}\\ &=\frac{y_1}{1+y_2^2}2f_{X_1X_2}(y_1,y_2)=\frac{y_1}{1+y_2^2}\frac{1}{\pi\sigma^2}\mathrm{e}^{-\frac{y_1^2}{2\sigma^2}}. \end{aligned} \tag{3.186}$$

Finally, we have:

$$f_{Y_1Y_2}(y_1,y_2)=\begin{cases}\dfrac{y_1}{1+y_2^2}\dfrac{1}{\pi\sigma^2}\mathrm{e}^{-\frac{y_1^2}{2\sigma^2}} & \text{for } y_1\geq 0,\\ 0 & \text{for } y_1<0.\end{cases} \tag{3.187}$$

3.4.3 Generalization for N Variables

Given N random variables

$$X_1,X_2,\ldots,X_N, \tag{3.188}$$

with the joint density function $f_{X_1X_2,\ldots,X_N}(x_1,x_2,\ldots,x_N)$, are transformed into new random variables

$$Y_1,Y_2,\ldots,Y_N, \tag{3.189}$$

where an unique transformation of the input variables (3.189) is defined as:

$$\begin{aligned} Y_1 &= g_1(X_1, \ldots, X_N) \\ Y_2 &= g_2(X_1, \ldots, X_N) \\ &\vdots \\ Y_N &= g_N(X_1, \ldots, X_N) \end{aligned} \tag{3.190}$$

Similarly as in (3.157), we have:

$$\mathrm{d}x_1, \ldots, \mathrm{d}x_N = \frac{\mathrm{d}y_1, \ldots, \mathrm{d}y_N}{J(x_1, \ldots, x_N)}, \tag{3.191}$$

where J is the Jacobian of the transformation (3.190), defined as:

$$J(x_1, \ldots, x_N) = \begin{vmatrix} \frac{\partial g_1}{\partial x_1} & \cdots & \frac{\partial g_1}{\partial x_N} \\ \vdots & \vdots & \vdots \\ \frac{\partial g_N}{\partial x_1} & \cdots & \frac{\partial g_N}{\partial x_N} \end{vmatrix}. \tag{3.192}$$

Similarly, like (3.159), using (3.190)–(3.192), we get:

$$f_{Y_1 Y_2, \ldots, Y_N}(y_1, y_2, \ldots, y_N) = \frac{f_{X_1 X_2, \ldots, X_N}(x_1, x_2, \ldots, x_N)}{|J(x_1, x_2, \ldots, x_N)|}\Bigg|_{\substack{x_1 = g_1^{-1}(y_1, y_2, \ldots, y_N) \\ x_2 = g_2^{-1}(y_1, y_2, \ldots, y_N) \\ \vdots \\ x_N = g_N^{-1}(y_1, y_2, \ldots, y_N)}}. \tag{3.193}$$

3.5 Characteristic Functions

3.5.1 Definition

The *joint characteristic function* of two random variables X_1 and X_2, denoted as $\phi_{X_1 X_2}(\omega_1, \omega_2)$, is defined as the expected value of the complex function $\mathrm{e}^{j(\omega_1 X_1 + \omega_2 X_2)}$,

$$\phi_{X_1 X_2}(\omega_1, \omega_2) = E\left\{\mathrm{e}^{j(\omega_1 X_1 + \omega_2 X_2)}\right\}. \tag{3.194}$$

According to (3.87), the expression (3.194) is equal to:

$$\phi_{X_1 X_2}(\omega_1, \omega_2) = \int_{-\infty}^{\infty} \int_{-\infty}^{\infty} \mathrm{e}^{j(\omega_1 x_1 + \omega_2 x_2)} f_{X_1 X_2}(x_1, x_2) \mathrm{d}x_1 \, \mathrm{d}x_2. \tag{3.195}$$

The obtained expression (3.195) can be interpreted with the exception of the exponent sign, as a two-dimensional Fourier transform of the joint PDF

$f_{X_1X_2}(x_1, x_2)$. This means that using the inverse two-dimensional Fourier transform, one can obtain the joint PDF from its joint characteristic function, as shown in the following expression:

$$f_{X_1X_2}(x_1, x_2) = \frac{1}{(2\pi)^2} \int_{-\infty}^{\infty} \int_{-\infty}^{\infty} \phi_{X_1X_2}(\omega_1, \omega_2) \mathrm{e}^{-j(\omega_1 x_1 + \omega_2 x_2)} \mathrm{d}\omega_1 \, \mathrm{d}\omega_2. \tag{3.196}$$

If random variables X_1 and X_2 are independent, then their joint density is equal to the product of the marginal densities. Thus the joint characteristic function (3.195) becomes:

$$\phi_{X_1X_2}(\omega_1, \omega_2) = \int_{-\infty}^{\infty} \int_{-\infty}^{\infty} \mathrm{e}^{j\omega_1 x_1} \, \mathrm{e}^{j\omega_2 x_2} f_{X_1}(x_1) f_{X_2}(x_2) \mathrm{d}x_1 \, \mathrm{d}x_2 =$$

$$\int_{-\infty}^{\infty} \mathrm{e}^{j\omega_1 x_1} f_{X_1}(x_1) \mathrm{d}x_1 \int_{-\infty}^{\infty} \mathrm{e}^{j\omega_2 x_2} f_{X_2}(x_2) \mathrm{d}x_2 = \phi_{X_1}(\omega_1) \phi_{X_2}(\omega_2). \tag{3.197}$$

Therefore, for the independent random variables, the joint characteristic function is equal to the product of the *marginal characteristic functions*. The reverse is also true, that is if the joint characteristic function is equal to the product of marginal characteristic functions, then the corresponding random variables are independent.

The marginal characteristic functions are obtained by making ω_1 or ω_2 equal to zero in the joint characteristic function, as demonstrated in (3.198):

$$\begin{aligned} \phi_{X_1}(\omega_1) &= \phi_{X_1X_2}(\omega_1, 0), \\ \phi_{X_2}(\omega_2) &= \phi_{X_1X_2}(0, \omega_2). \end{aligned} \tag{3.198}$$

3.5.2 *Characteristic Function of the Sum of the Independent Variables*

Consider the sum of two random variables X_1 and X_2,

$$Y = X_1 + X_2. \tag{3.199}$$

The characteristic function of the variable Y, according to (2.386) and the definition (3.194), is equal to:

$$\phi_Y(\omega) = E\{\mathrm{e}^{j\omega(X_1+X_2)}\} = \phi_{X_1X_2}(\omega, \omega). \tag{3.200}$$

If the random variables X_1 and X_2 are independent, then the joint characteristic function of their sum is equal to the product of the marginal characteristic functions,

$$\phi_Y(\omega) = \phi_{X_1}(\omega)\phi_{X_2}(\omega). \tag{3.201}$$

This result can be applied for the sum of N independent random variables X_i,

$$Y = \sum_{i=1}^{N} X_i. \tag{3.202}$$

The joint characteristic function of (3.202) is

$$\phi_Y(\omega) = \prod_{i=1}^{N} \phi_{X_i}(\omega). \tag{3.203}$$

Example 3.5.1 Find the characteristic function of the random variable Y, where

$$Y = aX_1 + bX_2 \tag{3.204}$$

and where X_1, and X_2 are independent.

Solution Using (2.386) and the definition of the characteristic function (3.194) and the condition of independence (3.201), we have:

$$\phi_Y(\omega) = E\{e^{j\omega(aX_1+bX_2)}\} = E\{e^{j\omega aX_1}\}E\{e^{j\omega bX_2}\} = \phi_{X_1}(a\omega)\phi_{X_2}(b\omega). \tag{3.205}$$

Example 3.5.2 The random variables X_1 and X_2 are independent. Find the joint characteristic function of the variables X and Y, as given in the following equations:

$$\begin{aligned} X &= X_1 + 2X_2, \\ Y &= 2X_1 + X_2. \end{aligned} \tag{3.206}$$

Solution From the definition (3.194), and using (3.206), we have:

$$\begin{aligned} \phi_{XY}(\omega_1, \omega_2) &= E\{e^{j(\omega_1 X + \omega_2 Y)}\} = E\{e^{j\omega_1(X_1+2X_2)+j\omega_2(2X_1+X_2)}\} \\ &= E\{e^{jX_1(\omega_1+2\omega_2)+jX_2(2\omega_1+\omega_2)}\}. \end{aligned} \tag{3.207}$$

Knowing that the random variables X_1 and X_2 are independent from (3.207) and (3.201), we arrive at:

$$\begin{aligned} \phi_{XY}(\omega_1, \omega_2) &= E\{e^{jX_1(\omega_1+2\omega_2)}\}E\{e^{jX_2(2\omega_1+\omega_2)}\} \\ &= \phi_{X_1}(\omega_1 + 2\omega_2)\phi_{X_2}(2\omega_1 + \omega_2). \end{aligned} \tag{3.208}$$

3.5.3 Moment Theorem

Using the moment theorem for one random variable, as an analogy, we arrive at the moment theorem that finds joint moments m_{nk} from the joint characteristic function, as

$$m_{nk} = (-j)^{n+k} \frac{\partial^{n+k} \phi_{X_1X_2}(\omega_1, \omega_2)}{\partial \omega_1{}^n \partial \omega_2{}^k} \bigg|_{\substack{\omega_1 = 0 \\ \omega_2 = 0}} . \tag{3.209}$$

3.5.4 PDF of the Sum of Independent Random Variables

The random variable Y is equal to the sum of the independent random variables X_1 and X_2,

$$Y = X_1 + X_2. \tag{3.210}$$

The PDF of the variable Y (see (2.387)) is expressed by its characteristic function as,

$$f_Y(y) = \frac{1}{2\pi} \int_{-\infty}^{\infty} \phi_Y(\omega) \mathrm{e}^{-j\omega y} \, \mathrm{d}\omega. \tag{3.211}$$

Placing (3.201) into (3.211), we have:

$$f_Y(y) = \frac{1}{2\pi} \int_{-\infty}^{\infty} \phi_{X_1}(\omega) \phi_{X_2}(\omega) \mathrm{e}^{-j\omega y} \, \mathrm{d}\omega. \tag{3.212}$$

By applying the relation (2.386) to the characteristic function $\phi_{X_1}(\omega)$ in (3.212), and interchanging the order of the integrations, we arrive at:

$$\begin{aligned} f_Y(y) &= \frac{1}{2\pi} \int_{-\infty}^{\infty} \phi_{X_2}(\omega) \mathrm{e}^{-j\omega y} \, \mathrm{d}\omega \int_{-\infty}^{\infty} f_{X_1}(x_1) \mathrm{e}^{j\omega x_1} \, \mathrm{d}x_1 \\ &= \frac{1}{2\pi} \int_{-\infty}^{\infty} f_{X_1}(x_1) \mathrm{d}x_1 \int_{-\infty}^{\infty} \phi_{X_2}(\omega) \mathrm{e}^{-j\omega(y - x_1)} \, \mathrm{d}\omega. \end{aligned} \tag{3.213}$$

Using (2.387), we get:

$$\frac{1}{2\pi}\int_{-\infty}^{\infty}\phi_{X_2}(\omega)e^{-j\omega(y-x_1)}\,d\omega = f_{X_2}(y-x_1), \tag{3.214}$$

From (3.213) and (3.214), we arrive at:

$$f_Y(y) = \int_{-\infty}^{\infty} f_{X_1}(x_1)f_{X_2}(y-x_1)dx_1. \tag{3.215}$$

This expression can be rewritten as:

$$f_Y(y) = \int_{-\infty}^{\infty} f_{X_2}(x_2)f_{X_1}(y-x_2)dx_2. \tag{3.216}$$

The expressions (3.215) and (3.216) present the convolution of the PDFs of random variables X_1 and X_2, and can be presented as:

$$f_Y(y) = f_{X_1}(x_1) * f_{X_2}(x_2) = f_{X_2}(x_2) * f_{X_1}(x_1), \tag{3.217}$$

where * stands for the convolution operation.

This result can be generalized for the sum of N independent variables X_i,

$$Y = \sum_{i=1}^{N} X_i, \tag{3.218}$$

$$f_Y(y) = f_{X_1}(x_1) * f_{X_2}(x_2) * \cdots * f_{X_N}(x_N). \tag{3.219}$$

Example 3.5.3 Two resistors R_1 and R_2 are in a serial connection (Fig. 3.16a). Each of them randomly changes its value in a uniform way for 10% about its nominal value of 1,000 Ω. Find the PDF of the equivalent resistor R,

$$R = R_1 + R_2 \tag{3.220}$$

if both resistors have uniform density in the interval [900, 1,100] Ω.

Solution Denote the equivalent resistor R as a random variable X and the particular resistors R_1 and R_2, as the random variables X_1 and X_2. Then,

$$X = X_1 + X_2. \tag{3.221}$$

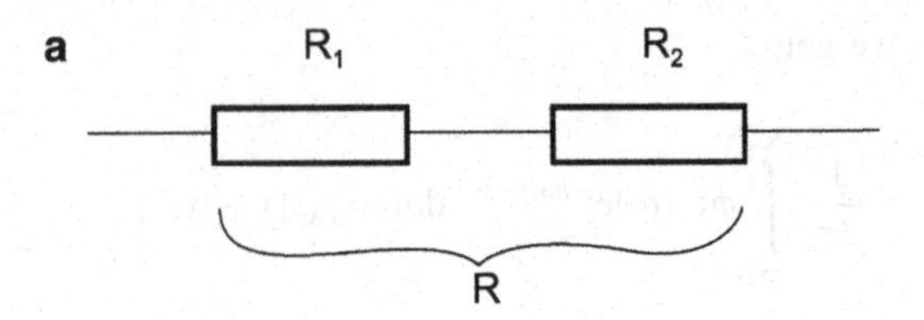

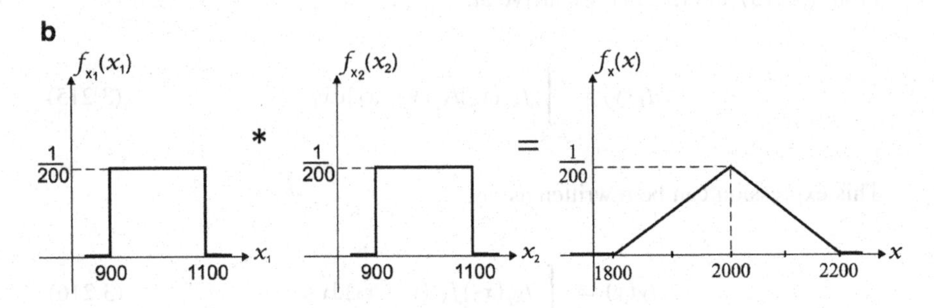

Fig. 3.16 Convolution of uniform PDFs

From (3.217), the PDF of the random variable X is equal to the convolution of the PDFs of the random variables X_1 and X_2. In this case, it is convenient to present the convolution graphically, as shown in Fig. 3.16.

3.6 Numerical Exercises

Exercise 3.1 The joint random variables X_1 and X_2 are defined in a circle of a radius $r = 2$, as shown in Fig. 3.17. Their joint PDF is constant inside the circle. Find and plot the joint PDF and the marginal PDFs. Determine whether or not the random variables X_1 and X_2 are independent.

Answer The area A in Fig. 3.17 is:

$$A = r^2\pi = 4\pi. \tag{3.222}$$

The volume below the joint density is the height of the cylinder which, according to (3.37) must be unity, is shown in Fig. 3.18.

The joint density is:

$$f_{X_1X_2}(x_1, x_2) = \begin{cases} 1/4\pi & \text{for} \quad x_1^2 + x_2^2 \le 4, \\ 0 & \text{otherwise}. \end{cases} \tag{3.223}$$

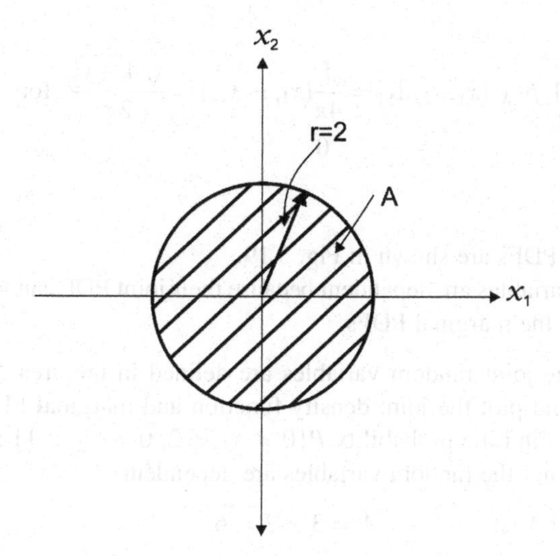

Fig. 3.17 Range in Exercise 3.1

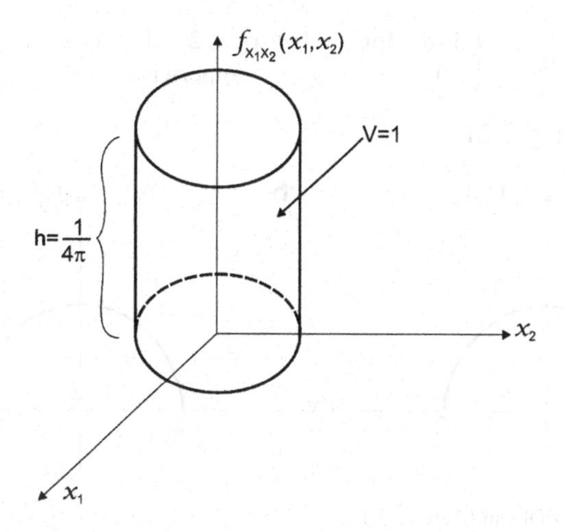

Fig. 3.18 Joint PDF in Exercise 3.1

From (3.223), we have:

$$x_{1_{1,2}} = \pm\sqrt{4 - x_2^2}. \tag{3.224}$$

$$x_{2_{1,2}} = \pm\sqrt{4 - x_1^2}. \tag{3.225}$$

Using (3.42a, 3.42b), and (3.223)–(3.225), we get the marginal densities:

$$f_{X_1}(x_1) = \begin{cases} \int\limits_{x_{2_1}}^{x_{2_2}} f_{X_1X_2}(x_1, x_2)\mathrm{d}x_2 = \dfrac{1}{4\pi}(x_{2_1} - x_{2_2}) = \dfrac{\sqrt{4 - x_1^2}}{2\pi} & \text{for} \quad |x_1| \leq 2, \\ 0 & \text{otherwise}, \end{cases} \tag{3.226}$$

$$f_{X_2}(x_2) = \begin{cases} \int\limits_{x_{1_1}}^{x_{1_2}} f_{X_1X_2}(x_1,x_2)\mathrm{d}x_1 = \frac{1}{4\pi}(x_{1_1} - x_{1_2}) = \frac{\sqrt{4-x_2^2}}{2\pi} & \text{for} \quad |x_2| \leq 2, \\ 0 & \text{otherwise.} \end{cases} \tag{3.227}$$

The marginal PDFs are shown in Fig. 3.19.

The random variables are dependent because their joint PDF cannot be presented as the product of the marginal PDFs.

Exercise 3.2 The joint random variables are defined in the area A, as shown in Fig. 3.20. Find and plot the joint density function and marginal PDFs if the joint PDF is constant. Find the probability $P\{0 < x_1 < 2, 0 < x_2 < 1\}$ as well. Determine whether or not the random variables are dependent.

Answer The area A is:

$$A = 3 \times 2 = 6. \tag{3.228}$$

From (3.36), the joint PDF is:

$$f_{X_1X_2}(x_1,x_2) = \begin{cases} 1/6 & \text{for} \quad -1 < x_1 < 2, -1 < x_2 < 1, \\ 0 & \text{otherwise.} \end{cases} \tag{3.229}$$

This is shown in Fig. 3.21.

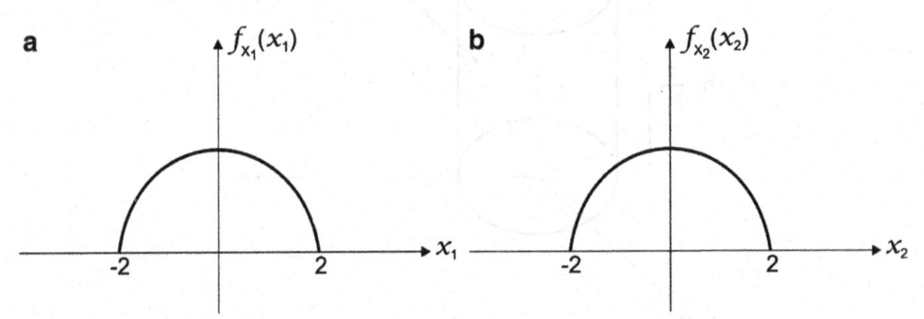

Fig. 3.19 Marginal PDFs in Exercise 3.1

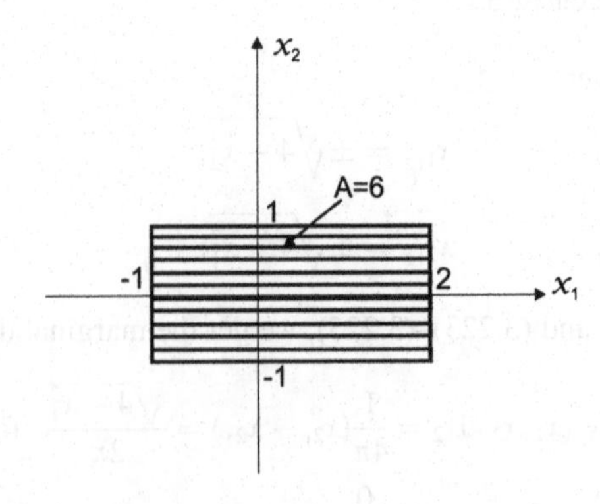

Fig. 3.20 Range of variables in Exercise 3.2

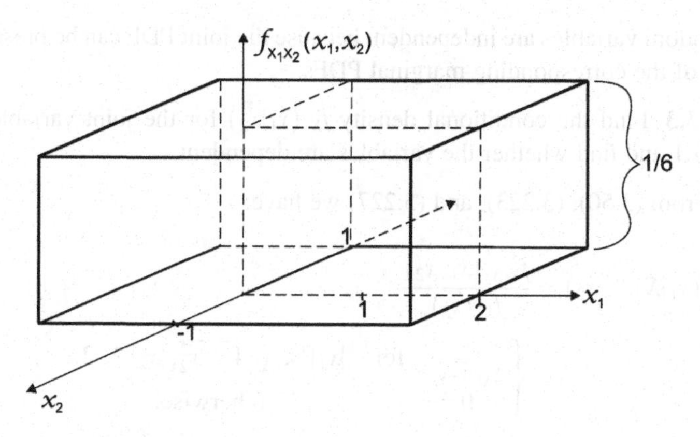

Fig. 3.21 Joint PDF in Exercise 3.2

The marginal PDFs are

$$f_{X_1}(x_1) = \begin{cases} \int_{-1}^{1} 1/6\,\mathrm{d}x_2 = 1/3 & \text{for} \quad -1 < x_1 < 2, \\ 0 & \text{otherwise.} \end{cases} \tag{3.230}$$

$$f_{X_2}(x_2) = \begin{cases} \int_{-1}^{2} 1/6\,\mathrm{d}x_1 = 1/2 & \text{for} \quad -1 < x_2 < 1, \\ 0 & \text{otherwise.} \end{cases} \tag{3.231}$$

The marginal densities are shown in Fig. 3.22.
The desired probability is:

$$P\{0 < x_1 < 2, 0 < x_2 < 1\} = \int_0^2 \int_0^1 1/6\,\mathrm{d}x_1\,\mathrm{d}x_2 = 1/3. \tag{3.232}$$

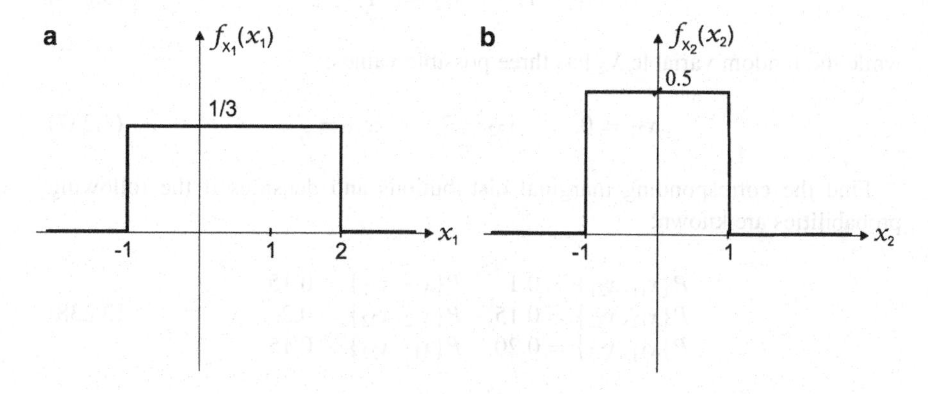

Fig. 3.22 Marginal densities in Exercise 3.2

The random variables are independent because the joint PDF can be presented as a product of the corresponding marginal PDFs.

Exercise 3.3 Find the conditional density $f_{X_1}(x_1|x_2)$ for the joint variables from Exercise 3.1 and find whether the variables are dependent.

Answer From (3.50), (3.223), and (3.227) we have:

$$\begin{aligned} f_{X_1}(x_1|X_2 = x_2) &= \frac{f_{X_1X_2}(x_1,x_2)}{f_{X_2}(x_2)} \\ &= \begin{cases} \frac{1}{2\sqrt{4-x_2^2}} & \text{for} \quad |x_1| < \sqrt{4-x_2^2}, |x_2| < 2, \\ 0 & \text{otherwise.} \end{cases} \end{aligned} \tag{3.233}$$

The conditional density (3.233) is different from $f_{X_1}(x_1)$, given in (3.226), thus confirming that the variables are dependent.

Exercise 3.4 Find the conditional density $f_{X_1}(x_1|X_2 = x_2)$ for the joint density from Exercise 3.2 and find whether the variables are independent.

Answer From (3.50), (3.223), and (3.227), we have:

$$f_{X_1}(x_1|X_2 = x_2) = \frac{f_{X_1X_2}(x_1,x_2)}{f_{X_2}(x_2)} = \begin{cases} \frac{1/6}{1/2} = 1/3 & \text{for} \quad -1 < x_1 < 2, \\ 0 & \text{otherwise.} \end{cases} \tag{3.234}$$

The result (3.234) shows that

$$f_{X_1|X_2}(x_1|x_2) = f_{X_1}(x_1), \tag{3.235}$$

thus confirming that the variables X_1 and X_2 are independent.

Exercise 3.5 In a two-dimensional discrete random variable (X_1 X_2), the random variable X_1 has two possible values:

$$x_{11} = 1, \qquad x_{12} = -1, \tag{3.236}$$

while the random variable X_2 has three possible values:

$$x_{21} = 0, \qquad x_{22} = 2, \qquad x_{23} = 5. \tag{3.237}$$

Find the corresponding marginal distributions and densities if the following probabilities are known:

$$\begin{aligned} &P\{x_{11}, x_{21}\} = 0.1, \quad P\{x_{12}, x_{21}\} = 0.15. \\ &P\{x_{11}, x_{22}\} = 0.15, \quad P\{x_{12}, x_{22}\} = 0.25. \\ &P\{x_{11}, x_{23}\} = 0.20, \quad P\{x_{12}, x_{22}\} = 0.15. \end{aligned} \tag{3.238}$$

Answer From (3.238), we have:

$$P\{X_1 = x_{11}\} = \sum_{j=1}^{3} P\{X_1 = x_{11}, X_2 = x_{2j}\} = 0.1 + 0.15 + 0.2 = 0.45.$$

$$P\{X_1 = x_{12}\} = \sum_{j=1}^{3} P\{X_1 = x_{12}, X_2 = x_{2j}\} = 0.15 + 0.25 + 0.15 = 0.55. \quad (3.239)$$

The marginal distribution and density are, respectively:

$$F_{X_1}(x_1) = 0.45u(x_1 - 1) + 0.55u(x_1 + 1), \quad (3.240)$$

$$f_{X_1}(x_1) = 0.45\delta(x_1 - 1) + 0.55\delta(x_1 + 1). \quad (3.241)$$

Similarly,

$$P\{X_2 = x_{21}\} = \sum_{j=1}^{2} P\{X_1 = x_{1j}, X_2 = x_{21}\} = 0.1 + 0.15 = 0.25,$$

$$P\{X_2 = x_{22}\} = \sum_{j=1}^{2} P\{X_1 = x_{1j}, X_2 = x_{22}\} = 0.15 + 0.25 = 0.4,$$

$$P\{X_2 = x_{23}\} = \sum_{j=1}^{2} P\{X_1 = x_{1j}, X_2 = x_{23}\} = 0.2 + 0.15 = 0.4, \quad (3.242)$$

$$F_{X_2}(x_2) = 0.25u(x_2) + 0.4u(x_2 - 2) + 0.35u(x_2 - 5). \quad (3.243)$$

$$f_{X_2}(x_2) = 0.25\delta(x_2) + 0.4\delta(x_2 - 2) + 0.35\delta(x_2 - 5). \quad (3.244)$$

Exercise 3.6 Find the conditional distributions,

$$F_{X_1}(x_1|X_2 = x_{21} = 0), \quad \text{and} \quad F_{X_2}(x_2|X_1 = x_{11} = 1), \quad (3.245)$$

for the random variables from Exercise 3.5.

Answer From (3.46), we have:

$$F_{X_1}(x_1|X_2 = x_{21} = 0) = \frac{\sum_{i=1}^{2} P\{X_1 = x_{1i}, X_2 = x_{21} = 0\}u(x_1 - x_{1i})}{P\{X_2 = x_{21} = 0\}}. \quad (3.246)$$

From (3.238), we have:

$$P\{X_1 = 1, X_2 = 0\} = 0.1, \qquad P\{X_1 = -1, X_2 = 0\} = 0.15. \quad (3.247)$$

From (3.242), it follows:

$$P\{X_2 = 0\} = 0.25. \tag{3.248}$$

Finally, from (3.246)–(3.248) we get the desired conditional density:

$$F_{X_1}(x_1|X_2 = x_{21} = 0) = 0.1/0.25u(x_1 - 1) + 0.15/0.25u(x_1 + 1). \tag{3.249}$$

Similarly, we have:

$$\begin{aligned} F_{X_2}(x_2|X_1 = x_{11} = 1) = 0.1/0.45u(x_2) + 0.15/0.45u(x_2 - 2) \\ + 0.2/0.45u(x_2 - 5). \end{aligned} \tag{3.250}$$

Exercise 3.7 Two-dimensional random variable (X_1, X_2), has a uniform joint PDF in the area A, as shown in Fig. 3.23. Find the marginal PDFs.

Answer Knowing that the area $A = 1/2$, the corresponding joint PDF is:

$$f_{X_1X_2}(x_1, x_2) = \begin{cases} 2 & \text{for} \quad (x_1, x_2) \in A, \\ 0 & \quad \text{otherwise.} \end{cases} \tag{3.251}$$

The marginal density is:

$$f_{X_1}(x_1) = \int_{-\infty}^{\infty} f_{X_1X_2}(x_1, x_2)\mathrm{d}x_2. \tag{3.252}$$

From Fig. 3.23, we see that x_2 changes from 0 to x_1, where x_1 is in the interval [0, 1]:

$$f_{X_1}(x_1) = \int_{-\infty}^{\infty} f_{X_1X_2}(x_1, x_2)\mathrm{d}x_2 = \int_{0}^{x_1} 2\mathrm{d}x_2 = 2x_1. \tag{3.253}$$

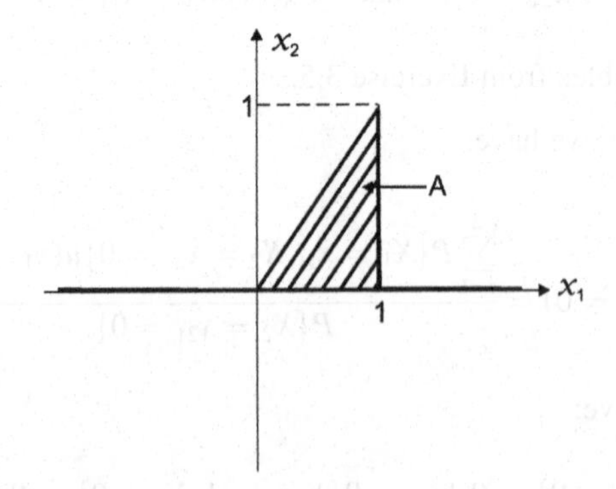

Fig. 3.23 Range in Exercise 3.7

Therefore,

$$f_{X_1}(x_1) = \begin{cases} 2x_1 & \text{for } 0 < x_1 < 1, \\ 0 & \text{otherwise.} \end{cases} \tag{3.254}$$

Similarly, we have:

$$f_{X_2}(x_2) = \int_{-\infty}^{\infty} f_{X_1X_2}(x_1, x_2)\mathrm{d}x_1 = \int_{0}^{x_2} 2\mathrm{d}x_1 = 2x_2 \tag{3.255}$$

and

$$f_{X_2}(x_2) = \begin{cases} 2x_2 & \text{for } 0 < x_2 < 1, \\ 0 & \text{otherwise.} \end{cases} \tag{3.256}$$

Exercise 3.8 The random variables X_1 and X_2 are related as:

$$X_2 = -2X_1 + 5. \tag{3.257}$$

Determine whether or not the random variables are orthogonal and correlated if,

$$E\{X_1\} = 2, \quad \text{and} \quad \sigma^2_{X_1} = 1. \tag{3.258}$$

Answer The random variables are orthogonal if

$$E\{X_1X_2\} = R_{X_1X_2} = 0. \tag{3.259}$$

From (3.257), we have:

$$R_{X_1X_2} = E\{(-2X_1 + 5)X_1\} = -2E\{X_1^2\} + 5E\{X_1\}. \tag{3.260}$$

From (3.258), it follows:

$$\sigma^2_{X_1} = 1 = E\{X_1^2\} - E\{X_1\}^2 = E\{X_1^2\} - 4 \tag{3.261}$$

and

$$E\{X_1^2\} = 5. \tag{3.262}$$

Finally,

$$R_{X_1X_2} = -2 \times 5 + 5 \times 2 = 0. \tag{3.263}$$

This result confirms that the variables X_1 and X_2 are orthogonal.

From (3.257), we can see that there is a linear relation between the random variables X_1 and X_2. It results in a maximum correlation factor. Due to the coefficient -2, the correlation is negative and the correlation factor is equal to -1. This is confirmed in the following numerical calculation.

From (3.130), the correlation coefficient is:

$$\begin{aligned}\rho_{X_1X_2} &= \frac{C_{X_1X_2}}{\sigma_{X_1}\sigma_{X_2}} = \frac{\overline{(X_1-2)(X_2-\overline{X_2})}}{\sigma_{X_1}\sigma_{X_2}} \\ &= \frac{\overline{X_1X_2} - 2\overline{X_2} - \overline{X_1}\,\overline{X_2} + 2\overline{X_2}}{\sigma_{X_1}\sigma_{X_2}} = \frac{\overline{X_1X_2} - \overline{X_1}\,\overline{X_2}}{\sigma_{X_1}\sigma_{X_2}}.\end{aligned} \tag{3.264}$$

From (3.259) and (3.263), we have:

$$\overline{X_1X_2} = 0. \tag{3.265}$$

From (3.257), we find the mean value and the variance of the variable X_2:

$$\begin{aligned}\overline{X_2} &= -2\overline{X_1} + 5 = 1, \\ \sigma^2_{X_2} &= 4\sigma^2_{X_1} = 4.\end{aligned} \tag{3.266}$$

Finally, placing (3.265) and (3.266) into (3.264), we arrive at:

$$\rho_{X_1X_2} = \frac{-2\times 1}{1\times 2} = -1. \tag{3.267}$$

Exercise 3.9 The random variable X is uniform in the interval $[0, 2\pi]$. Show that the random variables Y and Z are dependent and uncorrelated, if

$$Y = \sin X, \qquad Z = \cos X. \tag{3.268}$$

Answer From (3.268), we have: $Y^2 + Z^2 = 1.$ (3.269)

This indicates that the variables are dependent. We also notice that the dependence is squared, that is, it does not contain any degree of linear relation. Therefore, the correlation is zero and the random variables are uncorrelated. This is confirmed in the following calculation.

The mean values and the autocorrelation are:

$$\begin{aligned}\overline{Y} &= \frac{1}{2\pi}\int_0^{2\pi} \sin x\,\mathrm{d}x = 0, \\ \overline{Z} &= \frac{1}{2\pi}\int_0^{2\pi} \cos x\,\mathrm{d}x = 0, \\ \overline{YZ} &= \frac{1}{2\pi}\int_0^{2\pi} \sin x\cos x\,\mathrm{d}x = \frac{1}{4\pi}\int_0^{2\pi} \sin 2x\,\mathrm{d}x = 0.\end{aligned} \tag{3.270}$$

Using (3.270) in the middle, the covariance is:

$$C_{YZ} = \overline{(Y - \overline{Y})(Z - \overline{Z})} = \overline{YZ} = 0. \tag{3.271}$$

From here, it follows that the random variables are uncorrelated.

Exercise 3.10 The discrete random variables X_1 and X_2 have the possible values: -1, 0, 1. Find the mean values for the variables X_1 and X_2, if it is known:

$$\begin{aligned}
&P\{X_1=1,X_2=1\}=0, && P\{X_1=1,X_2=0\}=1/15, && P\{X_1=1,X_2=-1\}=4/15,\\
&P\{X_1=0,X_2=1\}=2/15, && P\{X_1=0,X_2=0\}=2/15, && P\{X_1=0,X_2=-1\}=1/15,\\
&P\{X_1=-1,X_2=1\}=0, && P\{X_1=-1,X_2=0\}=1/15, && P\{X_1=-1,X_2=-1\}=4/15.
\end{aligned} \tag{3.272}$$

Answer In order to find the mean values for X_1 and X_2, we need the corresponding probabilities $P(X_{1i})$ and $P(X_{2j})$, $i = 1, \ldots, 3$; $j = 1, \ldots, 3$.

From (1.67), we have:

$$\begin{aligned}
P(X_1 = 1) = P\{X_1 = 1, X_2 = 1\} + P\{X_1 = 1, X_2 = 0\}&\\
+ P\{X_1 = 1, X_2 = -1\} = 1/15 + 4/15 = 5/15 = 1/3.&
\end{aligned} \tag{3.273}$$

$$\begin{aligned}
P(X_1 = 0) &= P\{X_1 = 0, X_2 = 1\} + P\{X_1 = 0, X_2 = 0\} + P\{X_1 = 0, X_2 = -1\}\\
&= 2/15 + 2/15 + 1/15 = 5/15 = 1/3.
\end{aligned} \tag{3.274}$$

$$\begin{aligned}
P(X_1=-1) &= P\{X_1=-1,X_2=1\} + P\{X_1=-1,X_2=0 + P\{X_1=-1,X_2=-1\}\}\\
&= 1/15+4/15=5/15=1/3.
\end{aligned} \tag{3.275}$$

$$\begin{aligned}
P(X_2 = 1) = P\{X_2 = 1, X_1 = 1\} + P\{X_2 = 1, X_1 = 0\}&\\
+P\{X_2 = 1, X_1 = -1\} = 2/15.&
\end{aligned} \tag{3.276}$$

$$\begin{aligned}
P(X_2 = 0) &= P\{X_2 = 0, X_1 = 1\} + P\{X_2 = 0, X_1 = 0\} + P\{X_2 = 0, X_1 = -1\}\\
&= 1/15 + 2/15 + 1/15 = 4/15.
\end{aligned} \tag{3.277}$$

$$\begin{aligned}
P(X_2 = -1) = P\{X_2 = -1, X_1 = 1\} + P\{X_2 = -1, X_1 = 0\}&\\
+ P\{X_2 = -1, X_1 = -1\} = 4/15 + 1/15 + 4/15 = 9/15.&
\end{aligned} \tag{3.278}$$

From (3.273) to (3.275), we get:

$$E\{X_1\} = 1\,P\{X_1=1\} + 0\,P\{X_1=0\} + (-1)\,P\{X_1=-1\} = 1/3 - 1/3 = 0. \tag{3.279}$$

Similarly from (3.276) to (3.278), we obtain:

$$E\{X_2\} = 1\,P\{X_2 = 1\} + 0\,P\{X_2 = 0\} + (-1)\,P\{X_2 = -1\}$$
$$= 2/15 - 9/15 = -7/15. \quad (3.280)$$

Exercise 3.11 The discrete random variable X_1 has 0 and 1 as its discrete values, whereas X_2 has 0, 1, and -1 as its discrete values. Find the mean value and the variance of the random variable X, if

$$X = 2X_1 + X_2^2. \quad (3.281)$$

The corresponding probabilities are:

$$P\{X_1=0,X_2=0\}=0.2, \quad P\{X_1=0,X_2=1\}=0, \quad P\{X_1=0,X_2=-1\}=0.1,$$
$$P\{X_1=1,X_2=0\}=0.3, \quad P\{X_1=1,X_2=1\}=0.2, \quad P\{X_1=1,X_2=-1\}=0.2. \quad (3.282)$$

Answer The variance of the random variable X is:

$$\sigma_X^2 = \overline{X^2} - \overline{X}^2 = \overline{(2X_1+X_2^2)^2} - \overline{(2X_1+X_2^2)}^2. \quad (3.283)$$

From (3.281), we get the mean value of the variable X:

$$\overline{X} = \overline{2X_1+X_2^2} = \sum_{i=1}^{2}\sum_{j=1}^{3}(2x_{1i}+x_{2j}^2)P\{X_1=x_{1i},X_2=x_{2j}\}$$
$$= 0.1 + 2\times 0.3 + (2+1)\times 0.2 + (2+1)\times 0.2 = 0.1+0.6+0.6+0.6 = 1.9. \quad (3.284)$$

Similarly, the mean squared value of X is:

$$\overline{X^2} = \overline{(2X_1+X_2^2)^2} = \sum_{i=1}^{2}\sum_{j=1}^{3}(2x_{1i}+x_{2j}^2)^2P\{X_1=x_{1i},X_2=x_{2j}\}$$
$$= 0.1 + 1.2 + 1.8 + 1.8 = 4.9. \quad (3.285)$$

Placing (3.284) and (3.285) into (3.283), we arrive at:

$$\sigma_X^2 = \overline{X^2} - \overline{X}^2 = 4.9 - 1.9^2 = 1.29. \quad (3.286)$$

Exercise 3.12 The random variables X_1 and X_2 are independent and have the density functions

$$f_{X_1}(x_1) = \mathrm{e}^{-x_1}u(x_1),$$
$$f_{X_2}(x_2) = \mathrm{e}^{-x_2}u(x_2). \tag{3.287}$$

Determine whether or not the random variables

$$Y_1 = X_1 + X_2 \tag{3.288}$$

and

$$Y_2 = \frac{X_1}{X_1 + X_2} \tag{3.289}$$

are independent.

Answer From (3.288) to (3.289), we write:

$$y_1 = x_1 + x_2,$$
$$y_2 = \frac{x_1}{x_1 + x_2}. \tag{3.290}$$

The random variables X_1 and X_2 are independent, and their joint PDF from (3.287) is:

$$f_{X_1X_2}(x_1, x_2) = f_{X_1}(x_1)f_{X_2}(x_2) = \begin{cases} \mathrm{e}^{-(x_1+x_2)} & \text{for} \quad x_1 \geq 0, x_2 \geq 0, \\ 0 & \text{otherwise.} \end{cases} \tag{3.291}$$

Using (3.289), we can present the joint density (3.291) in the following form:

$$f_{X_1X_2}(y_1, y_2) = \begin{cases} \mathrm{e}^{-y_1} & \text{for} \quad y_1 \geq 0, 0 \leq y_2 \leq 1, \\ 0 & \text{otherwise.} \end{cases} \tag{3.292}$$

The Jacobian of the transformation (3.290) is:

$$J = \begin{vmatrix} 1 & 1 \\ \frac{x_2}{(x_1+x_2)^2} & \frac{-x_1}{(x_1+x_2)^2} \end{vmatrix} = -\frac{1}{x_1 + x_2} = -\frac{1}{y_1}. \tag{3.293}$$

The joint density of Y_1 and Y_2 is obtained from (3.159), (3.291), and (3.293) as:

$$f_{Y_1Y_2}(y_1, y_2) = \frac{f_{X_1X_2}(y_1, y_2)}{|J(y_1, y_2)|} = \frac{\mathrm{e}^{-y_1}}{\left|-\frac{1}{y_1}\right|}$$
$$= \begin{cases} y_1\,\mathrm{e}^{-y_1} & \text{for} \quad y_1 \geq 0, 0 \leq y_2 \leq 1, \\ 0 & \text{otherwise.} \end{cases} \tag{3.294}$$

The joint density (3.294) can be rewritten as:

$$f_{Y_1Y_2}(y_1, y_2) = f_{Y_1}(y_1)f_{Y_2}(y_2), \tag{3.295}$$

where

$$\begin{aligned} f_{Y_1}(y_1) &= y_1\,\mathrm{e}^{-y_1}, \quad &&\text{for } y_1 \geq 0, \\ f_{Y_2}(y_2) &= 1, &&\text{for } 0 \leq y_2 \leq 1. \end{aligned} \tag{3.296}$$

Therefore, from (3.295) it follows that the random variables Y_1 and Y_2 are also independent.

Exercise 3.13 The random variables X_1 and X_2 are independent and have the following density functions:

$$f_{X_1}(x_1) = \begin{cases} \mathrm{e}^{-x_1} & \text{for} \quad x_1 > 0 \\ 0 & \text{otherwise} \end{cases}, \quad f_{X_2}(x_2) = \begin{cases} \mathrm{e}^{-x_2} & \text{for} \quad x_2 > 0 \\ 0 & \text{otherwise} \end{cases}. \tag{3.297}$$

Find the PDF of the random variable X, if

$$X = \frac{X_1}{X_2}. \tag{3.298}$$

Solution We have two input random variables and one output variable. In order to apply the expression (3.159) we must first introduce the auxiliary variable Y,

$$Y = X_1. \tag{3.299}$$

Now we have the transformation defined in the following two equations:

$$\begin{aligned} x &= \frac{x_1}{x_2}, \\ y &= x_1. \end{aligned} \tag{3.300}$$

This system of equations (3.300) has one solution:

$$x_{11} = y, \quad \text{and} \quad x_{21} = y/x. \tag{3.301}$$

The Jacobian of the transformation (3.300) is:

$$J = \begin{vmatrix} \frac{1}{x_2} & -\frac{x_1}{x_2^2} \\ 1 & 0 \end{vmatrix} = \frac{x_1}{x_2^2} = \frac{y}{y^2/x^2} = \frac{x^2}{y}. \tag{3.302}$$

Using (3.159), the joint density of the variables X and Y is:

$$f_{XY}(x,y) = \frac{f_{X_1X_2}(x,y)}{|J(x,y)|} = \frac{y}{x^2}\mathrm{e}^{-y}\,\mathrm{e}^{-y/x}, \quad y > 0, \quad x > 0. \tag{3.303}$$

The desired density is obtained from (3.42a):

$$f_X(x) = \int_0^{\infty} f_{XY}(x,y)\mathrm{d}y = \int_0^{\infty} \frac{y}{x^2}\mathrm{e}^{-y(1+1/x)}\mathrm{d}y = \frac{1}{x^2}\int_0^{\infty} y\,\mathrm{e}^{-ay}\,\mathrm{d}y, \tag{3.304}$$

where

$$a = 1 + 1/x. \tag{3.305}$$

Using integral 1 from Appendix A and from (3.304) and (3.305), we get:

$$f_X(x) = \frac{1}{x^2}\frac{x^2}{(x+1)^2} = \frac{1}{(x+1)^2};\; x \geq 0 \tag{3.306}$$

Exercise 3.14 Find the PDF of the random variable X,

$$X = X_1 + a\cos X_2, \tag{3.307}$$

if the joint density of the variables X_1 and X_2 is known.

Answer After introducing the auxiliary random variable $Y = X_2$, we have the following equations:

$$\begin{aligned} x &= x_1 + a\cos x_2, \\ y &= x_2. \end{aligned} \tag{3.308}$$

The Jacobian of the transformation (3.308) is:

$$J = \begin{vmatrix} 1 & -a\sin x_2 \\ 0 & 1 \end{vmatrix} = 1. \tag{3.309}$$

The joint density of the variables X and Y is:

$$f_{XY}(x,y) = f_{X_1X_2}(x_1,x_2) = f_{X_1X_2}(x - a\cos y, y). \tag{3.310}$$

From here,

$$f_X(x) = \int_{-\infty}^{\infty} f_{X_1X_2}(x - a\cos y, y)\mathrm{d}y. \tag{3.311}$$

Exercise 3.15 The random variables X_1 and X_2 are independent. Find the PDF of the random variable X, if

$$X = \frac{X_1}{X_2}. \tag{3.312}$$

The marginal densities are:

$$f_{X_1}(x_1) = \frac{x_1}{\alpha^2} \mathrm{e}^{-\frac{x_1^2}{2\alpha^2}} u(x_1), \qquad f_{X_2}(x_2) = \frac{x_2}{\beta^2} \mathrm{e}^{-\frac{x_2^2}{2\beta^2}} u(x_2). \tag{3.313}$$

Answer The auxiliary variable $Y = X_2$ is introduced. The corresponding set of the transformation equations

$$\begin{aligned} x &= x_1/x_2, \\ y &= x_2, \end{aligned} \tag{3.314}$$

has the unique solution $x_1 = xy$ and $x_2 = y$. The Jacobian of the transformation (3.314) is

$$J = \begin{vmatrix} \frac{1}{x_2} & -\frac{x_1}{x_2^2} \\ 0 & 1 \end{vmatrix} = \frac{1}{x_2} = \frac{1}{y}. \tag{3.315}$$

The joint density function of the variables X and Y is given as:

$$f_{XY}(x, y) = \frac{f_{X_1X_2}(x, y)}{|J(x, y)|} = \frac{yx_1x_2}{\alpha^2\beta^2} \mathrm{e}^{-\frac{1}{2}\left(\frac{x_1^2}{\alpha^2} + \frac{x_2^2}{\beta^2}\right)} = \frac{xy^3}{\alpha^2\beta^2} \mathrm{e}^{-\frac{y^2}{2}\left(\frac{x^2}{\alpha^2} + \frac{1}{\beta^2}\right)}. \tag{3.316}$$

Denoting

$$a = \frac{1}{2}\left(\frac{x^2}{\alpha^2} + \frac{1}{\beta^2}\right), \tag{3.317}$$

from (3.316) and (3.317), we have:

$$f_X(x) = \int_0^\infty f_{XY}(x, y)\,\mathrm{d}y = \frac{x}{\alpha^2\beta^2} \int_0^\infty y^3 \mathrm{e}^{-ay^2}\,\mathrm{d}y. \tag{3.318}$$

Using the integral 4 from Appendix A and (E.2.15.6) we have:

$$f_X(x) = \frac{x}{\alpha^2\beta^2} \frac{1}{2a^2} = \frac{2\alpha^2}{\beta^2} \frac{x}{\left(x^2 + \frac{\alpha^2}{\beta^2}\right)^2}; \ x \geq 0 \tag{3.319}$$

Exercise 3.16 The random variables X_1 and X_2 are independent and uniform in the intervals [1, 2] and [3, 5], respectively (as shown in Fig. 3.24). Find the density of their sum:

$$X = X_1 + X_2. \tag{3.320}$$

Answer The density of the sum of the independent random variables is equal to the convolution of their densities. This result can be easily obtained graphically, as shown in Fig. 3.25.

Exercise 3.17 The random variable X_1 uniformly takes values around its nominal value 100 in the interval [100 ± 10%]. Similarly, the variable X_2 changes uniformly in the interval [200 ± 10%]. Find the probability that the sum of the variables $X = X_1 + X_2$ is less than 310, if the variables X_1 and X_2 are independent.

Answer The density of the sum is the convolution of the corresponding densities and is obtained graphically, as shown in Fig. 3.26. The desired probability is presented in the shaded area, and is equal to:

$$P\{X < 310\} = 1 - (330 - 310)/(2 \times 40) = 1 - 1/4 = 3/4. \tag{3.321}$$

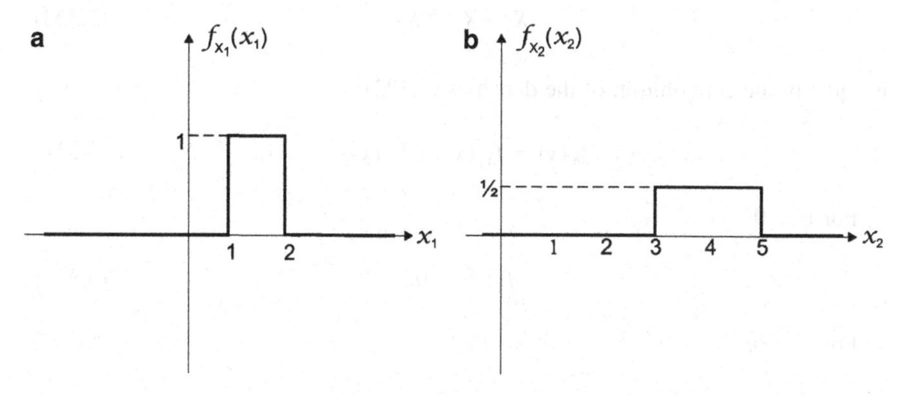

Fig. 3.24 Densities of the random variables X_1 and X_2

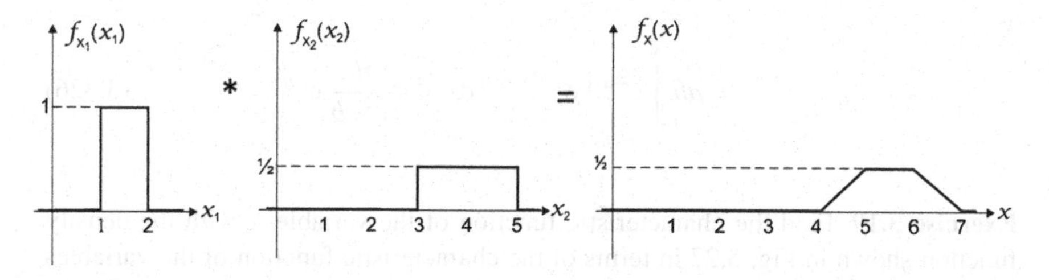

Fig. 3.25 Convolution of densities

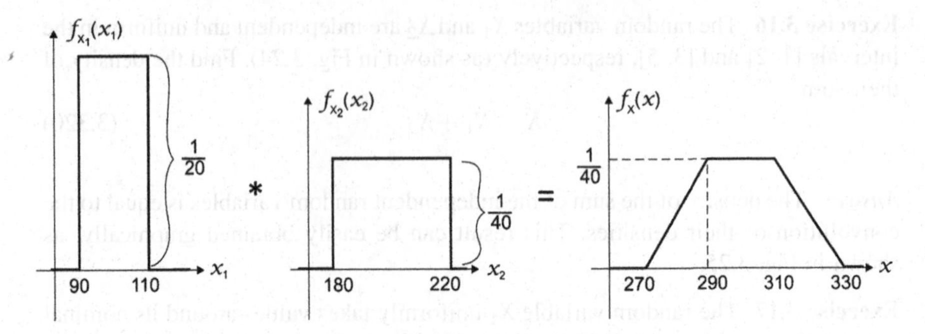

Fig. 3.26 Convolution of the uniform PDFs

Exercise 3.18 The random variables X_1 and X_2 are independent. Find the density of their sum if the corresponding densities are given as:

$$f_{X_1}(x_1) = a\,\mathrm{e}^{-ax_1}u(x_1), \qquad f_{X_2}(x_2) = b\,\mathrm{e}^{-bx_2}u(x_2), \tag{3.322}$$

where a and b are constants.

Solution The density of the sum

$$X = X_1 + X_2 \tag{3.323}$$

is equal to the convolution of the densities (3.322):

$$f_X(x) = f_{X_1}(x_1) * f_{X_2}(x_2). \tag{3.324}$$

For $x < 0$,

$$f_X(x) = 0. \tag{3.325}$$

For $x > 0$,

$$\begin{aligned} f_X(x) &= \int_{-\infty}^{\infty} f_{X_1}(x_1) f_{X_2}(x - x_1)\mathrm{d}x_1 \\ &= ab \int_{x}^{\infty} \mathrm{e}^{-ax_1}\,\mathrm{e}^{-b(x-x_1)}\,\mathrm{d}x_1 = \frac{ab}{a-b}\,\mathrm{e}^{-ax}. \end{aligned} \tag{3.326}$$

Exercise 3.19 Find the characteristic function of the variable X with the density function shown in Fig. 3.27 in terms of the characteristic function of the variables X_1 and X_2. The variable X is the sum of X_1 and X_2, and the variables X_1 and X_2 are independent.

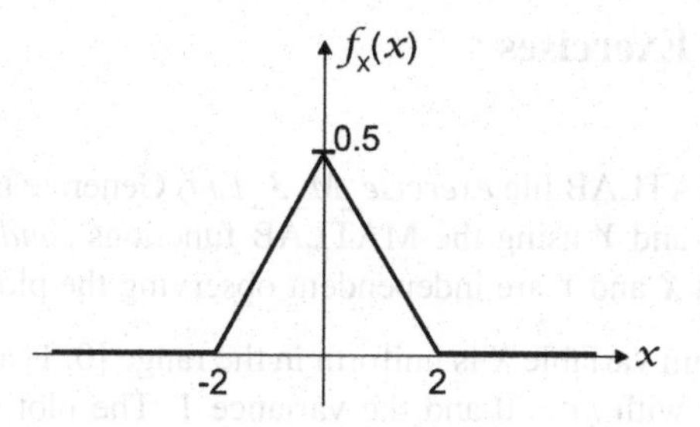

Fig. 3.27 PDF of the variable X

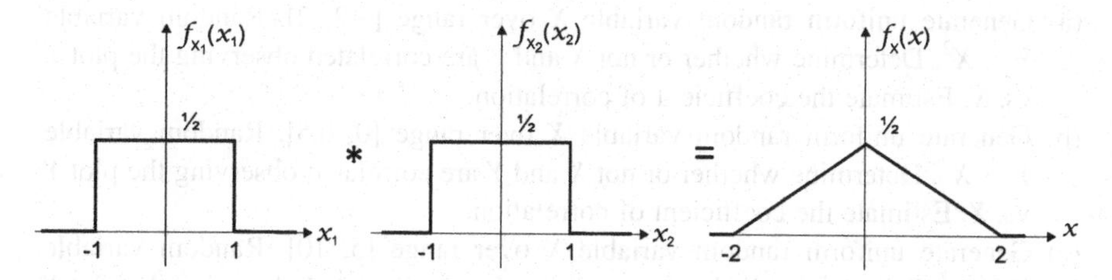

Fig. 3.28 Convolution of PDFs

Answer The characteristic function of the sum of independent variables is equal to the product of their characteristic functions:

$$\phi_X(\omega) = \phi_{X_1}(\omega)\phi_{X_2}(\omega), \tag{3.327}$$

where

$$\phi_{X_1}(\omega) = \int_{x_1} e^{j\omega x_1} f_{X_1}(x_1)\mathrm{d}x_1, \qquad \phi_{X_2}(\omega) = \int_{x_2} e^{j\omega x_2} f_{X_2}(x_2)\mathrm{d}x_2. \tag{3.328}$$

The PDF from Fig. 3.27 is obtained by the convolution of the densities of the variables X_1 and X_2, as shown in Fig. 3.28.

The characteristic functions are:

$$\phi_{X_1}(\omega) = \int_{-1}^{1} \frac{1}{2} e^{j\omega x_1}\,\mathrm{d}x_1 = \phi_{X_2}(\omega) = \int_{-1}^{1} \frac{1}{2} e^{j\omega x_2}\,\mathrm{d}x_2 = \frac{\sin\omega}{\omega}. \tag{3.329}$$

Finally, from (3.327) and (3.329) we have:

$$\phi_X(\omega) = \left(\frac{\sin\omega}{\omega}\right)^2. \tag{3.330}$$

3.7 MATLAB Exercises

Exercise M.3.1 (MATLAB file *exercise_M_3_1.m*) Generate uniform and normal random variables X and Y using the MATLAB functions *rand* and *randn*. Determine whether or not X and Y are independent observing the plot Y vs. X.

Solution The random variable X is uniform in the range [0, 1] and the variable Y is a Gaussian variable with $m = 0$ and the variance 1. The plot shown in Fig. 3.29 indicates that the variables are independent.

Exercise M.3.2 (MATLAB file *exercise_M_3_2.m*)

(a) Generate uniform random variable X over range $[-2, 2]$. Random variable $Y = X^2$. Determine whether or not X and Y are correlated observing the plot Y vs. X. Estimate the coefficient of correlation.
(b) Generate uniform random variable X over range [0, 0.5]. Random variable $Y = X^2$. Determine whether or not X and Y are correlated observing the plot Y vs. X. Estimate the coefficient of correlation.
(c) Generate uniform random variable X over range [5, 10]. Random variable $Y = X^2$. Determine whether or not X and Y are correlated observing the plot Y vs. X. Estimate the coefficient of correlation.
(d) Generate uniform random variable X over range $[-10,-5]$. Random variable $Y = X^2$. Determine whether or not X and Y are correlated observing the plot Y vs. X. Estimate the coefficient of correlation.

Solution

(a) The plot in Fig. 3.30a indicates that the variables are dependent but uncorrelated.

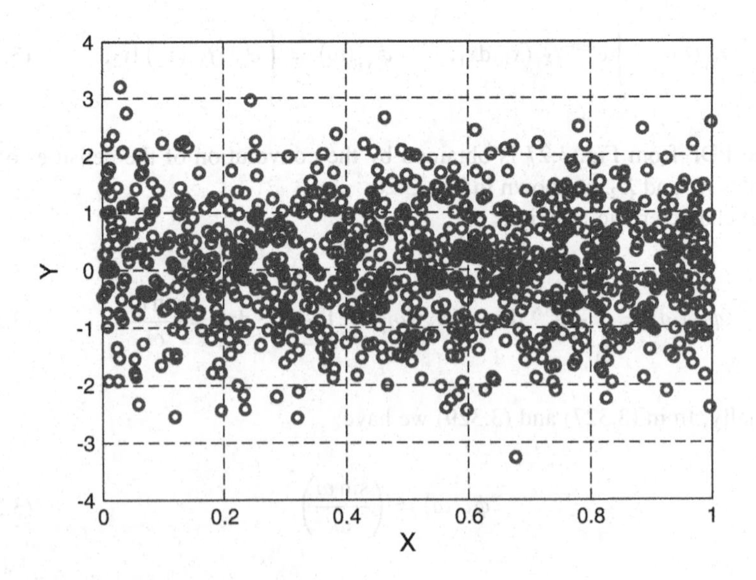

Fig. 3.29 Independent random variables

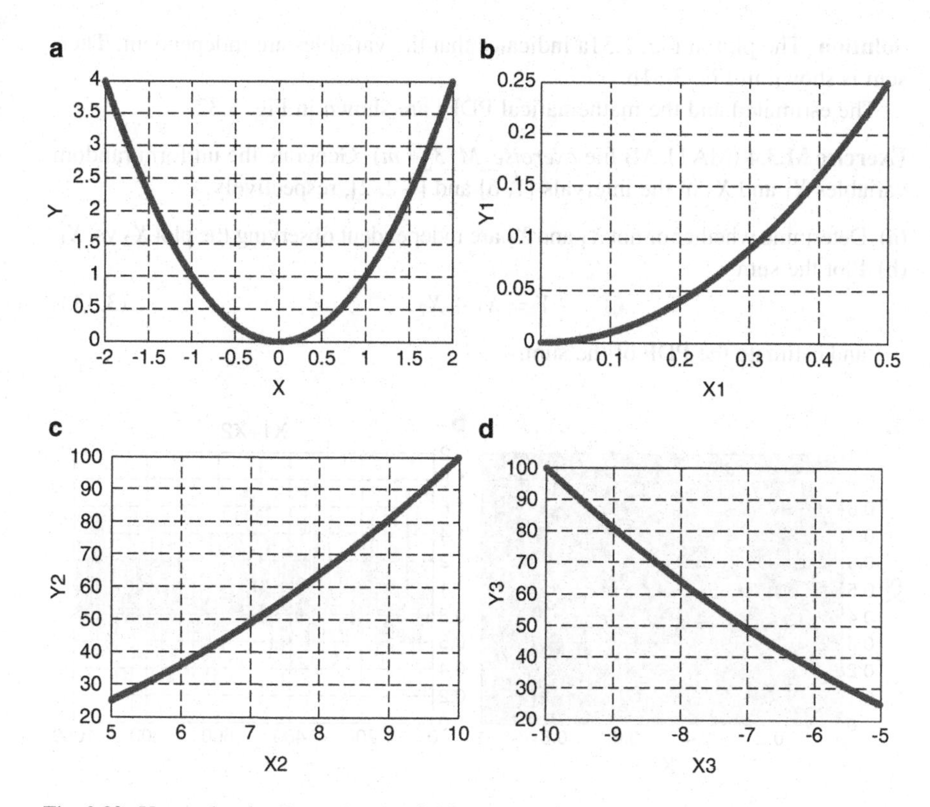

Fig. 3.30 Uncorrelated and correlated variables

The coefficient of correlation is estimated using the MATLAB file *corrcoef.m*. The estimated coefficient of correlation is −0.0144.

(b) The variables are correlated as shown in Fig. 3.30b. The estimated coefficient of correlation is 0.9679.

(c) The variables are correlated as shown in Fig. 3.30c. The estimated coefficient of correlation is 0.9964.

(d) The variables are negative correlated as shown in Fig. 3.30d. The estimated coefficient of correlation is −0.9964.

Exercise M.3.3 (MATLAB file *exercise_M_3_3.m*) Generate uniform random variables X_1 and X_2 using *rand.m*.

(a) Determine whether or not X_1 and X_2 are independent observing the plot X_2 vs. X_1.
(b) Plot the sum

$$Y = X_1 + X_2. \tag{3.331}$$

and estimate the PDF of the sum.

(c) Find the mathematical PDF of Y as the convolution of the corresponding PDFs,

$$f_Y(y) = f_{X_1}(x_1) * f_{X_2}(x_2). \tag{3.332}$$

Solution The plot in Fig. 3.31a indicates that the variables are independent. Their sum is shown in Fig. 3.31b.

The estimated and the mathematical PDFs are shown in Fig. 3.32.

Exercise M.3.4 (MATLAB file *exercise_M_3_4.m*) Generate the uniform random variables X_1 and X_2 in the intervals [1, 6] and [−2, 2], respectively.

(a) Determine whether or not X_1 and X_2 are independent observing the plot X_2 vs. X_1.
(b) Plot the sum

$$Y = X_1 + X_2. \tag{3.333}$$

and estimate the PDF of the sum.

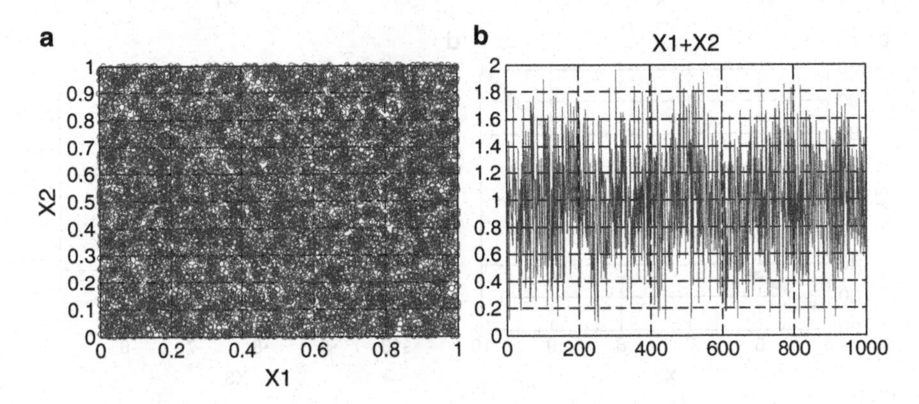

Fig. 3.31 Independent variables and their sum in Exercise M.3.3

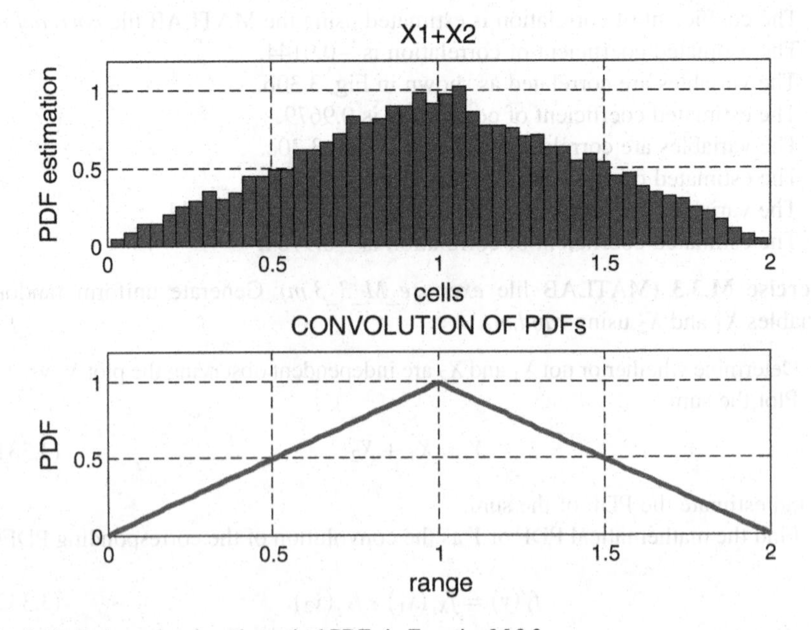

Fig. 3.32 Estimated and mathematical PDFs in Exercise M.3.3

(c) Find the mathematical PDF of Y as the convolution of the corresponding PDFs,

$$f_Y(y) = f_{X_1}(x_1) * f_{X_2}(x_2). \tag{3.334}$$

Solution The plot in Fig. 3.33a indicates that the variables are independent. Their sum is shown in Fig. 3.32b.

The estimated and the mathematical PDFs are shown in Fig. 3.34.

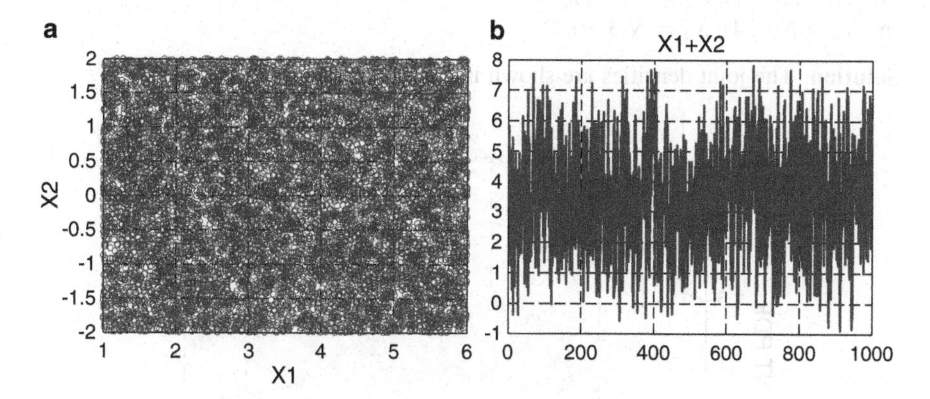

Fig. 3.33 Independent variables and their sum in Exercise M.3.4

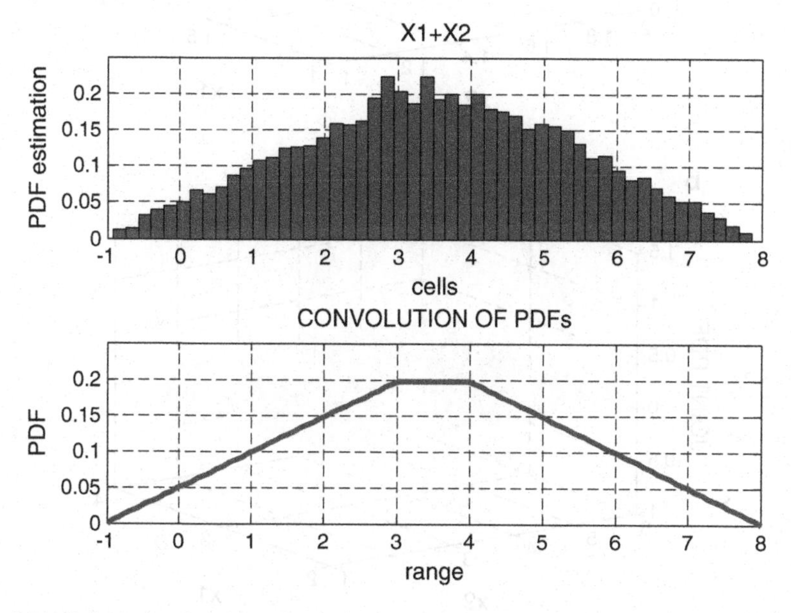

Fig. 3.34 Estimated and mathematical PDFs in Exercise M.3.4

Exercise M.3.5 (MATLAB file *exercise_M_3_5.m*) Plot the joint density for independent uniform random variables X_1 and X_2:

(a) X_1 and X_2 are in the intervals [1, 2] and [1, 2], respectively.
(b) X_1 and X_2 are in the intervals [2, 4] and [1, 6], respectively.

Solution The joint densities are shown in Fig. 3.35.

Exercise M.3.6 (MATLAB file *exercise_M_3_6.m*) Plot the joint density for independent normal random variables X_1 and X_2:

(a) $X_1 = N(0, 1)$; $X_2 = N(0, 1)$.
(b) $X_1 = N(4, 4)$; $X_2 = N(3, 9)$.

Solution The joint densities are shown in Fig. 3.36.

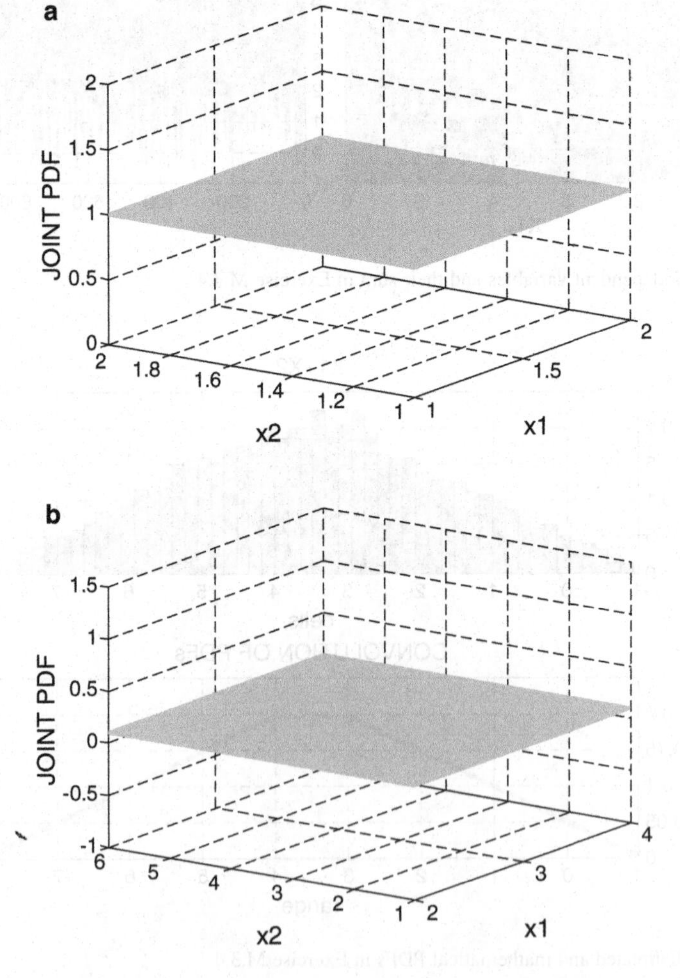

Fig. 3.35 Joint PDFs (**a**) $1 \leq X_1 \leq 2$; $1 \leq X_2 \leq 2$ (**b**) $2 \leq X_1 \leq 4$; $1 \leq X_2 \leq 6$

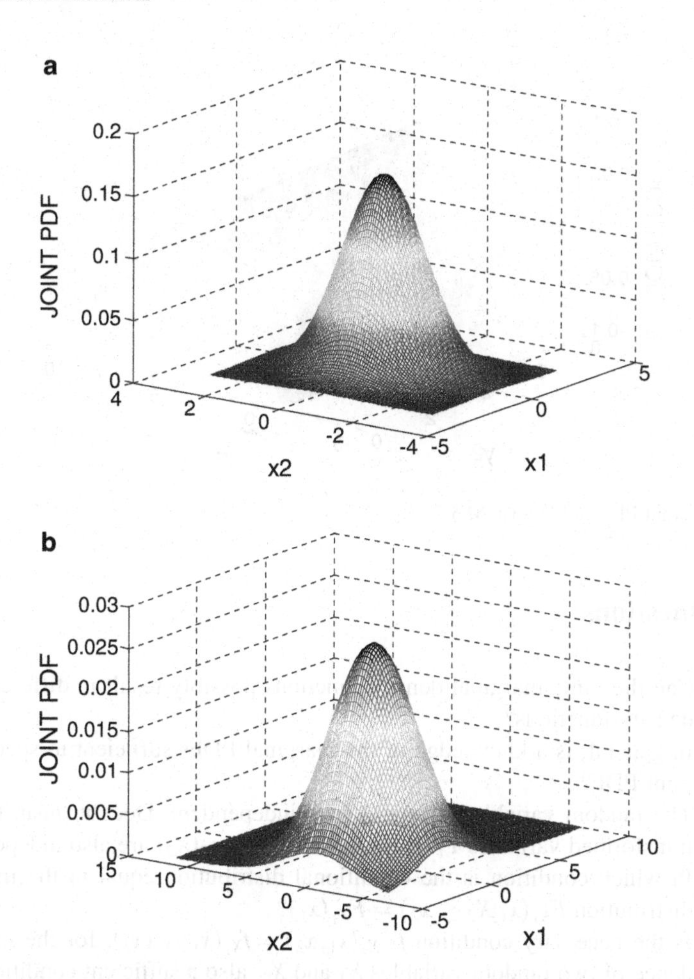

Fig. 3.36 Joint PDFs (**a**) $X_1 = N(0, 1)$; $X_2 = N(0, 1)$ (**b**) $X_1 = N(4, 4)$; $X_2 = N(3, 9)$

Exercise M.3.7 (MATLAB file *exercise_M_3_7.m*) Plot the joint density for the variables Y_1 and Y_2 from Example 3.4.1:

$$f_{Y_1Y_2}(y_1, y_2) = \begin{cases} \frac{y_1}{2\pi\sigma^2} e^{-\frac{y_1^2}{2\sigma^2}} & \text{for} \quad y_1 \geq 0; 0 \leq y_2 \leq 2\pi, . \\ 0 & \text{otherwise.} \end{cases} \tag{3.335}$$

Solution The joint density is shown in Fig. 3.37.

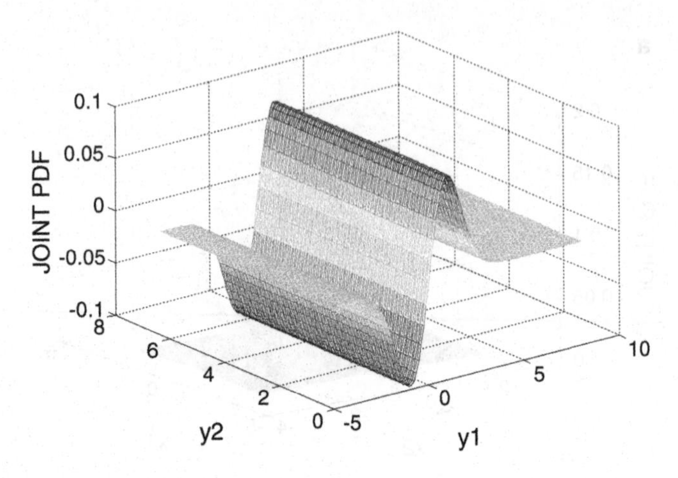

Fig. 3.37 Joint PDF in Exercise M.3.7

3.8 Questions

Q.3.1. Can the same marginal density functions possibly result in different joint density functions?

Q.3.2. In general, is a knowledge of the marginal PDFs sufficient to specify the joint PDF?

Q.3.3. The random variables X_1 and X_2 are independent. Does it mean that the transformed variables $Y_1 = g_1(X_1)$ and $Y_2 = g_2(X_2)$ are also independent?

Q.3.4. In which condition is the conditional distribution equal to the marginal distribution $F_{X_1}(x_1|X_2 \leq x_2) = F_{X_1}(x_1)$?

Q.3.5. Is the necessary condition $f_{X_1X_2}(x_1, x_2) = f_{X_1}(x_1)f_{X_2}(x_2)$, for the independence of two random variables X_1 and X_2, also a sufficient condition?

Q.3.6. Is the following true?

$$P\{a < X_1 \leq b, c < X_2 \leq d\} \neq F_{X_1X_2}(b, d) - F_{X_1X_2}(a, c). \tag{3.336}$$

Q.3.7. Is that the following probability true?

$$P\{x_1 < X_1 \leq x_1 + \mathrm{d}x_1, x_2 < X_2 \leq x_2 + \mathrm{d}x_2\} = f_{X_1X_2}(x_1, x_2)\mathrm{d}x_1\,\mathrm{d}x_2. \tag{3.337}$$

Q.3.8. For three random variables X_1, X_2, and X_3 we have:

$$f_{X_1X_2}(x_1, x_2) = f_{X_1}(x_1)f_{X_2}(x_2), \tag{3.338}$$

$$f_{X_1X_3}(x_1, x_3) = f_{X_1}(x_1)f_{X_3}(x_3), \tag{3.339}$$

$$f_{X_2X_3}(x_2,x_3) = f_{X_2}(x_2)f_{X_3}(x_3). \quad (3.340)$$

Are the random variables X_1, X_2, and X_3 independent?

Q.3.9. Is it possible for joint random variables to be of different types? For example, one discrete random variable and one continuous random variable?

Q.3.10. When is the correlation equal to the covariance?

Q.3.11. Is the coefficient of correlation zero, if random variables X and Y are related as in $Y = X^2$?

Q.3.12. Why is the covariance not zero if the variables are correlated?

3.9 Answers

A.3.1. Yes. As illustrated in the following example, two discrete random variables X_1 and X_2 have the following possible values:

$$(X_1 = 1, X_2 = 1), (X_1 = 1, X_2 = 0), (X_1 = 0, X_2 = 1), (X_1 = 0, X_2 = 0). \quad (3.341)$$

We consider two cases:

$$\begin{aligned} \text{(a)}\quad & P\{X_1 = 1, X_2 = 1\} = P\{X_1 = 0, X_2 = 0\} = 1/8, \\ & P\{X_1 = 1, X_2 = 0\} = P\{X_1 = 0, X_2 = 1\} = 3/8. \end{aligned} \quad (3.342)$$

$$\begin{aligned} \text{(b)}\quad & P\{X_1 = 1, X_2 = 1\} = P\{X_1 = 0, X_2 = 0\} = P\{X_1 = 1, X_2 = 0\} \\ & = P\{X_1 = 0, X_2 = 1\} = 1/4. \end{aligned} \quad (3.343)$$

The joint density function in case (a) is found to be:

$$\begin{aligned} f_{X_1X_2}(x_1,x_2) = {} & 1/8[\delta(x_1 - 1)\delta(x_2 - 1) + \delta(x_1)\delta(x_2)] \\ & + 3/8[\delta(x_1 - 1)\delta(x_2) + \delta(x_1)\delta(x_2 - 1)]. \end{aligned} \quad (3.344)$$

Similarly, in case (b), we have:

$$\begin{aligned} f_{X_1X_2}(x_1,x_2) = {} & 1/4[\delta(x_1 - 1)\delta(x_2 - 1) + \delta(x_1)\delta(x_2) + \delta(x_1 - 1)\delta(x_2) \\ & + \delta(x_1)\delta(x_2 - 1)]. \end{aligned} \quad (3.345)$$

Note that the joint PDFs (3.344) and (3.345) are different.
Next we find the marginal densities.

For case (a), we have:

$$f_{X_1}(x_1) = \int_{-\infty}^{\infty} f_{X_1X_2}(x_1,x_2)\mathrm{d}x_2 = 1/8[\delta(x_1-1)+\delta(x_1)] + 3/8[\delta(x_1-1)+\delta(x_1)]$$
$$= 1/2[\delta(x_1-1)+\delta(x_1)],$$
$$f_{X_2}(x_2) = \int_{-\infty}^{\infty} f_{X_1X_2}(x_1,x_2)\mathrm{d}x_1 = 1/8[\delta(x_2-1)+\delta(x_2)] + 3/8[\delta(x_2-1)+\delta(x_2)]$$
$$= 1/2[\delta(x_2-1)+\delta(x_2)]. \tag{3.346}$$

Similarly, in case (b), we get:

$$f_{X_1}(x_1) = \int_{-\infty}^{\infty} f_{X_1X_2}(x_1,x_2)\mathrm{d}x_2 = 1/4[\delta(x_1-1)+\delta(x_1)+\delta(x_1-1)+\delta(x_1)]$$
$$= 1/2[\delta(x_1-1)+\delta(x_1)] \tag{3.347}$$
$$f_{X_2}(x_2) = \int_{-\infty}^{\infty} f_{X_1X_2}(x_1,x_2)\mathrm{d}x_1 = 1/4[\delta(x_2-1)+\delta(x_2)+\delta(x_2-1)+\delta(x_2)]$$
$$= 1/2[\delta(x_2-1)+\delta(x_2)].$$

Note that the marginal densities (3.346) and (3.347) are equal in both cases, but the joint densities are different.

Therefore, the same marginal density functions may result in different joint density functions.

A.3.2. In the previous example, we concluded that knowledge of the marginal density functions does not provide all of the information about the relations of the random variables.

In case (a),

$$f_{X_1}(x_1)f_{X_2}(x_2) = 1/4[\delta(x_1-1)\delta(x_2-1) + \delta(x_1-1)\delta(x_2) + \delta(x_1)\delta(x_2-1) + \delta(x_1)\delta(x_2)] \neq f_{X_1X_2}(x_1,x_2). \tag{3.348}$$

This result shows us that if variables X_1 and X_2 are dependent, it is not possible to obtain the joint density from the marginal densities.

However, in case (b) we have:

$$f_{X_1}(x_1)f_{X_2}(x_2) = 1/4[\delta(x_1-1)\delta(x_2-1) + \delta(x_1-1)\delta(x_2) + \delta(x_1)\delta(x_2-1) + \delta(x_1)\delta(x_2)] = f_{X_1X_2}(x_1,x_2), \tag{3.349}$$

which demonstrates that the random variables are independent.

This result shows us that if the random variables are independent, we can obtain the joint density from the marginal densities.

A.3.3. Let us suppose that

$$P\{x_1 < X_1 \leq x_1 + \mathrm{d}x_1, x_2 < X_2 \leq x_2 + \mathrm{d}x_2\} = P\{y_1 < Y_1 \leq y_1 + \mathrm{d}y_1, y_2 < Y_2 \leq y_2 + \mathrm{d}y_2\}. \tag{3.350}$$

From (3.350), we have:

$$f_{X_1X_2}(x_1, x_2)\mathrm{d}x_1\,\mathrm{d}x_2 = f_{Y_1Y_2}(y_1, y_2)\mathrm{d}y_1\,\mathrm{d}y_2. \tag{3.351}$$

Knowing that the random variables X_1 and X_2 are independent, we get:

$$f_{X_1}(x_1)f_{X_2}(x_2)\mathrm{d}x_1\,\mathrm{d}x_2 = f_{Y_1Y_2}(y_1, y_2)\mathrm{d}y_1\,\mathrm{d}y_2. \tag{3.352}$$

The joint distribution of Y_1 and Y_2 is:

$$\begin{aligned} F_{Y_1Y_2}(y_1,y_2) &= \int_{-\infty}^{y_1}\int_{-\infty}^{y_2} f_{Y_1Y_2}(y_1,y_2)\mathrm{d}y_1\,\mathrm{d}y_2 = \int_{-\infty}^{g(x_1)}\int_{-\infty}^{g(x_2)} f_{X_1X_2}(x_1,x_2)\mathrm{d}x_1\,\mathrm{d}x_2 \\ &= \int_{-\infty}^{g(x_1)}\int_{-\infty}^{g(x_2)} f_{X_1}(x_1)f_{X_2}(x_2)\mathrm{d}x_1\,\mathrm{d}x_2 = \int_{-\infty}^{y_1} f_{X_1}(x_1)\mathrm{d}x_1 \int_{-\infty}^{y_2} f_{X_2}(x_2)\mathrm{d}x_2 = F_{Y_1}(y_1)F_{Y_2}(y_2). \end{aligned} \tag{3.353}$$

The joint distribution function is equal to the product of the marginal distributions. As a consequence, the variables Y_1 and Y_2 are independent.

A.3.4. The conditional distribution can be rewritten as:

$$F_{X_1}(x_1|X_2 \leq x_2) = \frac{P\{X_1 \leq x_1, X_2 \leq x_2\}}{P\{X_2 \leq x_2\}} = \frac{F_{X_1X_2}(x_1, x_2)}{F_{X_2}(x_2)}. \tag{3.354}$$

If the random variables X_1 and X_2 are independent, then

$$F_{X_1X_2}(x_1, x_2) = F_{X_1}(x_1)F_{X_2}(x_2), \tag{3.355}$$

resulting in:

$$F_{X_1}(x_1|X_2 \leq x_2) = \frac{F_{X_1}(x_1)F_{X_2}(x_2)}{F_{X_2}(x_2)} = F_{X_1}(x_1). \tag{3.356}$$

If the random variables X_1 and X_2 are independent, then the conditional distribution of variable X_1, given the other variable X_2, is equal to the marginal distribution of X_1.

A.3.5. The joint density function for the independent random variables X_1 and X_2 can be written as (see Q.3.3. and [THO71, p. 71]):

$$f_{X_1X_2}(x_1,x_2) = g_1(x_1)g_2(x_2). \tag{3.357}$$

The marginal densities can be expressed as:

$$\begin{aligned} f_{X_1}(x_1) &= \int_{-\infty}^{\infty} f_{X_1X_2}(x_1,x_2)\mathrm{d}x_2 = \int_{-\infty}^{\infty} g_1(x_1)g_2(x_2)\mathrm{d}x_2 \\ &= g_1(x_1)\int_{-\infty}^{\infty} g_2(x_2)\mathrm{d}x_2 = g_1(x_1)K_1, \quad \int_{-\infty}^{\infty} g_2(x_2)\mathrm{d}x_2 = K_1, \end{aligned} \tag{3.358}$$

$$\begin{aligned} f_{X_2}(x_2) &= \int_{-\infty}^{\infty} f_{X_1X_2}(x_1,x_2)\mathrm{d}x_1 = \int_{-\infty}^{\infty} g_1(x_1)g_2(x_2)\mathrm{d}x_1 \\ &= g_2(x_2)\int_{-\infty}^{\infty} g_1(x_1)\mathrm{d}x_1 = g_2(x_2)K_2, \quad \int_{-\infty}^{\infty} g_1(x_1)\mathrm{d}x_1 = K_2. \end{aligned} \tag{3.359}$$

From (3.358) and (3.359), we have:

$$\begin{aligned} g_1(x_1) &= f_{X_1}(x_1)/K_1 \\ g_2(x_2) &= f_{X_2}(x_2)/K_2. \end{aligned} \tag{3.360}$$

Using (3.357) and (3.360), we get:

$$f_{X_1X_2}(x_1,x_2) = \frac{f_{X_1}(x_1)f_{X_2}(x_2)}{K_1K_2}. \tag{3.361}$$

From (3.37), (3.357) and (3.361), we have:

$$\begin{aligned} 1 &= \int_{-\infty}^{\infty}\int_{-\infty}^{\infty} f_{X_1X_2}(x_1,x_2)\mathrm{d}x_1\,\mathrm{d}x_2 \\ &= \int_{-\infty}^{\infty}\int_{-\infty}^{\infty} g_1(x_1)g_2(x_2)\mathrm{d}x_1\,\mathrm{d}x_2 = \frac{1}{K_1K_2}\int_{-\infty}^{\infty}\int_{-\infty}^{\infty} f_{X_1}(x_1)f_{X_2}(x_2)\mathrm{d}x_1\,\mathrm{d}x_2. \end{aligned} \tag{3.362}$$

From (3.362), it follows:

$$K_1K_2 = 1. \tag{3.363}$$

Finally, from (3.357) and (3.363) we arrive at:

$$
\begin{aligned}
g_1(x_1) &= f_{X_1}(x_1), \\
g_2(x_2) &= f_{X_2}(x_2),
\end{aligned}
\tag{3.364}
$$

which confirms that the condition is also a sufficient condition for the independence.

A.3.6. Yes. It is true, as explained in the following:
The event

$$A = \{a < X_1 \le b, c < X_2 \le d\}. \tag{3.365}$$

can be presented as a difference between the events A_1 and A_2, as shown in Fig. 3.38.

$$A_1 = \{a < X_1 \le b, X_2 \le d\}, A_2 = \{a < X_1 \le b, X_2 \le c\}, \tag{3.366}$$

$$P\{A\} = P\{A_1\} - P\{A_2\}. \tag{3.367}$$

Placing (3.365) and (3.366) into (3.367), we get:

$$P\{a < X_1 \le b, c < X_2 \le d\} = P\{a < X_1 \le b, X_2 \le d\} - P\{a < X_1 \le b, X_2 \le c\}. \tag{3.368}$$

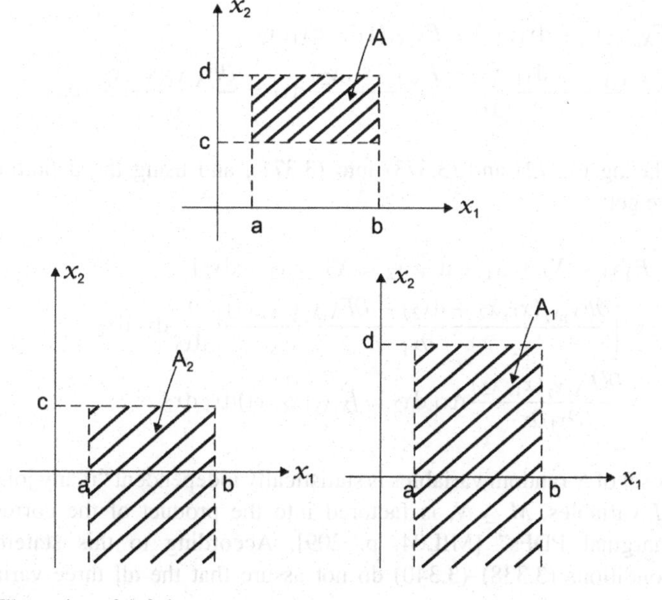

Fig. 3.38 Illustration of A.3.6

Using (3.9), we can rewrite (3.368) in the following form:

$$P\{a < X_1 \le b, c < X_2 \le d\} = F_{X_1X_2}(b,d) - F_{X_1X_2}(a,d) - [F_{X_1X_2}(b,c) - F_{X_1X_2}(a,c)]. \tag{3.369}$$

Finally, from (3.369) we obtain (3.15):

$$P\{a < X_1 \le b, c < X_2 \le d\} = F_{X_1X_2}(b,d) - F_{X_1X_2}(a,d) - F_{X_1X_2}(b,c) + F_{X_1X_2}(a,c). \tag{3.370}$$

A.3.7. Yes. Using (3.15), the probability (3.337) is expressed as:

$$\begin{aligned} &P\{x_1 < X_1 \le x_1 + \mathrm{d}x_1, x_2 < X_2 \le x_2 + \mathrm{d}x_2\} \\ &= [F_{X_1X_2}(x_1 + \mathrm{d}x_1, x_2 + \mathrm{d}x_2) - F_{X_1X_2}(x_1, x_2 + \mathrm{d}x_2)] \\ &- [F_{X_1X_2}(x_1 + \mathrm{d}x_1, x_2) - F_{X_1X_2}(x_1, x_2)]. \end{aligned} \tag{3.371}$$

The expression in the first bracket of (3.371) can be rewritten as:

$$\begin{aligned} &F_{X_1X_2}(x_1 + \mathrm{d}x_1, x_2 + \mathrm{d}x_2) - F_{X_1X_2}(x_1, x_2 + \mathrm{d}x_2) \\ &= \frac{F_{X_1X_2}(x_1 + \mathrm{d}x_1, x_2 + \mathrm{d}x_2) - F_{X_1X_2}(x_1, x_2 + \mathrm{d}x_2)}{\mathrm{d}x_1}\mathrm{d}x_1 \\ &= \frac{\partial F_{X_1X_2}(x_1, x_2 + \mathrm{d}x_2)}{\partial x_1}\mathrm{d}x_1. \end{aligned} \tag{3.372}$$

Similarly, we have:

$$\begin{aligned} &- F_{X_1X_2}(x_1 + \mathrm{d}x_1, x_2) + F_{X_1X_2}(x_1 + \mathrm{d}x_1, x_2) \\ &= \frac{F_{X_1X_2}(x_1 + \mathrm{d}x_1, x_2) - F_{X_1X_2}(x_1, x_2)}{\mathrm{d}x_1}\mathrm{d}x_1 = \frac{\partial F_{X_1X_2}(x_1, x_2)}{\partial x_1}\mathrm{d}x_1. \end{aligned} \tag{3.373}$$

Placing (3.372) and (3.373) into (3.371), and using the definition (3.27), we get:

$$\begin{aligned} &P\{x_1 < X_1 \le x_1 + \mathrm{d}x_1, x_2 < X_2 \le x_2 + \mathrm{d}x_2\} \\ &= \left[\frac{\partial F_{X_1X_2}(x_1, x_2 + \mathrm{d}x_2) - \partial F_{X_1X_2}(x_1, x_2)}{\partial x_1}\right]\frac{1}{\mathrm{d}x_2}\mathrm{d}x_1\,\mathrm{d}x_2 \\ &= \frac{\partial^2 F_{X_1X_2}(x_1, x_2)}{\partial x_1 \partial x_2}\mathrm{d}x_1\,\mathrm{d}x_2 = f_{X_1X_2}(x_1, x_2)\mathrm{d}x_1\,\mathrm{d}x_2. \end{aligned} \tag{3.374}$$

A.3.8. A set of N random variables is statistically independent "if any joint PDF of M variables, $M \le N$, is factored into the product of the corresponding marginal PDFs" [MIL04, p. 209]. According to this statement, the conditions (3.338)–(3.340) do not assure that the all three variables are independent because the condition is satisfied only for $M < N = 3$. Therefore, the condition for $M = N = 3$ should be added:

$$f_{X_1X_2}(x_1, x_2, x_3) = f_{X_1}(x_1)f_{X_2}(x_2)f_{X_3}(x_3). \tag{3.375}$$

However, the condition (3.375) itself includes the conditions (3.338)–(3.340). As a consequence, the condition (3.375) is a necessary and sufficient condition for the independence of three random variables.

A.3.9. Yes. However, in this case, it is more comfortable to work with the corresponding joint probabilities rather than either joint PDFs or distributions [LEO94, pp. 206–207]. The following example, adapted from [LEO94, pp. 206–207], illustrates the concept.

The input to the communication channel (Fig. 3.39) is a discrete random variable X with the discrete values: U and $-U$ (the polar signal) and the corresponding probabilities $P\{U\} = P\{-U\} = 1/2$. The uniform noise N over the range $[-a, a]$ is added to the signal X, resulting in the output of the channel $Y = X + N$. Find the probabilities

$$\begin{aligned} &P\{(X = U) \cap (Y \le 0)\} \\ &P\{(X = -U) \cap (Y \ge 0)\}. \end{aligned} \tag{3.376}$$

Using the conditional probabilities we can rewrite the probabilities (3.376) as:

$$\begin{aligned} P\{(X = U) \cap (Y \le 0)\} &= P\{Y \le 0|X = U\}\, P\{X = U\} = P\{Y \le 0|X = U\}{}^1\!/_2. \\ P\{(X = -U) \cap (Y \ge 0)\} &= P\{Y \ge 0|X = -U\}\, P\{X = -U\} = P\{Y \ge 0|X = -U\}{}^1\!/_2 \end{aligned} \tag{3.377}$$

The input random variable X is a discrete random variable, while the output random variable Y is a continuous random variable. The conditional variables are also continuous: $Y_1 = Y|U$ and $Y_2 = Y|-U$.

The PDFs of the noise and the variables Y_1 and Y_2 are shown in Fig. 3.40.

From Fig. 3.40, we get:

$$P\{Y \le 0|X = U\} = |U - a|/2a = P\{Y \ge 0|X = -U\} = |a - U|/2a. \tag{3.378}$$

Finally, from (3.377) and (3.378), we have:

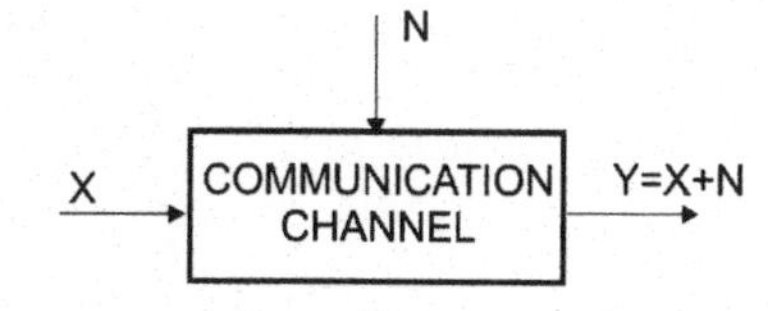

Fig. 3.39 Communication channel with added noise

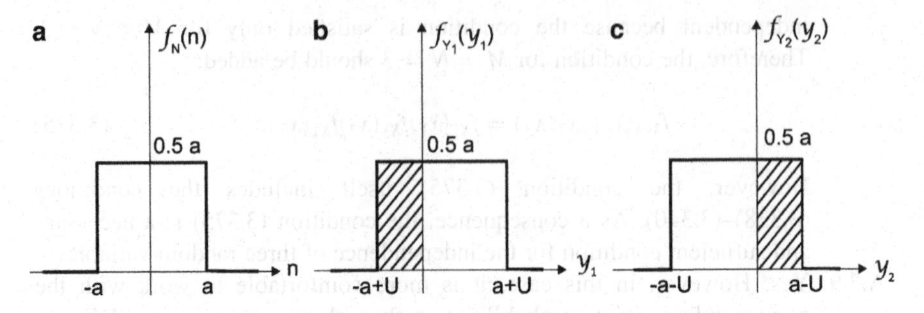

Fig. 3.40 PDFs of (**a**) noise (**b**) r.v. Y_1, and (**c**) r.v. Y_2

$$P\{(X = U) \cap (Y \leq 0)\} = P\{(X = U) \cap (Y \geq 0)\} = |a - U|/4a. \tag{3.379}$$

A.3.10. Recall that from (3.110) and (3.123):

$$C_{X_1X_2} = \overline{X_1X_2} - \overline{X_1}\,\overline{X_2} = R_{X_1X_2} - \overline{X_1}\,\overline{X_2}. \tag{3.380}$$

From (3.380), it follows that the covariance and correlation are equal if either mean value of X_1 or X_2–or of both of them–is zero.

A.3.11. It is not necessarily be zero. It depends on the range of the variable X. See Example 3.3.4.

A.3.12. Let us recall of the definition of covariance (3.121):

$$C_{X_1X_2} = E\{(X_1 - \overline{X_1})(X_2 - \overline{X_2})\}. \tag{3.381}$$

When variables X_1 and X_2 are correlated, they have either the same tendency (positive correlation), or the opposite tendency (negative correlation). Therefore, if there is a positive correlation between them, the dissipation of the random variable X_1 around its mean value will have the same sign (in a majority of cases) as the dissipation of the random variable X_2 around its mean value. As a consequence, the expected value of the product of $(X_1 - E\{X_1\})$ and $(X_2 - E\{X_2\})$ cannot equal zero, and the covariance is not zero.

Similarly, if variables X_1 and X_2 have a negative correlation, the signs of the dissipations of the values of the random variables will be opposite. As a consequence, the expected value of the product of $(X_1 - E\{X_1\})$ and $(X_2 - E\{X_2\})$ cannot equal zero, and the covariance is not zero.

Chapter 4
Normal Random Variable

4.1 Normal PDF

4.1.1 Definition

The most important density function is the normal PDF which is present in almost all science and scientific techniques. This is why the PDF is called normal, because it presents a normal behavior for a lot of random appearances. It is also known as a Gaussian random variable in honor of famous mathematician K.F. Gauss who used it in his important works in the theory of probability. However, it was first defined by the mathematician De Moivre in 1733.

The random variables which are described using a normal PDF are called *normal* or *Gaussian random variables*.

The normal PDF of the random variable X with a range of x is defined as:

$$f_X(x) = \frac{1}{\sqrt{2\pi}b} \mathrm{e}^{-\frac{(x-a)^2}{2b^2}}, \quad -\infty < x < \infty, \tag{4.1}$$

where a and b are constants, $b > 0$, and $-\infty < a < \infty$.

Below we find the meaning of the constants a and b.

To this end, we find the mean value of the r.v. X using (2.261) and (4.1):

$$E\{X\} = m_X = \int_{-\infty}^{\infty} x f_X(x) \mathrm{d}x = \frac{1}{\sqrt{2\pi}b} \int_{-\infty}^{\infty} x\, \mathrm{e}^{-\frac{(x-a)^2}{2b^2}} \mathrm{d}x$$

$$= \frac{1}{\sqrt{2\pi}b} \int_{-\infty}^{\infty} (x-a+a) \mathrm{e}^{-\frac{(x-a)^2}{2b^2}} \mathrm{d}x$$

G.J. Dolecek, *Random Signals and Processes Primer with MATLAB*,
DOI 10.1007/978-1-4614-2386-7_4, © Springer Science+Business Media New York 2013

$$= \frac{1}{\sqrt{2\pi}b} \int_{-\infty}^{\infty} (x-a)\mathrm{e}^{-\frac{(x-a)^2}{2b^2}} \mathrm{d}x + \frac{1}{\sqrt{2\pi}b} \int_{-\infty}^{\infty} a\,\mathrm{e}^{-\frac{(x-a)^2}{2b^2}} \mathrm{d}x. \tag{4.2}$$

One can easily see that the first term in (4.2) can be expressed as:

$$\frac{1}{\sqrt{2\pi}b} \int_{-\infty}^{\infty} (x-a)\mathrm{e}^{-\frac{(x-a)^2}{2b^2}} \mathrm{d}x = \frac{1}{\sqrt{2\pi}b} \int_{0}^{\infty} (x-a)\mathrm{e}^{-\frac{(x-a)^2}{2b^2}} \mathrm{d}x - \frac{1}{\sqrt{2\pi}b} \int_{0}^{\infty} (x-a)\mathrm{e}^{-\frac{(x-a)^2}{2b^2}} \mathrm{d}x = 0. \tag{4.3}$$

Similarly, using the PDF property (2.83) and (4.1), we have:

$$\frac{1}{\sqrt{2\pi}b} \int_{-\infty}^{\infty} a\,\mathrm{e}^{-\frac{(x-a)^2}{2b^2}} \mathrm{d}x = a \frac{1}{\sqrt{2\pi}b} \int_{-\infty}^{\infty} \mathrm{e}^{-\frac{(x-a)^2}{2b^2}} \mathrm{d}x = a. \tag{4.4}$$

From (4.1) to (4.4), we get:

$$E\{X\} = m = a. \tag{4.5}$$

Next we find the variance of the normal random variable X, using (2.344) and (4.1):

$$\sigma^2 = \int_{-\infty}^{\infty} (x-m)^2 f_X(x)\mathrm{d}x = \frac{1}{\sqrt{2\pi}b} \int_{-\infty}^{\infty} (x-m)^2 \mathrm{e}^{-\frac{(x-m)^2}{2b^2}} \mathrm{d}x. \tag{4.6}$$

By introducing the variable u

$$u = \frac{x-m}{\sqrt{2}b} \tag{4.7}$$

in (4.6) we arrive at:

$$\sigma^2 = \frac{2b^2}{\sqrt{\pi}} \int_{-\infty}^{\infty} u^2\,\mathrm{e}^{-u^2}\,\mathrm{d}u = \frac{4b^2}{\sqrt{\pi}} \int_{0}^{\infty} u^2\,\mathrm{e}^{-u^2}\,\mathrm{d}u. \tag{4.8}$$

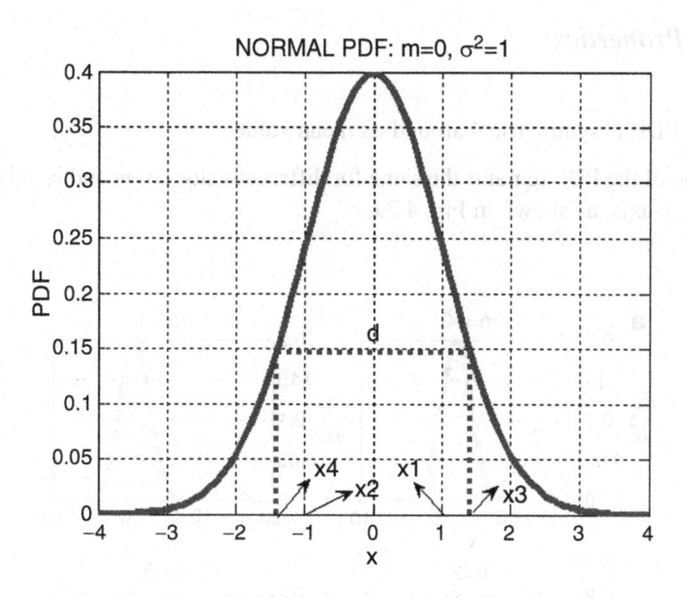

Fig. 4.1 Normal PDF

Using integral 4 in Appendix A, we have:

$$\sigma^2 = \frac{4b^2}{\sqrt{\pi}} \int_0^{\infty} u^2\, \mathrm{e}^{-u^2}\, \mathrm{d}u = \frac{4b^2}{\sqrt{\pi}} \frac{\sqrt{\pi}}{4} = b^2. \tag{4.9}$$

Therefore, the constant b is the standard deviation of the normal random variable.

Using the results obtained in (4.5) and (4.9), the PDF (4.1) can be rewritten in its more common form as:

$$f_X(x) = \frac{1}{\sqrt{2\pi}\sigma} \mathrm{e}^{-\frac{(x-m)^2}{2\sigma^2}}, \quad -\infty<x<\infty \tag{4.10}$$

As a consequence, the normal PDF is completely determined by its two first moments and, consequently, it is often done in the simple form:

$$f_X(x) = N(m, \sigma^2). \tag{4.11}$$

The density is shown in Fig. 4.1 for $m = 0$ and $\sigma^2 = 1$.

4.1.2 Properties

P.1 The PDF is symmetrical around its mean value.

The shape of the PDF remains the same for different values of m and is only shifted along the x-axis, as shown in Fig. 4.2a, b.

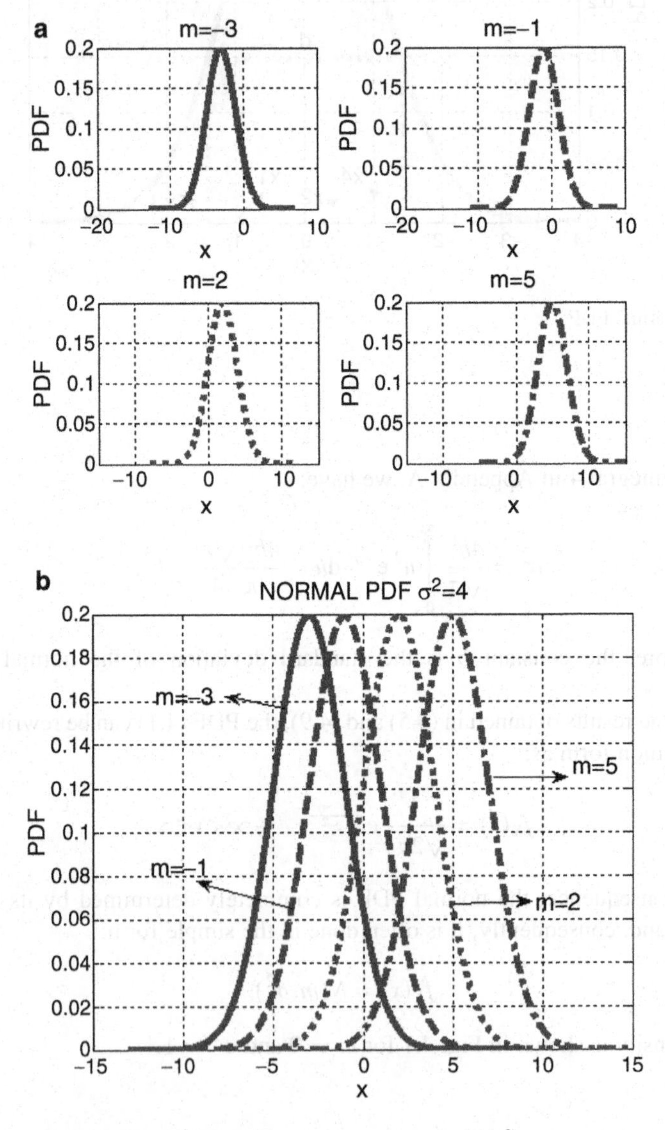

Fig. 4.2 Normal PDFs for different values of m and the variance $\sigma^2 = 4$

P.2 The PDF has a maximum value in $x = m$ that is equal to:

$$f_X(x)_{\max} = \frac{1}{\sqrt{2\pi}\sigma}. \tag{4.12}$$

P.3 The width d in the PDF is defined as the distance between the points x_3 and x_4 (see Fig. 4.1),

$$d = x_3 - x_4, \tag{4.13}$$

where

$$f_X(x_3) = f_X(x_4) = \mathrm{e}^{-1} f_X(x)_{\max} = \frac{\mathrm{e}^{-1}}{\sqrt{2\pi}\sigma}. \tag{4.14}$$

From (4.10) and (4.14), it follows that

$$\frac{(x-m)^2}{2\sigma^2} = 1, \tag{4.15}$$

resulting in:

$$x_{3,4} = m \pm \sqrt{2}\sigma \tag{4.16}$$

and

$$d = 2\sqrt{2}\sigma. \tag{4.17}$$

From (4.12) and (4.17), it follows that the variance affects the shape of the PDF. By increasing the variance, the maximum value of the PDF decreases and its width increases. The same can be concluded from the property of the PDF to keep the area under the PDF equal to 1. So, in increasing the maximum value of the PDF, the width must decrease in order for the area under the PDF to remain equal to 1.

The normal densities for different variances and mean values equal to 5 are shown in Fig. 4.3.

P.4 The distance between the mean value and the inflexion points is equal to two standard deviations 2σ, where the inflexion points are values of x for which the second derivative of the PDF is equal to zero,

$$\frac{\mathrm{d}^2 f_X(x)}{\mathrm{d}x^2} = \frac{1}{\sqrt{2\pi}\sigma^3}\left[\frac{(x-m)^2}{\sigma^2} - 1\right] = 0, \tag{4.18}$$

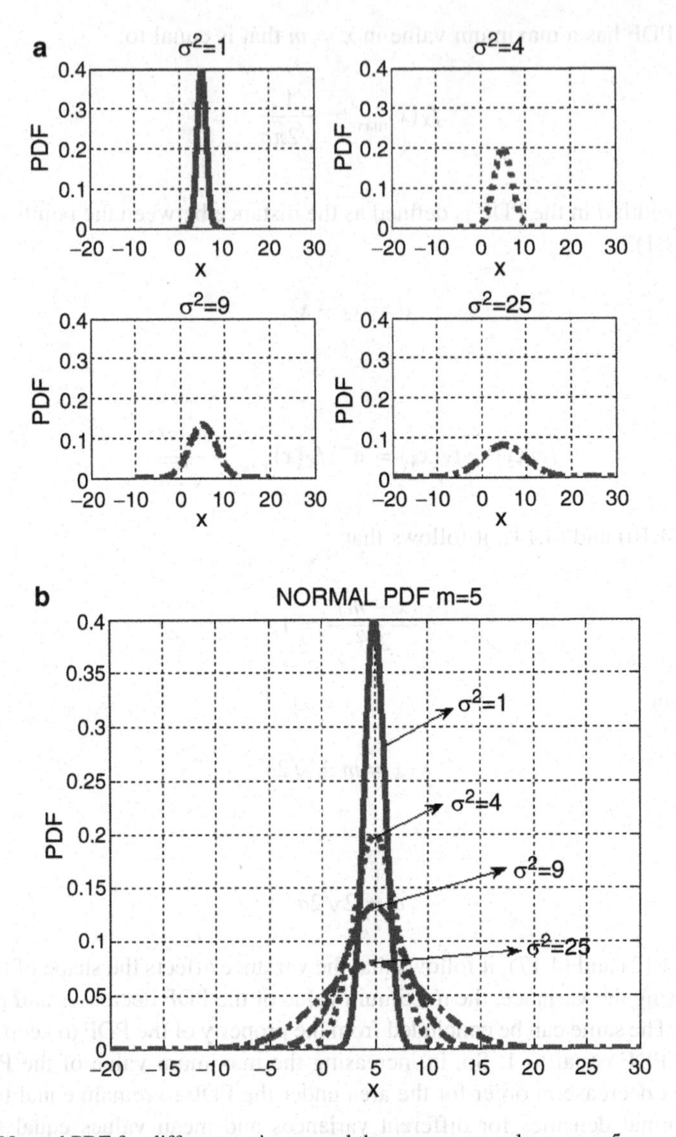

Fig. 4.3 Normal PDF for different variances and the same mean values, $m = 5$

and the third derivative of the PDF is not equal to zero for any x. From (4.18) the inflexion points are:

$$x_{1,2} = m \pm \sigma, \tag{4.19}$$

resulting in (see Fig. 4.1):

$$x_1 - x_2 = 2\sigma. \tag{4.20}$$

P.5 By increasing x, the PDF approaches to zero.

Theoretically,

$$\lim_{x \to \pm\infty} f_X(x) = 0. \tag{4.21}$$

However, the PDF practically approaches zero for finite values of x, as shown in Sect. 4.2.

4.2 Normal Distribution

4.2.1 Definition

Using the definition (2.10) and (4.10), we define the normal distribution as:

$$F_X(x) = P\{X \leq x\} = \int_{-\infty}^{x} f_X(x)\mathrm{d}x = \frac{1}{\sqrt{2\pi}\sigma} \int_{-\infty}^{x} \mathrm{e}^{-\frac{(x-m)^2}{2\sigma^2}} \mathrm{d}x. \tag{4.22}$$

This distribution is shown in Fig. 4.4 for $m = 0$ and $\sigma^2 = 1$.

Next we consider how the different values of the mean value and variance affect the shape of the normal distribution.

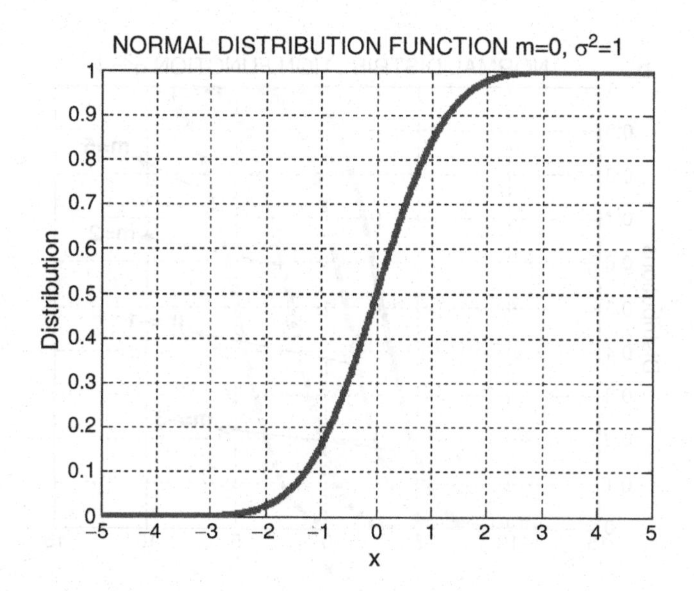

Fig. 4.4 Normal distribution

First, we consider the distributions of normal random variables that have equal variance and different mean values. To illustrate this, Fig. 4.5 shows the distribution functions for the normal variables from Fig. 4.2. The distributions are presented together in Fig. 4.5b. Note that the distributions are equal to 0.5 for $x = m$ and that all have the same shape and are displaced in x-axis.

Second, we consider distributions of normal variables that have equal mean values and different variances. As an example, Fig. 4.6a shows the distributions for the normal variables from Fig. 4.3, that have different variances and a mean value

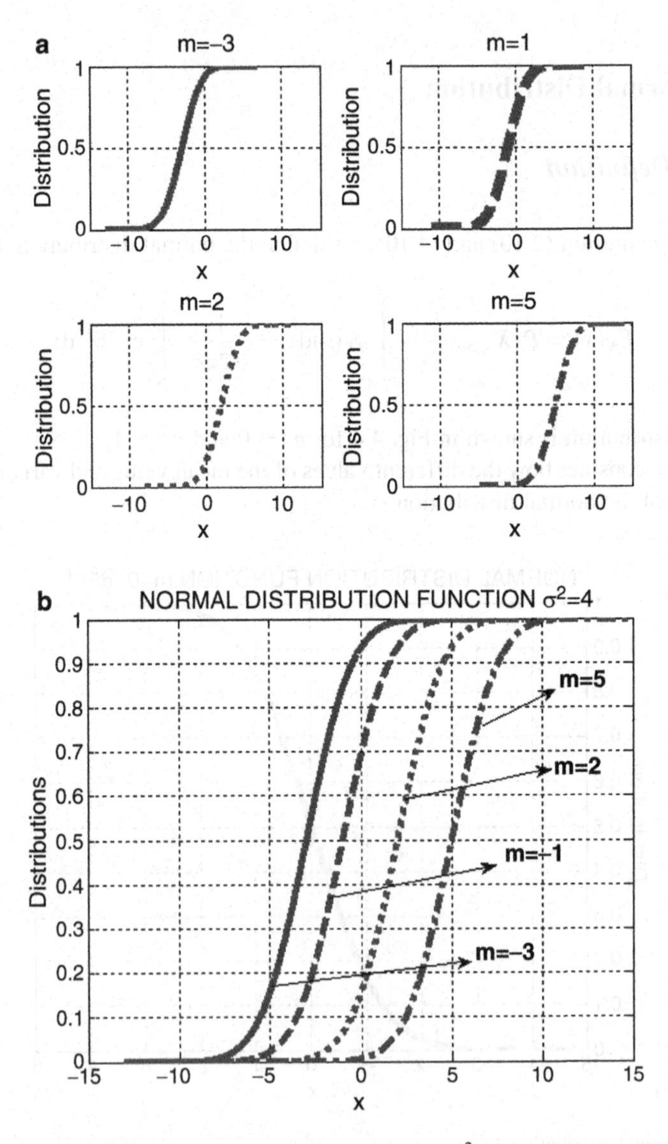

Fig. 4.5 Distribution functions for different values of m and $\sigma^2 = 4$

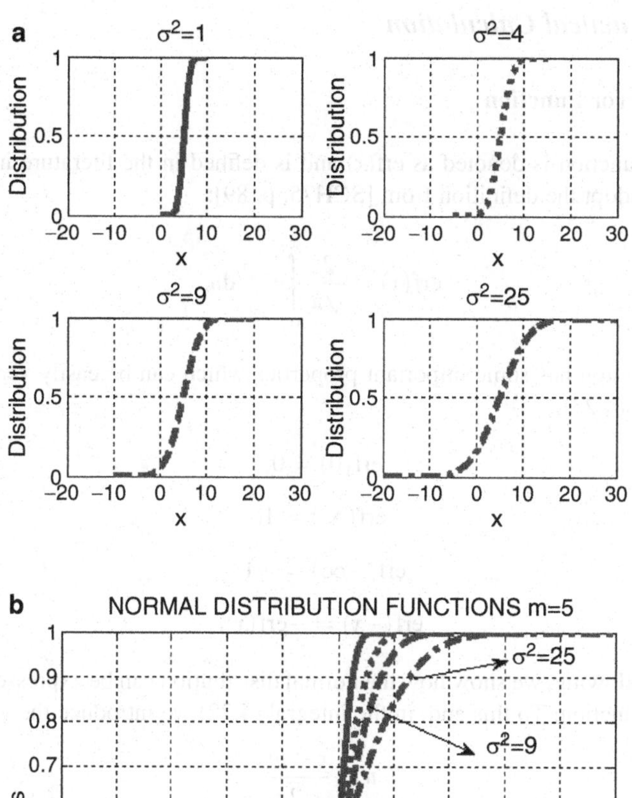

Fig. 4.6 Distribution functions for different values of σ^2 and $m = 5$

$m = 5$. Figure 4.6b shows all of these distributions in one plot. Note that all these distributions intersect at $x = 5$, and have different shapes.

The calculation of a distribution function (4.22) requires the calculation of the integral for each value of x and the given values of both the mean value and the variance. A more elegant way of using the special functions is described in the following section.

4.2.2 *Practical Calculation*

4.2.2.1 Error Function

The error function is denoted as erf(x) and is defined in the literature in different ways. We adopt the definition from [SCH75, p. 89]:

$$\operatorname{erf}(x) = \frac{2}{\sqrt{\pi}} \int_0^x \mathrm{e}^{-u^2}\, \mathrm{d}u. \tag{4.23}$$

This function has some important properties which can be easily verified from the integral (4.23):

$$\operatorname{erf}(0) = 0, \tag{4.24a}$$

$$\operatorname{erf}(\infty) = 1, \tag{4.24b}$$

$$\operatorname{erf}(-\infty) = -1, \tag{4.24c}$$

$$\operatorname{erf}(-x) = -\operatorname{erf}(x). \tag{4.24d}$$

In the following, we show how the normal distribution can be expressed in terms of the erf function. To this end, in the integral (4.22) we introduce the variable

$$u = \frac{x - m}{\sqrt{2}\sigma} \tag{4.25}$$

resulting in:

$$F_X(x) = \frac{1}{\sqrt{\pi}} \int_{-\infty}^{u} \mathrm{e}^{-u^2}\, \mathrm{d}u = \frac{1}{2}\left[\frac{2}{\sqrt{\pi}} \int_{-\infty}^{0} \mathrm{e}^{-u^2}\, \mathrm{d}u + \frac{2}{\sqrt{\pi}} \int_{0}^{u} \mathrm{e}^{-u^2}\, \mathrm{d}u\right]$$
$$= \frac{1}{2}\left[-\frac{2}{\sqrt{\pi}} \int_{0}^{-\infty} \mathrm{e}^{-u^2}\, \mathrm{d}u + \frac{2}{\sqrt{\pi}} \int_{0}^{u} \mathrm{e}^{-u^2}\, \mathrm{d}u\right]. \tag{4.26}$$

Applying the definition (4.23) and property (4.24c), the first term in (4.26) becomes:

$$-\frac{2}{\sqrt{\pi}} \int_{0}^{-\infty} \mathrm{e}^{-u^2}\, \mathrm{d}u = -\operatorname{erf}(-\infty) = -(-1) = 1. \tag{4.27}$$

Similarly, from (4.23) and (4.25), it follows:

$$\frac{2}{\sqrt{\pi}}\int_{0}^{u} \mathrm{e}^{-u^2}\,\mathrm{d}u = \mathrm{erf}(u) = \mathrm{erf}\left(\frac{x-m}{\sqrt{2}\sigma}\right). \tag{4.28}$$

Finally, from (4.26) to (4.28), we have:

$$F_X(x) = \frac{1}{2}\left[1 + \mathrm{erf}\left(\frac{x-m}{\sqrt{2}\sigma}\right)\right]. \tag{4.29}$$

Using (4.29), the probability $P\{X \le x_1\}$ can be written as:

$$P\{X \le x_1\} = F_X(x_1) = \frac{1}{2}\left[1 + \mathrm{erf}\left(\frac{x_1-m}{\sqrt{2}\sigma}\right)\right]. \tag{4.30}$$

Similarly, from (2.46) and (4.29), we have:

$$\begin{aligned} P\{x_1 \le X \le x_2\} &= F_X(x_2) - F_X(x_1) \\ &= \frac{1}{2}\left[1 + \mathrm{erf}\left(\frac{x_2-m}{\sqrt{2}\sigma}\right)\right] - \frac{1}{2}\left[1 + \mathrm{erf}\left(\frac{x_1-m}{\sqrt{2}\sigma}\right)\right] \\ &= \frac{1}{2}\left[\mathrm{erf}\left(\frac{x_2-m}{\sqrt{2}\sigma}\right) - \mathrm{erf}\left(\frac{x_1-m}{\sqrt{2}\sigma}\right)\right]. \end{aligned} \tag{4.31}$$

We often need to find the probability that the normal variable is around its mean value in the interval $[m - k\sigma, m + k\sigma]$, where k is any real number. Using (4.24d) and (4.31), where,

$$\begin{aligned} x_2 &= m + k\sigma, \\ x_1 &= m - k\sigma, \end{aligned}$$

we find the desired probability as:

$$\begin{aligned} P\{m - k\sigma \le X \le m + k\sigma\} &= \frac{1}{2}\left[\mathrm{erf}\left(\frac{m + k\sigma - m}{\sqrt{2}\sigma}\right) - \mathrm{erf}\left(\frac{m - k\sigma - m}{\sqrt{2}\sigma}\right)\right] \\ &= \mathrm{erf}\left(\frac{k}{\sqrt{2}}\right). \end{aligned} \tag{4.33}$$

Example 4.2.1 Given a normal random variable with a mean value of 5 and the mean squared value of 89, find the probability that the random variable is less than 10 and that is in the interval $[-2, 8]$ calculating the erf function using the MATLAB file erf(x).

Solution The variance and the standard deviation are:

$$\sigma^2 = E\{X^2\} - (E\{X\})^2 = 89 - 25 = 64, \sigma = 8.$$

From (4.30) we have:

$$P\{X \leq 10\} = F_X(10) = \frac{1}{2}\left[1 + \text{erf}\left(\frac{5}{8\sqrt{2}}\right)\right] = 0.734, \tag{4.35}$$

where

$$\text{erf}\left(\frac{5}{8\sqrt{2}}\right) = 0.468. \tag{4.36}$$

From (4.31), we have:

$$\begin{aligned} P\{-2 \leq X \leq 8\} &= \frac{1}{2}\left[\text{erf}\left(\frac{3}{8\sqrt{2}}\right) - \text{erf}\left(\frac{-7}{8\sqrt{2}}\right)\right] \\ &= \frac{1}{2}\left[\text{erf}\left(\frac{3}{8\sqrt{2}}\right) + \text{erf}\left(\frac{7}{8\sqrt{2}}\right)\right] \\ &= \frac{1}{2}[0.2923 + 0.6184] = 0.4554. \end{aligned} \tag{4.37}$$

4.2.2.2 Complementary Error Function

Complementary error function is denoted as erfc and is related to error function as:

$$\text{erfc}(x) = 1 - \text{erf}(x). \tag{4.38}$$

From (4.23) and (4.24b), it follows:

$$\text{erfc}(x) = \frac{2}{\sqrt{\pi}}\int_0^{\infty} \mathrm{e}^{-u^2}\,\mathrm{d}u - \frac{2}{\sqrt{\pi}}\int_0^{x} \mathrm{e}^{-u^2}\,\mathrm{d}u = \frac{2}{\sqrt{\pi}}\int_x^{\infty} \mathrm{e}^{-u^2}\,\mathrm{d}u. \tag{4.39}$$

From (4.30) and (4.38), we have:

$$\begin{aligned} P\{X \leq x_1\} = F_X(x_1) &= \frac{1}{2}\left[1 + 1 - \text{erfc}\left(\frac{x_1 - m}{\sqrt{2}\sigma}\right)\right] \\ &= 1 - \frac{1}{2}\text{erfc}\left(\frac{x_1 - m}{\sqrt{2}\sigma}\right). \end{aligned} \tag{4.40}$$

Similarly, the probability (4.31) can be expressed in terms of complementary error function as:

$$\begin{aligned} P\{x_1 \le X \le x_2\} &= F_X(x_2) - F_X(x_1) \\ &= \frac{1}{2}\left[1 - \operatorname{erfc}\left(\frac{x_2 - m}{\sqrt{2}\sigma}\right) - 1 + \operatorname{erfc}\left(\frac{x_1 - m}{\sqrt{2}\sigma}\right)\right] \\ &= \frac{1}{2}\left[\operatorname{erfc}\left(\frac{x_1 - m}{\sqrt{2}\sigma}\right) - \operatorname{erfc}\left(\frac{x_2 - m}{\sqrt{2}\sigma}\right)\right]. \end{aligned} \tag{4.41}$$

The probability (4.33) is:

$$P\{m - k\sigma \le X \le m + k\sigma\} = \operatorname{erf}\left(\frac{k}{\sqrt{2}}\right) = 1 - \operatorname{erfc}\left(\frac{k}{\sqrt{2}}\right). \tag{4.42}$$

Example 4.2.2 Find the probabilities from Example 4.2.1 in terms of the erfc function, where the erfc function is calculated using the MATLAB file erfc(x).

Solution From (4.40) and Example 4.2.1, we have:

$$\begin{aligned} P\{X \le 10\} &= 1 - \frac{1}{2}\operatorname{erfc}\left(\frac{x_1 - m}{\sqrt{2}\sigma}\right) \\ &= 1 - \frac{1}{2}\operatorname{erfc}\left(\frac{5}{8\sqrt{2}}\right) = 1 - \frac{1}{2}0.532 = 0.734. \end{aligned} \tag{4.43}$$

From (4.41), we get:

$$\begin{aligned} &P\{-2 \le X \le 8\} \\ &= \frac{1}{2}\left[\operatorname{erfc}\left(\frac{-7}{8\sqrt{2}}\right) - \operatorname{erfc}\left(\frac{3}{8\sqrt{2}}\right)\right] = \frac{1}{2}[1.6184 - 0.7077] = 0.4554. \end{aligned} \tag{4.44}$$

4.2.2.3 *Q* Function

Consider the probability that the values of a normal random variable are outside of the interval $m + k\sigma$,

$$P\{X > m + k\sigma\} = \int_{m+k\sigma}^{\infty} f_X(x)\mathrm{d}x = \frac{1}{\sqrt{2\pi}\sigma}\int_{m+k\sigma}^{\infty} \mathrm{e}^{-\frac{(x-m)^2}{2\sigma^2}}\mathrm{d}x. \tag{4.45}$$

Introducing the variable

$$u = \frac{x - m}{\sigma} \tag{4.46}$$

into the integral (4.45), we arrive at:

$$P\{X > m + k\sigma\} = \frac{1}{\sqrt{2\pi}} \int_{k}^{\infty} e^{-\frac{u^2}{2}} \mathrm{d}u = Q(k). \tag{4.47}$$

The integral (4.47) represents the area under the tail of the normal PDF outside of $m + k\sigma$. The values of the integral (4.47) for a given value of k are given in tables in literature.

Due to the symmetrical property of the normal PDF, it stands:

$$P\{X>m + k\sigma\} = P\{X<m - k\sigma\} = Q(k). \tag{4.48}$$

Let us express the probability (4.30) in terms of $Q(k)$. To this end, we write:

$$x_1 = m + k\sigma \quad \text{for} \quad x_1>m \tag{4.49}$$

and

$$x_1 = m - k\sigma \quad \text{for} \quad x_1<m. \tag{4.50}$$

From here,

$$k = \frac{|x_1 - m|}{\sigma}, \tag{4.51}$$

and the desired probability is:

$$P\{X \leq x_1\} = 1 - P\{X>x_1\}. \tag{4.52}$$

From (4.48) and (4.51), we get:

$$P\{X \leq x_1\} = 1 - P\{X>x_1\} = 1 - Q(k) = 1 - Q\left(\frac{|x_1 - m|}{\sigma}\right). \tag{4.53}$$

Similarly, the probability (4.31) can be calculated from (2.45) and (4.53) as:

$$\begin{aligned} P\{x_1<X \leq x_2\} &= 1 - Q(k_1) - Q(k_2) \\ &= 1 - Q\left(\frac{|x_1 - m|}{\sigma}\right) - Q\left(\frac{|x_2 - m|}{\sigma}\right). \end{aligned} \tag{4.54}$$

Using (4.54), the probability (4.33) can be expressed in terms of the function $Q(k)$ as:

$$P\{m - k\sigma<X \leq m + k\sigma\} = 1 - 2Q(k). \tag{4.55}$$

Example 4.2.3 Find the probabilities of Example 4.2.1 in terms of Q function.

Solution From (4.53), it follows:

$$P\{X \leq 10\} = 1 - P\{X>10\} = 1 - Q(k) = 1 - Q\left(\frac{5}{8}\right). \tag{4.56}$$

Using (4.54), we arrive at:

$$P\{-2<X \leq 8\} = 1 - Q\left(\frac{7}{8}\right) - Q\left(\frac{3}{8}\right). \tag{4.57}$$

4.2.2.4 Relation Between Different Functions

Sometimes it is necessary to express the probability (given in terms of one function) in terms of a different function.

From (4.42) and (4.55), we write:

$$\operatorname{erf}\left(\frac{x}{\sqrt{2}}\right) = 1 - 2Q(x). \tag{4.58}$$

From here, we have:

$$Q(x) = \frac{1}{2}\left[1 - \operatorname{erf}\left(\frac{x}{\sqrt{2}}\right)\right]. \tag{4.59}$$

$$\operatorname{erf}(x) = 1 - 2Q\left(\sqrt{2}x\right). \tag{4.60}$$

Similarly, from (4.42) and (4.55), we have:

$$1 - \operatorname{erfc}\left(\frac{x}{\sqrt{2}}\right) = 1 - 2Q(x), \tag{4.61}$$

resulting in:

$$Q(x) = \frac{1}{2}\operatorname{erfc}\left(\frac{x}{\sqrt{2}}\right) \tag{4.62}$$

and

$$\operatorname{erfc}(x) = 2Q\left(\sqrt{2}x\right). \tag{4.63}$$

The relations (4.38), (4.59), (4.60), (4.62), and (4.63) are summarized in Table 4.1.

Table 4.1 The relations between erf(x), erfc(x), and $Q(x)$

	erf(x)	erfc(x)	$Q(x)$
erf(x)	–	$\text{erf}(x) = 1 - \text{erfc}(x)$	$\text{erf}(x) = 1 - 2Q(\sqrt{2}x)$
erfc(x)	$\text{erfc}(x) = 1 - \text{erf}(x)$	–	$\text{erfc}(x) = 2Q(\sqrt{2}x)$
$Q(x)$	$Q(x) = \frac{1}{2}\left[1 - \text{erf}\left(\frac{x}{\sqrt{2}}\right)\right]$	$Q(x) = \frac{1}{2}\text{erfc}\left(\frac{x}{\sqrt{2}}\right)$	–

Example 4.2.4 Using Table 4.1, verify that the results in Example 4.2.3 correspond to the results in Examples 4.2.1 and 4.2.2.

Solution From (4.56), we have:

$$P\{X \leq 10\} = 1 - Q\left(\frac{5}{8}\right). \tag{4.64}$$

From Table 4.1 we express $Q(x)$ in terms of erf function:

$$Q\left(\frac{5}{8}\right) = \frac{1}{2}\left[1 - \text{erf}\left(\frac{5/8}{\sqrt{2}}\right)\right] = \frac{1}{2}[1 - 0.468] = 0.266. \tag{4.65}$$

Placing (4.65) into (4.64) we get:

$$P\{X \leq 10\} = 1 - Q\left(\frac{5}{8}\right) = 1 - 0.266 = 0.734, \tag{4.66}$$

which has the same result as (4.35) and (4.43).

Similarly, from (4.57) and Table 4.1, we write:

$$\begin{aligned} P\{-2 < X \leq 8\} &= 1 - Q\left(\frac{7}{8}\right) - Q\left(\frac{3}{8}\right) \\ &= 1 - \frac{1}{2}\text{erfc}\left(\frac{7/8}{\sqrt{2}}\right) - \frac{1}{2}\text{erfc}\left(\frac{3/8}{\sqrt{2}}\right) \\ &= 1 - \frac{1}{2}0.3816 - \frac{1}{2}0.7077 = 0.4554. \end{aligned} \tag{4.67}$$

4.2.3 The "3 σ Rule"

Consider the probability that the random variable X is in the interval of k standard deviations around its mean value, as given in (4.33):

$$P\{m - k\sigma \leq X \leq m + k\sigma\} = \text{erf}\left(\frac{k}{\sqrt{2}}\right). \tag{4.68}$$

Table 4.2 The "3 σ rule"

k	$P\{m - k\sigma \leq X \leq m + k\sigma\}$
0.5	0.3829
1	0.6827
1.5	0.8664
2	0.9545
2.5	0.9876
3	0.9973
3.5	0.9995
4	0.9999

Table 4.2 presents the values of probabilities (4.68) for different values of k. From Table 4.2, we note that for $k = 3$ we have:

$$P\{m - 3\sigma \leq X \leq m + 3\sigma\} = 0.9973, \tag{4.69}$$

or, that 99.73% of all values of the normal variable are in symmetrical intervals of 3 standard deviations around the mean value. This is known as the "3 σ rule." This rule states that the absolute value of the dissipation of a normal random variable from its mean value effectively does not pass 3 standard deviations.

4.3 Transformation of Normal Random Variable

All transformations discussed in Sect. 2.6 can easily be applied to the normal variable.

4.3.1 *Monotone Transformation*

Let us first consider the linear transformation.

4.3.1.1 Linear Transformation

Let X be a normal random variable with mean value and variance m_X and σ_X^2, respectively, and

$$Y = aX + b, \tag{4.70}$$

where a and b are constants.

From (2.138), we have:

$$f_Y(y) = \frac{f_X(x)}{\left|\frac{dy}{dx}\right|}\Bigg|_{x=\frac{(y-b)}{a}} = \frac{\frac{1}{\sqrt{2\pi}\sigma_X} e^{-\frac{(x-m_X)^2}{2\sigma_X^2}}}{|a|}\Bigg|_{x=\frac{y-b}{a}} = \frac{1}{\sqrt{2\pi}|a|\sigma_X} e^{-\frac{\left(\frac{y-b}{a}-m_X\right)^2}{2\sigma_X^2}}$$

$$= \frac{1}{\sqrt{2\pi}|a|\sigma_X} e^{-\frac{(y-(b+am_X))^2}{2a^2\sigma_X^2}}. \tag{4.71}$$

Comparing the result (4.71) with the expression for the normal PDF, we can write:

$$f_Y(y) = \frac{1}{\sqrt{2\pi}\sigma_Y} e^{-\frac{(y-m_Y)^2}{2\sigma_Y^2}}, \quad -\infty<y<\infty, \tag{4.72}$$

where

$$m_Y = am_X + b,$$
$$\sigma_Y^2 = (a\sigma_X)^2 = a^2\sigma_X^2. \tag{4.73}$$

Observing (4.70), we can easily conclude that the parameters (4.73) can be obtained from (4.70), using (2.244) and (2.355):

$$m_Y = E\{Y\} = E\{aX+b\} = aE\{X\} + b = am_X + b,$$
$$\sigma_{aX+b}^2 = a^2\sigma_X^2. \tag{4.74}$$

The following rule stands:

The linear transformation of the normal random variable results in a normal variable with the mean value and variance given in (4.73).

Example 4.3.1 Find the PDF of the random variable $Y = -3X + 2$, where X is the normal random variable with a mean value and variance equal to -1 and 4, respectively.

Solution The transformed random variable Y is also a normal random variable with the parameters obtained from (4.73):

$$m_Y = -3m_X + 2 = -3(-1) + 2 = 5,$$
$$\sigma_{-3X+2}^2 = (-3)^2\sigma_X^2 = 9 \times 4 = 36. \tag{4.75}$$

From here, the PDF of the random variable Y is as follows:

$$f_Y(y) = \frac{1}{\sqrt{2\pi}6} e^{-\frac{(y-5)^2}{72}} \quad \text{for} \quad -\infty<y<\infty. \tag{4.76}$$

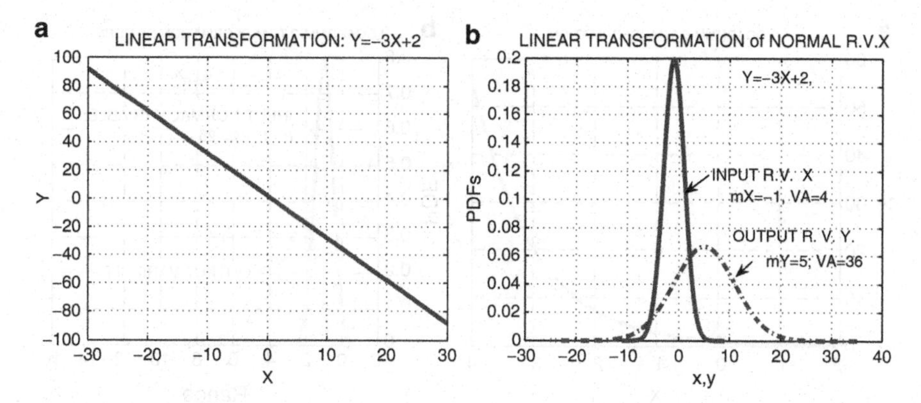

Fig. 4.7 Linear transformation of a normal random variable. (**a**) Linear transformation. (**b**) Input and output PDFs

Figure 4.7a shows the transformation and Fig. 4.7b shows the input and the transformed PDFs.

4.3.1.2 Nonlinear Transformation

We use the expression (2.138) to illustrate that the nonlinear monotone transformation of a normal variable does not result in a normal variable.

Example 4.3.2 In this example, consider the logarithmic transformation

$$Y = \mathrm{e}^{X} \tag{4.77}$$

of a normal variable with parameters m_X and σ_X^2.

The transformed PDF according to (2.138) is:

$$f_Y(y) = \left.\frac{f_X(x)}{\left|\frac{\mathrm{d}y}{\mathrm{d}x}\right|}\right|_{x=\ln y} = \frac{1}{\sqrt{2\pi}\sigma_X y}\mathrm{e}^{-\frac{(\ln y - m_X)^2}{2\sigma_X^2}}; \quad y>0 \tag{4.78}$$

The obtained PDF is lognormal and will be examined in more detail in the next chapter. The transformation as well as input and output PDFs are shown in Fig. 4.8a, b, respectively.

4.3.2 Nonmonotone Transformation

The nonmonotone transformation of a normal random variable changes the variable such that the resulting variable is not normal, as shown in the following example.

Example 4.3.3 We consider the absolute value of the normal variable with a zero mean value and a variance of 1.

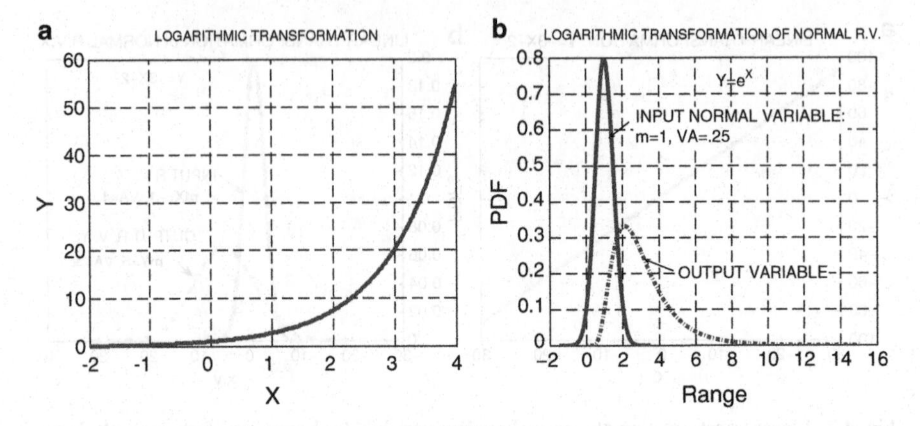

Fig. 4.8 Logarithmic transformation. (**a**) Transformation. (**b**) PDFs

Solution The transformation is given as:

$$Y = |X| = \begin{cases} X & \text{for} \quad X>0, \\ -X & \text{for} \quad X<0. \end{cases} \tag{4.79}$$

From here

$$\left|\frac{dy}{dx}\right| = 1. \tag{4.80}$$

Using (2.170) and (4.80), we have:

$$f_Y(y) = f_X(-x) + f_X(x)|_{y=|x|} = 2f_X(y) = \frac{2}{\sqrt{2\pi}}e^{-\frac{y^2}{2}}, \quad y>0. \tag{4.81}$$

The transformation as well as the input and output PDFs are shown in Fig. 4.9a, b, respectively.

Next example illustrates how the discrete random variable is obtained from the transformation of a normal random variable.

Example 4.3.4 The transformation of a normal random variable X with a mean value equal to 0 and a variance equal to 1, is given as:

$$Y = \begin{cases} 0.3 & \text{for} \quad X>0, \\ -0.3 & \text{for} \quad X<0. \end{cases} \tag{4.82}$$

Find the PDF of the random variable Y.

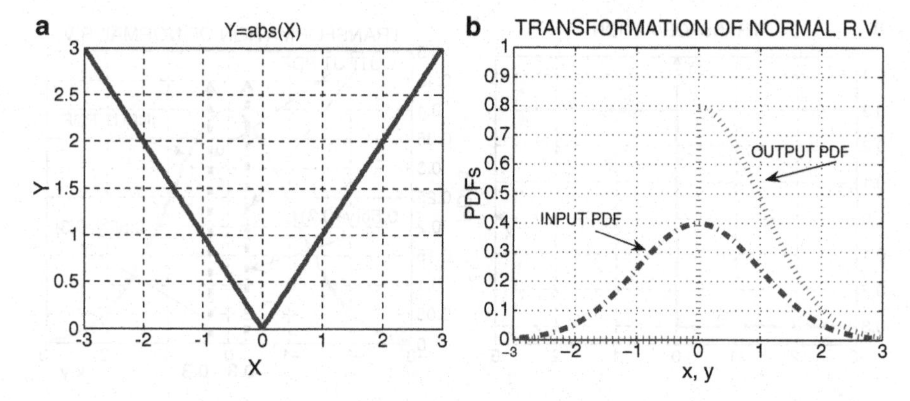

Fig. 4.9 Absolute value of a normal random variable. (**a**) Transformation. (**b**) Input and output PDFs

Solution The random variable Y takes only two discrete values: -0.3 and 0.3. Consequently, the random variable Y is a discrete random variable and its PDF will have two delta functions:

$$f_Y(y) = P\{Y = 0.3\}\delta(y - 0.3) + P\{Y = -0.3\}\delta(y + 0.3). \tag{4.83}$$

From (4.82), the random variable Y has the value 0.3 if the random variable X is positive. Therefore,

$$P\{Y = 0.3\} = P\{X>0\} = 0.5. \tag{4.84}$$

Similarly,

$$P\{Y = -0.3\} = P\{X<0\} = 0.5. \tag{4.85}$$

From (4.83) to (4.85), we have:

$$f_Y(y) = 0.5\delta(y - 0.3) + 0.5\delta(y + 0.3). \tag{4.86}$$

The input and output PDFs along with the transformation are shown in Fig. 4.10.

A normal random variable can also be transformed into a mixed random variable, as shown in the following example.

Example 4.3.5 The normal random variable has a mean value equal to 0, and a variance equal to 1. Find the PDF of the random variable Y, if

$$Y = \begin{cases} 1 & \text{for} \quad X>1 \\ X & \text{for} \quad -1<X<1 \\ -1 & \text{for} \quad X<-1 \end{cases}, \tag{4.87}$$

as shown in Fig. 4.11a.

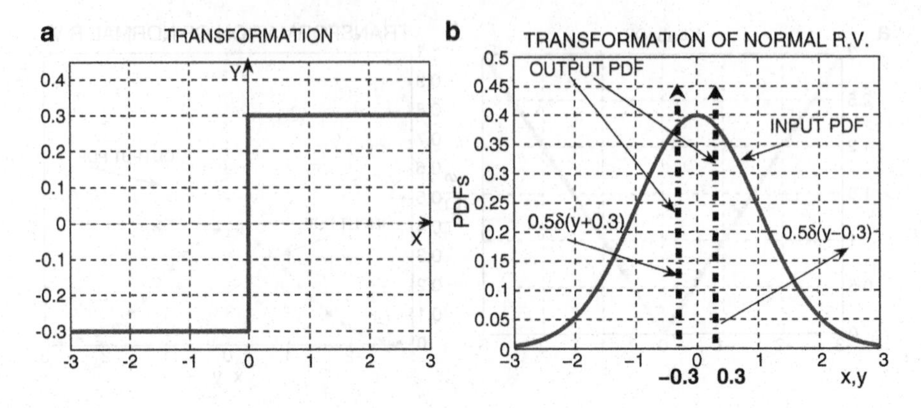

Fig. 4.10 Illustration of the transformation of a normal r.v. to a discrete r.v. in Example 4.3.4 (**a**) Transformation (**b**) Input and output PDFs

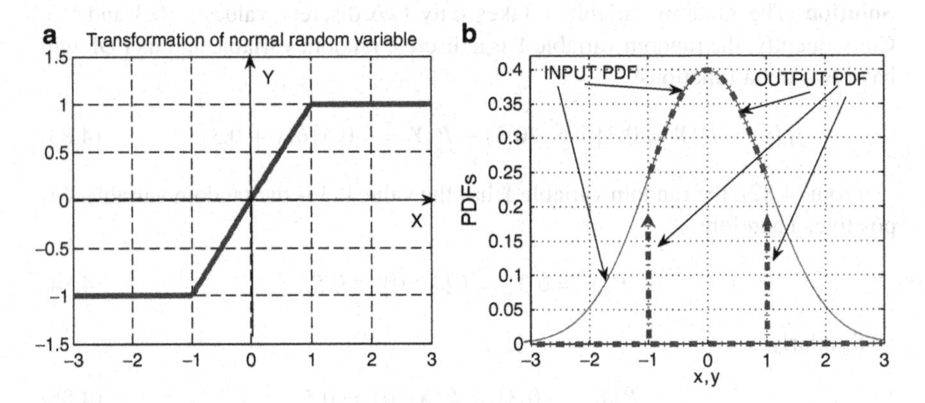

Fig. 4.11 Illustration of transformation of the normal r.v. to a mixed r.v. in Example 4.3.5 (**a**) Transformation (**b**) Input and output PDFs

Solution The random variable Y is continuous in the range $[-1, 1]$ and takes two discrete values -1 and 1.

For $-1 < X < 1$

$$f_Y(y) = f_X(x)|_{x=y} = \frac{1}{\sqrt{2\pi}}e^{-\frac{y^2}{2}}, \quad -1<y<1. \tag{4.88}$$

For $X > 1$, the random variable takes the discrete value 1.

$$P\{Y = 1\} = P\{X>1\}. \tag{4.89}$$

Similarly, for $X < 1$, the random variable takes the discrete value -1.

$$P\{Y = -1\} = P\{X < -1\}. \tag{4.90}$$

According to the normal PDF symmetry, the probability (4.89) is equal to (4.90). Using (4.47) and (4.59), we have:

$$P\{X > 1\} = P\{X < -1\} = Q(1) = 0.5\left(1 - \mathrm{erf}\left(\frac{1}{\sqrt{2}}\right)\right) = 0.1587. \tag{4.91}$$

Therefore, from (4.88) to (4.91) the output PDF is given as:

$$f_Y(y) = \frac{1}{\sqrt{2\pi}} \mathrm{e}^{-\frac{y^2}{2}} + 0.1587\delta(y-1) + 0.1587\delta(y+1) \quad \text{for} \quad -1 \le y \le 1. \tag{4.92}$$

The transformation and the corresponding PDFs are shown in Fig. 4.11a and b, respectively.

4.4 How to Generate a Normal Variable in MATLAB?

In this section, we consider the generation of a normal variable in MATLAB, using an arbitrary mean value and variance. However, the standard MATLAB file *randn.m* generates the normal random variable with the mean value 0 and variance 1, as shown in the following.

Example 4.4.1 Generate a normal random variable with a length of $N = 10{,}000$, a mean value 0, and a variance of 1. Estimate its PDF.

Solution The variable and the corresponding estimated PDF are shown in Fig. 4.12.

In order to generate the normal random variable with an arbitrary mean value and variance, let us recall the linear transformation of the normal random variable from Sect. 4.3, where it has been shown that the linear transformation

$$Y = aX + b \tag{4.93}$$

of the normal random variable X results in a normal random variable with a mean value and variance equal to:

$$\begin{aligned} m_Y &= a m_X + b, \\ \sigma_Y^2 &= a^2 \sigma_X^2. \end{aligned} \tag{4.94}$$

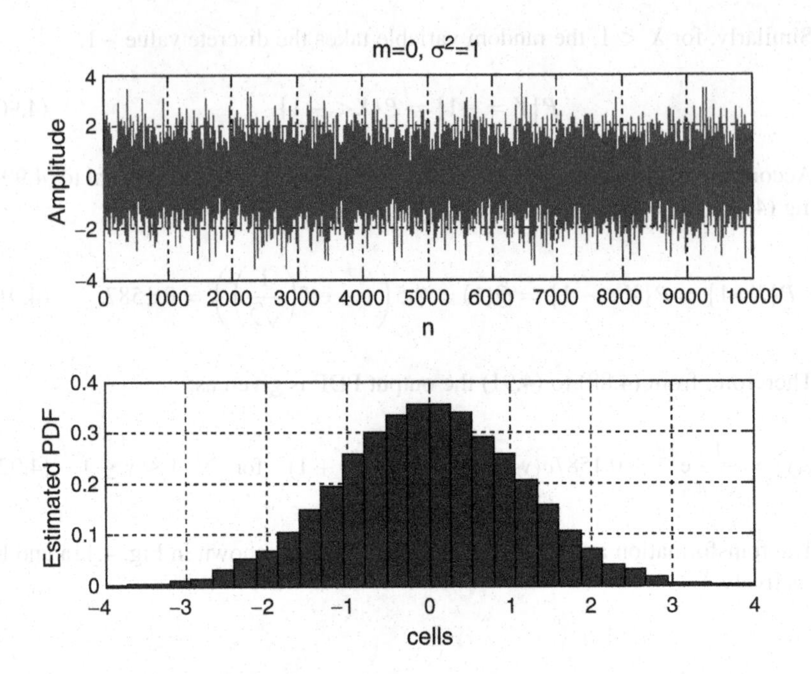

Fig. 4.12 Normal random variable and its estimated PDF

Consider that the variable X (generated using the MATLAB file *randn.m*) is a normal random variable with a mean value of 0 and a variance of 1. From (4.94), it follows:

$$m_Y = b; \qquad \sigma_Y^2 = a^2. \tag{4.95}$$

From (4.93) and (4.95), we get a linear transformation that has to be applied to the normal variable X in order to obtain a normal variable with the desired mean value and variance (m_Y and σ_Y^2, respectively) of length N

$$Y = \sigma_Y X + m_Y. \tag{4.96}$$

Using (4.93) the MATLAB file $r =$ *fnorm(N,VA,ME)* generates the desired normal variable r with a length of N and the given variance VA and mean value ME.

Example 4.4.2 Generate normal variables with the variance $\sigma^2 = 4$ and four different mean values: −6, −2, 3, 8. Estimate the corresponding PDFs.

Solution The generated variables are shown in Fig. 4.13a, and the corresponding estimated densities are shown in Fig. 4.13b. Note that the estimated densities have the same shape and are only shifted along the x-axis. Similarly, according to the "3σ rule" all signals have the same range around their mean values.

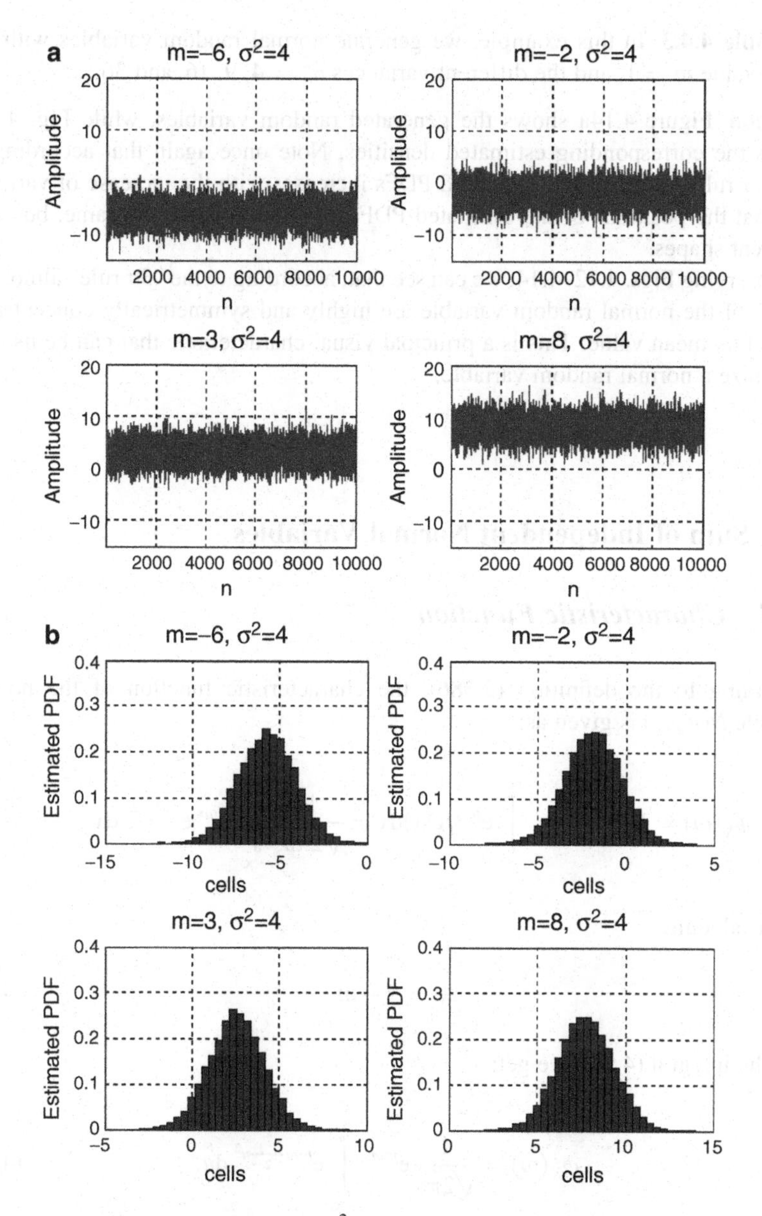

Fig. 4.13 Normal random variables with $\sigma^2 = 4$ and different mean values. (**a**) Normal random variables. (**b**) Estimated PDFs

Example 4.4.3 In this example, we generate normal random variables with the mean value $m = 1$, and the different variances $\sigma^2 = 4, 9, 16$, and 36.

Solution Figure 4.14a shows the generated random variables, while Fig. 4.14b shows the corresponding estimated densities. Note once again that according to the "3σ rule" range of the estimated PDFs increases with the increase of variance and that the positions of the estimated PDFs on the x-axis are the same, but with different shapes.

Observing Figs. 4.12–4.14, we can see that, according to the "3σ rule" almost all values of the normal random variable are highly and symmetrically concentrated around its mean value. This is a principal visual characteristic that can be used to recognize a normal random variable.

4.5 Sum of Independent Normal Variables

4.5.1 Characteristic Function

According to the definition (2.386), the characteristic function of the normal variable $N(m,\sigma^2)$ is given as:

$$\phi_X(\omega) = E\{\mathrm{e}^{j\omega X}\} = \int_{-\infty}^{\infty} \mathrm{e}^{j\omega x} f_X(x)\mathrm{d}x = \frac{1}{\sqrt{2\pi}\sigma} \int_{-\infty}^{\infty} \mathrm{e}^{j\omega x}\, \mathrm{e}^{-\frac{(x-m)^2}{2\sigma^2}}\, \mathrm{d}x. \tag{4.97}$$

Introducing

$$u = x - m. \tag{4.98}$$

into the integral (4.97), we get:

$$\phi_X(\omega) = \frac{1}{\sqrt{2\pi}\sigma} \mathrm{e}^{j\omega m} \int_{-\infty}^{\infty} \mathrm{e}^{j\omega u - \frac{1}{2\sigma^2}u^2}\, \mathrm{d}u. \tag{4.99}$$

Using integral 6 from Appendix A $\left(\int_{-\infty}^{\infty} \mathrm{e}^{-a^2x^2+bx}\, \mathrm{d}x = \frac{\sqrt{\pi}}{a} \mathrm{e}^{\frac{b^2}{4a^2}} \right)$, we arrive at:

$$\phi_X(\omega) = \mathrm{e}^{j\omega m - \frac{\omega^2\sigma^2}{2}}. \tag{4.100}$$

Figure 4.15 shows the absolute values of characteristic functions for $m = 0, 2$, and -3, and $\sigma^2 = 1, 4$, and 16.

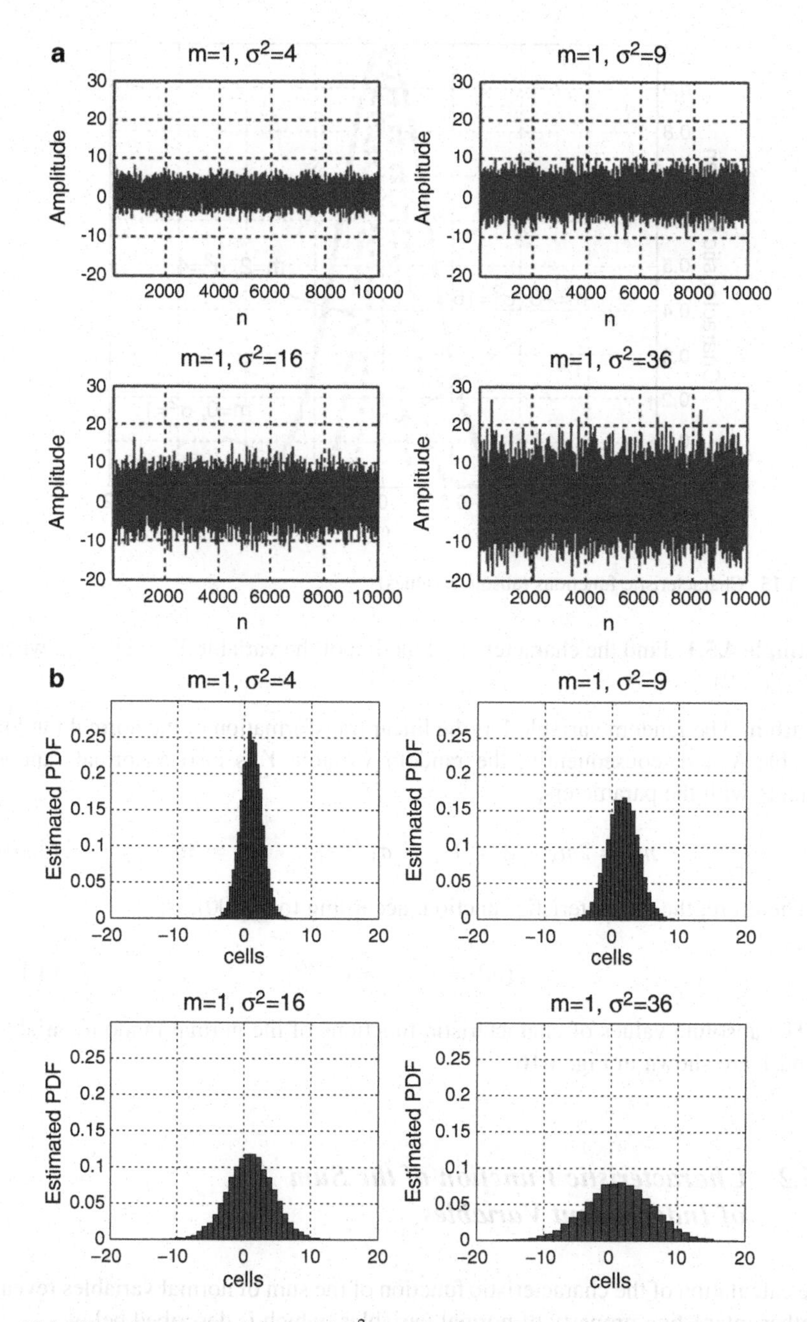

Fig. 4.14 Normal random variables with $\sigma^2 = 4, 9, 16$, and 36 and $m = 1$. (**a**) Normal variables. (**b**) Estimated PDFs

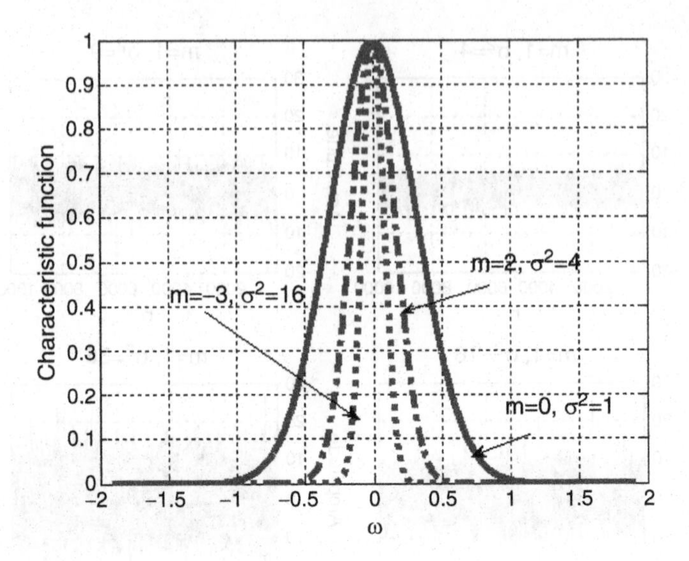

Fig. 4.15 Characteristic functions (absolute values)

Example 4.5.1 Find the characteristic function of the variable $Y = 2X - 2$, where $X = N(1, 9)$.

Solution The random variable Y is the linear transformation of the normal random variable X and, consequently, the random variable Y is also a normal random variable with the parameters,

$$m_Y = 2m_X - 2 = 0, \qquad \sigma_X^2 = 9, \qquad \sigma_Y^2 = 36. \tag{4.101}$$

Therefore, the characteristic function, according to (4.100), is:

$$\phi_Y(\omega) = \mathrm{e}^{-\frac{\omega^2 36}{2}} = \mathrm{e}^{-18\omega^2}. \tag{4.102}$$

The absolute values of characteristic functions of the normal random variables X and Y are shown in Fig. 4.16.

4.5.2 *Characteristic Function of the Sum of Independent Variables*

The calculation of the characteristic function of the sum of normal variables reveals another interesting property of normal variables, which is described below.

Consider two independent normal random variables $X_1 = N(m_1, \sigma_1^2)$ and $X_2 = N(m_2, \sigma_2^2)$. Denote their sum as the variable X,

$$X = X_1 + X_2. \tag{4.103}$$

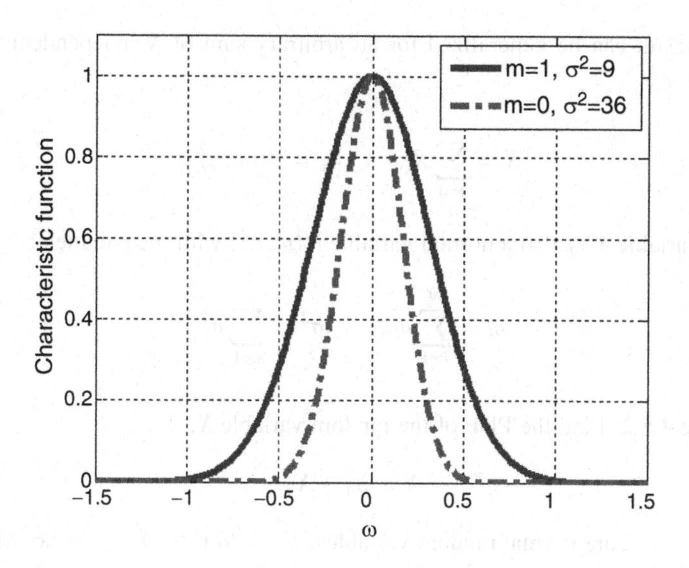

Fig. 4.16 Characteristic functions for normal variables *X* and *Y* (absolute values)

The mean value and the variance of variable *X* are shown below (see (3.95) and (3.128)):

$$m = m_1 + m_2, \qquad \sigma^2 = \sigma_1^2 + \sigma_2^2. \tag{4.104}$$

The characteristic function of the sum of independent variables is equal to the product of the characteristic functions:

$$\phi(\omega) = \phi_1(\omega)\phi_2(\omega). \tag{4.105}$$

The characteristic functions of the normal variables X_i, $i = 1, 2$, according to (4.100), is:

$$\phi_{X_i}(\omega) = e^{j\omega m_i - \frac{\omega^2 \sigma_i{}^2}{2}}, \quad i = 1, 2. \tag{4.106}$$

From (4.103) to (4.106) we have:

$$\phi_X(\omega) = e^{j\omega m_1 - \frac{\omega^2 \sigma_1{}^2}{2}} e^{j\omega m_2 - \frac{\omega^2 \sigma_2{}^2}{2}} = e^{j\omega(m_1+m_2) - \frac{\omega^2(\sigma_1{}^2+\sigma_2{}^2)}{2}} = e^{j\omega m - \frac{\omega^2\sigma^2}{2}}. \tag{4.107}$$

Comparing (4.107) with the expression for the characteristic function of a normal variable (4.100), we can see that the expressions are equal, thus indicating that *the sum of normal random variables is also a normal random variable.*

This result can be generalized for an arbitrary sum of N independent random variables:

$$X = \sum_{i=1}^{N} X_i, \qquad X_i = N(m_i, \sigma_i^2). \tag{4.108}$$

The variable X is also a normal variable $N(m, \sigma^2)$ with the parameters:

$$m = \sum_{i=1}^{N} m_i, \qquad \sigma^2 = \sum_{i=1}^{N} \sigma_i^2. \tag{4.109}$$

Example 4.5.2 Find the PDF of the random variable X,

$$X = X_1 + X_2. \tag{4.110}$$

where $i = 1, 2$ are normal random variables: $X_1 = N(1, 5)$, $X_2 = N(2.6, 8)$.

Solution According to the observations in this section, the random variable X is also a normal random variable with the parameters:

$$m = 1 + 2.6 = 3.6, \qquad \sigma^2 = 5 + 8 = 13. \tag{4.111}$$

The corresponding signals and PDFs are shown in Fig. 4.17.

4.5.3 *Sum of Linear Transformations of Independent Normal Random Variables*

Consider independent normal random variables $X_i(m_i,{\sigma_i}^2)$. The linear transformation

$$Y_i = a_i X_i + b_i \tag{4.112}$$

results in independent normal random variables Y_i. Their sum denoted as Y,

$$Y = \sum_{i=1}^{N} Y_i = \sum_{i=1}^{N} a_i X_i + b_i \tag{4.113}$$

is also normal variable with parameters

$$m_Y = \sum_{i=1}^{N} a_i m_i + b_i, \qquad \sigma_Y^2 = \sum_{i=1}^{N} a_i^2 \sigma_i^2. \tag{4.114}$$

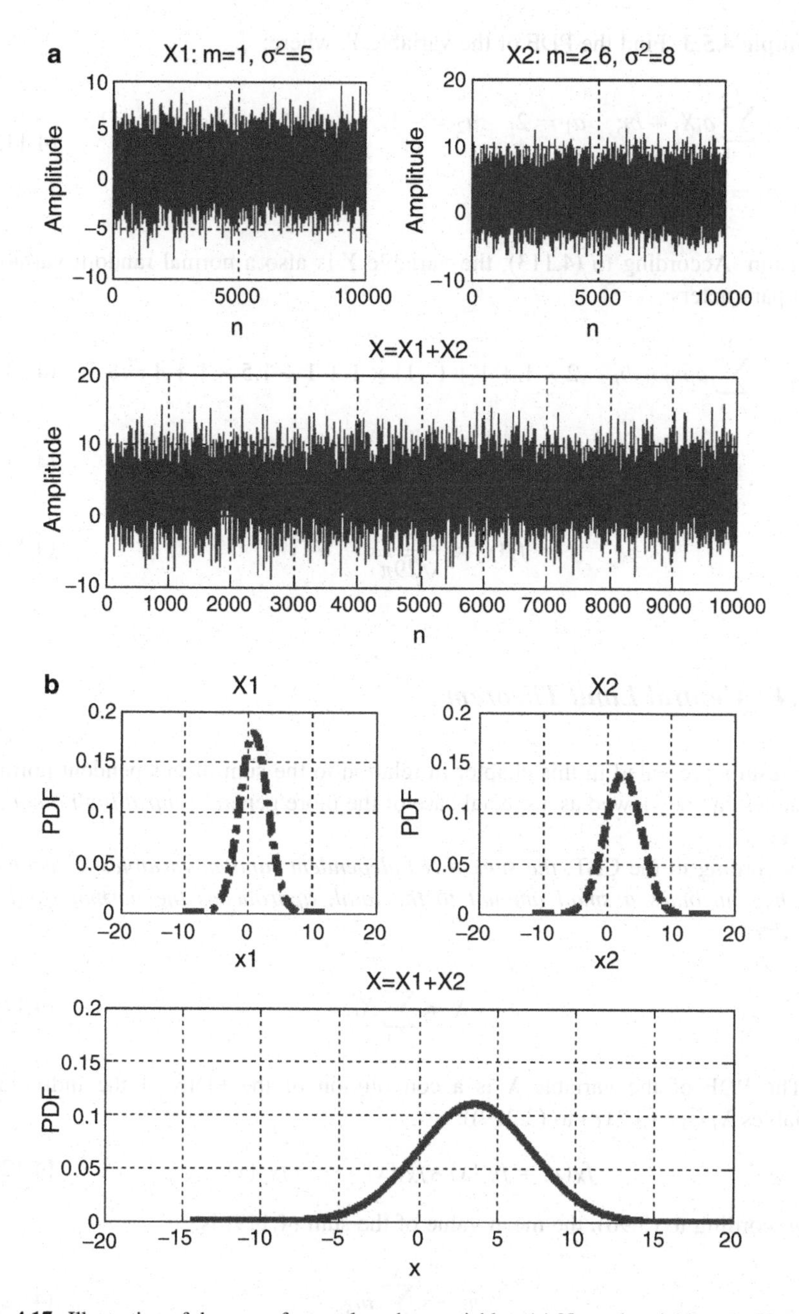

Fig. 4.17 Illustration of the sum of normal random variables. (**a**) Normal variables. (**b**) PDFs

Example 4.5.3 Find the PDF of the variable Y, where:

$$Y=\sum_{i=1}^{3} a_i X_i + b_i; \quad a_1=2, \quad a_2=-1, \quad a_3=1.5; \quad b_1=b_2=1,$$
$$b_3=4; \quad X_i=N(1,2). \tag{4.115}$$

Solution According to (4.113), the variable Y is also a normal random variable with parameters:

$$m_Y=\sum_{i=1}^{3} a_i m_i + b_i = 2\times 1+1+(-1)\times 1+1+1.5\times 1+4=8.5. \tag{4.116}$$

$$\sigma_Y^2=\sum_{i=1}^{3} a_i^2\sigma_i^2 = 4\times 2+1\times 2+2.25\times 2=14.5. \tag{4.117}$$

$$f_Y(y)=\frac{1}{\sqrt{29\pi}}\mathrm{e}^{-\frac{(y-8.5)^2}{29}}. \tag{4.118}$$

4.5.4 *Central Limit Theorem*

The results presented in this chapter in relation to the sum of independent normal variables can be viewed as a special case of the more general *central limit theorem* (CLT).

According to the CLT, *the sum of N independent random variables X_i, each of which contributes a small amount to the total, approaches the normal random variable.*

$$X=\sum_{i=1}^{N} X_i. \tag{4.119}$$

The PDF of the variable X is a convolution of the PDFs of the individual variables X_i (see (3.218) to (3.219)):

$$f_X(x)=f_{X_1}(x)*f_{X_2}(x)*\cdots*f_{X_N}(x). \tag{4.120}$$

According to (3.96), the mean value of the sum (4.119) is:

$$m_X=\sum_{i=1}^{N} m_{X_i}. \tag{4.121}$$

Similarly, using the result (3.129) we find the variance of the variable X_i in (4.119):

$$\sigma_X^2=\sum_{i=1}^{N} \sigma_{X_i}^2. \tag{4.122}$$

According to the CLT and using (4.121) and (4.122), we have:

$$f_X(x) \rightarrow \frac{1}{\sqrt{2\pi}\sigma_X} \mathrm{e}^{-\frac{(x-m_X)^2}{2\sigma_X^2}}, \text{ as } N \rightarrow \infty. \tag{4.123}$$

The proof of the CLT can be found, for example, in [LEO94, pp. 287–288].

The CLT can be interpreted as a property of the convolutions (4.120) involving a large number of positive functions. This result is independent of the type of the random variables in the sum, supposing that the participation of each variable in the sum is small in comparison with the total sum. The value of N in (4.119) must be a large number in order to obtain a good approximation to the normal variable. However, if the PDFs are "reasonably concentrated near mean value," [PAP65, p. 267], then a close approximation to (4.123) is obtained even for moderate values of N. To see more discussion of the normal approximation for a finite number of N, see [BRE69, pp. 117–119].

The CLT shows why the normal random variables are so important in different applications. There are many situations in which a random occurrence can be considered the result of many independent occurrences where the participation of each occurrence in the total sum is small. For example, the resulting error of a given measurement can be viewed as the result of many independent random occurrences where the participation of each of them in the total error is small, and can thus be considered a normal random variable.

The CLT is especially important in communications, where the noise which is added in the channel to the signal, is the result of different sources, and can thus be presented as normal noise.

Similarly, thermal noise is the result of random independent movement of electrons and can thus be presented as a normal noise.

In some special cases, the CLT can also be applied to dependent random variables [PEE93, p. 118].

Finally, we must mention that a more adequate name for the CLT would be the theorem of normal convergence. However, the word "central" is useful to remind us that the PDF converges to the normal PDF around the center of the mean value and more errors are expected at the tails of the PDF.

4.6 Jointly Normal Variables

4.6.1 Two Jointly Normal Variables

Consider two normal random variables, $X_1 = N(m_1, \sigma_1^2)$ and $X_2 = N(m_2, \sigma_2^2)$. The variables X_1 and X_2 are jointly normal if their joint density function is given as:

$$f_{X_1,X_2}(x_1,x_2) = \frac{1}{2\pi\sigma_1\sigma_2\sqrt{1-\rho_{1,2}^2}} e^{-\frac{1}{2(1-\rho_{1,2}^2)}\left[\frac{(x_1-m_1)^2}{\sigma_1^2} + \frac{(x_2-m_2)^2}{\sigma_2^2} - 2\rho_{1,2}\frac{(x_1-m_1)(x_2-m_2)}{\sigma_1\sigma_2}\right]}, \tag{4.124}$$

where $\rho_{1,2}$ is the coefficient of correlation.

The expression (4.124) shows some interesting properties of normal random variables:

P.1 The joint density function (4.124) is completely determined by first two moments of marginal variables: the mean value and the mean squared value (note that the variance and the correlation coefficient are given by (2.338) and (3.130), respectively).

P.2 If the random variables are jointly normal, then the marginal variables are also normal.

P.3 If normal random variables X_1 and X_2 are not correlated (i.e., the coefficient of correlation is equal to zero), then they are also independent,

$$\rho_{1,2} = 0. \tag{4.125}$$

From (4.124) and (4.125), we arrive at:

$$\begin{aligned} f_{X_1,X_2}(x_1,x_2) &= \frac{1}{2\pi\sigma_1\sigma_2} \mathrm{e}^{-\frac{1}{2}\left[\frac{(x_1-m_1)^2}{\sigma_1^2}+\frac{(x_2-m_2)^2}{\sigma_2^2}\right]} \\ &= \frac{1}{\sqrt{2\pi}\sigma_1} \mathrm{e}^{-\frac{(x_1-m_1)^2}{2\sigma_1^2}} \frac{1}{\sqrt{2\pi}\sigma_2} \mathrm{e}^{-\frac{(x_2-m_2)^2}{2\sigma_2^2}} = f_{X_1}(x_1) f_{X_2}(x_2). \end{aligned} \tag{4.126}$$

From (4.126), we see that the joint density function is equal to the product of the corresponding densities indicating that the random variables X_1 and X_2 are independent.

Therefore, if the normal random variables are not correlated, they are also independent. This is exceptional and holds only for normal random variables. Previously, in Chap. 3, it was stated that the condition of noncorrelation is weaker than that of independence and that, in general, if random variables are uncorrelated they can be either dependent or independent.

Example 4.6.1 The joint density of the random variables X_1 and X_2 is given as:

$$f_{X_1,X_2}(x_1,x_2) = \frac{1}{12\pi} \mathrm{e}^{-\frac{1}{2}\left[\frac{x_1^2}{4}+\frac{(x_2-2)^2}{9}\right]}. \tag{4.127}$$

Find the PDF of the random variable

$$X = X_1 - X_2. \tag{4.128}$$

Solution From (4.127), we note that X_1 and X_2 are independent normal random variables with the corresponding mean values and variances,

$$m_1 = 0, \quad m_2 = 2, \quad \sigma_1^2 = 4, \quad \sigma_2^2 = 9. \tag{4.129}$$

From (4.128), it follows that X is also a normal random variable with parameters:

$$m = m_1 - m_2 = -2, \qquad \sigma^2 = 4 + 9 = 13. \tag{4.130}$$

Therefore, the PDF of the variable X is given as:

$$f_x(x) = \frac{1}{\sqrt{26\pi}} e^{-x^2/26}. \tag{4.131}$$

4.6.2 N-*Jointly Normal Random Variables*

Two jointly normal random variables can be extended to N-dimensional case, thus leading to N-jointly normal random variables. From (4.124), we can conclude that the corresponding expression for the N-dimensional case would be very complex. It is much simpler to write the N-dimensional density in its matrix form.

To this end, consider the column vector $\boldsymbol{X}$ of random variables, such that its transpose is a $(1 \times N)$ row vector given as:

$$\boldsymbol{X}^{\mathrm{T}} = [X_1, \ldots, X_N], \tag{4.132}$$

where upper index T means transpose.

The parameters of the normal vector (4.132) are the mean vector $\boldsymbol{m_X}$ and the $(N \times N)$ covariance matrix $\mathbf{C_X}$, given in the following expression:

$$\boldsymbol{m}_{\boldsymbol{X}}^{\mathrm{T}} = E\{\boldsymbol{X}^{\mathrm{T}}\} = [m_1, \ldots, m_N], \tag{4.133}$$

where m_i is the mean value of the random variable X_i.

The covariance matrix $\mathbf{C_X}$ is an $(N \times N)$ matrix of variances and covariances,

$$\mathbf{C_X} = E\left\{(\boldsymbol{X} - \boldsymbol{m_X})(\boldsymbol{X} - \boldsymbol{m_X})^{\mathrm{T}}\right\} = \begin{bmatrix} \sigma_1^2 & C_{1,2} & \ldots & C_{1,N} \\ C_{2,1} & \sigma_2^2 & \ldots & C_{2,N} \\ \ldots & \ldots & \ldots & \ldots \\ C_{N,1} & C_{N,2} & \ldots & \sigma_N^2 \end{bmatrix}, \tag{4.134}$$

where σ_i^2 is the variance of the random variable X_i, and $C_{i,j}$ is the covariance of the random variables X_i and X_j. Because of $C_{i,j} = C_{j,i}$ the covariance matrix is symmetric. Denoting the determinant and the inverse matrix of the matrix (4.134) as $|\mathbf{C_X}|$ and $\mathbf{C_X^{-1}}$, respectively, we have the PDF of N-jointly normal variables:

$$f_{\mathbf{X}}(\boldsymbol{X}) = \frac{1}{(2\pi)^{N/2}\sqrt{|\mathbf{C_X}|}} e^{\dfrac{-(\boldsymbol{X} - \boldsymbol{m_X})^{\mathrm{T}}\mathbf{C_X^{-1}}(\boldsymbol{X} - \boldsymbol{m_X})}{2}}. \tag{4.135}$$

This expression can be rewritten in the following form:

$$f_{\mathbf{X}}(\boldsymbol{X}) = \frac{1}{(2\pi)^{N/2}\sqrt{|\mathbf{C_X}|}} \mathrm{e}^{-\left[\frac{1}{2|\mathbf{C_X}|}\sum_i \sum_j (x_i - m_i)(x_j - m_j)\Delta_{i,j}\right]}; \quad i,j = 1,\ldots,N, \tag{4.136}$$

where $\Delta_{i,j}$ is the cofactor of the matrix $\mathbf{C_X}$.

Example 4.6.2 In this example, we demonstrate how the PDF of two jointly normal random variables (4.124) can be derived from the general expression (4.136).

Solution The covariance matrix $\mathbf{C_X}$ is

$$\mathbf{C_X} = \begin{bmatrix} \sigma_1^2 & C_{1,2} \\ C_{2,1} & \sigma_2^2 \end{bmatrix}, \tag{4.137}$$

where the covariances are:

$$C_{1,2} = C_{2,1} = E\{X_1 - m_1)(X_2 - m_2)\} = \rho_{1,2}\sigma_1\sigma_2. \tag{4.138}$$

The determinant $|\mathbf{C_X}|$ is given as:

$$|\mathbf{C_X}| = \begin{vmatrix} \sigma_1^2 & C_{1,2} \\ C_{2,1} & \sigma_2^2 \end{vmatrix} = \sigma_1^2\sigma_2^2 - C_{1,2}^2. \tag{4.139}$$

Using (4.138), the determinant (4.139) is expressed as:

$$|\mathbf{C_X}| = \sigma_1^2\sigma_2^2(1 - \rho_{1,2}^2). \tag{4.140}$$

Its cofactors are:

$$\Delta_{1,1} = \sigma_2^2; \quad \Delta_{2,2} = \sigma_1^2; \quad \Delta_{1,2} = \Delta_{2,1} = -C_{1,2} = -C_{2,1} = -\rho_{1,2}\sigma_1\sigma_2. \tag{4.141}$$

Placing (4.138)–(4.141) into (4.136), we get:

$$\begin{aligned} f_{X_1,X_2}(x_1,x_2) &= \frac{1}{2\pi\sigma_1\sigma_2\sqrt{1-\rho_{1,2}^2}} \exp\left(-\frac{1}{2(1-\rho_{1,2}^2)}\left[\frac{(x_1-m_1)^2\sigma_2^2}{\sigma_1^2\sigma_2^2} + \frac{(x_2-m_2)^2\sigma_1^2}{\sigma_1^2\sigma_2^2} - 2\rho_{1,2}\sigma_1\sigma_2\frac{(x_1-m_1)(x_2-m_2)}{\sigma_1^2\sigma_2^2}\right]\right) \\ &= \frac{1}{2\pi\sigma_1\sigma_2\sqrt{1-\rho_{1,2}^2}} \exp\left(-\frac{1}{2(1-\rho_{1,2}^2)}\left[\frac{(x_1-m_1)^2}{\sigma_1^2} + \frac{(x_2-m_2)^2}{\sigma_2^2} - 2\rho_{1,2}\frac{(x_1-m_1)(x_2-m_2)}{\sigma_1\sigma_2}\right]\right), \end{aligned} \tag{4.142}$$

which is the same expression as in (4.124).

4.7 Summary of Properties

Normal variables have several useful properties, which are summarized here.

P.1 A normal PDF is completely determined from its two first moments and consequently, it is often given in the simple form

$$f_X(x) = N(m, \sigma^2),$$

where m is its mean value and σ^2 is its variance.

P.2 The PDF is symmetrical around its mean value.

P.3 By increasing x, the PDF approaches to zero. Theoretically,

$$f_x(x) \to 0, \text{ as } x \to \pm\infty$$

P.4 The "3 σ rule" states that the absolute value of the dissipation of a normal random variable from its mean value practically does not overpass its 3 standard deviations.

P.5 The linear transformation $Y = aX + b$ of the normal random variable results in a normal variable with parameters

$$m_Y = am_X + b,$$
$$\sigma_Y^2 = (a\sigma_X)^2 = a^2\sigma_X^2.$$

P.6 The sum of N independent normal random variables

$$X = \sum_{i=1}^{N} X_i, \quad X_i = N(m_i, \sigma_i^2),$$

is also a normal variable $N(m, \sigma^2)$ with corresponding parameters:

$$m = \sum_{i=1}^{N} m_i, \quad \sigma^2 = \sum_{i=1}^{N} \sigma_i^2.$$

P.7 The joint normal density function is completely determined by its first two moments, i.e., mean value and the mean squared value.

P.8 If random variables are jointly normal, then the marginal variables are also normal variables.

P.9 If random variables are not correlated, it follows that they are also independent.

4.8 Numerical Exercises

Exercise 4.1 The normal random variable X has the parameters

$$\overline{X} = 5; \qquad \overline{X^2} = 81. \tag{4.143}$$

Find the probability that the random variable is in the interval [4, 20] using the erf and Q functions.

Answer Using (4.143) the variance of the variable X is:

$$\sigma^2 = \overline{X^2} - \overline{X}^2 = 81 - 5^2 = 56. \tag{4.144}$$

Using (4.31) and using the MATLAB function erf, we have:

$$\begin{aligned} P\{x_1 \leq X \leq x_2\} = P\{4 \leq X \leq 20\} &= \frac{1}{2}\left[\operatorname{erf}\left(\frac{x_2 - m}{\sqrt{2}\sigma}\right) - \operatorname{erf}\left(\frac{x_1 - m}{\sqrt{2}\sigma}\right)\right] \\ &= \frac{1}{2}\left[\operatorname{erf}\left(\frac{20 - 5}{\sqrt{2 \times 56}}\right) - \operatorname{erf}\left(\frac{4 - 5}{\sqrt{2 \times 56}}\right)\right] \\ &= \frac{1}{2}[0.9550 - (-0.1063)] = 0.5306. \end{aligned} \tag{4.145}$$

The desired probability (Fig. 4.18) can also be expressed as:

$$P\{4 \leq X \leq 20\} = P\{X>4\} - P\{X>20\} = 1 - Q(k_1) - Q(k_2). \tag{4.146}$$

From the definition of the Q function (4.47) and (4.146), we have:

$$4 = m - k_1\sigma; \qquad 20 = m + k_2\sigma. \tag{4.147}$$

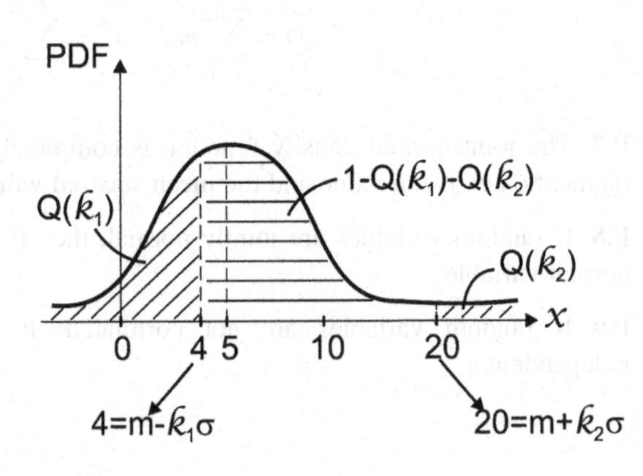

Fig. 4.18 The probability and Q function

From (4.147), (4.143) and (4.144), we arrive at:

$$\begin{aligned} k_1 &= 0.1336, \\ k_2 &= 2.0045. \end{aligned} \tag{4.148}$$

From (4.146) and (4.148), it follows:

$$P\{4 \le X \le 20\} = 1 - Q(0.1336) - Q(2.0045). \tag{4.149}$$

From (4.59), we have:

$$Q(0.1336) = \frac{1}{2}\left[1 - \operatorname{erf}\left(\frac{0.1336}{\sqrt{2}}\right)\right] = 0.4469. \tag{4.150}$$

$$Q(2.0045) = \frac{1}{2}\left[1 - \operatorname{erf}\left(\frac{2.0045}{\sqrt{2}}\right)\right] = 0.0225. \tag{4.151}$$

Finally, from (4.146), (4.150), and (4.151), we get:

$$\begin{aligned} P\{4 \le X \le 20\} &= 1 - Q(0.1336) - Q(2.0045) \\ &= 0.5531 - 0.0225 = 0.5306, \end{aligned} \tag{4.152}$$

which is the same result as we had in (4.145).

Exercise 4.2 Find the symmetrical interval of $k\sigma$ around the mean value $m = 16$ for the normal random variable X with a variance equal to 16, for which the probability that the variable is in the interval is equal to 0.683:

$$P\{16 - k \times 4 \le X \le 16 + k \times 4\} = 0.683. \tag{4.153}$$

Answer From (4.33), the probability that the random variable is in the symmetrical interval around its mean value is:

$$P\{16 - k \times 4 \le X \le 16 + k \times 4\} = \operatorname{erf}\left(\frac{k}{\sqrt{2}}\right) = 0.683. \tag{4.154}$$

From here, using the MATLAB file *erfinv.m,* we get:

$$\operatorname{erfinv}(0.683) = 0.7076 \tag{4.155}$$

or equivalently

$$\frac{k}{\sqrt{2}} = 0.7076 \tag{4.156}$$

resulting in $k = 0.9984 \approx 1$. Therefore, the symmetrical interval is [16 – 4, 16 + 4] = [12, 20]; this is shown in Fig. 4.19.

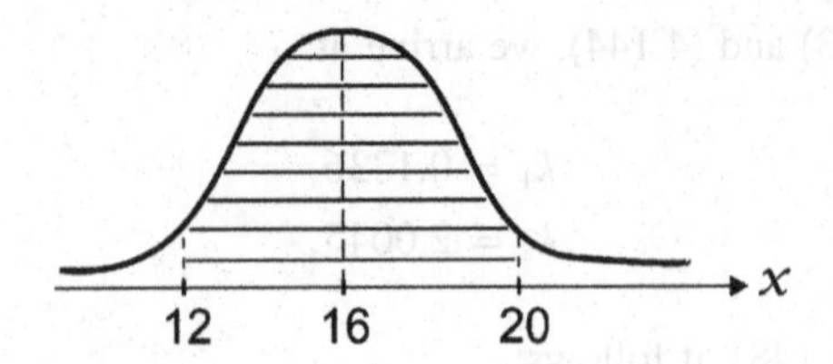

Fig. 4.19 Probability in Exercise 4.2

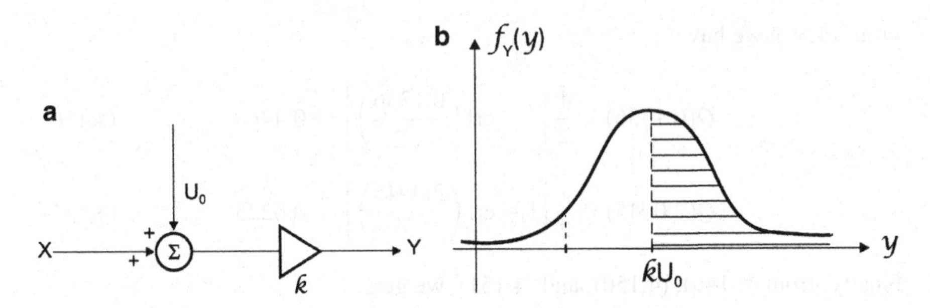

Fig. 4.20 Illustration of Exercise 4.4

Exercise 4.3 For a normal random noise X with a variance equal to 9 and a mean value equal to zero, find the value c so that we find

$$|X|<c, \tag{4.157}$$

99% of the time.

Answer Considering the mean value $m = 0$ and the standard deviation $\sigma = 3$ while using (4.31) it follows:

$$P\{|X|<c\} = P\{-c<X<c\} = 0.99 = \text{erf}\left(c/3\sqrt{2}\right). \tag{4.158}$$

Using *erfinv.m* we have:

$$\text{erfinv}(0.99) = 1.8214, \tag{4.159}$$

resulting in:

$$c = 3\sqrt{2} \times 1.8214 = 7.7275. \tag{4.160}$$

Exercise 4.4 A DC component U_0 is added to a normal noise with a mean value and variance equal to 0, and σ^2, respectively. Next, the sum is amplified k-times, as shown in Fig. 4.20a.

(a) Find the probability that the amplitude of the output noise Y will be more than kU_0.
(b) Determine whether or not this probability depends on the variance of the input noise X.

Answer

(a) The output noise is the linear transformation of the input noise X, and thus is also a normal noise.

$$Y = (X + U_0)k = Xk + U_0k. \quad (4.161)$$

Knowing that the mean or expected value of X is zero, the mean value of the normal random variable Y becomes:

$$E\{Y\} = E\{Xk + U_0k\} = E\{Xk\} + E\{U_0k\} = E\{U_0k\} = U_0k. \quad (4.162)$$

The normal random variable is symmetrical around its mean value, resulting in:

$$P\{X > kU_0\} = 1/2. \quad (4.163)$$

(b) The result (4.163) does not depend on the value of the variance, as also shown in Fig. 4.21b.

Exercise 4.5 Find the density of the output random variable Y if the transformation of the input normal random variable $X = N(0, 1)$ it is given in Fig. 4.21a.

Answer The input normal random variable is symmetrical around its mean value which is zero, resulting in:

$$P\{X < 0\} = P\{X > 0\} = 1/2. \quad (4.164)$$

The output random variable is a discrete random variable taking only the values U and $-U$,

$$P\{Y = U\} = P\{X > 0\} = 1/2; \qquad P\{Y = -U\} = P\{X < 0\} = 1/2. \quad (4.165)$$

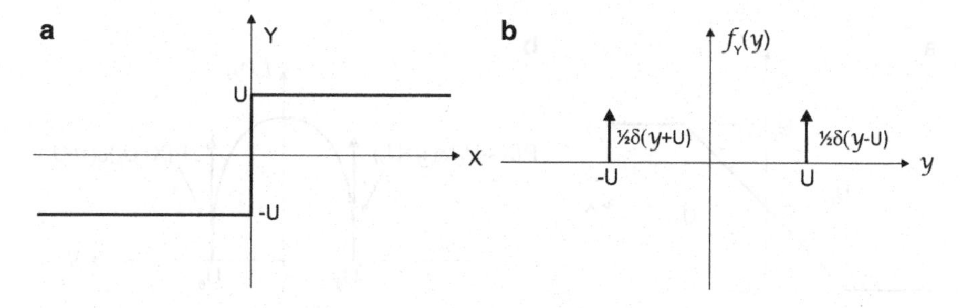

Fig. 4.21 Transformation of normal variable and the output density function

The density function of the output random variable is given as:

$$f_Y(y) = 0.5\delta(y - U) + 0.5\delta(y + U), \tag{4.166}$$

and it is shown in Fig. 4.21b.

Exercise 4.6 Find the PDF of the output random variable Y where the input random variable $X = N(0, 1)$ and the output variable is obtained from the output of the limiter, as shown in Fig. 4.22a.

Answer The output random variable Y is mixed, having two discrete values U_0 and $-U_0$, and is continuous for $\{-U_0 < X < U_0\}$. The discrete values have the following probabilities:

$$P\{Y = U_0\} = P\{Y = -U_0\} = 0.5\left[1 + \text{erf}(U_0/\sqrt{2})\right]. \tag{4.167}$$

For the values of the input variable $\{-U_0 < X < U_0\}$, the output random variable Y is equal to the input variable X, thus having the same PDF:

$$f_Y(y) = f_X(x) = \frac{1}{\sqrt{2\pi}}\mathrm{e}^{-\frac{y^2}{2}} \quad \text{for} \quad -U_0 < Y < U_0. \tag{4.168}$$

From (4.167) and (4.168), the corresponding PDF of the variable Y is:

$$f_Y(y) = \begin{cases} P\{Y = -U_0)\delta(y + U_0) & \text{for} \quad Y = -U_0 \\ \frac{1}{\sqrt{2\pi}}\mathrm{e}^{-\frac{y^2}{2}} & \text{for} \quad -U_0 < Y < U_0 \\ P\{Y = U_0)\delta(y - U_0) & \text{for} \quad Y = U_0 \end{cases}. \tag{4.169}$$

The PDF (4.169) is shown in Fig. 4.22b.

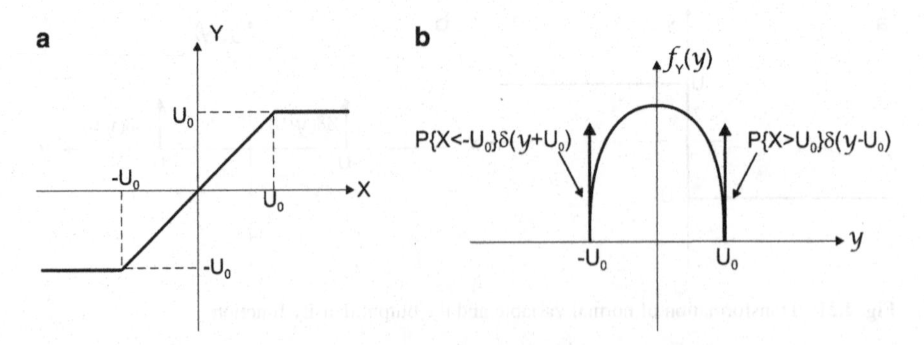

Fig. 4.22 Transformation of a normal variable and the output density function

Exercise 4.7 A normal noise with a zero mean and a variance of σ^2 enter into a circuit with the characteristics shown in Fig. 4.23a. Find the expression and plot the PDF of the noise in the circuit's output.

Answer For the negative values of the input noise the output noise will be equal to zero,

$$P\{Y=0\}=P\{X<0\}. \tag{4.170}$$

The PDF of the input noise is symmetrical around zero, resulting in:

$$P\{Y=0\}=P\{X<0\}=1/2. \tag{4.171}$$

For the positive values of the input noise, the output noise Y will be equal to the input noise thus having

$$f_Y(y)=f_X(x)=\frac{1}{\sqrt{2\pi\sigma^2}}\mathrm{e}^{-\frac{y^2}{2\sigma^2}} \quad \text{for} \quad y>0. \tag{4.172}$$

From (4.170) to (4.172), we have:

$$f_Y(y)=\begin{cases} 1/2\delta(y) & \text{for} \quad y=0 \\ \frac{1}{\sqrt{2\pi\sigma^2}}\mathrm{e}^{-\frac{y^2}{2\sigma^2}} & \text{for} \quad y>0 \end{cases}. \tag{4.173}$$

The density is shown in Fig. 4.23b.

Exercise 4.8 Find the characteristic function of the random variable Y at the output of the block shown in Fig. 4.24, if the input random variable X is the normal variable with the zero mean value and the variance $\sigma^2=1$.

Answer The output random variable Y is a discrete random variable with the values 0 and 1.

$$P\{Y=1\}=P\{X<1.5]. \tag{4.174}$$

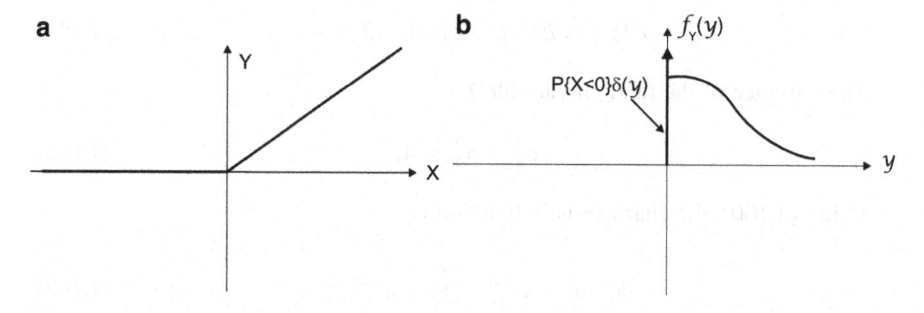

Fig. 4.23 The circuit characteristic and the output PDF

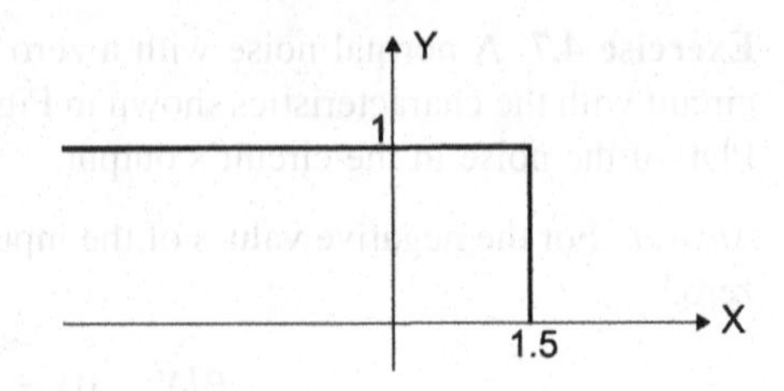

Fig. 4.24 The characteristic of the transformation block

$$P\{Y=0\}=P\{X>1.5\}=1-P\{Y=1\}. \tag{4.175}$$

From (4.30), we have

$$P\{X\leq 1.5\}=F_X(1.5)=\frac{1}{2}\left[1+\operatorname{erf}\left(\frac{1.5}{\sqrt{2}}\right)\right]=0.9332. \tag{4.176}$$

From (4.174) to (4.176), we have:

$$P\{Y=1\}=0.9332, \qquad P\{Y=0\}=0.0668. \tag{4.177}$$

From (2.419), we have:

$$\phi_Y(\omega)=\sum_i \mathrm{e}^{j\omega y_i}P(Y=y_i)=\mathrm{e}^{j\omega\times 1}P\{Y=1\}+\mathrm{e}^{j\omega\times 0}P\{Y=0\}. \tag{4.178}$$

Using (4.174)–(4.178), we arrive at:

$$\phi_Y(\omega)=0.9332\,\mathrm{e}^{j\omega}+0.0668. \tag{4.179}$$

Exercise 4.9 Find the characteristic function of the random variable

$$Y=X-2, \tag{4.180}$$

where X is the normal random variable $X=N(1,4)$.

Answer The random variable Y is the linear transformation of the normal random variable X and thus itself is a normal random variable with the parameters,

$$E\{Y\}=E\{X\}-2=1-2=-1. \tag{4.181}$$

The variance of the random variable Y is:

$$\sigma_Y^2=\sigma_X^2=4. \tag{4.182}$$

Using (4.100), the characteristic function is:

$$\phi_Y(\omega)=\mathrm{e}^{j\omega m_Y-\frac{\omega^2\sigma_Y^2}{2}}=\mathrm{e}^{-j\omega-2\omega^2}. \tag{4.183}$$

Exercise 4.10 The independent normal random variables X_1 and X_2 are related with the random variable Y, as shown in Fig. 4.25. Find the variance for the random variable Y if the parameters of the variables X_1 and X_2 are $m_1 = 2$, $\sigma_1^2 = 4$ and $m_2 = 0$, $\sigma_2^2 = 5$, respectively. (Use Q function.)

Answer The random variable X is a sum of independent normal random variables and has the following parameters:

$$\sigma^2 = \sigma_1^2 + \sigma_2^2 = 9,$$
$$m = m_1 + m_2 = 2.$$

The variable X is also a normal random variable, $X = N(2, 9)$.

The output random variable Y is a discrete random variable with the discrete values 2 and 0. The corresponding probabilities are:

$$P\{Y = 2\} = P\{X<8\}; \qquad P\{Y = 0\} = P\{X>8\}. \tag{4.185}$$

Using (4.53), we have:

$$\begin{aligned} P\{Y = 2\} &= P\{X \leq 8\} = 1 - P\{X>8\} = 1 - Q(k) \\ &= 1 - Q\left(\frac{|8-2|}{3}\right) = 1 - Q(2), \\ P\{Y = 0\} &= 1 - P\{Y = 2\} = Q(2). \end{aligned} \tag{4.186}$$

Using (4.186), the variance of the variable Y is obtained as:

$$\sigma_Y^2 = \overline{Y^2} - \overline{Y}^2 = 4(1 - Q(2)) - [2(1 - Q(2)]^2 = 4Q(2) - 4[Q(2)]^2. \tag{4.187}$$

The function Q is related to the erf function as shown in Table 4.1, resulting in:

$$\begin{aligned} \sigma_Y^2 &= \frac{4}{2}\left[1 - \text{erf}\left(\frac{2}{\sqrt{2}}\right)\right] - \frac{4}{2}\left[1 - \text{erf}\left(\frac{2}{\sqrt{2}}\right)\right]^2 \\ &= 0.091 - 0.0041 = 0.0869. \end{aligned} \tag{4.188}$$

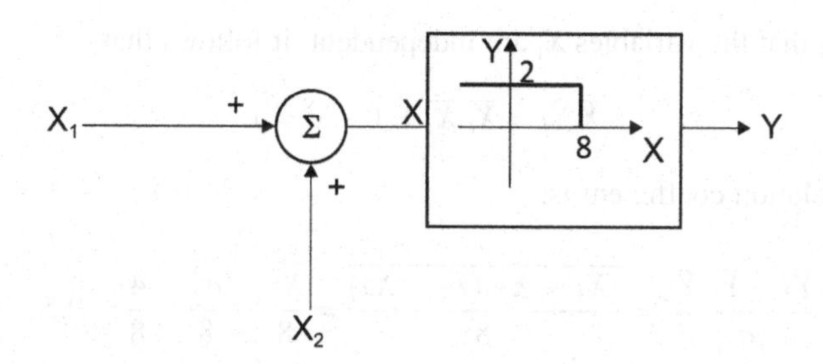

Fig. 4.25 Random variables X_1, X_2, and Y

Exercise 4.11 The joint density function of the normal random variables X_1 and X_2 is given as:

$$f_{X_1X_2}(x_1,x_2)=\frac{1}{300\pi\sqrt{0.75}}\mathrm{e}^{-\frac{1}{1.5}\left[\frac{(x_1-5)^2}{100}-\frac{(x_1-5)x_2}{150}+\frac{x_2^2}{225}\right]}. \tag{4.189}$$

Find the correlation coefficient and the mean value of the variable

$$X=aX_1+bX_2. \tag{4.190}$$

Answer Comparing the general expression (4.124) and (4.189), we have:

$$1.5=2(1-\rho^2)\quad\text{and}\quad\rho=0.5. \tag{4.191}$$

The mean values of the variables X_1 and X_2 are 5 and zero, respectively. From (4.190), we have:

$$E\{X\}=a\{X_1\}+bE\{X_2\}=5a+0=5a. \tag{4.192}$$

Exercise 4.12 Three independent normal random variables have the following parameters:

$$m_i=0,\qquad\sigma_i^2=4,\quad i=1,2,3. \tag{4.193}$$

Find the coefficient of correlation for variables Y_1 and Y_2 if

$$Y_1=X_1+X_2,\qquad Y_2=X_2-X_3. \tag{4.194}$$

Answer The random variables Y_1 and Y_2 are the sums of normal random variables and consequently, they are also normal with the parameters,

$$\begin{aligned}E\{Y_1\}&=E\{X_1\}+E\{X_2\}=0,\\E\{Y_2\}&=E\{X_2\}-E\{X_3\}=0,\\\sigma_{Y_1}^2&=\sigma_1^2+\sigma_2^2=8,\\\sigma_{Y_2}^2&=\sigma_2^2+\sigma_3^2=8.\end{aligned} \tag{4.195}$$

Knowing that the variables X_i are independent, it follows that

$$\overline{X_iX_j}=\overline{X_i}\,\overline{X_j}=0,\quad i\neq j. \tag{4.196}$$

The correlation coefficient is:

$$\rho=\frac{\overline{Y_1Y_2}-\overline{Y_1}\,\overline{Y_2}}{\sigma_{Y_1}\sigma_{Y_2}}=\frac{\overline{(X_1+X_2)(X_2-X_3)}}{8}=\frac{\overline{X_2^2}}{8}=\frac{\sigma_2^2}{8}=\frac{4}{8}=0.5. \tag{4.197}$$

Exercise 4.13 The normal random variable X has a zero mean value and a variance of σ^2. Find the mean value of the random variable Y

$$Y = 2 + 3X^2 - 3X^3. \tag{4.198}$$

Answer From (4.198), the mean value of Y is given as:

$$\overline{Y} = 2 + 3\overline{X^2} - 3\overline{X^3}. \tag{4.199}$$

A mean squared value of the random variable X is equal to the variance σ^2. Additionally, the third moment is equal to zero because the density function is symmetrical around zero.

Therefore, from (4.199) we have:

$$\overline{Y} = 2 + 3\sigma^2. \tag{4.200}$$

Exercise 4.14 The random variable X is a normal random variable $N(2, 4)$ with the probability 0.6, and it is a normal random variable $N(-2, 9)$ with the probability 0.4. Find the PDF of the random variable X.

Answer The PDF of the random variable X is given as:

$$f_X(x) = 0.6\frac{1}{2\sqrt{2\pi}}\mathrm{e}^{-\frac{(x-2)^2}{8}} + 0.4\frac{1}{3\sqrt{2\pi}}\mathrm{e}^{-\frac{(x+2)^2}{18}}. \tag{4.201}$$

Let us verify that (4.201) is a PDF. To this end, the condition (2.83) must be satisfied:

$$\int_{-\infty}^{\infty} f_X(x)\mathrm{d}x = 0.6\frac{1}{2\sqrt{2\pi}}\int_{-\infty}^{\infty}\mathrm{e}^{-\frac{(x-2)^2}{8}}\mathrm{d}x + 0.4\frac{1}{3\sqrt{2\pi}}\int_{-\infty}^{\infty}\mathrm{e}^{-\frac{(x+2)^2}{18}}\mathrm{d}x = 0.6 \times 1 + 0.4 \times 1 = 1;$$

$$\frac{1}{2\sqrt{2\pi}}\int_{-\infty}^{\infty}\mathrm{e}^{-\frac{(x-2)^2}{8}}\mathrm{d}x = 1; \quad \frac{1}{3\sqrt{2\pi}}\int_{-\infty}^{\infty}\mathrm{e}^{-\frac{(x+2)^2}{18}}\mathrm{d}x = 1. \tag{4.202}$$

Exercise 4.15 The random variables X and Y are independent. Their corresponding density functions are:

$$f_X(x) = \frac{1}{4\sqrt{2\pi}}\mathrm{e}^{-\frac{(x-3)^2}{32}}, \quad f_Y(y) = \begin{cases} 1/3 & \text{for} \quad -1 \le x \le 2 \\ 0 & \text{otherwise} \end{cases}. \tag{4.203}$$

Find the mean values of the sum and the product of variables X and Y.

Answer The mean value of the sum of random variables X and Y is:

$$E\{X+Y\} = E\{X\} + E\{Y\}. \tag{4.204}$$

From (4.203), it follows:

$$E\{X\} = 3. \tag{4.205}$$

The mean value of the uniform variable Y is at the point of symmetry of its PDF, i.e.,

$$E\{Y\} = 0.5. \tag{4.206}$$

Placing (4.205) and (4.206) into (4.204), we get:

$$E\{X+Y\} = 3 + 0.5 = 3.5. \tag{4.207}$$

The random variables X and Y are independent, and as a consequence we have:

$$E\{XY\} = E\{X\}E\{Y\}. \tag{4.208}$$

Placing (4.205) and (4.206) into (4.208) we arrive at:

$$E\{XY\} = 1.5. \tag{4.209}$$

Exercise 4.16 N messages are transmitted over a channel. The times necessary to transmit the messages are the independent variables X_i with equal mean values $m_i = m$ and equal variances $\sigma_i^2 = \sigma^2$.

(a) Find the density of a random time in which all messages could be transmitted.
(b) Find the approximate total time needed to transmit all N messages.

Answer
(a) According to the CLT, the total time needed to transmit all messages is approximately equal to the normal random variable

$$X = \sum_{i=1}^{N} X_i \approx N(m_X, \sigma_X^2), \tag{4.210}$$

where

$$m_X = \sum_{i=1}^{N} m_i = Nm; \qquad \sigma_X^2 = \sum_{i=1}^{N} \sigma_i^2 = N\sigma^2. \tag{4.211}$$

The approximated total time necessary to transmit all N messages is $m_X + 3\sigma_X$.

$$X \approx m_X + 3\sigma_X = Nm + 3\sigma\sqrt{N}. \tag{4.212}$$

Exercise 4.17 A joint probability density of the random variables X and Y is given as:

$$f_{XY}(x, y) = k\,e^{-\frac{(x-1)^2}{8} - \frac{(y+1)^2}{18}}. \tag{4.213}$$

Find the constant k and find the probability

$$P\{-1<X<1, -2<Y<4\}. \tag{4.214}$$

Answer From (4.213), we can see that the random variables X and Y are independent normal random variables with the following parameters:

$$m_X = 1, \qquad \sigma_X^2 = 4; \qquad m_Y = -1, \qquad \sigma_Y^2 = 9. \tag{4.215}$$

The joint density (4.213) can be written as:

$$f_{XY}(x, y) = f_X(x) f_Y(y) = \frac{1}{2\sqrt{2\pi}} e^{-\frac{(x-1)^2}{8}} \frac{1}{3\sqrt{2\pi}} e^{-\frac{(y+1)^2}{18}}. \tag{4.216}$$

By comparing (4.213) and (4.216), we find that $k = 1/(12\pi)$.

Due to the independence of the random variables X and Y, the joint probability (4.214) can be written as:

$$P\{(-1 \le X \le 1) \cap (-2 \le Y \le 4)\} = P\{-1<X<1\} P\{-2<Y<4\}. \tag{4.217}$$

From (4.31), we calculate:

$$\begin{aligned} P\{-1 \le X \le 1\} &= \frac{1}{2}\left[\operatorname{erf}\left(\frac{1-1}{2\sqrt{2}}\right) - \operatorname{erf}\left(\frac{-1-1}{2\sqrt{2}}\right)\right] \\ &= \frac{1}{2}\operatorname{erf}\left(\frac{1}{\sqrt{2}}\right) = 0.3413, \end{aligned} \tag{4.218}$$

$$\begin{aligned} P\{-2 \le Y \le 4\} &= \frac{1}{2}\left[\operatorname{erf}\left(\frac{4+1}{3\sqrt{2}}\right) - \operatorname{erf}\left(\frac{-2+1}{3\sqrt{2}}\right)\right] \\ &= \frac{1}{2}\left[\operatorname{erf}\left(\frac{5}{3\sqrt{2}}\right) + \operatorname{erf}\left(\frac{1}{3\sqrt{2}}\right)\right] = 0.5828. \end{aligned} \tag{4.219}$$

Finally, from (4.217) to (4.219), we obtain:

$$\begin{aligned} P\{(-1 \le X \le 1) \cap (-2 \le Y \le 4)\} &= P\{-1 \le X \le 1\} P\{-2 \le Y \le 4\} \\ &= 0.3413 \times 0.5828 = 0.1989. \end{aligned} \tag{4.220}$$

Exercise 4.18 The normal random variables X_1 and X_2 have the mean values

$$E\{X_1\} = 4, \qquad E\{X_2\} = 2, \tag{4.221}$$

and the covariance matrix

$$\mathbf{C} = \begin{vmatrix} 4 & 3/4 \\ 3/4 & 9 \end{vmatrix}. \tag{4.222}$$

Find the joint density function.

Answer From (4.222), we have:

$$\sigma_1^2 = 4, \qquad \sigma_2^2 = 9, \qquad C_{1,2} = C_{2,1} = 3/4. \tag{4.223}$$

The correlation coefficient is:

$$\rho = \frac{C_{1,2}}{\sigma_1\sigma_2} = \frac{3/4}{2 \times 3} = 1/8 = 0.125. \tag{4.224}$$

From (4.124), the joint density is:

$$f_{X_1,X_2}(x_1, x_2) = \frac{1}{11.906\pi}\mathrm{e}^{-0.5079\left[\frac{(x_1-4)^2}{4}+\frac{(x_2-2)^2}{9}-0.25\frac{(x_1-4)(x_2-2)}{6}\right]}. \tag{4.225}$$

Exercise 4.19 Consider a Gaussian random variable with $m = 0$ and $\sigma^2 = 1$, given in Fig. 4.26.

$$f_X(x) = \frac{1}{\sqrt{2\pi}}\mathrm{e}^{-\frac{x^2}{2}}. \tag{4.226}$$

The condition event A is defined such that the random variable X is positive,

$$A = \{X > 0\}. \tag{4.227}$$

Find $E\{X|A\}$.

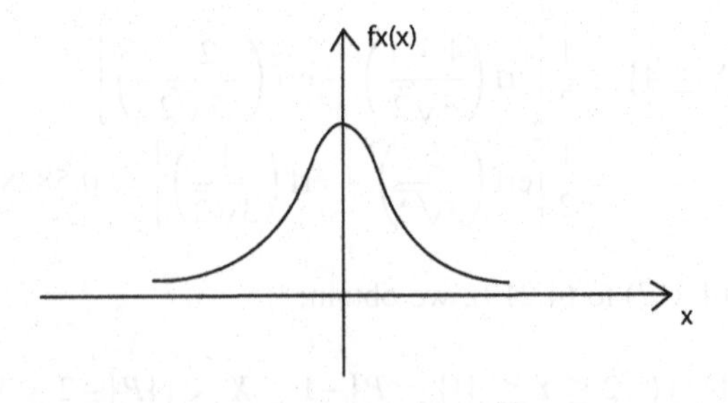

Fig. 4.26 Gaussian probability density in Example 4.19

Answer Observing Fig. 4.26, we can easily conclude that

$$P\{X>0\} = 1/2. \tag{4.228}$$

From (2.305), (2.115), and (4.228), we have:

$$E\{X|A\} = \frac{\int_0^{\infty} x f_X(x)\mathrm{d}x}{1/2} = \sqrt{\frac{2}{\pi}} \int_0^{\infty} x\,\mathrm{e}^{-\frac{x^2}{2}}\,\mathrm{d}x. \tag{4.229}$$

The integral on the right side is calculated using integral 4 from Appendix A for $n = 1$ and $a = 1/2$, and the expression for Gamma function in Appendix C:

$$\int_0^{\infty} x\,\mathrm{e}^{-\frac{x^2}{2}}\,\mathrm{d}x = \frac{\Gamma\left(\frac{1+1}{2}\right)}{2\left(\frac{1}{2}\right)^{(1+1)/2}} = 1, \tag{4.230}$$

resulting in:

$$E\{X|A\} = \sqrt{\frac{2}{\pi}}. \tag{4.231}$$

4.9 MATLAB Exercises

Exercise M.4.1 (MATLAB file: *exercise_M_4_1.m*). Generate 50,000 samples of the normal random variable $N(1, 9)$ and estimate its density and distribution functions.

Solution The normal random variable is shown in Fig. 4.27a for the first 10,000 samples. The corresponding PDF and distribution are given in Fig. 4.27b, c, respectively. The estimated PDF is obtained using 30 cells and is shown in Fig. 4.27d.

Exercise M.4.2 (MATLAB file: *exercise_M_4_2.m*). Generate 50,000 values of two normal random variables with equal variances $\sigma_1^2 = \sigma_2^2 = 12$, and different mean values $m_1 = 0$, $m_2 = 10$. Find the corresponding PDFs in order to demonstrate how the mean value affects the normal random variable and its PDF.

Solution The normal variables are shown in Fig. 4.28. Note that the variables are only moved along the amplitude-axis. The corresponding densities are given in Fig. 4.29, indicating that the shape of both PDFs is the same, but that they are positioned differently on the x-axis.

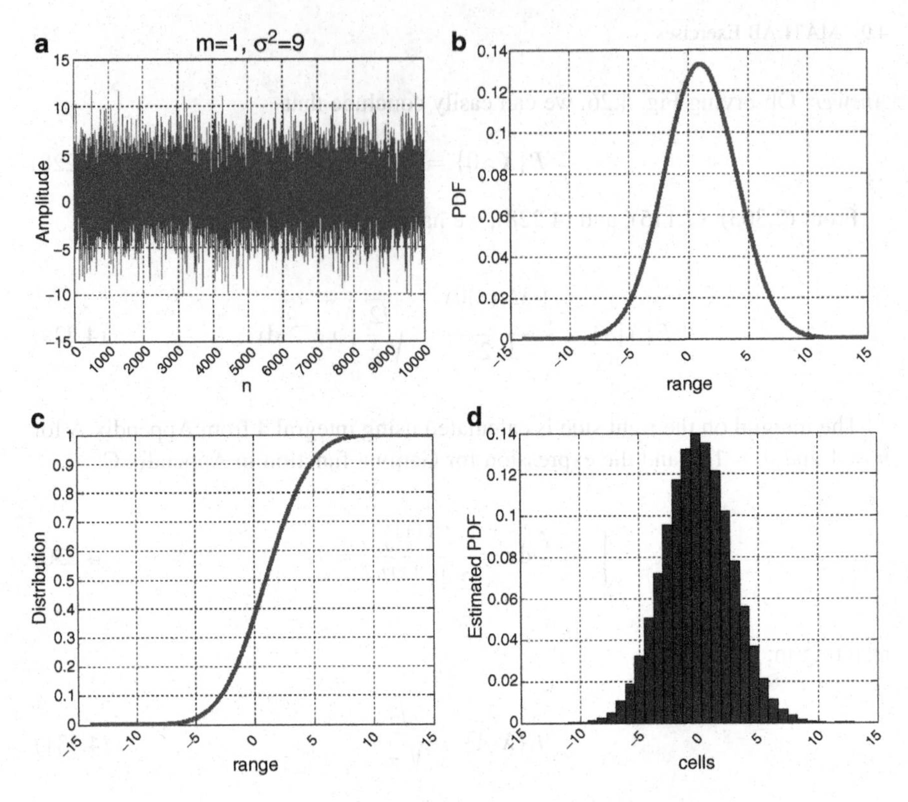

Fig. 4.27 (**a**) Normal variable. (**b**) PDF. (**c**) Distribution. (**d**) Estimated PDF

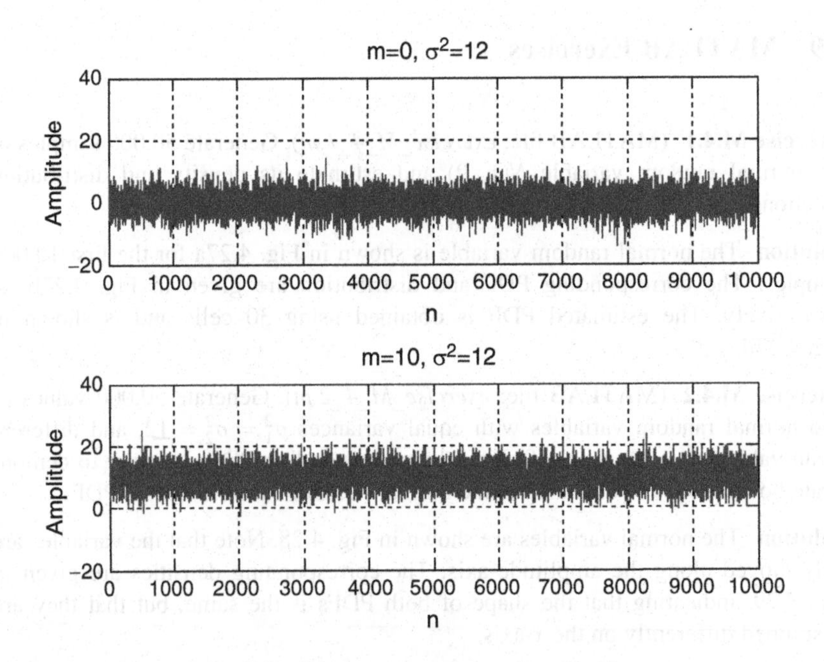

Fig. 4.28 Random variables with different mean values and equal variances

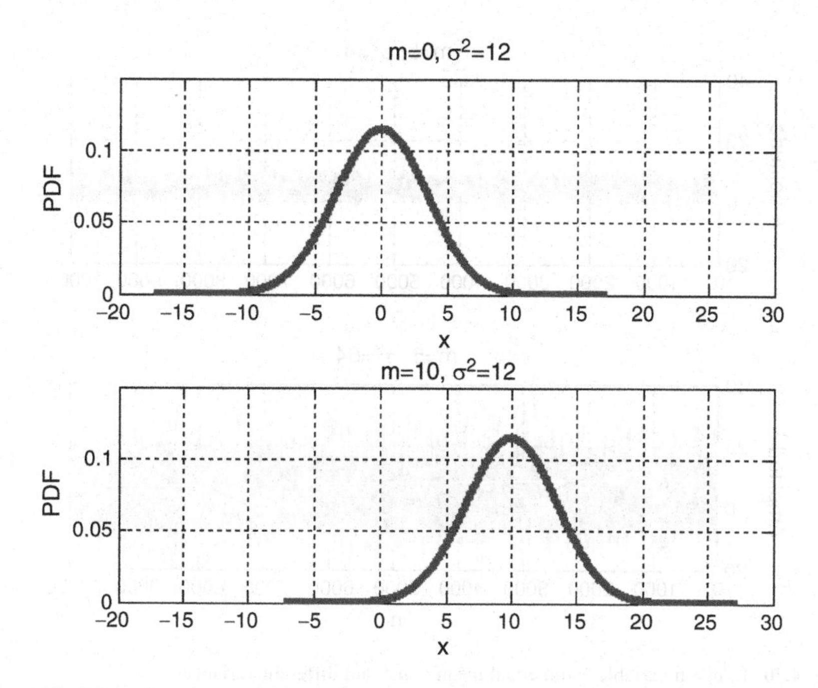

Fig. 4.29 PDFs of random variables with different mean values and equal variances

Exercise M.4.3 (MATLAB file: *exercise_M_4_3.m*). In this exercise, we show how different variances change the normal variables and their corresponding densities. To this end, generate 50,000 values of two normal random variables with equal mean values $m_1 = m_2 = 5$ and different variances $\sigma_1^2 = 4$, $\sigma_2^2 = 64$.

Solution The corresponding random variables and the PDFs are shown in Figs. 4.30 and 4.31, respectively. Note that the positions of the PDFs on the x-axis are the same but have different shapes. By increasing the variance, the maximum value of the PDF decreases and the width increases.

Exercise M.4.4 (MATLAB file: *exercise_M_4_4.m*) Generate a normal random variable X with a mean value equal to 1 and a variance equal to 4. Plot the corresponding PDF and distribution and find the corresponding probabilities:

(a) $P_1 = P\{X < 3\}$
(b) $P_2 = \{X < 0\}$

Solution
(a) The first 1,000 values of the random variable along with the value of 3 are shown in Fig. 4.32, where the limit of 3 on the values for calculating the probability P_1 is denoted with a dotted line. The corresponding probability is:

$$P_1 = 0.84135 \tag{4.232}$$

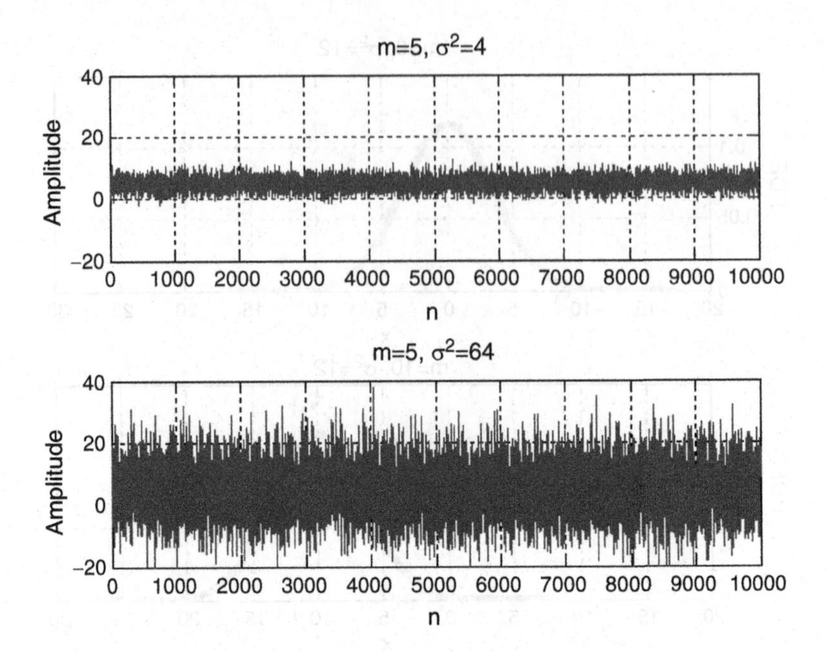

Fig. 4.30 Random variables with equal mean value and different variances

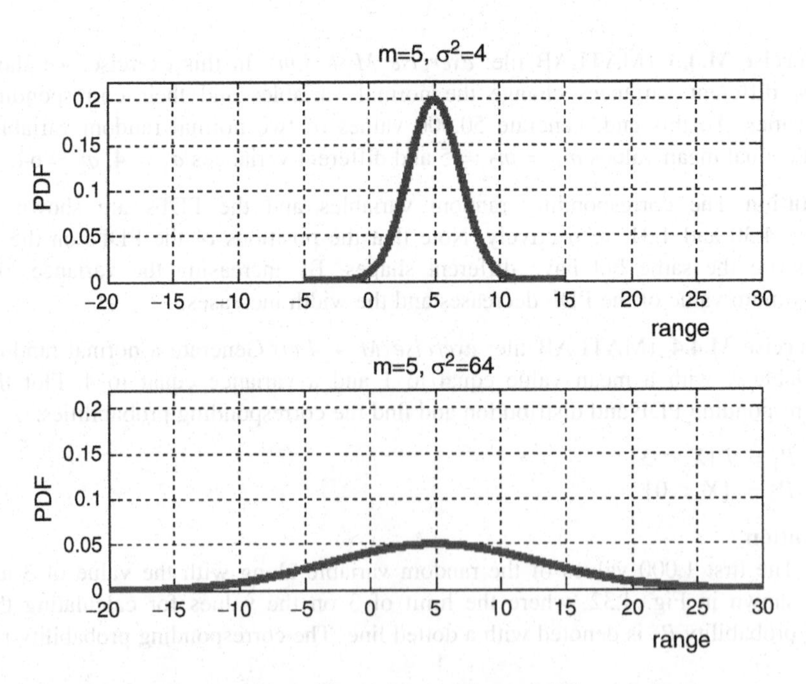

Fig. 4.31 PDF of random variables with equal mean value and different variances

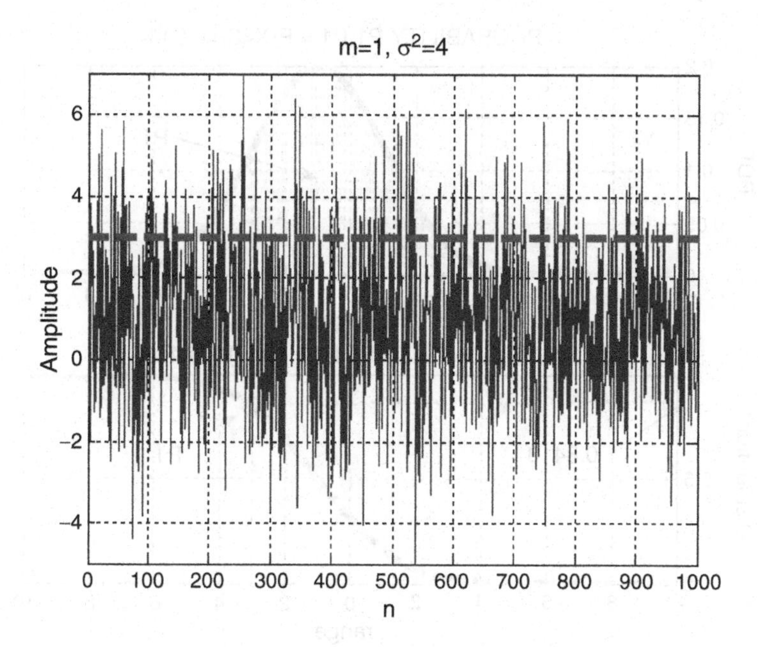

Fig. 4.32 Random variables in Exercise M.4.4

(b) The probability P_1 is illustrated in Fig. 4.33, indicating that the probability is the area below the PDF in the range $[-\infty, 3]$ and also it is the value of the distribution function for $x = 3$.
The probability P_2 is:

$$P_2 = 0.3085. \tag{4.233}$$

The probability P_2 is related with the PDF and distribution as shown in Fig. 4.34.

Exercise M.4.5 (MATLAB file: *exercise_M_4_5.m*). Find the probability that the random variable $X = N(2, 9)$ is in the interval $[-1, 5]$ and show this probability in the PDF and distribution plots.

Solution The random variable along with the interval (dotted line), where the desired probability being looked for, is shown in Fig. 4.35.
The desired probability is obtained from the erf function:

$$P_{21} = 0.1587, \tag{4.234}$$

$$P_{22} = 0.8413, \tag{4.235}$$

$$P_{22} - P_{21} = 0.682. \tag{4.236}$$

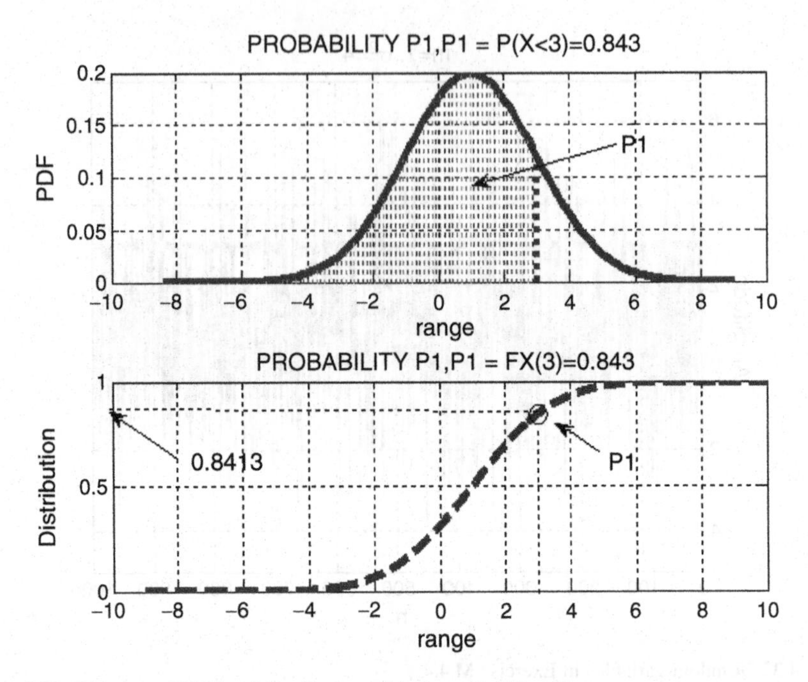

Fig. 4.33 Relation of PDF and distribution with the probability P_1

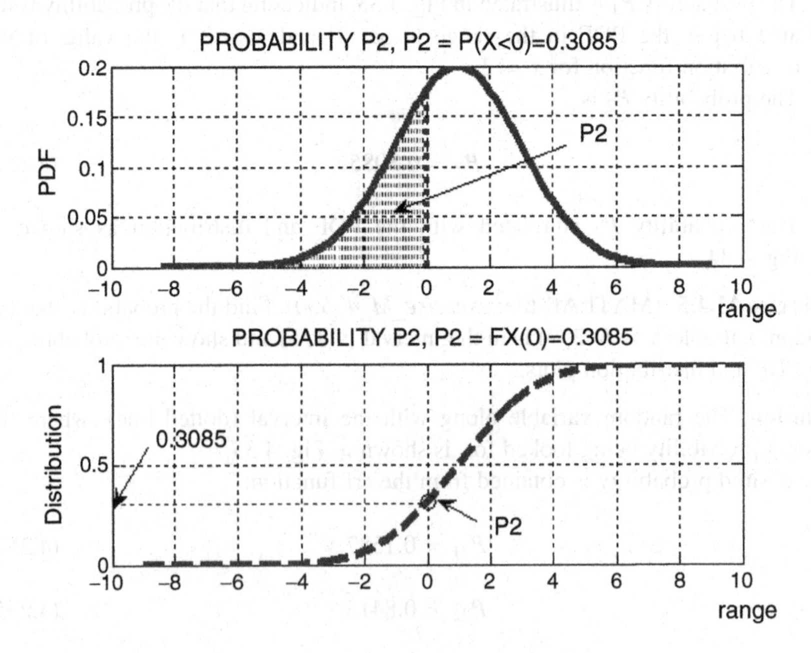

Fig. 4.34 Relation of PDF and distribution with the probability P_2

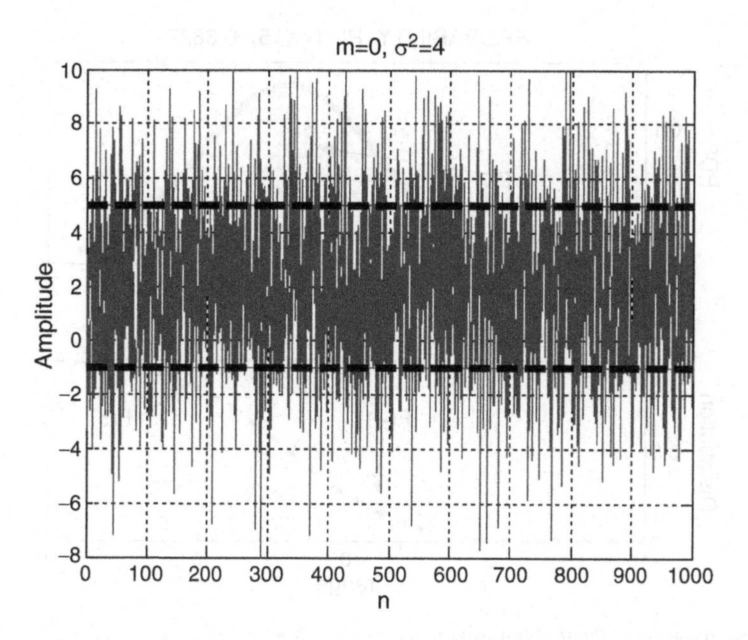

Fig. 4.35 Random variable and the interval for probability calculation

This probability is illustrated in Fig. 4.36, as the area below the PDF in the interval $[-1, 5]$. Similarly, the probability is presented in the distribution function as the difference of $F_X(5)$ and $F_X(-1)$.

Exercise M.4.6 (MATLAB file: *exercise_M_4_6.m*) Show that the linear transformation $Y = 3X + 3$ of the normal random variable $X = N(1, 1)$ is also a normal random variable.

Solution The input and output random variables along with the transformation are shown in Fig. 4.37.

The estimated PDFs are shown in Fig. 4.38.

The estimated PDFs show that both random variables are normal variables. The mean value and variance of the output variable Y are:

$$m_Y = 3m_X + 3 = 6; \qquad \sigma_Y^2 = 3^2\sigma_X^2 = 9. \tag{4.237}$$

Exercise M.4.7 (MATLAB file: *exercise_M_4_7.m*) Show that the sum of two independent normal random variables:

$$X_1 = N(-1, 4) \text{ and } X_2 = N(5, 8) \tag{4.238}$$

is also a normal random variable.

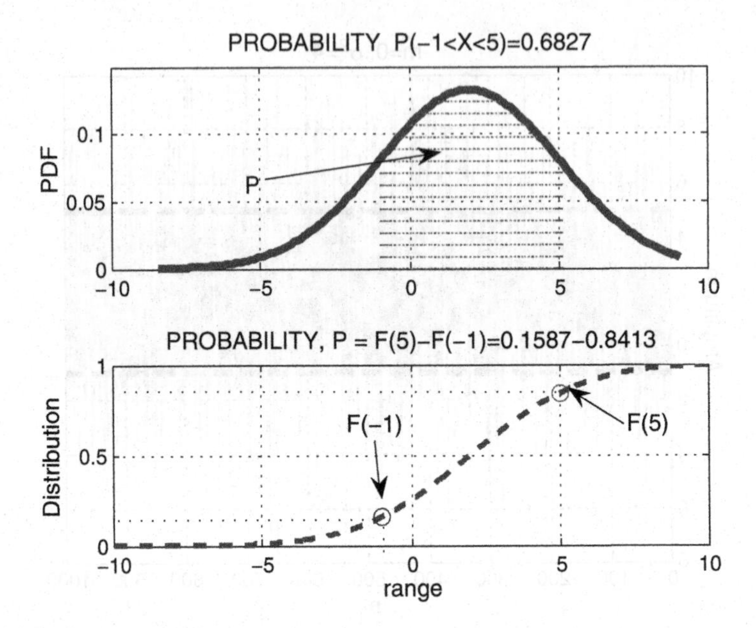

Fig. 4.36 Probability, PDF, and distribution

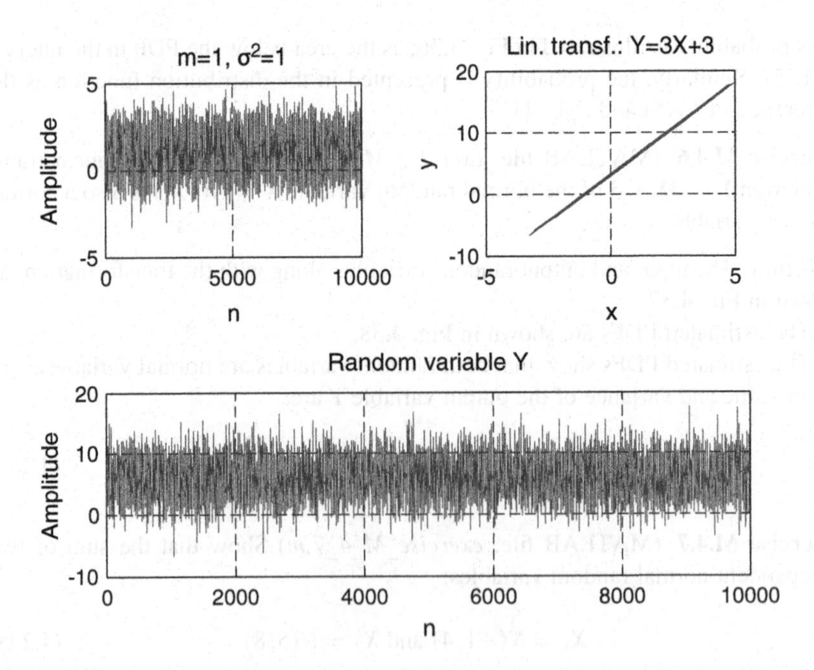

Fig. 4.37 Input variable X, output variable Y, and the corresponding linear transformation

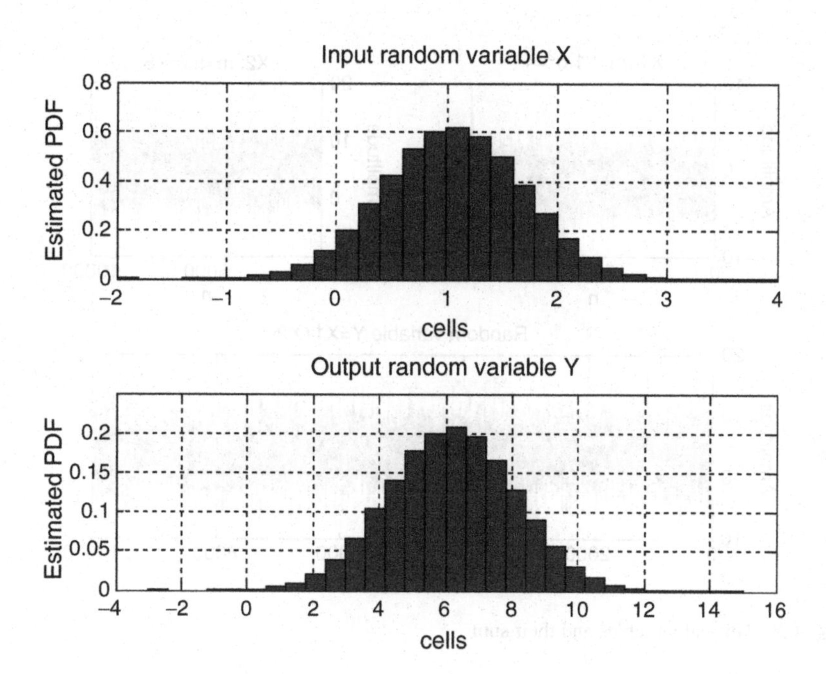

Fig. 4.38 Input and output estimated PDFs

Solution The random variables X_1 and X_2 and their sum are shown in Fig. 4.39. Observing the random variable Y, we can easily conclude that the variable Y is also a normal random variable. This is confirmed by estimating the PDFs of the variables, X_1, X_2, and Y (shown in Fig. 4.40).

Exercise M.4.8 (MATLAB file: *exercise_M_4_8.m*) Show the CLT considering the sum of five uniform random variables in the interval [1, 4].

Solution We generate a uniform random variable and present the sum of 2, 3, 4, and 5, uniform variables in Fig. 4.41, along with the corresponding PDF estimations.

The uniform random variable has the mean value and the variance:

$$m_i = 2.5, \qquad \sigma_i^2 = 0.75, \quad i = 1, \ldots, 5 \tag{4.239}$$

The sum of five uniform random variables Y has a mean value and variance of:

$$\begin{aligned} Y &= \sum_{i=1}^{5} X_i; \\ m_Y &= \sum_{i=1}^{5} m_i = 2.5 \times 5 = 12.5; \\ \sigma_Y^2 &= \sum_{i=1}^{5} \sigma_i^2 = 0.75 \times 5 = 3.75. \end{aligned} \tag{4.240}$$

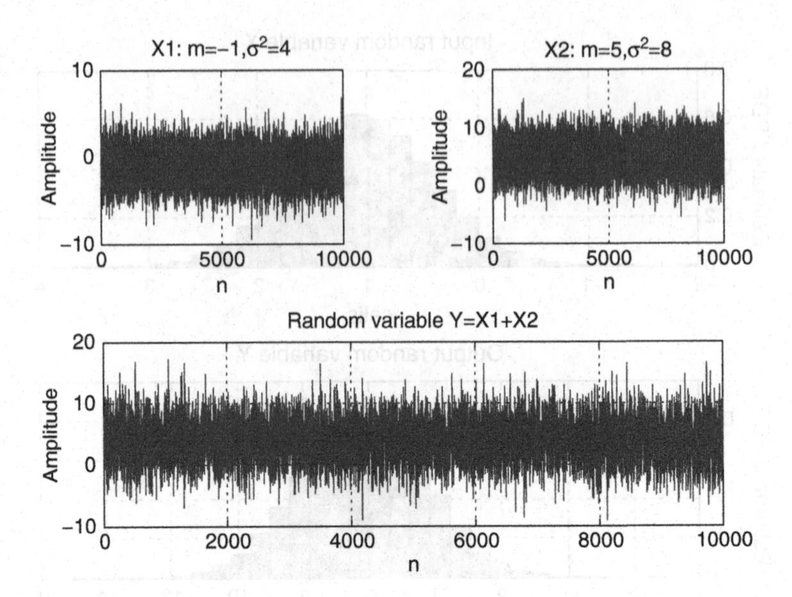

Fig. 4.39 Normal variables and their sum

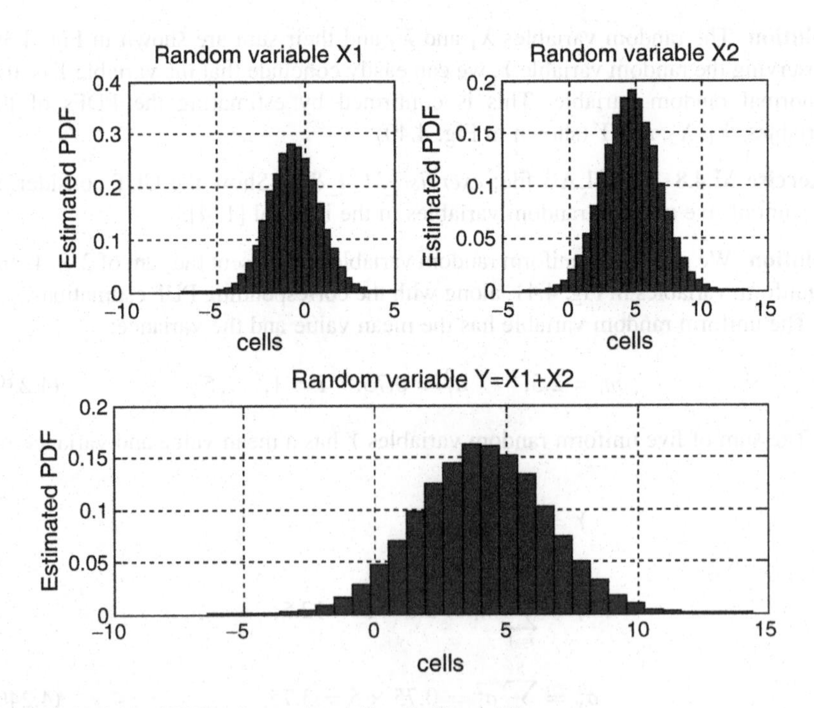

Fig. 4.40 Estimated PDFs of normal variables X_1, X_2, and Y

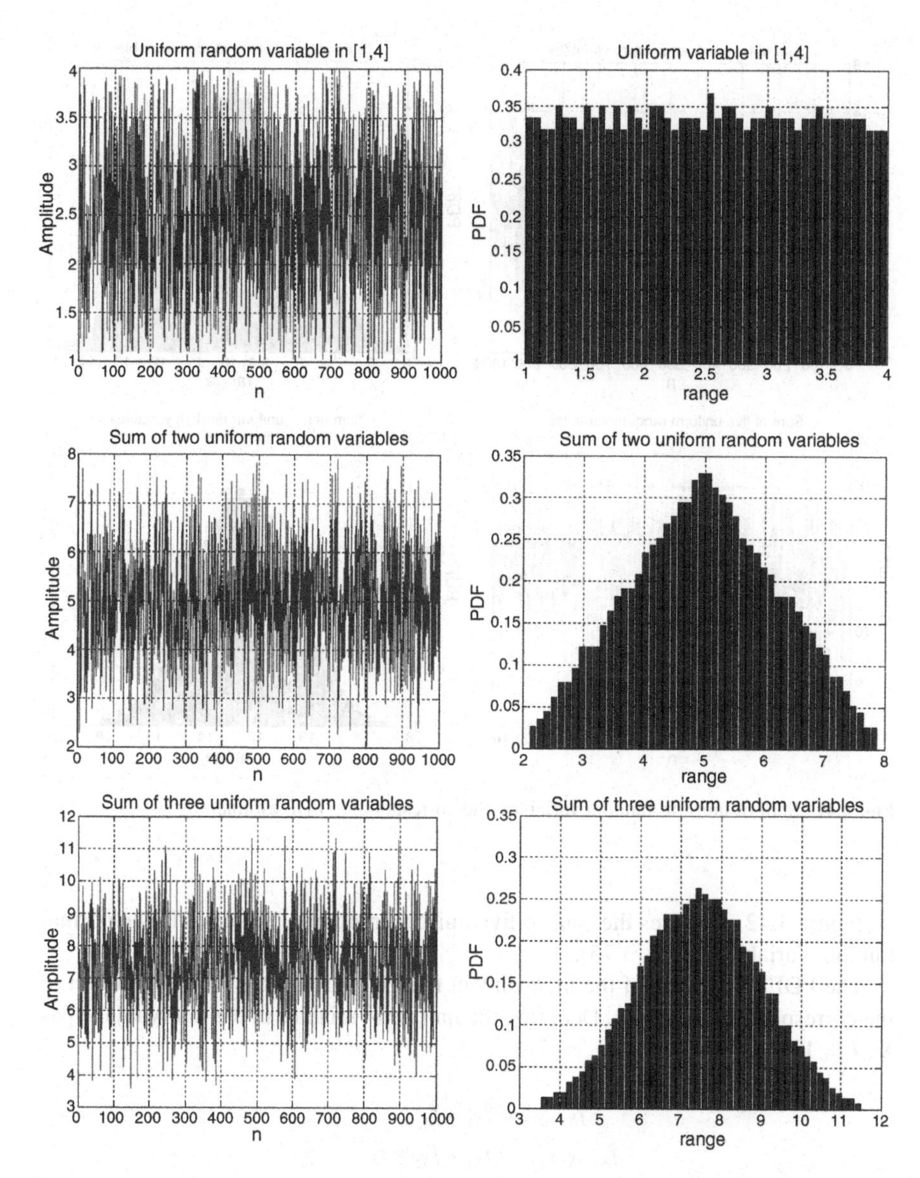

Fig. 4.41 (continued)

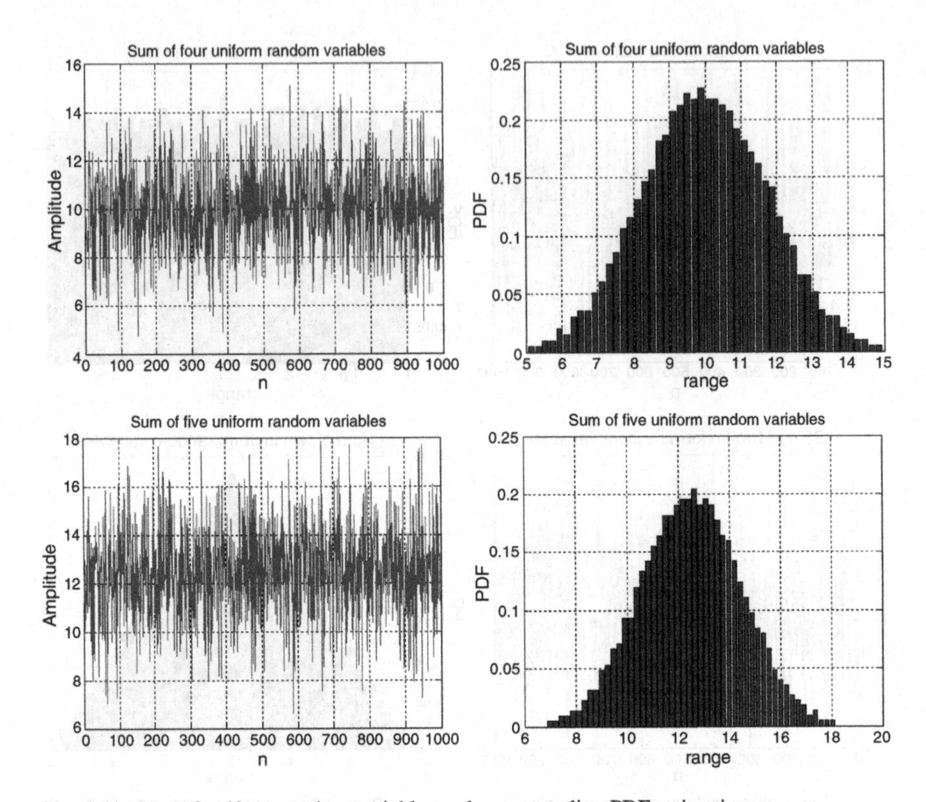

Fig. 4.41 Sum of uniform random variables and corresponding PDF estimations

Figure 4.42 compares the sum of five uniform random variables and the normal random variable $N(12.5, 3.75)$.

The PDF of the sum of the independent random variables is the convolution of the corresponding densities. Denoting the uniform variables in the interval [1, 4] as X_i, $i = 1, \ldots, 5$, we have:

$$\begin{aligned} f_{X_1+X_2} &= f_{X_1} * f_{X_2}, \\ f_{X_1+X_2+X_3} &= f_{X_1} * f_{X_2} * f_{X_3}, \\ f_{X_1+X_2+X_3+X_4} &= f_{X_1} * f_{X_2} * f_{X_3} * f_{X_4}, \\ f_{X_1+X_2+X_3+X_4+X_5} &= f_{X_1} * f_{X_2} * f_{X_3} * f_{X_4} * f_{X_5}, \end{aligned} \tag{4.241}$$

where * stands for the correlation operation.

Figure 4.43 shows the corresponding densities obtained by convolution of uniform densities.

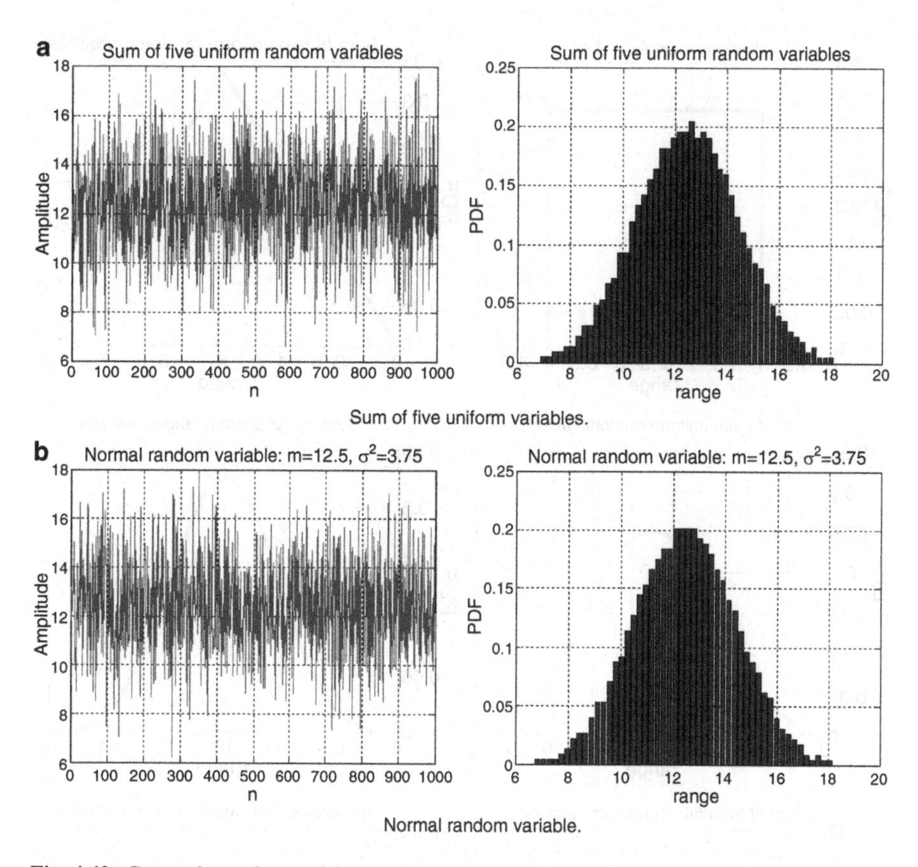

Fig. 4.42 Comparison of sum of five uniform r.v. (**a**) and normal r.v. (**b**)

4.10 Questions

Q.4.1. Why does a linear transformation of both, normal and uniform random variables result in a normal and uniform variables, respectively?

Q.4.2. Why is the sum of normal independent variables also normal in contrast to the sum of uniform independent variables, which is not a uniform variable?

Q.4.3. What happens to the normal PDF as the variance of the normal variable is approaching to zero?

Q.4.4. Two normal variables have different mean values and equal variances. What is the difference in their PDFs?

Q.4.5. Two normal variables have different variances and equal mean values. What is the difference in their PDFs?

Q.4.6. What is the result of the convolution of two normal PDFs?

Q.4.7. What is a unique property of the characteristic function of a normal variable?

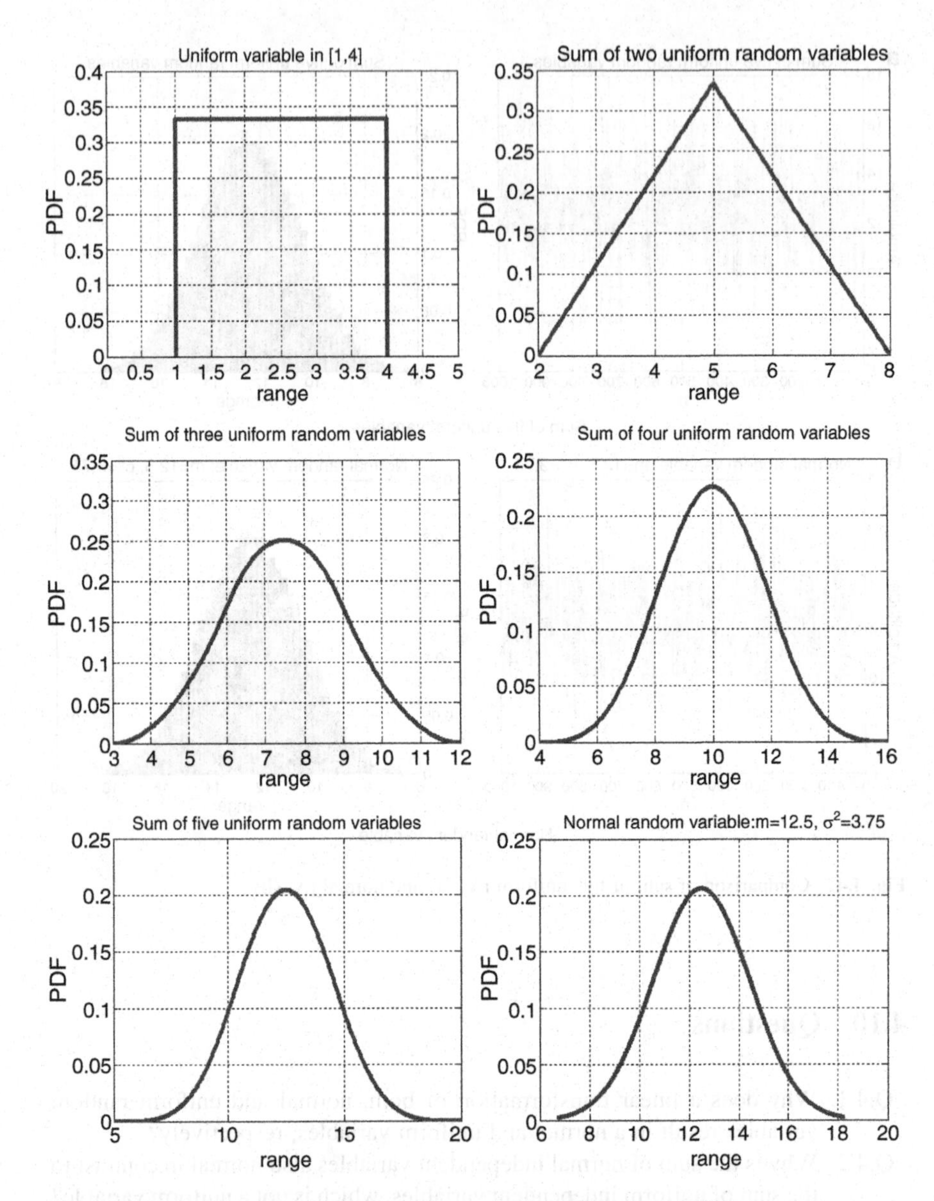

Fig. 4.43 PDFs of the sum of uniform random variables and the normal PDF

Q.4.8. The normal random variable X has a mean value of zero and a variance of 1. Which of the two events, $A = \{|X| \leq 0.5\}$ and $B = \{|X| \geq 0.5\}$, is more probable?

Q.4.9. Is it possible to obtain a single normal variable using a transformation of a single uniform random variable?

Q.4.10. What is the practical usefulness of the CLT because if N is approaching infinity, the variance is also approaching to infinity?

Q.4.11. Does the CLT also stand for a discrete random variables; knowing that PDFs of discrete random variables have delta impulses?

4.11 Answers

A.4.1. Consider a linear transformation

$$Y = aX + b \tag{4.242}$$

of a random variable X, where a and b are constants.

The linear transformation (4.242) changes the mean value and the variance of the variable X:

$$\begin{aligned} m_Y &= am_X + b, \\ \sigma_Y^2 &= a^2\sigma_X^2. \end{aligned} \tag{4.243}$$

Each normal variable is determined only by its two parameters: a mean value and a variance. Therefore, the linearly transformed normal random variable is also a normal variable with new parameters (4.243).

The uniform variable X, defined in the range $[a_1, a_2]$, is equivalently defined by using the length of its range $\Delta_X = a_2 - a_1$, and its mean value m_X – which is in the middle of the range-as shown in Fig. 4.44a.

The length of the range Δ_X is related to the variance as:

$$\sigma_X^2 = \frac{\Delta_X^2}{12}. \tag{4.244}$$

Therefore, a uniform variable, like a normal variable, is also uniquely defined with its mean value and variance.

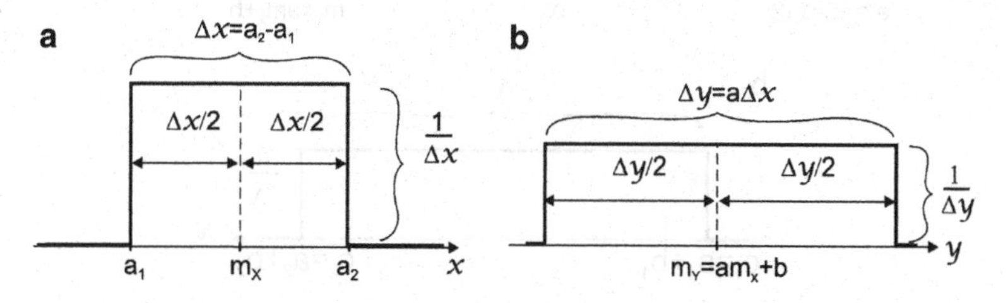

Fig. 4.44 Linear transformation of uniform random variable

From (4.243). we have:

$$\sigma_Y^2 = \frac{\Delta_Y^2}{12} = a^2 \frac{\Delta_X^2}{12} \tag{4.245}$$

resulting in:

$$\Delta_Y = a\Delta_X. \tag{4.246}$$

The mean value of the random variable Y is determined by the first equation (4.243), resulting in the uniform PDF of the variable Y, as shown in Fig. 4.44b.

A.4.2. The mean value m of a sum of independent (or dependent) random variables X_1 and X_2, is equal to the sum of the corresponding mean values m_1 and m_2:

$$m = m_1 + m_2. \tag{4.247}$$

Similarly, the variance σ^2 of the sum of independent random variables X_1 and X_2, is equal to the sum of the corresponding variances ${\sigma_1}^2$ and ${\sigma_2}^2$:

$$\sigma^2 = \sigma_1^2 + \sigma_2^2. \tag{4.248}$$

Knowing that a normal random variable is uniquely defined with its mean value and variance, it follows that the sum X will also be a normal variable with the new parameters (4.247) and (4.248).

Let the uniform random variables X_1 and X_2 be uniform within the range $[a_1, a_2]$ and $[b_1, b_2]$, respectively (as shown in Fig. 4.45a).

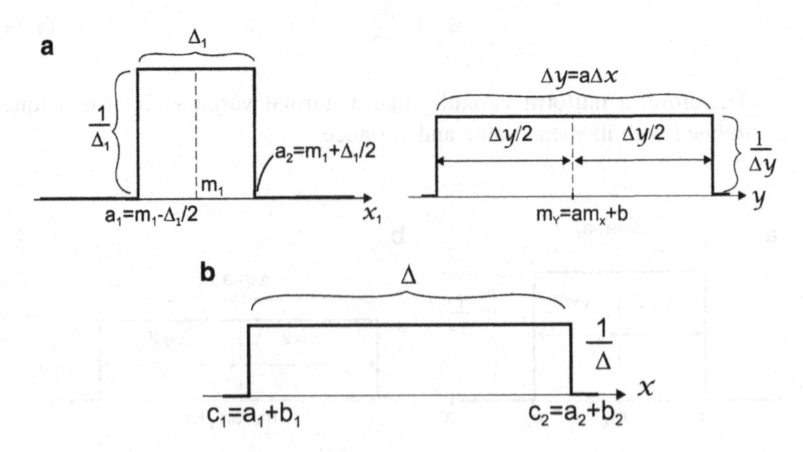

Fig. 4.45 Sum of uniform random variables

Their sum X will be a random variable in the range $[c_1, c_2]$, where

$$c_1 = a_1 + b_1 \tag{4.249}$$

and

$$c_2 = a_2 + b_2. \tag{4.250}$$

Let us further suppose that the random variable X is also a uniform random variable in the range $[c_1, c_2]$, as shown in Fig. 4.45b.

From (4.249), we write:

$$c_1 = a_1 + b_1 = m_1 - \frac{\Delta_1}{2} + m_2 - \frac{\Delta_2}{2} = m - \frac{\Delta_1 + \Delta_2}{2}. \tag{4.251}$$

Similarly, from (4.250) we have:

$$c_2 = a_2 + b_2 = m_1 + \frac{\Delta_1}{2} + m_2 + \frac{\Delta_2}{2} = m + \frac{\Delta_1 + \Delta_2}{2}. \tag{4.252}$$

From (4.251) and (4.252), it follows:

$$c_2 - c_1 = \Delta = \frac{\Delta_1 + \Delta_2}{2} + \frac{\Delta_1 + \Delta_2}{2} = \Delta_1 + \Delta_2, \tag{4.253}$$

or

$$\Delta^2 = (\Delta_1 + \Delta_2)^2. \tag{4.254}$$

From the other side, using (4.248), we have:

$$\frac{\Delta^2}{12} = \frac{\Delta_1^2 + \Delta_2^2}{12}, \tag{4.255}$$

or equivalently

$$\Delta^2 = \Delta_1^2 + \Delta_2^2. \tag{4.256}$$

Note that two different conditions (4.254) and (4.256) cannot be satisfied at the same time and as a consequence, the assumption that the variable X is uniform was wrong. Therefore, the sum of two uniform random variables is not a uniform variable.

A.4.3. The normal PDF becomes a delta function with an area of 1 at the mean value m of the random variable; thus indicating that a normal variable becomes a deterministic value m with the probability of 1.

A.4.4. Both PDFs will have the same shape. The only difference is positions of the PDFs on the x-axis. Each PDF is centered on its mean value.

A.4.5. Both PDFs are centered on the same mean value, but have different shapes.

A.4.6. The PDF of the sum of two independent random variables is equal to the convolution of their PDFs.
On the other hand, knowing that the sum of independent normal random variables is itself a normal random variable, it follows that a convolution of two normal PDFs is also a normal PDF.

A.4.7. The characteristic function of a normal variable has the shape of a normal PDF.

A.4.8. The probability of event A is:

$$P\{A\} = P\{|X| \le 0.5\} = P\{-0.5 \le X \le 0.5\} = \text{erf}\left(\frac{0.5}{\sqrt{2}}\right) = 0.3829. \quad (4.257)$$

The probability of event B is:

$$\begin{aligned} P\{B\} &= P\{|X| \ge 0.5\} = P\{-0.5 \ge X \ge 0.5\} = 2P\{X \ge 0.5\} \\ &= 1 - P\{A\} = 0.6171. \end{aligned} \quad (4.258)$$

Therefore, event B is more probable.

A.4.9. It is not possible [KOM87, pp. 103–104]. However, the transformation, known as a Box–Muller transformation, transforms a pair of independent uniform variables into a pair of independent normal random variables [MIL04, pp. 190–191], [KOM87, pp. 102–104], as described below.
The random variables X_1 and X_2 are both independent and uniform over [0,1]:

$$f_{X_1}(x_1) = \begin{cases} 1 & \text{for } 0 < x_1 \le 1, \\ 0 & \text{otherwise}. \end{cases} \quad (4.259)$$

$$f_{X_2}(x_2) = \begin{cases} 1 & \text{for } 0 < x_2 \le 1, \\ 0 & \text{otherwise}. \end{cases} \quad (4.260)$$

Two new random variables Y_1 and Y_2 are obtained by the following transformation:

$$Y_1 = \sqrt{-2\ln(X_1)}\cos(2\pi X_2), \quad (4.261)$$

$$Y_2 = \sqrt{-2\ln(X_1)}\sin(2\pi X_2). \quad (4.262)$$

The joint density of variables Y_1 and Y_2 is found to be:

$$f_{Y_1Y_2}(y_1,y_2)=\frac{f_{X_1X_2}(y_1,y_2)}{|J(y_1,y_2)|}. \tag{4.263}$$

The Jacobian is calculated by taking derivatives of y_1 and y_2 with respect to x_1 and x_2:

$$J=\begin{vmatrix}\frac{\partial y_1}{\partial x_1} & \frac{\partial y_1}{\partial x_2}\\ \frac{\partial y_2}{\partial x_1} & \frac{\partial y_2}{\partial x_2}\end{vmatrix}=\begin{vmatrix}-\frac{\cos(2\pi x_2)}{x_1\sqrt{-2\ln(x_1)}} & -2\pi\sqrt{-2\ln(x_1)}\sin(2\pi x_2)\\ -\frac{\sin(2\pi x_2)}{x_1\sqrt{-2\ln(x_1)}} & 2\pi\sqrt{-2\ln(x_1)}\cos(2\pi x_2)\end{vmatrix}=-\frac{2\pi}{x_1}. \tag{4.264}$$

Since x_1 is always positive (see (4.259)), then

$$|J|=\frac{2\pi}{x_1}. \tag{4.265}$$

From (4.261) and (4.262), we find:

$$x_1=\mathrm{e}^{-\frac{y_1^2+y_2^2}{2}}. \tag{4.266}$$

Placing (4.266) into (4.264) from (4.263), we arrive at:

$$f_{Y_1Y_2}(y_1,y_2)=\frac{1}{2\pi}\mathrm{e}^{-\frac{y_1^2+y_2^2}{2}}=\frac{1}{\sqrt{2\pi}}\mathrm{e}^{-\frac{y_1^2}{2}}\frac{1}{\sqrt{2\pi}}\mathrm{e}^{-\frac{y_2^2}{2}}=f_{Y_1}(y_1)f_{Y_2}(y_2). \tag{4.267}$$

Therefore, the transformation of two independent uniform random variables produces two independent standard normal variables (with a mean of zero and a variance of 1).

A.4.10. It is useful because of the fact that the sum X may have a PDF that is closely approximated to a normal PDF for finite number N [PEE93, p. 119].

A.4.11. In this case, the practical usefulness comes from the fact that the areas of delta functions approximate Gaussian PDF curve. For more details see [PAP65, pp. 268–272]. Also see Chap. 5.

Chapter 5
Other Important Random Variables

5.1 Lognormal Random Variable

5.1.1 Density Function

A *lognormal variable* X is obtained through the exponential transformation of a normal random variable Y that has a mean value of m_Y and a variance of σ_Y^2,

$$X = \mathrm{e}^Y, \tag{5.1}$$

or

$$x = \mathrm{e}^y, \quad x>0, \tag{5.2}$$

where x and y are ranges of the variables X and Y, respectively.

From (5.2), we have only one solution:

$$y = \ln x. \tag{5.3}$$

The relation (5.3) explains why the name "lognormal" comes for the variable X.

The PDF of the variable X from (5.1) is obtained using (2.140):

$$f_X(x) = \left.\frac{f_Y(y)}{\left|\frac{\mathrm{dx}}{\mathrm{dy}}(y)\right|}\right|_{y=\ln x}. \tag{5.4}$$

The derivative of the transformation (5.2) is:

$$\frac{\mathrm{d}x}{\mathrm{d}y} = \mathrm{e}^y = x, \quad x>0. \tag{5.5}$$

G.J. Dolecek, *Random Signals and Processes Primer with MATLAB*,
DOI 10.1007/978-1-4614-2386-7_5,

The normal variable Y has the following PDF:

$$f_Y(y) = \frac{1}{\sqrt{2\pi}\sigma_Y} e^{-\frac{(y-m_Y)^2}{2\sigma_Y^2}}. \tag{5.6}$$

Placing (5.6) and (5.5) into (5.4), we obtain the density function of the lognormal variable:

$$\begin{aligned} f_X(x) &= \frac{1}{\sqrt{2\pi}\sigma_Y x} e^{-\frac{(\ln x - m_Y)^2}{2\sigma_Y^2}} && \text{for} \quad x \geq 0, \\ f_X(x) &= 0 && \text{for} \quad x < 0 \end{aligned} \tag{5.7}$$

Note that m_Y and σ_Y^2 respectively, are the mean value and variance of the normal random variable Y, and not of the lognormal variable X. Figure 5.1 shows the lognormal variables (a) and densities (b), for different values of σ_Y and for $m_Y = 0.01$. For small values of σ_Y, ($\sigma_Y \leq 0.3$), the lognormal PDF becomes more symmetric, while for $\sigma_Y > 0.3$, the PDF becomes asymmetric.

5.1.2 Distribution Function

Using the definitions (2.61) and (5.7), the distribution function of a lognormal variable is:

$$F_X(x) = P\{X \leq x\} = \int_{-\infty}^{x} f_X(x)dx = \int_{-\infty}^{x} \frac{1}{\sqrt{2\pi\sigma_Y^2}x} e^{-\frac{(\ln x - m_Y)^2}{2\sigma_Y^2}} dx. \tag{5.8}$$

However, we can express the distribution (5.8) in terms of the normal distribution and thus make possible the use of erf function to more easily derive the lognormal distribution, as shown in the following equation:

$$F_X(x) = P\{X \leq x\} = \int_{-\infty}^{\ln x} f_Y(y)dy = \int_{-\infty}^{\ln x} \frac{1}{\sqrt{2\pi\sigma_Y^2}} e^{-\frac{(y-m_Y)^2}{2\sigma_Y^2}} dy. \tag{5.9}$$

Using the expressions for erf function (4.29) we arrive at:

$$F_X(x) = \begin{cases} \frac{1}{2}\left[1 + \operatorname{erf}\left(\frac{\ln x - m_Y}{\sqrt{2}\sigma_Y}\right)\right] & \text{for} \quad x \geq 0, \\ 0 & \text{otherwise.} \end{cases} \tag{5.10}$$

As an example, Fig. 5.2 shows the distribution function for $m_Y = 0.01$ and $\sigma_Y = 1$.

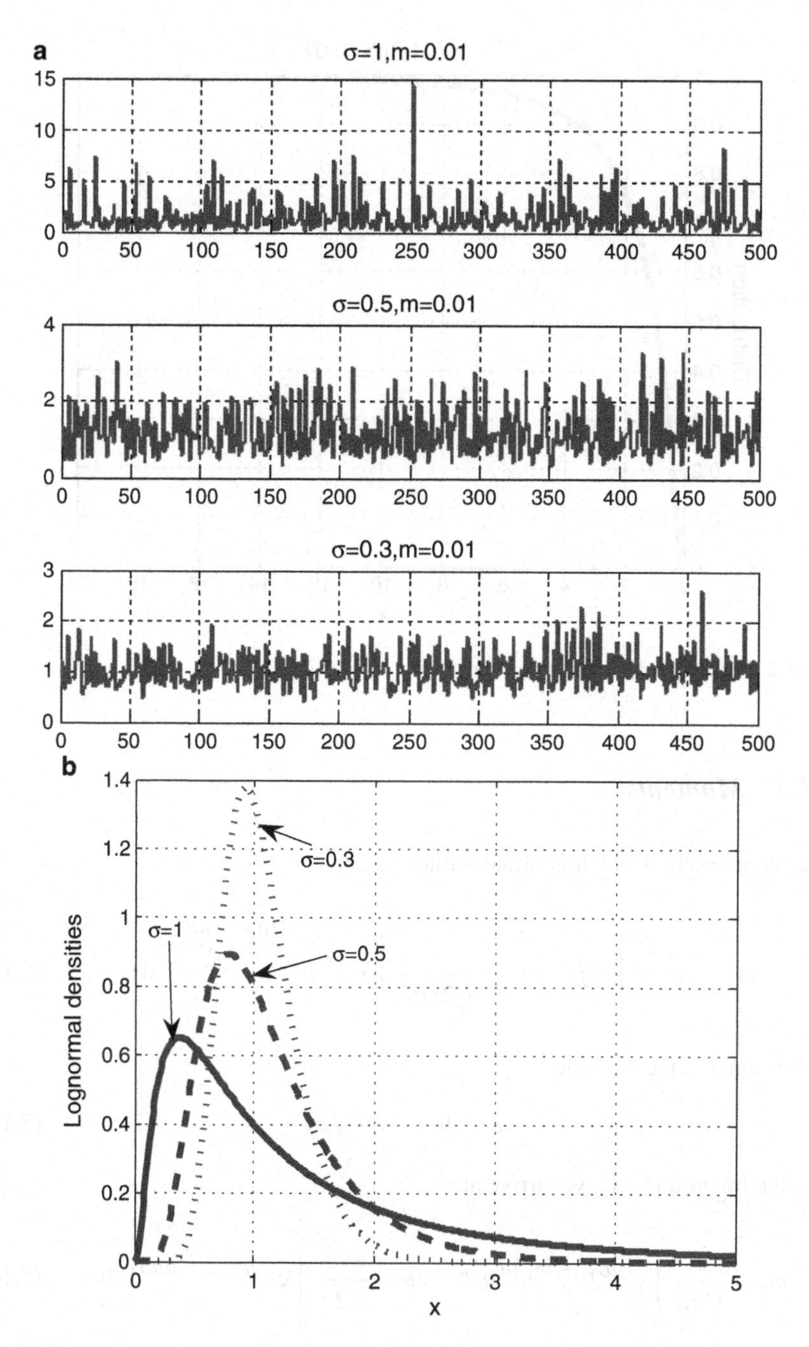

Fig. 5.1 Lognormal densities for different values of σ_Y and $m_Y = 0.01$. (**a**) Lognormal variables. (**b**) Densities

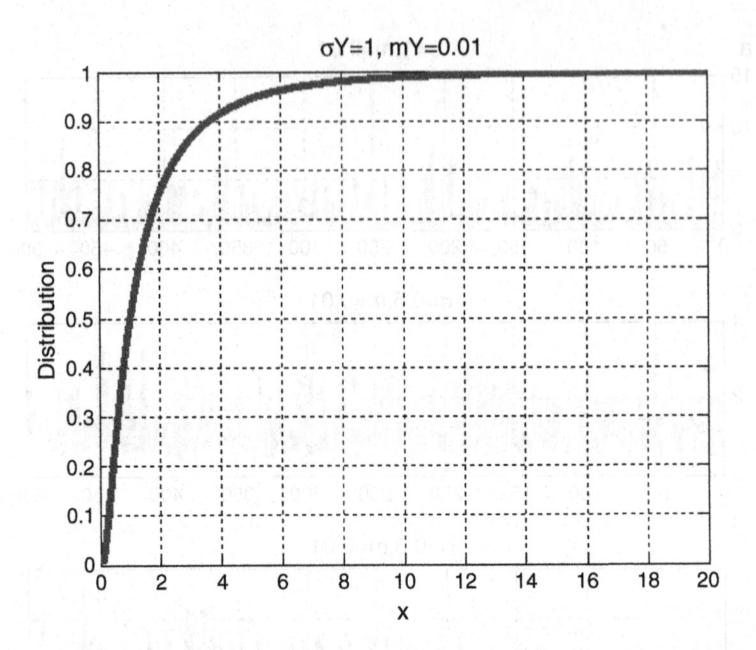

Fig. 5.2 Lognormal distribution function

5.1.3 Moments

The nth moment of the lognormal variable X is:

$$m_n = \overline{X^n} = \int_0^\infty x^n f_X(x)\mathrm{d}x = \frac{1}{\sqrt{2\pi}\sigma_Y}\int_0^\infty x^{n-1}\,\mathrm{e}^{-\frac{(\ln x - m_Y)^2}{2\sigma_Y^2}}\mathrm{d}x. \tag{5.11}$$

Introducing the variable

$$u = (\ln x - m_Y)/\sigma_Y \tag{5.12}$$

into the integral (5.11), we arrive at:

$$m_n = \frac{1}{\sqrt{2\pi}}\int_0^\infty \mathrm{e}^{(u\sigma_Y + m_Y)n}\,\mathrm{e}^{-u^2/2}\,\mathrm{d}u = \frac{\mathrm{e}^{nm_Y}}{\sqrt{2\pi}}\int_0^\infty \mathrm{e}^{-u^2/2 + un\sigma_Y}\,\mathrm{d}u. \tag{5.13}$$

Using the integral 6 from Appendix A, we finally arrive at:

$$m_n = \mathrm{e}^{nm_Y + n^2\sigma_Y^2/2}. \tag{5.14}$$

From (5.14), taking $n = 1$ and $n = 2$, we obtain the mean and the mean squared values:

$$m_1 = m_X = \overline{X} = e^{m_Y + \sigma_Y^2/2}, \tag{5.15}$$

$$m_2 = \overline{X^2} = e^{2m_Y + 2\sigma_Y^2}. \tag{5.16}$$

The corresponding variance is obtained from (5.15) and (5.16):

$$\sigma_X^2 = m_2 - m_1^2 = e^{2m_Y + \sigma_Y^2}(e^{\sigma_Y^2} - 1). \tag{5.17}$$

5.1.4 What Does a Lognormal Variable Tell Us?

Consider a large number of independent random variables

$$Y_1, \ldots, Y_N, \tag{5.18}$$

and the random variable X, which is equal to the product of the variables (5.18)

$$X = \prod_{i=1}^{N} Y_i. \tag{5.19}$$

Taking the natural logarithm of both sides of (5.19), we get:

$$\ln X = \sum_{i=1}^{N} \ln Y_i. \tag{5.20}$$

According to the central limit theorem, the right side of (5.20) can be approximated as a normal random variable Y,

$$Y \approx \sum_{i=1}^{N} \ln Y_i. \tag{5.21}$$

Taking into account (5.21), the expression (5.20) becomes:

$$\ln X \approx Y. \tag{5.22}$$

This expression can be rewritten as:

$$X \approx e^Y. \tag{5.23}$$

Comparing (5.23) and (5.1), it follows that the random variable X from (5.23) is a lognormal variable which itself represents the product of a number of independent variables, as can be seen in (5.19).

As a result, as the sum of a large number of independent random variables approaches a normal random variable (central limit theorem) *the product of a large number of independent random variables gives rise to a lognormal variable.*

The lognormal PDF can be used to describe the life distribution of many semiconductor components, the failure of which is a result of cracks caused by material fatigue [KLA89, p. 51].

5.2 Rayleigh Random Variable

5.2.1 Density Function

The *Rayleigh* density function is defined as:

$$\begin{aligned} f_X(x) &= kx^{-k\frac{x^2}{2}}, \quad \text{for} \quad x \geq 0 \\ f_X(x) &= 0, \quad\quad\quad\; \text{for} \quad x < 0 \end{aligned} \tag{5.24}$$

where k is a positive constant.

The density function is also expressed using the parameter σ^2, which is related with the constant k as follows:

$$\sigma^2 = 1/k, \tag{5.25}$$

resulting in:

$$\begin{aligned} f_X(x) &= \frac{1}{\sigma^2}x^{-\frac{x^2}{2\sigma^2}}, \quad \text{for} \quad x \geq 0 \\ f_X(x) &= 0, \quad\quad\quad\quad\; \text{for} \quad x < 0 \end{aligned} \tag{5.26}$$

The Rayleigh random variable and the corresponding density for the parameter $\sigma^2 = 4$ are shown in Fig. 5.3.

The Rayleigh density is useful in describing the envelope of a narrowband normal noise. It is also important in the analysis of measurement errors [PEE93, p. 56].

Note that the random variable has only positive values and that the PDF is asymmetrical and has a maximum value for

$$x = \sigma = 1/\sqrt{k}. \tag{5.27}$$

$$f_X\left(\frac{1}{\sqrt{k}}\right) = f_X(\sigma) = \frac{1}{\sigma\sqrt{e}} = \sqrt{\frac{k}{e}}. \tag{5.28}$$

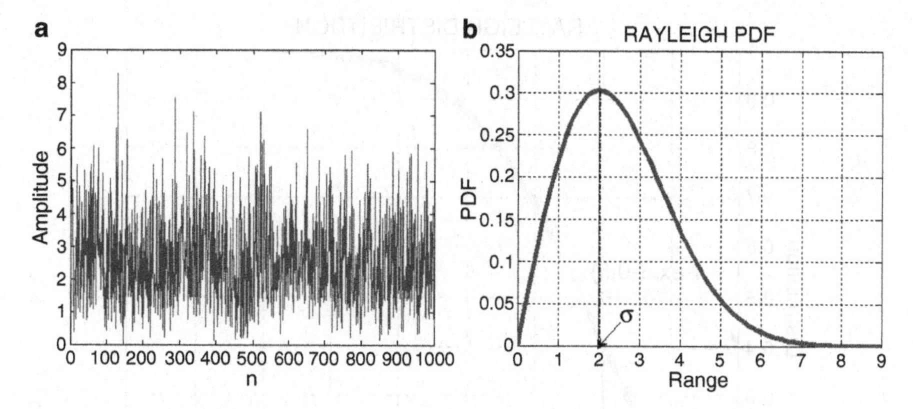

Fig. 5.3 Rayleigh variable and density, $\sigma^2 = 4$. (**a**) Rayleigh variable. (**b**) PDF

5.2.2 *Distribution Function*

The distribution function is obtained from (5.26) as:

$$F_X(x) = \int_{-\infty}^{x} f_X(x)\mathrm{d}x = \int_{0}^{x} \frac{1}{\sigma^2} x\,\mathrm{e}^{-\frac{x^2}{2\sigma^2}}\mathrm{d}x = 1 - \mathrm{e}^{-\frac{x^2}{2\sigma^2}}. \tag{5.29}$$

The distribution function for $\sigma^2 = 4$ is shown in Fig. 5.4.

Note that for $x = \sigma$, the distribution is

$$F_X(\sigma) = P\{X \leq \sigma\} = 1 - \mathrm{e}^{-0.5}. \tag{5.30}$$

Example 5.2.1 Find the probability that a Rayleigh signal is c times greater than the σ.

Solution From (5.29), we have:

$$P\{X > c\sigma\} = 1 - P\{X \leq c\sigma\} = 1 - F_X(c\sigma) = \mathrm{e}^{-c^2/2}. \tag{5.31}$$

Table 5.1 shows the probabilities (5.31) for different values of c.

5.2.3 *Moments*

The mean value and the mean squared value are obtained from (5.26) as:

$$E\{X\} = \frac{1}{\sigma^2} \int_{0}^{\infty} x^2\,\mathrm{e}^{-\frac{x^2}{2\sigma^2}}\,\mathrm{d}x, \tag{5.32}$$

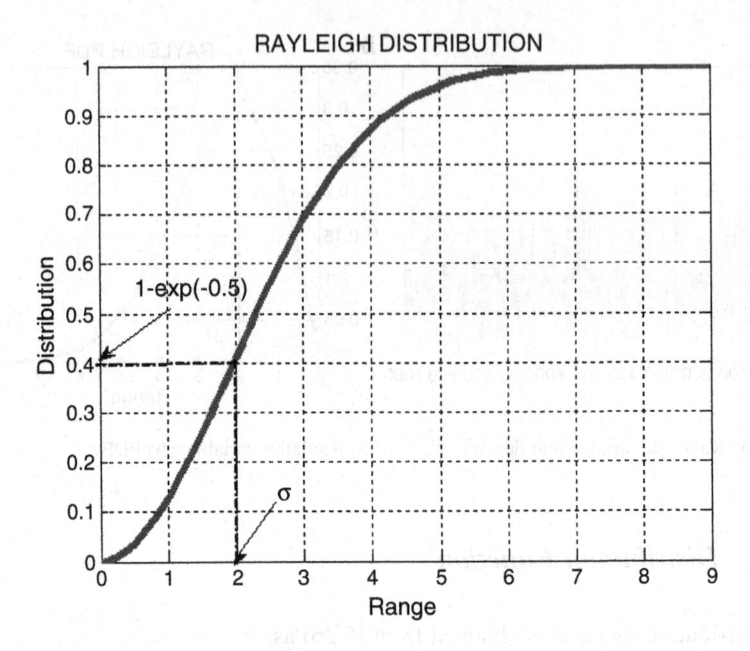

Fig. 5.4 Distribution

Table 5.1 The probabilities (5.31)

c	$P\{X > c\sigma\}$
1/5	0.9802
1/4	0.9692
1/3	0.9460
1/2	0.8825
1	0.6065
1.5	0.3247
2	0.135
3	0.011

$$E\{X^2\} = \frac{1}{\sigma^2} \int_0^\infty x^3 \, e^{-\frac{x^2}{2\sigma^2}} \, dx. \tag{5.33}$$

Using the integral 4 from Appendix A, we have:

$$E\{X\} = \sqrt{\frac{\pi}{2k}} = \sigma\sqrt{\frac{\pi}{2}} = 1.25\sigma, \tag{5.34}$$

$$E\{X^2\} = \frac{2}{k} = 2\sigma^2. \tag{5.35}$$

From (5.34) to (5.35), we find the variance to be:

$$\sigma_X^2 = \overline{X^2} - \overline{X}^2 = 2\sigma^2 - \frac{\sigma^2\pi}{2} = \sigma^2\left(2 - \frac{\pi}{2}\right) = 0.4292\sigma^2. \tag{5.36}$$

5.2.4 Relation of Rayleigh and Normal Variables

Consider two independent normal random variables X_1 and X_2 with equal mean values and variances:

$$\begin{aligned} E\{X_1\} = E\{X_2\} &= 0 \\ \sigma_{X_1}^2 = \sigma_{X_2}^2 &= \sigma^2. \end{aligned} \tag{5.37}$$

The joint density function is given as:

$$f_{X_1X_2}(x_1, x_2) = f_{X_1}(x_1)f_{X_2}(x_2) = \frac{1}{2\pi\sigma^2}\mathrm{e}^{-\frac{(x_1^2+x_2^2)}{2\sigma^2}}. \tag{5.38}$$

Next we introduce the polar coordinates, as shown in Fig. 5.5
The polar coordinates are X and θ and the transformation is defined as:

$$\begin{aligned} x &= \sqrt{x_1^2 + x_2^2}, \\ \mathrm{tg}\theta &= \frac{x_2}{x_1}. \end{aligned} \tag{5.39}$$

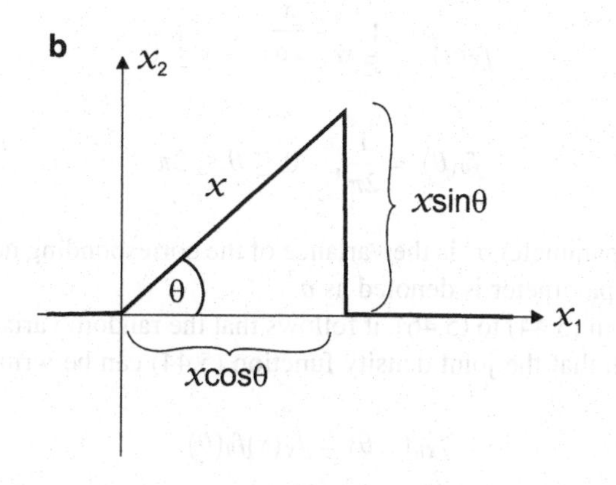

Fig. 5.5 Transformation X_1 and X_2 to polar coordinates X and θ

The joint output density function is

$$f_{X\theta}(x,\theta) = \frac{f_{X_1X_2}(x_1,x_2)}{|J(x_1,x_2)|}\Bigg|_{\substack{x_1 = f^{-1}(x,\theta) \\ x_2 = f^{-1}(x,\theta)}} \cdot \tag{5.40}$$

where J is a Jacobian of the transformation (5.39):

$$J = \begin{vmatrix} \dfrac{x_1}{\sqrt{{x_1}^2 + x_2^2}} & \dfrac{x_2}{\sqrt{{x_1}^2 + x_2^2}} \\ \dfrac{-x_2}{{x_1}^2 + x_2^2} & \dfrac{x_1}{{x_1}^2 + x_2^2} \end{vmatrix} = \frac{1}{\sqrt{{x_1}^2 + x_2^2}} = \frac{1}{x}. \tag{5.41}$$

Observe that in (5.39) x is always positive, such that

$$|J| = 1/x. \tag{5.42}$$

Placing (5.39) into (5.38), we get:

$$f_{X_1X_2}(x_1,x_2) = \frac{1}{2\pi\sigma^2}\mathrm{e}^{-\frac{(x_1^2 + x_2^2)}{2\sigma^2}} = \frac{1}{2\pi\sigma^2}\mathrm{e}^{-\frac{x^2}{2\sigma^2}}. \tag{5.43}$$

Placing (5.42) and (5.43) into (5.40), we get:

$$f_{X\theta}(x,\theta) = \frac{x}{2\pi\sigma^2}\mathrm{e}^{-\frac{x^2}{2\sigma^2}} = \frac{x}{\sigma^2}\mathrm{e}^{-\frac{x^2}{2\sigma^2}}\frac{1}{2\pi}. \tag{5.44}$$

We can easily recognize that in (5.44) the first term represents a Rayleigh and the second term a uniform density function:

$$f_X(x) = \frac{1}{\sigma^2}x\mathrm{e}^{-\frac{x^2}{2\sigma^2}}, \quad x \geq 0. \tag{5.45}$$

$$f_\theta(\theta) = \frac{1}{2\pi}, \quad 0 \leq \theta \leq 2\pi. \tag{5.46}$$

Note that the parameter σ^2 is the variance of the corresponding normal variables. This is why this parameter is denoted as σ^2.

Therefore, from (5.44) to (5.46), it follows that the random variables X and θ are independent such that the joint density function (5.44) can be written as:

$$f_{X\theta}(x,\theta) = f_X(x)f_\theta(\theta). \tag{5.47}$$

5.3 Rician Random Variable

5.3.1 *Relation of Rician, Rayleigh, and Normal Variables*

Consider two independent normal random variables X_1 and X_2 with equal mean values and variances, as defined in (5.37). Additionally, imagine that the constant A is added to the variable X_1, resulting in the normal variable $X_1{}'$

$$X_1{}' = X_1 + A. \tag{5.48}$$

The mean value of the variable $X_1{}'$ is equal to A.

$$E\{X_1{}'\} = A. \tag{5.49}$$

The joint density function is given as:

$$\begin{aligned} f_{X_1{}'X_2}(x_1{}', x_2) &= f_{X_1{}'}(x_1{}') f_{X_2}(x_2) = \frac{1}{2\pi\sigma^2} \mathrm{e}^{-\frac{(x_1{}' - A)^2}{2\sigma^2}} \mathrm{e}^{-\frac{x_2^2}{2\sigma^2}} \\ &= \frac{1}{2\pi\sigma^2} \mathrm{e}^{-\frac{(x_1{}' - A)^2 + x_2^2}{2\sigma^2}} = \frac{1}{2\pi\sigma^2} \mathrm{e}^{-\frac{x_1'^2 + x_2^2 - 2x_1' A + A^2}{2\sigma^2}}. \end{aligned} \tag{5.50}$$

Next we introduce the polar coordinates, as shown in Fig. 5.6.

The transformation of X_1 and X_2 into polar coordinates X and θ is defined as:

$$\begin{aligned} x &= \sqrt{x_1'^2 + x_2^2}, \\ \theta &= \tan^{-1}\left(\frac{x_2}{x_1'}\right). \end{aligned} \tag{5.51}$$

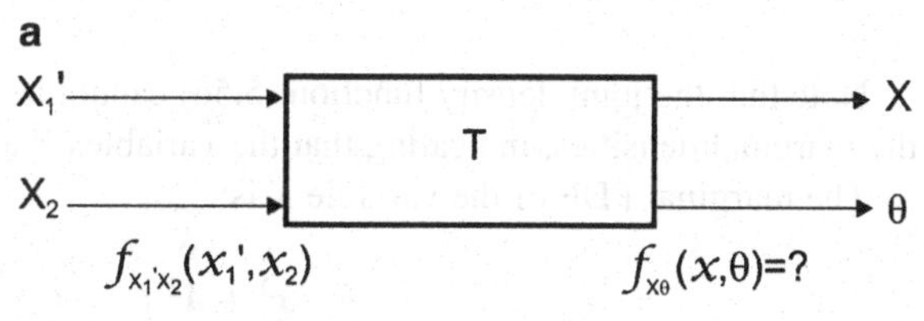

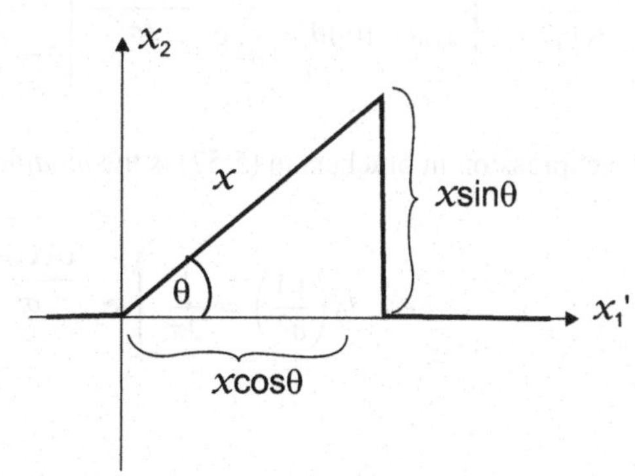

Fig. 5.6 Transformation of $X_1{}'$ and X_2 to polar coordinates X and θ

The joint output density function can be obtained from (5.50) and (5.51):

$$f_{X\theta}(x,\theta) = \frac{f_{X_1'X_2}(x_1',x_2)}{|J(x_1',x_2)|}\Bigg|_{\substack{x_1' = f^{-1}(x,\theta)\\ x_2 = f^{-1}(x,\theta)}}, \tag{5.52}$$

where J is a Jacobian of the transformation (5.51):

$$J = \begin{vmatrix} \frac{x_1'}{\sqrt{x_1'^2 + x_2^2}} & \frac{x_2}{\sqrt{x_1'^2 + x_2^2}} \\ \frac{-x_2}{x_1'^2 + x_2^2} & \frac{x_1}{x_1'^2 + x_2^2} \end{vmatrix} = \frac{1}{\sqrt{x_1'^2 + x_2^2}} = \frac{1}{x}. \tag{5.53}$$

Note that the Jacobian of (5.53) is positive for all values of x (see (5.51)). From Fig. 5.6, we have:

$$x_1' = x\cos\theta. \tag{5.54}$$

Placing (5.51) and (5.54) into (5.50), we arrive at:

$$f_{X_1'X_2}(x,\theta) = \frac{1}{2\pi\sigma^2}\mathrm{e}^{-\frac{x^2 - 2xA\cos\theta + A^2}{2\sigma^2}}. \tag{5.55}$$

Finally, from (5.52), (5.53), and (5.55), we get the joint density function of X and θ:

$$f_{X\theta}(x,\theta) = \frac{x}{2\pi\sigma^2}\exp\left(-\frac{x^2 - 2xA\cos\theta + A^2}{2\sigma^2}\right) \text{for } x \geq 0,\ -\pi \leq \theta \leq \pi. \tag{5.56}$$

Note that the joint density function (5.56) cannot be represented as a product of the marginal densities, indicating that the variables X and θ are dependent.

The marginal PDF of the variable X is:

$$f_X(x) = \int_{-\pi}^{\pi} f_{X\theta}(x,\theta)\mathrm{d}\theta = \frac{x}{\sigma^2}\mathrm{e}^{-\frac{x^2 + A^2}{2\sigma^2}}\left[\frac{1}{2\pi}\int_{-\pi}^{\pi}\mathrm{e}^{-\frac{xA\cos\theta}{\sigma^2}}\mathrm{d}\theta\right]. \tag{5.57}$$

The expression in brackets in (5.57) is the *modified zero-order Bessel function* I_0

$$I_0\left(\frac{xA}{\sigma^2}\right) = \frac{1}{2\pi}\int_{-\pi}^{\pi}\mathrm{e}^{-\frac{xA\cos\theta}{\sigma^2}}\mathrm{d}\theta. \tag{5.58}$$

Using the Bessel function (5.58), we get the PDF of the variable X as:

$$f_X(x) = \frac{x}{\sigma^2} \mathrm{e}^{-\frac{x^2 + A^2}{2\sigma^2}} I_0\left(\frac{xA}{\sigma^2}\right). \tag{5.59}$$

The PDF (5.59) is said to be a *Rician PDF*.

The Rician PDFs for different values of A and $\sigma^2 = 1$ are shown in Fig. 5.7.

For $A = 0$, the Bessel function $I_0 = 1$ and the Rician PDF becomes the Rayleigh PDF, as shown in Fig. 5.7. This is also obvious from (5.48) where the variable X_1' becomes X_1.

For higher values of A, the Rician PDF becomes more symmetrical and it approaches normal random variable.

5.4 Exponential Random Variable

5.4.1 Density and Distribution Function

The *exponential density function* is given as:

$$f_X(x) = \begin{cases} \lambda\, \mathrm{e}^{-\lambda x} & \text{for} \quad x \geq 0, \\ 0 & \text{otherwise}, \end{cases} \tag{5.60}$$

where λ is a constant parameter.

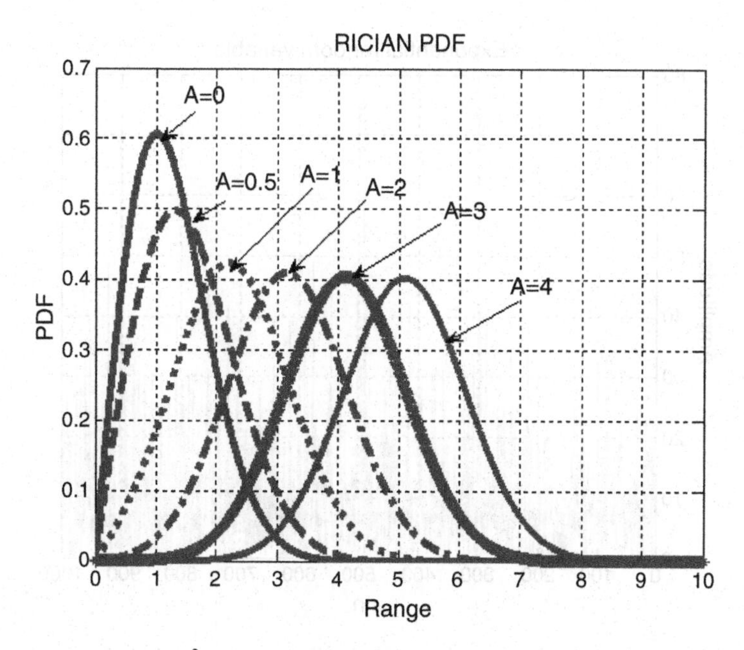

Fig. 5.7 Rician PDFs for $\sigma^2 = 1$ and different values A

The corresponding distribution function is:

$$F_X(x) = P\{X \leq x\} = \int_0^x f_X(x)\mathrm{d}x = \lambda \int_0^x \mathrm{e}^{-\lambda x}\,\mathrm{d}x. \tag{5.61}$$

From (5.61), we have:

$$F_X(x) = \begin{cases} 1 - \mathrm{e}^{-\lambda x} & \text{for} \quad x \geq 0, \\ 0 & \text{for} \quad x < 0. \end{cases} \tag{5.62}$$

Exercise E.2.13 shows that the exponential random variable X can be obtained by the following transformation of the uniform random variable Y in the interval [0, 1]:

$$X = -\frac{1}{\lambda}\ln(1 - Y). \tag{5.63}$$

An exponential variable with the parameter $\lambda = 0.1$, which is generated using (5.63), is shown in Fig. 5.8. The corresponding density and distribution functions are shown in Fig. 5.9a, b. Note that the low values of the random variable are much more probable than the high values.

Exponential random variables are used in the study of queuing systems to describe the time between arrivals of customers [MIL04, p. 64]. Additionally, the exponential density is useful in describing the fluctuations in signal strength received by radar from aircrafts [PEE93, p. 56].

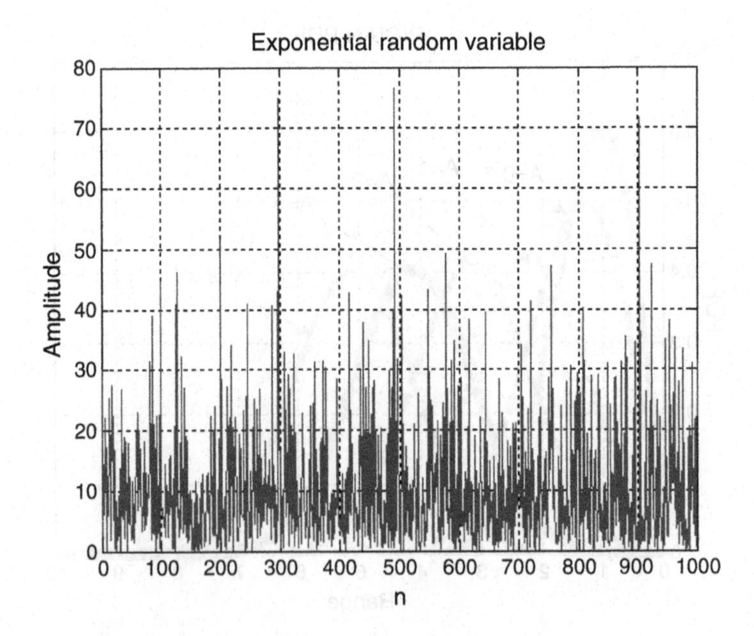

Fig. 5.8 Exponential variable with the parameter $\lambda = 0.1$

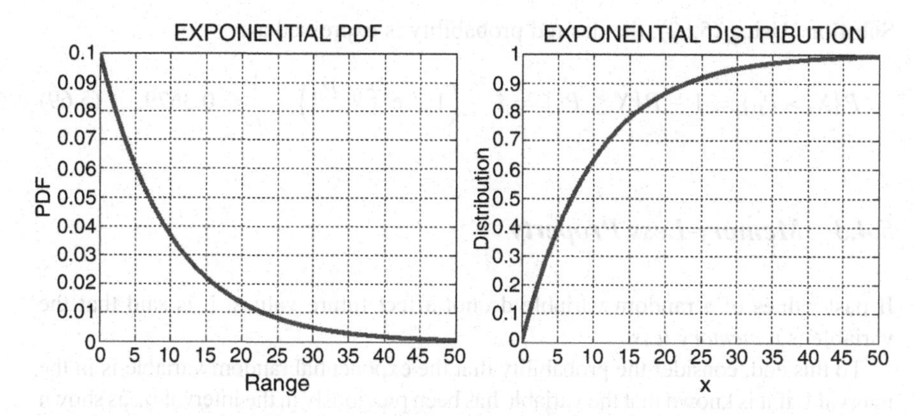

Fig. 5.9 Exponential PDF and distribution for the parameter $\lambda = 0.1$

5.4.2 *Characteristic Function and Moments*

Using the definition of the characteristic function, we have:

$$\phi_X(\omega) = \int_0^\infty e^{j\omega x} f_X(x)dx = \lambda \int_0^\infty e^{-x(\lambda - j\omega)}\, dx = \frac{\lambda}{\lambda - j\omega}. \tag{5.64}$$

Applying the moment theorem, we obtain the mean and mean squared values,

$$\overline{X} = \frac{1}{j} \frac{d\phi_X(\omega)}{d\omega}\bigg|_{\omega=0} = \frac{1}{\lambda}, \tag{5.65}$$

$$\overline{X^2} = \frac{1}{j^2} \frac{d^2\phi_X(\omega)}{d\omega^2}\bigg|_{\omega=0} = \frac{2}{\lambda^2}. \tag{5.66}$$

From (5.65) and (5.66), the variance is given as:

$$\sigma^2 = \overline{X^2} - \overline{X}^2 = \frac{2}{\lambda^2} - \frac{1}{\lambda^2} = \frac{1}{\lambda^2}. \tag{5.67}$$

Example 5.4.1 The power reflected from the aircraft into the radar is described using the exponential variable:

$$f_X(x) = \begin{cases} \frac{1}{P_0} e^{-x/P_0} & \text{for} \quad x \geq 0, \\ 0 & \quad\quad\quad \text{otherwise}, \end{cases} \tag{5.68}$$

where P_0 is the mean value of the received power.

Find the probability that the received power is larger than its average value.

Solution Using (5.62), the desired probability is expressed as:

$$P\{X > P_0\} = 1 - P\{X \leq P_0\} = 1 - \left(1 - e^{-P_0/P_0}\right) = \frac{1}{e} = 0.3679. \quad (5.69)$$

5.4.3 Memory-Less Property

If past values of a random variable do not affect future values, it is said that the variable is a *memory less*.

To this end, consider the probability that the exponential random variable is in the interval x, if it is known that the variable has been previously in the interval x_0, as shown in Fig. 5.10. This statement can be expressed in a mathematical form as in (5.70).

$$P\{X \leq x_0 + x_1 \mid X > x_0\} = \frac{P\{x_0 < X \leq x_0 + x_1\}}{P\{X > x_0\}} = \frac{F_X(x_0 + x_1) - F_X(x_0)}{1 - F_X(x_0)}. \quad (5.70)$$

Using (5.62), we arrive at:

$$P\{X \leq x_0 + x_1 \mid X > x_0\} = \frac{(1 - e^{-\lambda(x_0+x_1)}) - (1 - e^{-\lambda x_0})}{1 - (1 - e^{-\lambda x_0})} = 1 - e^{-\lambda x_1}. \quad (5.71)$$

The obtained result (5.71) indicates that the desired probability does not depend on the length of x_0, but on the length of x_1:

$$P\{X \leq x_0 + x_1 \mid X > x_0\} = P\{X \leq x_1\} = 1 - e^{-\lambda x_1}. \quad (5.72)$$

5.5 Variables Related with Exponential Variable

5.5.1 Laplacian Random Variable

A random variable is said to be a *Laplacian random variable* if its density function is defined as:

$$f_X(x) = k\,e^{-\lambda|x|}, \quad \text{for} -\infty < x < \infty, \quad (5.73)$$

where k is the constant and λ is the positive parameter.

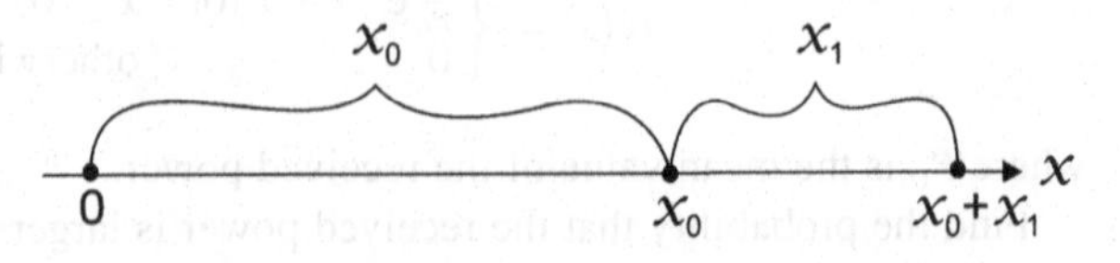

Fig. 5.10 Memory-less property of the exponential variable

Applying the condition that the area below a PDF is equal to the unity,

$$\int_{-\infty}^{\infty} f_X(x)\mathrm{d}x = \int_{-\infty}^{0} k\,\mathrm{e}^{\lambda x}\,\mathrm{d}x + \int_{0}^{\infty} k\,\mathrm{e}^{-\lambda x}\,\mathrm{d}x = \frac{2k}{\lambda} = 1, \tag{5.74}$$

the constant k is related with the parameter λ as:

$$k = \lambda/2. \tag{5.75}$$

Using (5.75), the density function (5.73) becomes

$$f_X(x) = \frac{\lambda}{2}\mathrm{e}^{-\lambda|x|}, \quad \text{for} - \infty<x<\infty. \tag{5.76}$$

The variable for $\lambda = 0.5$ is shown in Fig. 5.11a. Note that the density shown in Fig. 5.11b takes the form of a two-sided exponential PDF with a peak which is one-half of that of the exponential PDF.

The corresponding distribution function is

$$F_X(x) = \int_{-\infty}^{x} \frac{\lambda}{2}\mathrm{e}^{-\lambda|x|}\,\mathrm{d}x. \tag{5.77}$$

For $x \leq 0$, we have:

$$F_X(x) = \frac{\lambda}{2}\int_{-\infty}^{x} \mathrm{e}^{\lambda x}\,\mathrm{d}x = \frac{1}{2}\mathrm{e}^{\lambda x}. \tag{5.78}$$

For $x \geq 0$, it follows:

$$F_X(x) = \frac{\lambda}{2}\left[\int_{-\infty}^{0} \mathrm{e}^{\lambda x}\,\mathrm{d}x + \int_{0}^{x} \mathrm{e}^{-\lambda x}\,\mathrm{d}x\right] = 1 - \frac{1}{2}\mathrm{e}^{-\lambda x}. \tag{5.79}$$

From (5.78) and (5.79), we obtain:

$$F_X(x) = \begin{cases} 0.5\,\mathrm{e}^{\lambda x} & \text{for} \quad x \leq 0, \\ 1 - 0.5\,\mathrm{e}^{-\lambda x} & \text{for} \quad x \geq 0. \end{cases} \tag{5.80}$$

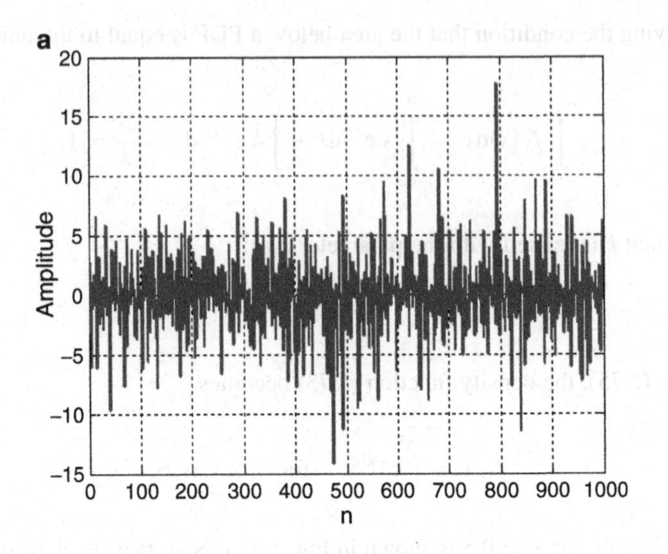

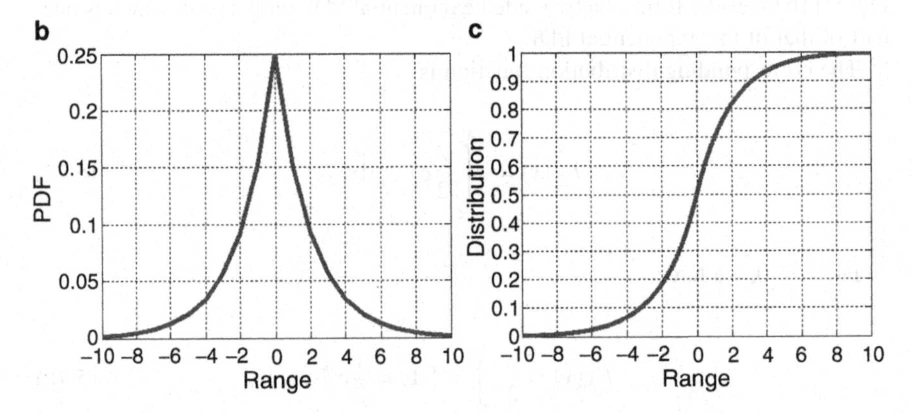

Fig. 5.11 Laplacian variable, density, and distribution for $\lambda = 0.5$. (**a**) Laplacian variable. (**b**) PDF. (**c**) Distribution

The distribution function for $\lambda = 0.5$ is shown in Fig. 5.11c.

In order to find the mean value and the variance, we compute the characteristic function:

$$\phi_X(\omega) = \int_{-\infty}^{\infty} e^{j\omega x} f_X(x)\mathrm{d}x = \frac{\lambda}{2}\left[\int_{-\infty}^{0} e^{\lambda x + j\omega x}\,\mathrm{d}x + \int_{0}^{\infty} e^{-\lambda x + j\omega x}\,\mathrm{d}x\right]$$
$$= \frac{\lambda^2}{\lambda^2 + \omega^2}. \tag{5.81}$$

The mean and mean squared values are obtained from the characteristic function (5.81) using the moment theorem:

$$\overline{X} = \frac{1}{j} \frac{\mathrm{d}\phi_X(\omega)}{\mathrm{d}\omega}\bigg|_{\omega=0} = 0. \tag{5.82}$$

$$\overline{X^2} = \frac{1}{j^2} \frac{\mathrm{d}^2\phi_X(\omega)}{\mathrm{d}\omega^2}\bigg|_{\omega=0} = \frac{2}{\lambda^2}. \tag{5.83}$$

From here the variance is:

$$\sigma^2 = \overline{X^2} - \overline{X}^2 = \frac{2}{\lambda^2}. \tag{5.84}$$

The Laplacian random variable is useful in modeling a speech signal [MIL04, p. 66].

5.5.2 Gamma and Erlang's Random Variables

5.5.2.1 Gamma Variable

As opposed to an exponential random variable with only one parameter, sometimes it is easier to use variables related with the exponential random variable but which have more parameters. One such variable is a *Gamma variable*.

Its density function is defined as:

$$f_X(x) = \begin{cases} \frac{\lambda^b}{\Gamma(b)} x^{b-1}\,\mathrm{e}^{-\lambda x} & \text{for} \quad x \geq 0, \\ 0 & \text{otherwise}, \end{cases} \tag{5.85}$$

where λ and b are positive parameters and $\Gamma(b)$ is the *Gamma function*, defined as:

$$\Gamma(b) = \int_0^\infty y^{b-1}\,\mathrm{e}^{-y}\,\mathrm{d}y. \tag{5.86}$$

The properties of Gamma function are given in Appendix C.

The Gamma densities for $\lambda = 1$, and the different values of the parameter b are shown in Fig. 5.12. Note that the Gamma variable becomes the exponential variable for $b = 1$.

The mean value of the Gamma variable is written as:

$$\overline{X} = \int_{0}^{\infty} x f_X(x)\mathrm{d}x = \int_{0}^{\infty} \frac{\lambda^b}{\Gamma(b)} x^b \,\mathrm{e}^{-\lambda x}\,\mathrm{d}x. \tag{5.87}$$

Using integral 1 from Appendix A, we obtain:

$$\overline{X} = \frac{\lambda^b}{\Gamma(b)} \frac{\Gamma(b+1)}{\lambda^{b+1}} = \frac{\lambda^b}{\Gamma(b)} \frac{b\Gamma(b)}{\lambda^{b+1}} = \frac{b}{\lambda}. \tag{5.88}$$

Similarly, the mean squared value is:

$$\overline{X^2} = \frac{\lambda^b}{\Gamma(b)} \frac{\Gamma(b+2)}{\lambda^{b+2}} = \frac{\lambda^b}{\Gamma(b)} \frac{b(b+1)\Gamma(b)}{\lambda^{b+2}} = \frac{b(b+1)}{\lambda^2}. \tag{5.89}$$

Finally, the variance is:

$$\sigma^2 = \frac{b^2 + b - b^2}{\lambda^2} = \frac{b}{\lambda^2}. \tag{5.90}$$

The Gamma variable is used in queuing theory [MIL04, p. 67].

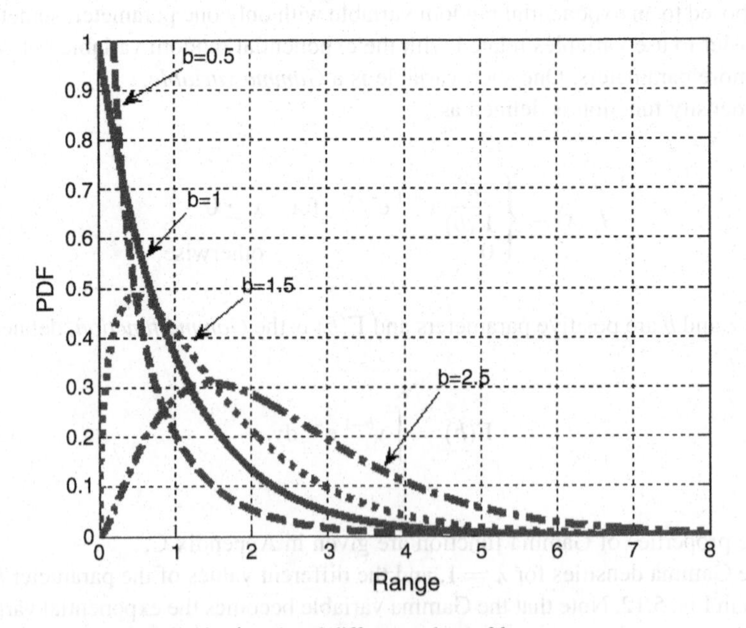

Fig. 5.12 Gamma densities for $\lambda = 1$ and different values of b

5.5.2.2 Erlang's Variable

If a parameter b of a Gamma variable is an integer $b = n$, then,

$$\Gamma(n) = (n-1)!, \tag{5.91}$$

and Gamma variable becomes an *Erlang's variable* with a density function of:

$$f_X(x) = \begin{cases} \dfrac{(\lambda x)^{n-1}}{(n-1)!} \lambda \, e^{-\lambda x} & \text{for} \quad x \geq 0, \\ 0 & \text{otherwise}. \end{cases} \tag{5.92}$$

The characteristic function is:

$$\phi(\omega) = \frac{\lambda^n}{(n-1)!} \int_0^{\infty} x^{n-1} \, e^{-x(\lambda - j\omega)} dx. \tag{5.93}$$

Using integral 1 from Appendix A, and (5.91), we have:

$$\phi(\omega) = \frac{\lambda^n}{(n-1)!} \frac{\Gamma(n)}{(\lambda - j\omega)^n} = \left[\frac{\lambda}{\lambda - j\omega}\right]^n. \tag{5.94}$$

Erlang's random variable is useful in the study of wire line telecommunication networks [MIL04, p. 67].

5.5.2.3 Sum of Independent Exponential Variables

Consider the sum of n independent exponential variables X_k with the same parameter λ,

$$X = \sum_{k=1}^{n} X_k. \tag{5.95}$$

The corresponding characteristic function is:

$$\phi_X(\omega) = \prod_{i=1}^{n} \phi_{X_k}(\omega). \tag{5.96}$$

From (5.64), we have the characteristic function of the exponential variable

$$\phi_{X_k}(\omega) = \frac{\lambda}{\lambda - j\omega}. \tag{5.97}$$

Placing (5.97) into (5.96), we get:

$$\phi_X(\omega) = \prod_{i=1}^{n} \frac{\lambda}{\lambda - j\omega} = \left[\frac{\lambda}{\lambda - j\omega}\right]^n. \tag{5.98}$$

Comparing (5.98) to the characteristic function of Erlang's variable (5.94), we can conclude that the variable X is an Erlang's variable:

The sum of independent exponential variables with equal parameters λ is the Erlang's variable.

Applying the central limit theorem to the sum of exponential variables (5.95), it follows that the Erlang's variable for high values of n approaches the normal variable, as shown in Fig. 5.13.

5.5.3 Weibull Random Variable

A *Weibull random variable* is also a two-parameter random variable with a density function of:

$$f_X(x) = \begin{cases} Kx^m \mathrm{e}^{-\frac{Kx^{m+1}}{m+1}} & \text{for} \quad x \geq 0, \\ 0 & \text{otherwise}, \end{cases} \tag{5.99}$$

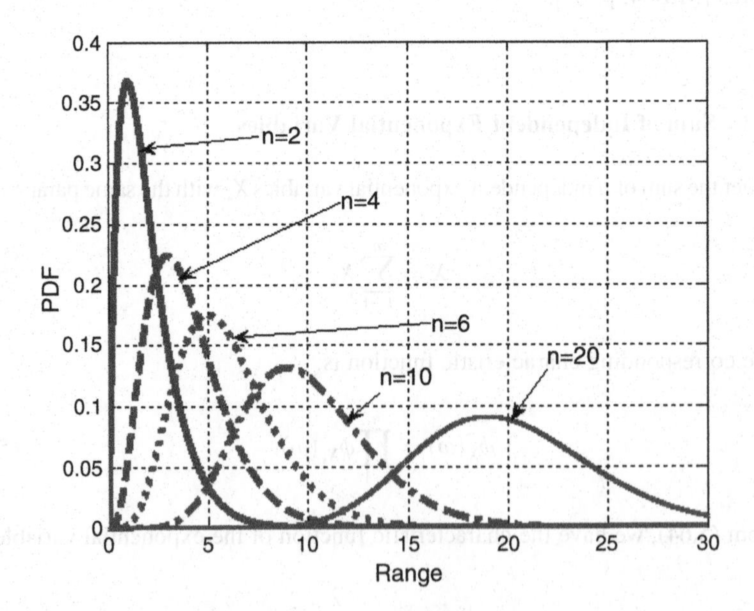

Fig. 5.13 Erlang's densities for different values of n and $\lambda = 1$

where K and m are constant parameters.

The Weibull densities are shown in Fig. 5.14 for $K = 1$ and for different values of m. For $m = 0$, the Weibull density becomes an exponential density; while for $m = 1$, it becomes a Rayleigh density.

5.6 Bernoulli and Binomial Random Variables

5.6.1 Bernoulli Experiments

An experiment is said to be a *Bernoulli experiment* if it has only two possible outcomes: either the *success* or *failure* of a given event A. We denote the failure of an event A as $\overline{A}$. The corresponding probabilities are denoted as:

$$P\{A\} = p, \tag{5.100}$$

$$P\{\overline{A}\} = 1 - p = q. \tag{5.101}$$

For example, a coin tossing experiment is a Bernoulli experiment.

A *Bernoulli random variable* is a random variable associated with a Bernoulli experiment, in which 1 is associated with the success of an event A, and 0 is associated with the failure of this event. Therefore, a Bernoulli random variable

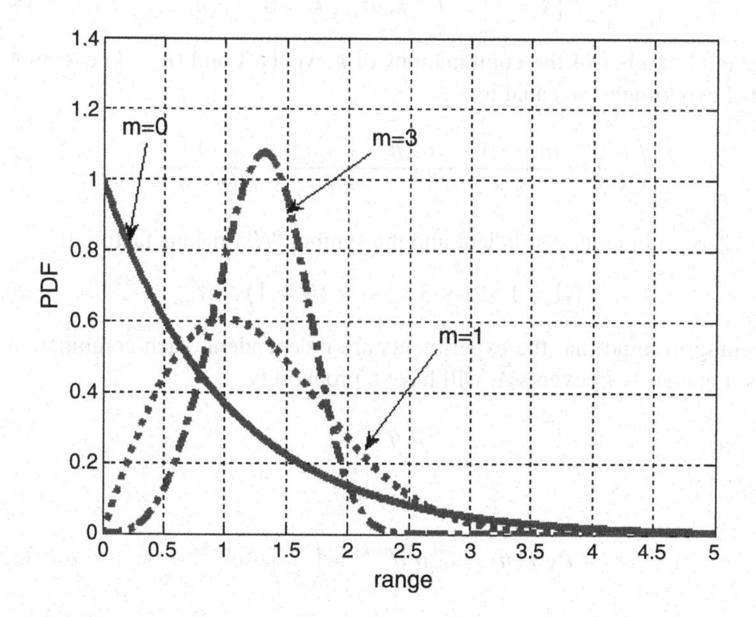

Fig. 5.14 Weibull PDFs for $K = 1$ and different values of m

is a discrete random variable with only two discrete values 1 and 0, that has the corresponding probabilities,

$$P\{X = 1\} = p. \tag{5.102}$$

$$P\{X = 0\} = 1 - p = q. \tag{5.103}$$

5.6.2 *What Is a Binomial Random Variable?*

Consider n independent random Bernoulli experiments. As a result of each experiment, either event A or its complement $\overline{A}$ occurs, with the corresponding probabilities (5.100) and (5.101). The outcomes of repeated trials are sequences of A and $\overline{A}$ where order of sequences A or $\overline{A}$ is arbitrary.

$$AAA\overline{A}\ldots\overline{A}AAA\ldots\overline{A}\,\overline{A}. \tag{5.104}$$

The number of A events in n repeated Bernoulli trials can be any integer, from 0 to n. A random variable associated with the number of the occurrences of event A is said to be a *binomial random variable*, with the parameters, n and p, where p is a probability of event A.

A binomial random variable X is a discrete random variable with the discrete values $x = 0, \ldots, n$. In order to describe this variable, we need the probability

$$P\{X = k\} = P_X(k; n), \quad k = 0, \ldots, n. \tag{5.105}$$

The total number of the combinations of k events A and $(n - k)$ events $\overline{A}$ in n repeated experiments is equal to

$$C_n^k = \binom{n}{k} = \frac{n(n-1), \ldots, (n-k+1)}{1 \times 2 \times 3 \times \cdots \times k} = \frac{n!}{k!(n-k)!}, \tag{5.106}$$

where C_n^k is a binomial coefficient and the symbol "$N!$" means factorial:

$$N! = 1 \times 2 \times 3 \times \cdots \times (N-1) \times N. \tag{5.107}$$

Keeping in mind that the experiments are independent, each combination of k events A and $(n - k)$ events $\overline{A}$, will have a probability,

$$p^k q^{n-k}, \tag{5.108}$$

resulting in:

$$P_X(k; n) = C_n^k p^k q^{n-k} = \binom{n}{k} p^k q^{n-k}. \tag{5.109}$$

This expression is called a *Bernoulli formula* and *Binomial probability law* [LEO94, p. 62].

5.6.3 Binomial Distribution and Density

From (5.108) the *binomial distribution* is given as:

$$F_X(x) = \sum_{k=0}^{n} \binom{n}{k} p^k q^{n-k} u(x-k) = \sum_{k=0}^{n} P_X(k;n) u(x-k). \tag{5.110}$$

The probabilities $P_X(k;n)$ are called the *probability mass function*, as indicated in Chap. 2.

Keeping in mind that a binomial variable is a discrete variable, its corresponding density function is:

$$f_X(x) = \sum_{k=0}^{n} \binom{n}{k} p^k q^{n-k} \delta(x-k) = \sum_{k=0}^{n} P_X(k,n) \delta(x-k). \tag{5.111}$$

The name "binomial" comes up from the probability (5.109) which itself represents the nth degree of the Newton binomial:

$$(p+q)^n = \sum_{k=0}^{n} C_n^k p^k q^{n-k}. \tag{5.112}$$

Taking into account (5.101), and using (5.109) from (5.112), we arrive at:

$$1^n = 1 = \sum_{k=0}^{n} C_n^k p^k q^{n-k} = \sum_{k=0}^{n} P_X(k,n). \tag{5.113}$$

The expression (5.113) proves that the binomial PDF satisfies the condition in (2.83).

Additionally, the expression (5.112) can be useful in calculating the characteristic function, as shown below.

5.6.4 Characteristic Functions and Moments

The characteristic function of a binomial variable is:

$$\begin{aligned}\phi_X(\omega) &= E\{e^{j\omega X}\} \\ &= \sum_{k=0}^{n} e^{j\omega k} P_X(k;n) = \sum_{k=0}^{n} e^{j\omega k} C_n^k p^k q^{n-k} = \sum_{k=0}^{n} C_n^k (p\, e^{j\omega})^k q^{n-k}. \end{aligned} \tag{5.114}$$

Table 5.2 Events and probabilities in Example 5.6.1

$k = 2$ events A in $n = 4$ events A and $\overline{A}$	Probability $p^k q^{n-k}$
$AA\overline{A}\,\overline{A}$	$0.5^2 0.5^2$
$A\overline{A}A\overline{A}$	$0.5^2 0.5^2$
$A\overline{A}\,\overline{A}A$	$0.5^2 0.5^2$
$\overline{A}\,\overline{A}AA$	$0.5^2 0.5^2$
$\overline{A}AA\overline{A}$	$0.5^2 0.5^2$
$\overline{A}A\overline{A}A$	$0.5^2 0.5^2$

Comparing (5.114) and (5.112), we get:

$$\phi_X(\omega) = (p\,\mathrm{e}^{j\omega} + q)^n. \tag{5.115}$$

The mean and the mean squared values are obtained by applying the moment's theorem:

$$E\{X\} = \frac{\mathrm{d}\phi_X(\omega)}{\mathrm{d}\omega}\frac{1}{j}\bigg|_{\omega=0} = \frac{jnp}{j} = np. \tag{5.116}$$

$$E\{X^2\} = \frac{\mathrm{d}^2\phi_X(\omega)}{\mathrm{d}\omega^2}\frac{1}{j^2}\bigg|_{\omega=0} = \frac{-n^2p^2 - npq}{j^2} = n^2p^2 + npq. \tag{5.117}$$

From here, the variance is

$$\sigma^2 = E\{X^2\} - E\{X\}^2 = n^2p^2 + npq - (np)^2 = npq. \tag{5.118}$$

Example 5.6.1 In $n = 4$ independent Bernoulli experiments A event occurs with a probability of $P\{A\} = p$. Find a binomial distribution and density which describes the number of occurrences of the A event. Also find the probability of the occurrence of two A events.

Solution From (5.106) we find the number of all possible occurrences of two events A for $n = 4$, $k = 2$ and $n - k = 2$, in n repeated Bernoulli trials:

$$C_4^2 = \frac{4!}{2!2!} = 6. \tag{5.119}$$

All six possible combinations, along with their corresponding probabilities, are shown in Table 5.2.

Note that the only difference is in the order of the occurrence of A events.

The probability that the variable X is equal to 2 is equal to the sum of the probabilities in the second column,

$$P\{X = 2\} = C_4^2 p^2 q^2 = 6\ 0.5^2\ 0.5^2 = 3/8. \tag{5.120}$$

In a similar way, we can find the corresponding probabilities for all other possible values of k, shown in Table 5.3.

Using the results from Tables 5.2 and 5.3, we have:

$$\begin{aligned} P_X(0;4) &= 1/16 \\ P_X(1;4) &= 4/16 = 1/4 \\ P_X(2;4) &= 3/8 \\ P_X(3;4) &= 4/16 = 1/4 \\ P_X(4;4) &= 1/16 \end{aligned} \qquad (5.121)$$

From (5.110), (5.111) and (5.121), the corresponding distribution and density functions are:

$$F_X(x) = \frac{1}{16}u(x) + \frac{1}{4}u(x-1) + \frac{3}{8}u(x-2) + \frac{1}{4}u(x-3) + \frac{1}{16}u(x-4). \qquad (5.122)$$

$$f_X(x) = \frac{1}{16}\delta(x) + \frac{1}{4}\delta(x-1) + \frac{3}{8}\delta(x-2) + \frac{1}{4}\delta(x-3) + \frac{1}{16}\delta(x-4). \qquad (5.123)$$

The distribution and density function are shown in Fig. 5.15. Note that the density function is symmetric, and that the mean value is at the point of symmetry,

$$E\{X\} = 2 = np = 4/2 = 2. \qquad (5.124)$$

In this example, the density function is symmetrical because the probabilities $p = q = 1/2$. However, in general, where $p \neq q$ the density function is not symmetrical, as shown in the following example.

Example 5.6.2 A code word has six pulses of "1" and four pulses of "0." As a result of the noise, each "1" pulse can be attenuated, becoming "0," with a probability of 0.04. Similarly, any "0" can be transformed into a false "1" with a probability of 0.1. It is assumed that the noise affects the "0" and "1," independently.

Table 5.3 Events and probabilities in Example 5.6.2

k	C_4^k	Events	Probabilities
0	1	$\bar{A}\,\bar{A}\,\bar{A}\,\bar{A}$	1/16
1	4	$A\,\bar{A}\,\bar{A}\,\bar{A}$	1/16
		$\bar{A}\,A\,\bar{A}\,\bar{A}$	1/16
		$\bar{A}\,\bar{A}\,A\,\bar{A}$	1/16
		$\bar{A}\,\bar{A}\,\bar{A}\,A$	1/16
3	4	$A\,A\,A\,\bar{A}$	1/16
		$A\,A\,\bar{A}\,A$	1/16
		$A\,\bar{A}\,A\,A$	1/16
		$\bar{A}\,A\,A\,A$	1/16
4	1	$A\,A\,A\,A$	1/16

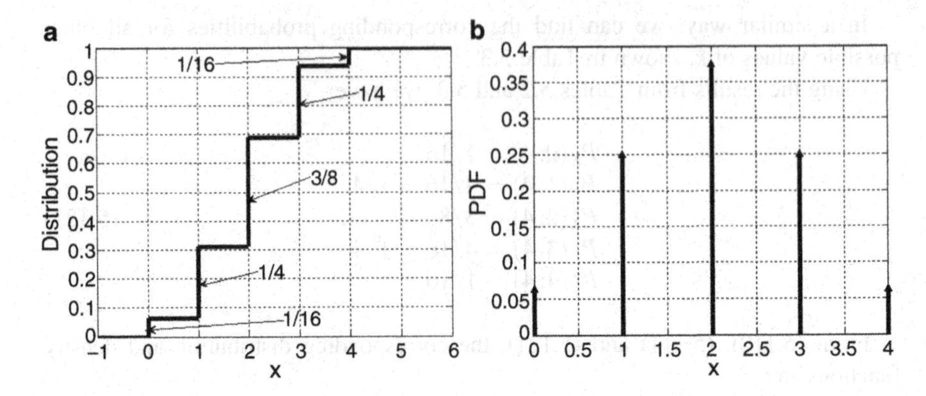

Fig. 5.15 Binomial distribution and density functions. (**a**) Distribution function. (**b**) Density function

(a) It is necessary to find the probability that exactly two "1" pulses become "0."
(b) Find the probability that no more than two false pulses appear ("0"s become "1"). Plot the density function of the corresponding binomial random variables which presents the number of false pulses.

Solution

(a) The number of lost "1" pulses in a code word is a binomial random variable in which $n = 6$ and $p = 0.04$

$$q = 1 - p = 1 - 0.04 = 0.96. \tag{5.125}$$

From (5.109), we have:

$$P\{X = 2\} = P_X(2; 6) = C_6^2 p^2 q^4 = \frac{6!}{2!4!} 0.04^2 0.96^4 = 0.0204. \tag{5.126}$$

The random variable X takes the discrete values $k = 0, \ldots, 6$, with the corresponding probabilities $p^k(1 - p)^{n-k}$.

(b) The number of possible false pulses in the code word is also a binomial random variable X in which $p = 0.1$, $n = 4$, and $k = 0, 1, \ldots, 4$. It follows that $q = 1 - 0.1 = 0.9$.
The probability that no more than two false pulses appear is given as:

$$\begin{aligned} P\{k \leq 2; 4\} &= \sum_{k=0}^{2} C_4^k 0.1^k 0.9^{n-k} = 0.6561 \\ &+ 0.2916 + 0.0486 = 0.9963. \end{aligned} \tag{5.127}$$

The corresponding probabilities are:

$$\begin{aligned} P_X(0;4) &= 0.6567 \\ P_X(1;4) &= 0.2916 \\ P_X(2;4) &= 0.0486\,. \\ P_X(3;4) &= 0.0036 \\ P_X(4;4) &= 0.001 \end{aligned} \tag{5.128}$$

The density function is given in Fig. 5.16.

5.6.5 *Approximation of Binomial Variable*

For high values of n, the calculation of the probabilities $P_X(k;n)$ is time consuming. Instead, it is helpful to use the asymptotic approximations for the binomial random variable.

We will consider two cases:

Case 1 The probability p is finite and n is very high.

Case 2 The probability p is very small, while n is very high, resulting in a finite value of np.

In this case, the approximation is the Poisson random variable, which will be considered in Sect. 5.7.

Let us now consider *Case 1*.

It follows that

$$np \gg 1 \tag{5.129}$$

and

$$npq \gg 1. \tag{5.130}$$

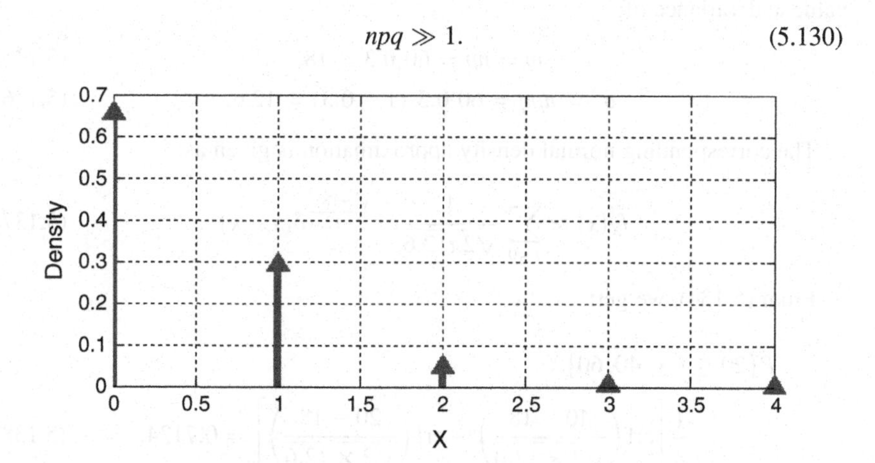

Fig. 5.16 Binomial density function

As a consequence most of the values of the random variable are concentrated around the mean value np, which is a well-known property of normal random variables. Therefore, in this case, the probabilities (5.109) can be approximated with a normal variable with the parameters:

$$\begin{aligned} m &= np \\ \sigma^2 &= npq. \end{aligned} \tag{5.131}$$

It follows:

$$f_X(x) = \sum_{k=0}^{n} C_n^k p^k q^{n-k} \delta(x-k) \approx \sum_{k=0}^{n} \frac{1}{\sqrt{2\pi npq}} \mathrm{e}^{-\frac{(k-np)^2}{2npq}} \delta(x-k). \tag{5.132}$$

Note that the normal density in (5.132) approximates the probability mass function (5.109)

In this way, we can use the functions shown in Sect. 4.2.2 to find the approximate probabilities for a binomial random variable:

$$P\{k \le k_1, n\} = \sum_{k=0}^{k_1} C_n^k p^k q^{n-k} \approx \frac{1}{2}\left[1 + \operatorname{erf}\frac{k_1 - m}{\sigma\sqrt{2}}\right]. \tag{5.133}$$

Similarly,

$$P\{k_2 \le k \le k_3, n\} = \sum_{k=k_2}^{k_3} C_n^k p^k q^{n-k} \approx \frac{1}{2}\left[\operatorname{erf}\frac{k_3 - m}{\sigma\sqrt{2}} - \operatorname{erf}\frac{k_2 - m}{\sigma\sqrt{2}}\right]. \tag{5.134}$$

Example 5.6.3 In $n = 60$ independent trials, a binary event A occurs with a probability of $p = 0.3$. Find the probability that event A occurs greater than or equal to 20 and less than or equal to 40 times.

Solution The number of occurrences of event A is a binomial variable with a mean value and variance of:

$$m = np = 60\ 0.3 = 18. \tag{5.135}$$

$$\sigma^2 = npq = 60\ 0.3\ (1 - 0.3) = 12.6. \tag{5.136}$$

The corresponding normal density approximation is given as:

$$f_X(x) \approx \sum_{k=0}^{n} \frac{1}{\sqrt{2\pi 12.6}} \mathrm{e}^{-\frac{(k-18)^2}{2\times 12.6}} \delta(x-k). \tag{5.137}$$

From (5.134), we get:

$$\begin{aligned} &P\{20 \le k \le 40, 60\} \\ &\approx \frac{1}{2}\left[\operatorname{erf}\left(\frac{40-18}{\sqrt{2\times 12.6}}\right) - \operatorname{erf}\left(\frac{20-18}{\sqrt{2\times 12.6}}\right)\right] = 0.7134. \end{aligned} \tag{5.138}$$

5.7 Poisson Random Variable

5.7.1 Approximation of Binomial Variable

Consider the asymptotic behavior of a binomial random variable if n is very high in value and the probability of p is very small such that the mean value of a binomial variable,

$$\overline{X} = \overline{k} = np \tag{5.139}$$

has a moderate value.

From (5.139), we write:

$$p = \frac{\overline{k}}{n}. \tag{5.140}$$

The obtained value (5.140) is placed into the expression $P_X(k;n)$ in (5.109), and n is assumed to go to ∞, resulting in:

$$\begin{aligned}
\lim_{n\to\infty} P_X(k,n) &= \lim_{n\to\infty} \binom{n}{k} \left(\frac{\overline{k}}{n}\right)^k \left(1-\frac{\overline{k}}{n}\right)^{n-k} = \lim_{n\to\infty} \frac{n!}{k!(n-k)!} \frac{\overline{k}^k}{n^k} \left(1-\frac{\overline{k}}{n}\right)^{n-k} \\
&= \frac{\overline{k}^k}{k!} \lim_{n\to\infty} \frac{n(n-1)\dots(n-k+1)}{n^k} \left(1-\frac{\overline{k}}{n}\right)^{-k} \left(1-\frac{\overline{k}}{n}\right)^n \\
&= \frac{\overline{k}^k}{k!} \lim_{n\to\infty} \frac{n}{n}\frac{n-1}{n}\dots\frac{n-k+1}{n} \left(1-\frac{\overline{k}}{n}\right)^{-k} \left(1-\frac{\overline{k}}{n}\right)^n \\
&= \frac{\overline{k}^k}{k!} \lim_{n\to\infty} \left(1-\frac{\overline{k}}{n}\right)^n
\end{aligned} \tag{5.141}$$

Knowing that

$$\lim_{n\to\infty} \left(1-\frac{\overline{k}}{n}\right)^n = e^{-\overline{k}}, \tag{5.142}$$

it follows that

$$\lim_{n\to\infty} P_X(k,n) = P\{X=k\} = \frac{\overline{k}^k}{k!} e^{-\overline{k}}. \tag{5.143}$$

The obtained expression for probability is called the *Poisson formula* and it is a probability mass function (see Sect. 2.2.1) of a *Poisson random variable*, where $\overline{k}$ is the parameter.

Therefore, the number of successes in Bernoulli trials is described using a Poisson random variable, if the number n of trials is large, and the probability p of success is small enough to make np of (5.139) of a moderate size. Because of the event's low probability of success the Poisson variable is also called *rare events variable*.

It is interesting to mention that the Poisson random variable was introduced by Simeon Poisson in 1837 in his book on the application of probability to verdicts in criminal and civil matters, [ROSS10, p. 143].

Practically, this approximation is used for

$$np<5. \tag{5.144}$$

Example 5.7.1 The probability that a call from any of 300 users comes to the telephone operator center is equal to 0.01. Find the probability that 10 calls come.

Solution The number $n = 300$ of users and $p = 0.01$, result in:

$$\overline{k} = np = 3. \tag{5.145}$$

From (5.143) and (5.145), we have:

$$P\{X = 4\} = \frac{\overline{k}^k}{4!}\mathrm{e}^{-\overline{k}} = \frac{3^4}{4!}\mathrm{e}^{-3} = 0.168. \tag{5.146}$$

5.7.2 Poisson Variable as a Counting Random Variable

A Poisson variable is not only used as a limiting case for a binomial variable. It is most commonly applied to a wide variety of counting-type applications in a queuing theory and communication networks.

If a time being observed has duration of τ, and the events being counted are known to occur at an average rate of λ,

$$\lambda = \frac{\overline{k}}{\tau}, \tag{5.147}$$

then the probability of the occurrence of k events in the time interval τ is a Poisson random variable with the parameter λ.

From (5.147),

$$\overline{k} = \lambda\tau. \tag{5.148}$$

Placing (5.148) into (5.143), we get the probability that k events occur in the interval τ,

$$P\{X = k, \tau\} = \frac{\bar{k}^k}{k!}\mathrm{e}^{-\bar{k}} = \frac{(\lambda\tau)^k}{k!}\mathrm{e}^{-\lambda\tau}. \tag{5.149}$$

The probability (5.149) is another form of the Poisson formula (5.143) and thus represents the probability mass function of a Poisson random variable.

Example 5.7.2 Telephone calls can come in to a telephone switching center with the frequency of 0.5 calls/min. Find the probability that exactly two calls come in, and the probability that no more than three calls come in.

Solution The number of calls that come in $\tau = 1$ min, is a Poisson random variable with the parameter $\lambda = 0.5$ calls/min.

From (5.149), we have:

$$P\{X = 2, \tau = 1 \text{ min}\} = \frac{0.5^2}{2!}\mathrm{e}^{-0.5} = 0.0758. \tag{5.150}$$

$$\begin{aligned} P\{X \le 3, \tau = 1 \text{ min}\} &= \sum_{k=0}^{3} P_X(k, \tau) \\ &= \frac{0.5^0}{0!}\mathrm{e}^{-0.5} + \frac{0.5^1}{1!}\mathrm{e}^{-0.5} + \frac{0.5^2}{2!}\mathrm{e}^{-0.5} \\ &\quad + \frac{0.5^3}{3!}\mathrm{e}^{-0.5} = 0.9982. \end{aligned} \tag{5.151}$$

5.7.3 Distribution and Density Functions

Using either (5.143) or (5.149), the corresponding distribution and density functions are:

$$\begin{aligned} F_X(x) &= \sum_{k=0}^{\infty} P\{X = k\}u(x - k) = \sum_{k=0}^{\infty} \frac{\bar{k}^k}{k!}\mathrm{e}^{-\bar{k}}\, u(x - k) \\ &= \sum_{k=0}^{\infty} \frac{(\lambda\tau)^k}{k!}\mathrm{e}^{-\lambda\tau}\, u(x - k), \end{aligned} \tag{5.152}$$

$$\begin{aligned} f_X(x) &= \sum_{k=0}^{\infty} P\{X = k\}\delta(x - k) = \sum_{k=0}^{\infty} \frac{\bar{k}^k}{k!}\mathrm{e}^{-\bar{k}}\, \delta(x - k) \\ &= \sum_{k=0}^{\infty} \frac{(\lambda\tau)^k}{k!}\mathrm{e}^{-\lambda\tau}\, \delta(x - k). \end{aligned} \tag{5.153}$$

We can easily verify that the condition (2.83) is satisfied.
From (5.147), we write

$$\sum_{k=0}^{\infty} P_X(k) = \sum_{k=0}^{\infty} \frac{(\lambda\tau)^k}{k!} \mathrm{e}^{-\lambda\tau} = \mathrm{e}^{-\lambda\tau} \sum_{k=0}^{\infty} \frac{(\lambda\tau)^k}{k!}. \tag{5.154}$$

Knowing that

$$\sum_{k=0}^{\infty} \frac{x^k}{k!} = \mathrm{e}^x, \tag{5.155}$$

from (5.154), we finally get:

$$\sum_{k=0}^{\infty} P_X(k) = \mathrm{e}^{-\lambda\tau} \mathrm{e}^{\lambda\tau} = 1. \tag{5.156}$$

5.7.4 Characteristic Function

We calculate a characteristic function in order to find the variance of the Poisson random variable.

$$\begin{aligned} \phi_X(\omega) = E\{\mathrm{e}^{j\omega X}\} &= \sum_{k=0}^{\infty} \mathrm{e}^{j\omega k} P\{X = k\} = \mathrm{e}^{-\lambda\tau} \sum_{k=0}^{\infty} \frac{(\lambda\tau)^k}{k!} \mathrm{e}^{j\omega k} \\ &= \mathrm{e}^{-\lambda\tau} \sum_{k=0}^{\infty} \frac{(\lambda\tau\, \mathrm{e}^{j\omega})^k}{k!}. \end{aligned} \tag{5.157}$$

Applying (5.155), we finally arrive at:

$$\phi_X(\omega) = \mathrm{e}^{-\lambda\tau} \mathrm{e}^{\mathrm{e}^{j\omega}\lambda\tau} = \mathrm{e}^{\lambda\tau(\mathrm{e}^{j\omega} - 1)} = \mathrm{e}^{\overline{k}(\mathrm{e}^{j\omega} - 1)}. \tag{5.158}$$

Using the moment theorem, we have:

$$E\{X\} = \frac{1}{j} \frac{\mathrm{d}\phi_X(\omega)}{\mathrm{d}\omega}\bigg|_{\omega=0} = \lambda\tau = \overline{k}. \tag{5.159}$$

This result is the same as (5.148).
The second moment,

$$E\{X^2\} = \frac{1}{j^2} \frac{\mathrm{d}^2\phi_X(\omega)}{\mathrm{d}\omega^2}\bigg|_{\omega=0} = (\lambda\tau)^2 + \lambda\tau = \overline{k}^2 + \overline{k}. \tag{5.160}$$

From (5.159) and (5.160), the variance is

$$\sigma^2 = \overline{X^2} - \overline{X}^2 = (\lambda\tau)^2 + \lambda\tau - (\lambda\tau)^2 = \lambda\tau = \overline{k}. \tag{5.161}$$

This result shows an interesting property of a Poisson variable:

The variance of a Poisson random variable is equal to its mean value.

Next section shows another interesting property of Poisson random variable.

5.7.5 Sum of Independent Poisson Variables

Consider two independent Poisson random variables X_1 and X_2 with mean values of $\overline{k_1}$ and $\overline{k_2}$, respectively:

$$X = X_1 + X_2. \tag{5.162}$$

The characteristic function of the sum (5.162) is equal to the product of the corresponding characteristic functions,

$$\phi_X(\omega) = \phi_{X_1}(\omega)\phi_{X_2}(\omega). \tag{5.163}$$

Using the expression for the characteristic function (5.158), we have:

$$\phi_X(\omega) = e^{\overline{k_1}(e^{j\omega} - 1)} e^{\overline{k_2}(e^{j\omega} - 1)} = e^{(\overline{k_1} + \overline{k_2})(e^{j\omega} - 1)} = e^{\overline{k}(e^{j\omega} - 1)}. \tag{5.164}$$

Comparing (5.158) and (5.164), we can conclude that the variable X is also a Poisson random variable and that the following is true:

The sum of independent Poisson variables is a Poisson variable.

The parameter of the X variable is,

$$\overline{k} = \overline{k_1} + \overline{k_2}, \tag{5.165}$$

where $\overline{k_1}$ and $\overline{k_2}$ are parameters (mean values) of the variables X_1 and X_2, respectively.

This result can be generalized to the sum of N independent Poisson random variables,

$$X = \sum_{i=1}^{N} X_i, \tag{5.166}$$

where

$$\overline{X} = \overline{k} = \sum_{i=1}^{N} \overline{X_i} = \sum_{i=1}^{N} \overline{k_i}. \tag{5.167}$$

According to the central limit theorem, the probabilities (5.149) for the sum (5.166) approximate Gaussian PDF curve. Consequently, if the Poisson random variable has a large mean value (5.167), this value can be considered to be a sum of a large number of corresponding mean values, as shown in (5.167). Taking this case, the probability (5.149) of a Poisson random variable can be computed using formulas from Chapter 4.2.2.

5.7.6 Poisson Flow of Events

A sequence of homogeneous events occurring one after another at random moments is called a *flow of events* or *calls*. Graphically, it can be presented as a set of points on time axis, as shown in Fig. 5.17.

The flow is called *ordinary*, if the probability of the occurrence of more than one event in the elementary interval $\Delta\tau$ is negligibly small compared to the probability of the occurrence of only one event. In a *stationary flow*, the probability that a certain number of events occurs in a given time interval depends only on the length of the time interval and is independent on the position of this interval on the time axis. In this case, the average number of events λ occurring per unit of time—called *intensity of flow*—is constant. If the number of events occurring in one interval of time does not depend on the number of events falling in any other nonoverlapping interval, the flow is said to be *without aftereffects*.

A stationary flow of events which is ordinary and without aftereffects is called an *elementary* or *Poisson flow*.

Next we show that the probability that k events occur in an arbitrary time interval τ of a Poisson flow is described using Poisson formula (5.149).

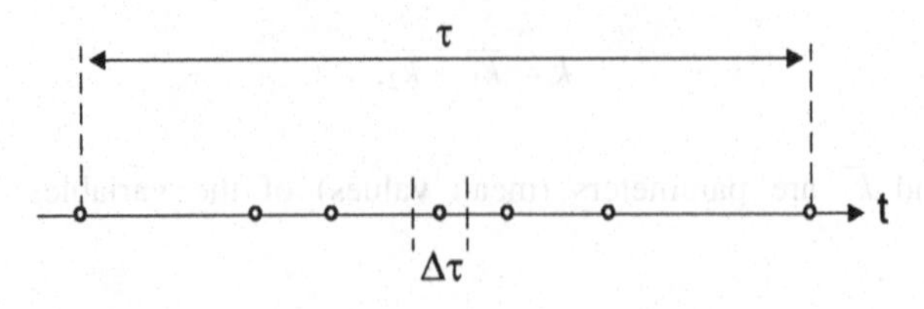

Fig. 5.17 Flow of events

Due to an ordinary property, in the elementary interval $\Delta\tau$ the event can either occur once or not occur. The probability of the occurrence is denoted as p, and of nonoccurrence as q, such that

$$p+q=1. \tag{5.168}$$

Due to stationary and nonaftereffects characteristics, the probability p is the same for any interval $\Delta\tau$ and depends only on the length $\Delta\tau$.

The expected number of events in an interval τ is:

$$p\times 1+q\times 0=p. \tag{5.169}$$

From here, the expected number of events in a unit of time is called the *intensity of flow*:

$$\lambda=\frac{p}{\Delta\tau}. \tag{5.170}$$

We are looking for the probability that k events occur in the time interval τ. To this end, the interval τ is divided into n equal subintervals $\Delta\tau$,

$$\Delta\tau=\frac{\tau}{n}. \tag{5.171}$$

In order to have k events in the interval τ, it is necessary that the events occur at k of total n elementary intervals $\Delta\tau$, where the events are mutually independents. The desired probability is described by the Bernoulli formula (5.109),

$$\begin{aligned} P\{k \text{ events in } \tau\} &\approx P\{\text{occurence of events in } k \text{ of } n \text{ elementary intervals } \Delta\tau\} \\ &= \binom{n}{k}p^k q^{n-k}, \end{aligned} \tag{5.172}$$

Instead of approximating we can obtain the exact expression (5.172) if

$$\Delta\tau\to 0 \quad \text{or} \quad n\to\infty. \tag{5.173}$$

Therefore, from (5.172), we have:

$$\begin{aligned} P\{k \text{ events in } \tau\} &= \lim_{n\to\infty}\binom{n}{k}p^k q^{n-k}=\lim_{n\to\infty}\binom{n}{k}\left(\frac{\lambda\tau}{n}\right)^k\left(1-\frac{\lambda\tau}{n}\right)^{n-k} \\ &= \frac{(\lambda\tau)^k}{k!}\lim_{n\to\infty}\frac{n(n-1),\ldots,(n-k+1)}{n^k}\left(1-\frac{\lambda\tau}{n}\right)^n\left(1-\frac{\lambda\tau}{n}\right)^{-k} \\ &= \frac{(\lambda\tau)^k}{k!}\lim_{n\to\infty}\left(1-\frac{\lambda\tau}{n}\right)^n. \end{aligned} \tag{5.174}$$

It is well known that

$$\lim_{n\to\infty}\left(1-\frac{\lambda\tau}{n}\right)^n = \mathrm{e}^{-\lambda\tau}. \tag{5.175}$$

Placing (5.175) into (5.174), we get:

$$P\{k \text{ events in } \tau\} = \frac{(\lambda\tau)^k}{k!}\mathrm{e}^{-\lambda\tau}. \tag{5.176}$$

This result confirms that *the probability that k events occur in an arbitrary time interval τ in a Poisson flow is described by a Poisson formula.*

Consider the random variable X which presents the random time between two subsequent Poisson events.

The probability

$$P\{X>x\} \tag{5.177}$$

presents the probability that in the interval [0, x] no events have occurred. From (5.176), this probability is equal to

$$P\{X>x\} = P\{k=0,x\} = \mathrm{e}^{-\lambda x}. \tag{5.178}$$

The distribution function $F_X(x)$ can be obtained from (5.178), as

$$F_X(x) = P\{X \le x\} = 1 - P\{X>x\} = 1 - \mathrm{e}^{-\lambda x}. \tag{5.179}$$

Comparing the result from (5.179) with (5.62), we can see that the distribution (5.179) is that of an exponential random variable. Consequently, the variable X is an exponential random variable.

This also means that the variable X has the following PDF:

$$f_X(x) = \frac{\mathrm{d}F_X(x)}{\mathrm{d}x} = \lambda\,\mathrm{e}^{-\lambda x}, \tag{5.180}$$

where λ is the intensity of the events defined in (5.170).

Therefore, *the time between the events in the Poisson flow of events is the exponential random variable.*

We can also show that if the time intervals between independent events are described using the exponential random variable, then a flow is said to be a Poisson flow.

We can conclude that there is a certain relationship between a constant intensity λ, a Poisson flow of events, and an exponential distribution of times between adjacent events. This relation is presented in Fig. 5.18.

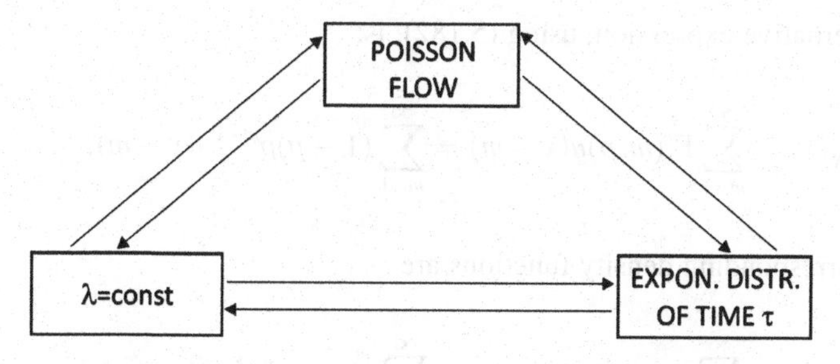

Fig. 5.18 Relation between Poisson flow, exponential time, and constant intensity λ

5.8 Geometric Random Variable

5.8.1 *What Is a Geometric Random Variable and Where Does This Name Come From?*

Consider a Bernoulli experiment where the possible outcomes are the events A (success) with a probability of p and the complementary event $\overline{A}$ (failure) with a probability $q = 1 - p$.

We are looking for the probability that event A occurs in mth experiment; that is, after $(m - 1)$ failures $(\overline{A})$.

Denoting this probability $\Gamma(m, p)$, we have:

$$\Gamma(m,p) = (1-p)^{m-1}p. \tag{5.181}$$

Similarly, one can find the probability of the first failure after $m - 1$ successes using a Bernoulli trials:

$$\Gamma'(m,p) = (1-p)p^{m-1}. \tag{5.182}$$

Note that the probabilities (5.181) and (5.182) are members of geometrical series and that is why the name *geometric* is used.

5.8.2 *Probability Distribution and Density Functions*

Let X be a discrete random variable taking all possible values of m (from 1 to ∞.)

Therefore, the distribution function is

$$F_X(x) = \sum_{m=1}^{\infty}\Gamma(m,p)u(x-m) = \sum_{m=1}^{\infty}(1-p)^{m-1}pu(x-m), \tag{5.183}$$

where $\Gamma(m,p)$is the probability mass function given in (5.181).

An alternative expression, using (5.182), is:

$$F'_X(x) = \sum_{m=1}^{\infty} \Gamma'(m,p)u(x-m) = \sum_{m=1}^{\infty} (1-p)p^{m-1}u(x-m). \tag{5.184}$$

The corresponding density functions are:

$$f_X(x) = \sum_{m=1}^{\infty} \Gamma(m,p)\delta(x-m) = \sum_{m=1}^{\infty} (1-p)^{m-1}p\delta(x-m) \tag{5.185}$$

and

$$f'_X(x) = \sum_{m=1}^{\infty} \Gamma'(m,p)\delta(x-m) = \sum_{m=1}^{\infty} (1-p)p^{m-1}\delta(x-m). \tag{5.186}$$

Next we will determine whether or not the sum of the probabilities (5.181) is equal to 1, as should be.

We have:

$$\sum_{m=1}^{\infty} \Gamma(m,p) = \sum_{m=1}^{\infty} p(1-p)^{m-1} = \frac{p(1-p)^{1-1}}{1-(1-p)} = \frac{p}{1-1+p} = 1. \tag{5.187}$$

Similarly,

$$\sum_{m=1}^{\infty} \Gamma'(m,p) = \sum_{m=1}^{\infty} p^{m-1}(1-p) = \frac{p^{1-1}(1-p)}{1-p} = 1. \tag{5.188}$$

5.8.3 *Characteristic Functions and Moments*

A characteristic function of a Geometric random variable is:

$$\begin{aligned}\phi_X(\omega) &= \sum_{m=1}^{\infty} e^{j\omega m} p(1-p)^{m-1} = p(1-p)^{-1} \sum_{m=1}^{\infty} e^{j\omega m}(1-p)^m \\ &= \frac{pq^{-1}e^{j\omega}q}{1-e^{j\omega}q} = \frac{pe^{j\omega}}{1-e^{j\omega}q}.\end{aligned} \tag{5.189}$$

Applying the moment's theorem we can obtain the first and second moments:

$$E\{X\} = \frac{1}{j} \frac{d\phi_x(\omega)}{d\omega}\bigg|_{\omega=0} = \frac{1}{p}, \tag{5.190}$$

$$E\{X^2\} = \frac{1}{j^2} \frac{d^2\phi_x(\omega)}{d\omega^2}\bigg|_{\omega=0} = \frac{1+q}{p^2}. \tag{5.191}$$

From (5.190) and (5.191), we get a variance:

$$\sigma^2 = E\{X^2\} - E\{X\}^2 = \frac{1+q}{p^2} - \frac{1}{p^2} = \frac{q}{p^2}. \tag{5.192}$$

Example 5.8.1 The message sent over the binary channel has symbols denoted as "0" and "1," where the probabilities of the occurrence of "0" or "1" are independent and equal. For example, a sequence of symbols is given as:

$$\ldots 000110111111000111111000001110000000011000\ldots.$$

It is necessary to find the distribution and density functions for the random variable which presents the number of identical symbols in a randomly chosen group of identical symbols in the message. Additionally, find the probability that will have at least $k = 5$ identical symbols in a group of identical symbols.

Solution The number of successive identical symbols in a group is a discrete random variable with discrete values of 1, 2, ..., m, ..., and corresponding probabilities of

$$\begin{aligned} P\{X=1\} &= 0.5 \\ P\{X=2\} &= 0.5^2 \\ \ldots & \quad \ldots \ldots \\ P\{X=m\} &= 0.5^m \\ \ldots & \quad \ldots \ldots \end{aligned} \tag{5.193}$$

It follows that the random variable X is a geometrical random variable with the distribution and density functions,

$$F_X(x) = \sum_{m=1}^{\infty} 0.5^m u(x-m), \tag{5.194}$$

$$f_X(x) = \sum_{m=1}^{\infty} 0.5^m \delta(x-m). \tag{5.195}$$

The distribution and density functions are shown in Fig. 5.19. From (5.190) the average number of identical symbols in a randomly chosen group is:

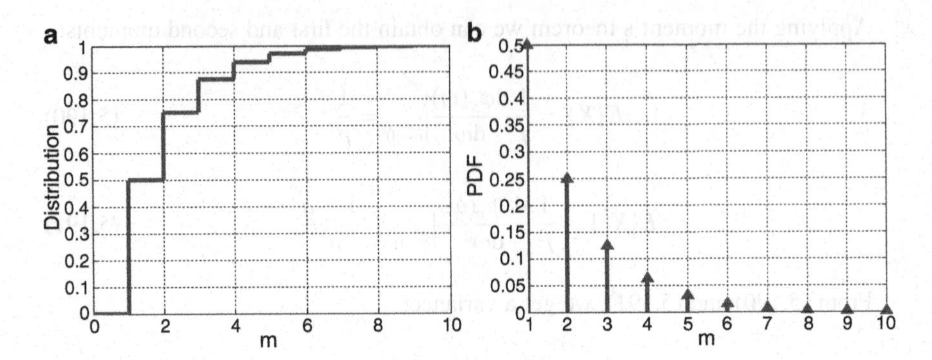

Fig. 5.19 Geometric random variable. (**a**) Distribution. (**b**) PDF

$$E\{X\} = \frac{1}{p} = 2. \tag{5.196}$$

The probability that the number of identical symbols is at least $k = 5$ is:

$$P\{X \geq 5\} = \sum_{m=5}^{\infty} 0.5^m = \frac{0.5^5}{1 - 0.5} = 0.5^4 = 0.0625. \tag{5.197}$$

5.8.4 The Memory-Less Property of Geometric Random Variable

A geometric random variable is a memory-less variable. That is, the future values of a random variable do not depend on the present value. In other words, like an exponential random variable, a *geometric random variable does not have aftereffects and it is said to be a memory-less variable.*

Therefore, a geometric random variable is a discrete counterpart of an exponential variable.

Imagine that the variable X represents the occurrence of an event A in $(n + m)$th experiment, if it is known that event A has not occurred in the first n experiments:

$$\begin{aligned} P\{X = m + n | X > n\} &= \frac{P\{X = m + n, X > n\}}{P\{X > n\}} = \frac{P\{X = m + n\}}{P\{X > n\}} \\ &= \frac{p(1-p)^{n+m-1}}{\sum_{m=n+1}^{\infty} p(1-p)^{m-1}} = \frac{p(1-p)^{n+m-1}}{p(1-p)^n/[1-(1-p)]} = p(1-p)^{m-1}. \end{aligned} \tag{5.198}$$

The obtained probability depends only on m and does not depend on n. This confirms the memory-less property.

5.9 Numerical Exercises

Exercise 5.1 Show that a Rayleigh variable has only positive values.

Answer The area below the PDF must be unity:

$$\frac{1}{\sigma^2}\int_{a}^{\infty} x\,\mathrm{e}^{-\frac{x^2}{2\sigma^2}}\,\mathrm{d}x = 1, \tag{5.199}$$

where a is the low integral limit which must be defined in order for the condition (5.199) to be satisfied.

We introduce the auxiliary variable y:

$$\frac{x^2}{2\sigma^2} = y. \tag{5.200}$$

From here,

$$x\,\mathrm{d}x/\sigma^2 = \mathrm{d}y, \tag{5.201}$$

Using (5.200) and (5.201), the integral (5.199) becomes:

$$\int_{a^2/2\sigma^2}^{\infty} \mathrm{e}^{-y}\,\mathrm{d}y = \mathrm{e}^{-y}\Big|_{\infty}^{a^2/2\sigma^2} = \mathrm{e}^{-a^2/2\sigma^2} = 1. \tag{5.202}$$

From here, it follows that $a = 0$. That is, in order to satisfy (5.199), a Rayleigh variable must have only positive values.

Exercise 5.2 Find the mode and median for a Rayleigh variable.

Answer Mode is a x value for which the PDF has its maximum value:

$$\frac{\mathrm{d}f_X(x)}{\mathrm{d}x} = \frac{\mathrm{d}}{\mathrm{d}x}\left(\frac{1}{\sigma^2}x\,\mathrm{e}^{-\frac{x^2}{2\sigma^2}}\right) = \frac{1}{\sigma^2}\mathrm{e}^{-\frac{x^2}{2\sigma^2}}\left(1 - \frac{x^2}{\sigma^2}\right) = 0. \tag{5.203}$$

From here $x = \sigma$, as shown in Fig. 5.3.

Median is a value of the random variable, for which $F_X(x) = 0.5$.

$$0.5 = F_X(x) = 1 - \mathrm{e}^{-\frac{x^2}{2\sigma^2}}. \tag{5.204}$$

From here, it follows:

$$0.5 = \mathrm{e}^{-\frac{x^2}{2\sigma^2}}. \tag{5.205}$$

From here,

$$\frac{x}{\sigma} = \sqrt{-2\ln(0.5)}. \tag{5.206}$$

From (5.206), the median is found to be:

$$x = 1.774\sigma. \tag{5.207}$$

Exercise 5.3 The random variable X is a Rayleigh variable with a parameter of σ. Find the variance of the random variable

$$Y = 2 + X^2. \tag{5.208}$$

Answer The variance of the random variable Y is

$$\sigma_Y^2 = \overline{Y^2} - \overline{Y}^2. \tag{5.209}$$

From (5.208), the mean and the mean squared values are:

$$\overline{Y} = 2 + \overline{X^2}, \tag{5.210}$$

$$\overline{Y^2} = \overline{(2 + X^2)^2} = 4 + \overline{X^4} + 4\overline{X^2}. \tag{5.211}$$

From (5.35), the mean squared value of Rayleigh variable X is

$$\overline{X^2} = 2\sigma^2. \tag{5.212}$$

The fourth moment $\overline{X^4}$ is obtained using the integral 4 from Appendix A,

$$\overline{X^4} = \frac{1}{\sigma^2}\int_0^{\infty} x^5 \mathrm{e}^{-\frac{x^2}{2\sigma^2}} \mathrm{d}x = \frac{1}{\sigma^2} 8\sigma^6 = 8\sigma^4. \tag{5.213}$$

Placing (5.212) into (5.210), we obtain:

$$\overline{Y} = 2 + 2\sigma^2. \tag{5.214}$$

Similarly,

$$\overline{Y^2} = 4 + \overline{X^4} + 4\overline{X^2} = 4 + 8\sigma^4 + 8\sigma^2. \tag{5.215}$$

Finally, from (5.209), (5.214), and (5.215), we get:

$$\sigma_Y^2 = 4 + 8\sigma^2 + 8\sigma^4 - (2 + 2\sigma^2)^2 = 4\sigma^4. \tag{5.216}$$

Exercise 5.4 The random voltage is described using a Rayleigh density function,

$$f_X(x) = \begin{cases} \frac{2}{5} x \, e^{-\frac{x^2}{5}} & \text{for} \quad x \geq 0, \\ 0 & \text{otherwise.} \end{cases} \tag{5.217}$$

Find the mean power of X, in a resistance of 1 Ω,

$$\overline{P} = \overline{X^2}. \tag{5.218}$$

Answer The mean squared value of X is

$$\overline{P} = \overline{X^2} = \frac{2}{5} \int_0^\infty x^2 x \, e^{-\frac{x^2}{5}} \, dx = \frac{2}{5} \int_0^\infty x^3 \, e^{-\frac{x^2}{5}} \, dx. \tag{5.219}$$

Using integral 4 from Appendix A, we arrive at:

$$\overline{P} = \frac{2}{5} \frac{5^2}{2} = 5 \, \text{W}. \tag{5.220}$$

Exercise 5.5 The power reflected from an aircraft to radar is described using the exponential random variable:

$$f_X(x) = \lambda \, e^{-\lambda x}, \tag{5.221}$$

where λ is a parameter.

Find the probability that the power received by radar is less than the mean power P_0.

Answer From (5.65), the mean power is

$$P_0 = \frac{1}{\lambda}. \tag{5.222}$$

The desired probability is:

$$P\{X < 1/\lambda\} = F_X(1/\lambda) = 1 - e^{-1} = 0.6321. \tag{5.223}$$

Exercise 5.6

(a) Find the probability that a random time between two consecutive Poisson events with a constant intensity of λ will be less than its mean value.
(b) Find the mean squared deviation from its mean value for the time interval τ in which k Poisson events occur.

Answer

(a) The time between two consecutive Poisson events is an exponential random variable X. From (5.65),

$$E\{X\} = 1/\lambda, \tag{5.224}$$

$$P\{X<1/\lambda\} = F_X(1/\lambda) = 1 - e^{-1} = 0.6321. \tag{5.225}$$

(b) A mean squared deviation from a mean value is a variance of random variable. The time interval in which k Poisson events occur is an Erlang's variable. From (5.90), the variance of a Gamma variable is

$$\sigma^2 = \frac{b}{\lambda^2}. \tag{5.226}$$

Replacing b with k in (5.226), we get the variance of Erlang's variable:

$$\sigma^2 = \frac{k}{\lambda^2}. \tag{5.227}$$

Exercise 5.7 A random variable X has the PDF

$$f_X(x) = e^{-x} u(x). \tag{5.228}$$

Find the PDF and mean value of the random variable Y if

$$Y = \begin{cases} aX & \text{for} \quad X \geq 0 \\ 0 & \quad\quad \text{otherwise;} \ a > 0. \end{cases} \tag{5.229}$$

Answer The PDF of the random variable Y is obtained from (2.138), as:

$$f_Y(y) = \frac{f_X(x)}{a} = \frac{f_X(y/a)}{a} = \frac{1}{a} e^{-\frac{y}{a}}, \quad \text{for} \quad y \geq 0. \tag{5.230}$$

$$f_Y(y) = 0 \quad \text{for} \quad y<0, \tag{5.231}$$

The variable X is an exponential variable for which $\lambda = 1$. As a consequence,

$$E\{X\} = 1/\lambda = 1. \tag{5.232}$$

The mean value of the variable Y is:

$$E\{Y\} = E\{aX\} = aE\{X\} = a. \tag{5.233}$$

Exercise 5.8 Find the PDF of the random variable Y if

$$Y = \begin{cases} a & \text{for } X \leq a \\ -a & \text{for } X > a \end{cases} \tag{5.234}$$

The random variable X has a PDF of

$$f_X(x) = \mathrm{e}^{-x}, \ x \geq 0 \tag{5.235}$$

Answer

$$P\{Y = a\} = P\{X \leq a\} = 1 - \mathrm{e}^{-a}, \tag{5.236}$$

$$P\{Y = -a\} = P\{X>a\} = 1 - P\{X \leq a\} = 1 - (1 - \mathrm{e}^{-a}) = \mathrm{e}^{-a}. \tag{5.237}$$

The random variable Y is discrete and has two discrete values as given in (5.236) and (5.237).

$$f_Y(y) = (1 - \mathrm{e}^{-a})\delta(y - a) + \mathrm{e}^{-a}\,\delta(y + a). \tag{5.238}$$

Exercise 5.9 A random variable Y, given an exponential random variable X with a parameter λ, has the following conditional PDF:

$$f_Y(y|x) = \begin{cases} x\,\mathrm{e}^{-xy} & \text{for } \ y \geq 0, \\ 0 & \text{otherwise.} \end{cases} \tag{5.239}$$

Find the joint density of the variables X and Y.

Answer A conditional PDF can be expressed in terms of the joint PDF and PDF of the variable X:

$$f_Y(y|x) = \frac{f_{YX}(y,x)}{f_X(x)}. \tag{5.240a}$$

From here,

$$f_{YX}(y,x) = f_Y(y|x)f_X(x) = \lambda\,\mathrm{e}^{-\lambda x}x\,\mathrm{e}^{-xy} = \lambda x\,\mathrm{e}^{-x(\lambda+y)}, \quad \text{for } \ x \geq 0, \ \ y \geq 0. \tag{5.240b}$$

$$f_{XY}(x,y) = 0, \quad \text{for} \quad x<0, \quad y<0. \tag{5.240c}$$

Exercise 5.10 For a given Poisson flow of events with an intensity of λ, form a new flow of events by placing a new event in the middle of the two successive Poisson events. For the random variable Y, which describes the time between two successive events in a new flow of events, find the PDF.

Answer The exponential random variable X describes the time between two successive Poisson events. The random variable Y is then:

$$Y = X/2, \tag{5.241}$$

$$f_Y(y) = 2f_X(2y) = 2\lambda\,e^{-2\lambda y}, \text{ for } y>0, \tag{5.242}$$

$$f_Y(y) = 0 \text{ for } y<0. \tag{5.243}$$

Exercise 5.11 Find the probability of error in a detecting binary signals "0" and "1" in an impulse noise X which is described using a Laplace PDF:

$$f_X(x) = 0.5\,e^{-|x|}, \quad -\infty<x<\infty. \tag{5.244}$$

The probability that signals "0" and "1" are sent is equal to ½.

Answer The received signal is:

$$v = s + x, \tag{5.245}$$

where the signal s is

$$s = \begin{cases} V, & \text{when ``1'' is sent,} \\ 0, & \text{when ``0'' is sent.} \end{cases} \tag{5.246}$$

Therefore, the received signal is:

$$v = \begin{cases} V + x, & \text{when ``1'' is sent,} \\ x, & \text{when ``0'' is sent.} \end{cases} \tag{5.247}$$

If "1" was sent, the corresponding PDF is:

$$f_{v|1}(v) = 0.5\,e^{-|v-V|}. \tag{5.248}$$

Similarly, if "0" was sent, the received signal is only impulse noise and the PDF is:

$$f_{v|0}(v) = 0.5\,e^{-|v|}. \tag{5.249}$$

The intersection of the PDFs corresponds to the value of $V/2$ for the received signal.

The following rule is applied:

When a received signal is higher than $V/2$, the decision is made that "1" was sent.

When a received signal is less than $V/2$, the decision is made that "0" was sent.

Consequently, error results if we decide either that "1" was sent when "0" was sent (P_e^0), or we decide that "0" was sent when "1" was sent (P_e^1). Due to the symmetry of the PDFs both are equally probable:

$$P_{\mathrm{e}}{}^{0} = P_{\mathrm{e}}{}^{1}, \tag{5.250}$$

$$P_{\mathrm{e}}^{0} = \int_{V/2}^{\infty} 0.5\,\mathrm{e}^{-v}\,\mathrm{d}v = 0.5\,\mathrm{e}^{-V/2}. \tag{5.251}$$

The total error is:

$$\begin{aligned} P_{\mathrm{e}} &= P(0)P_{\mathrm{e}}^{0} + P(1)P_{\mathrm{e}}^{1} = 0.5 \times 0.5\,\mathrm{e}^{-V/2} + 0.5 \times 0.5\,\mathrm{e}^{-V/2} \\ &= 0.5\,\mathrm{e}^{-V/2}. \end{aligned} \tag{5.252}$$

Exercise 5.12 The n random calls are made in a time interval [0, T]. What is the probability that k of n calls occur within the time interval τ (which is inside the interval T).

Answer Denote as the event as A if a call is received in τ.

The probability of event A can be expressed as a ratio of the time τ to the time T:

$$P\{A\} = \tau/T. \tag{5.253}$$

The probability that k of n calls received in the time interval τ is described using the Bernoulli formula:

$$P\{k;n\} = \binom{n}{k}\left(\frac{\tau}{T}\right)^{k}\left(1-\frac{\tau}{T}\right)^{n-k}. \tag{5.254}$$

Exercise 5.13 The binomial random variable X has the parameters $n = 5$ and a probability $p = 1/3$. Find the expected value of the random variable Y,

$$Y = \mathrm{e}^{-2X}. \tag{5.255}$$

Answer The probability mass function of the variable X is given with the Bernoulli formula:

$$P\{X=k;5\} = \binom{5}{k}\left(\frac{1}{3}\right)^{k}\left(\frac{2}{3}\right)^{5-k}. \tag{5.256}$$

The mean value of Y is:

$$E\{Y\} = \sum_{k=0}^{5} \mathrm{e}^{-2k} P\{X = k; 5\} = \sum_{k=0}^{5} \mathrm{e}^{-2k} \binom{5}{k} \left(\frac{1}{3}\right)^{k} \left(\frac{2}{3}\right)^{5-k}$$
$$= \sum_{k=0}^{5} \binom{5}{k} \left(\mathrm{e}^{-2}\frac{1}{3}\right)^{k} \left(\frac{2}{3}\right)^{5-k}. \tag{5.257}$$

Comparing (5.112) with (5.257), we can write:

$$E\{Y\} = \left(\frac{\mathrm{e}^{-2}}{3} + \frac{2}{3}\right)^{5} = 0.1827 \tag{5.258}$$

Exercise 5.14 If it is known that 20% of students did not pass an exam, find a probability that out of five randomly chosen students:

(a) One student did not pass exam.
(b) All five students passed the exam.
(c) At least three students passed the exam.

Answer Denote as a binary event A that a student did not pass the exam. The probability of event A is defined as:

$$P\{A\} = 0.2. \tag{5.259}$$

The probability that k out of n students did not pass the exam can be calculated using the Bernoulli formula.

(a) The probability that one out of five randomly chosen students did not pass the exam is:

$$P\{1; 5\} = \binom{5}{1} 0.2 \times 0.8^{4} = \frac{5!}{(5-1)!1!} 0.2 \times 0.8^{4} = 0.4096. \tag{5.260}$$

(b) The probability that all five randomly chosen students passed the exam (that is $k = 0$ students failed) is:

$$P\{0; 5\} = \binom{5}{0} 0.2^{0} \times 0.8^{5} = \frac{5!}{(5-0)!0!} 0.8^{5} = 0.8^{5} = 0.3277. \tag{5.261}$$

(c) In this case, 3, 4, or 5 students passed exam. This means that 2, 1, or 0 students did not pass the exam, or $k \leq 2$:

$$P\{k \leq 2; 5\} = \sum_{k=0}^{2} \binom{5}{k} 0.2^{k} \times 0.8^{5-k}$$
$$= \frac{5!}{(5-0)!0!} 0.8^{5} + \frac{5!}{(5-1)!1!} 0.2 \times 0.8^{4} + \frac{5!}{(5-2)!2!} 0.2^{2} \times 0.8^{3}$$
$$= 0.3277 + 0.4096 + 0.2048 = 0.9421. \tag{5.262}$$

Exercise 5.15 A manufacturer delivered 10,000 products. The probability that a product has a defect is 0.1. Find the probability that no more than 100 products have a defect.

Answer Event A is defined as an event in which a product has a defect.

$$P\{A\} = 0.1. \tag{5.263}$$

The probability that there will not be more than 100 products with a defect is:

$$P\{0 \leq k \leq 100; 10,000\} = \sum_{k=0}^{100} \binom{10,000}{k} (0.1)^k (0.9)^{10,000-k}. \tag{5.264}$$

We can clearly see that the calculation of the probability of (5.264) would be cumbersome. An easier way to approach this is to utilize an approximation of (5.264).

Note that

$$np = 10,000 \times 0.1 = 1,000. \tag{5.265}$$

This indicates that, according to (5.129), we can use a normal approximation (5.134) of (5.264), which leads to the following result:

$$P\{0 \leq k \leq 1,500\}$$
$$\approx \frac{1}{2}\left[\text{erf}\left(\frac{1,500 - 1,000}{\sqrt{2 \times 1,000 \times 0.9}}\right) - \text{erf}\left(\frac{0 - 1,000}{\sqrt{2 \times 1,000 \times 0.9}}\right)\right] \approx 1. \tag{5.266}$$

Exercise 5.16 A manufacturer delivered 3,000 products. The probability that a product has a defect is 0.001. Find the probability that there will be more than three products with defects.

Answer The desired probability can be calculated using the Bernoulli formula:

$$P\{k>3; 3,000\} = \sum_{k=3}^{3,000} \binom{3,000}{k} 10^{-3k}(1 - 10^{-3})^{3,000-k}. \tag{5.267}$$

We can see that an exact calculation of (5.267) is cumbersome. Instead, we will use the approximation of (5.267).

As opposed to Example 5.15, here the probability of failure is very small compared to the total number of products (3,000). As a result,

$$np = 3,000 \times 0.001 = 3. \tag{5.268}$$

This indicates that we can use a Poisson approximation (5.143):

$$P\{k>3; 3,000\} = 1 - P\{k \leq 3; 3,000\} = 1 - \sum_{k=0}^{3} \frac{3^k}{k!} e^{-3} = 1 - 0.6472$$
$$= 0.3528. \tag{5.269}$$

Exercise 5.17 An electronic system has 2,000 components. In a given period of time, the probability of failure of any one of its components is equal to 0.001 and is not dependent on the failure of any other components. Find the probability that exactly three components will fail.

Answer

$$P\{3; 2,000\} = \binom{2,000}{3} 10^{-9}(1 - 10^{-3})^{2,000-3}$$
$$= \frac{2,000!}{1,997!3!} 10^{-9}(1 - 10^{-3})^{1,997} = 0.1805. \tag{5.270}$$

Let us now find the approximated probability (5.270) using a Poisson approximation (5.143), finding that the following condition it is satisfied:

$$np = (2,000)(10)^{-3} = 2, \tag{5.271}$$

$$P\{k = 3\} = \frac{2^3}{3!} e^{-2} = 0.1804. \tag{5.272}$$

Note that (5.272) is a good approximation of (5.270).

Exercise 5.18 In a random sequence of binary symbols "0" and "1," see, for example,

$$\ldots 00000111010001111100 \ldots \tag{5.273}$$

the probabilities of "0" and "1" are equal and independent of one another.

A successive sequence of identical symbols forms groups, as shown below:

$$\ldots 00000 \quad 111 \quad 0 \quad 1 \quad 000 \quad 11111 \quad 00 \ldots . \tag{5.274}$$

There are a total of n groups in a sequence.

Define a random variable X as a number of equal symbols in ith group. Then define a random variable Y as a ratio of the number of symbols in all groups divided by the number of groups n. Find the mean value of Y.

Answer The number of symbols in a group is a geometric random variable described by the probability (5.181) where $p = 0.5$ (see Example 5.8.1). From (5.190), the mean value of X is:

$$E\{X\} = 1/p = 2. \tag{5.275}$$

The random variable

$$Y = \frac{\sum_{i=1}^{n} X_i}{n}. \tag{5.276}$$

From here,

$$E\{Y\} = \frac{nE\{X\}}{n} = E\{X\} = 2. \tag{5.277}$$

Exercise 5.19 In a Poisson flow of calls there is an average number of calls in 1 s, which is $\lambda = 100$ calls/s. Find the probability that at least 1 call appears in 1 ms.

Answer The number of calls in $\tau = 10^{-3}$ s is described using a Poisson formula (5.149):

$$\begin{aligned} P\{k \geq 1; 10^{-3}\,\mathrm{s}\} &= 1 - P\{k<1\} = 1 - P\{k = 0\} \\ &= \frac{(100 \times 10^{-3})^0}{0!} \mathrm{e}^{-100\times 10^{-3}} = 0.9048. \end{aligned} \tag{5.278}$$

Exercise 5.20 Find the probability that the time between two successive Poisson events will be less than its mean value. Also find the mean squared dissipation around the mean value of the random interval τ in which n events have occurred.

Answer The time between two successive Poisson events is an exponential random variable. Then from (5.62) and (5.65), we have:

$$P\{X \leq \overline{X}\} = F_X(\overline{X}) = 1 - \mathrm{e}^{-\lambda\frac{1}{\lambda}} = 1 - \mathrm{e}^{-1} = 0.6321. \tag{5.279}$$

The random time τ is an Erlang's random variable and so, from (5.90) and $b = n$, we have:

$$\sigma^2 = \frac{n}{\lambda^2}. \tag{5.280}$$

5.10 MATLAB Exercises

Exercise M.5.1 (MATLAB file: *exercise_M_5_1.m*). Generate the lognormal variable X for $N = 10{,}000$ samples using parameters of the corresponding normal variable Y, $m_Y = 1$ and $\sigma^2 = 0.5$. Estimate and plot the PDF and the distribution.

Solution The first 500 samples of the lognormal variable are shown in Fig. 5.20a. The estimated PDF is shown in Fig. 5.20b.
An estimation of the lognormal distribution and the mathematical lognormal distribution are shown in Fig. 5.21.

Exercise M.5.2 (MATLAB file: *exercise_M_5_2.m*). For the lognormal variable from Exercise M.5.1, find the probability P_1, that the random variable is less than $A = 3$ and the probability P_2, that the random variable is in the interval $[B_1, B_2]$, where $B_1 = 2$ and $B_2 = 5$. Illustrate both probabilities in the PDF and distribution plots.

Solution The probability $P_1 = 0.5555$ is illustrated in Fig. 5.22.
The probability $P_2 = 0.4735$ is illustrated in Fig. 5.23.

Exercise M.5.3 (MATLAB file: *exercise_M_5_3.m*). Show how the lognormal PDF becomes more asymmetric, considering that $\sigma = 0.4$, 0.6, 0.8, and 0.9.
Next consider that $\sigma = 0.2$, 0.1, 0.08, and 0.05, to show how the lognormal PDF approaches normal PDF.

Solution
First case: The values of the normal standard deviations are increased: $\sigma_1 = 0.4$, $\sigma_2 = 0.6$, $\sigma_3 = 0.8$, $\sigma_4 = 0.9$. Figure 5.24a shows the values of the variables, while Fig. 5.24b shows the corresponding estimated densities.

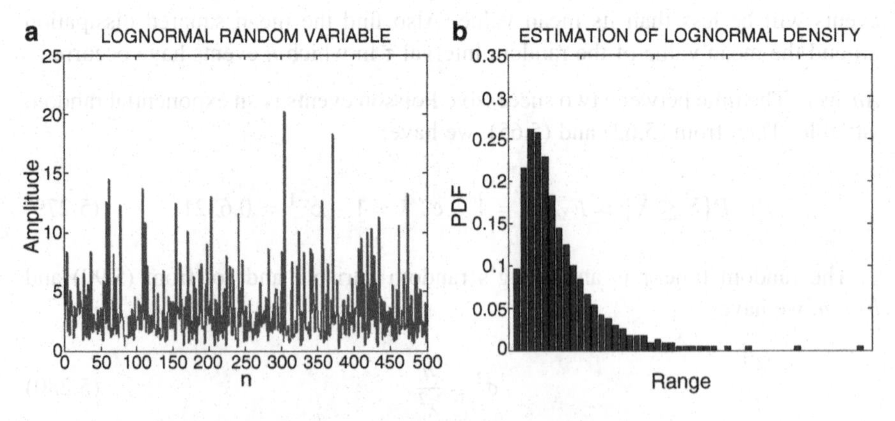

Fig. 5.20 Lognormal variable. (**a**) Lognormal variable. (**b**) Estimated PDF

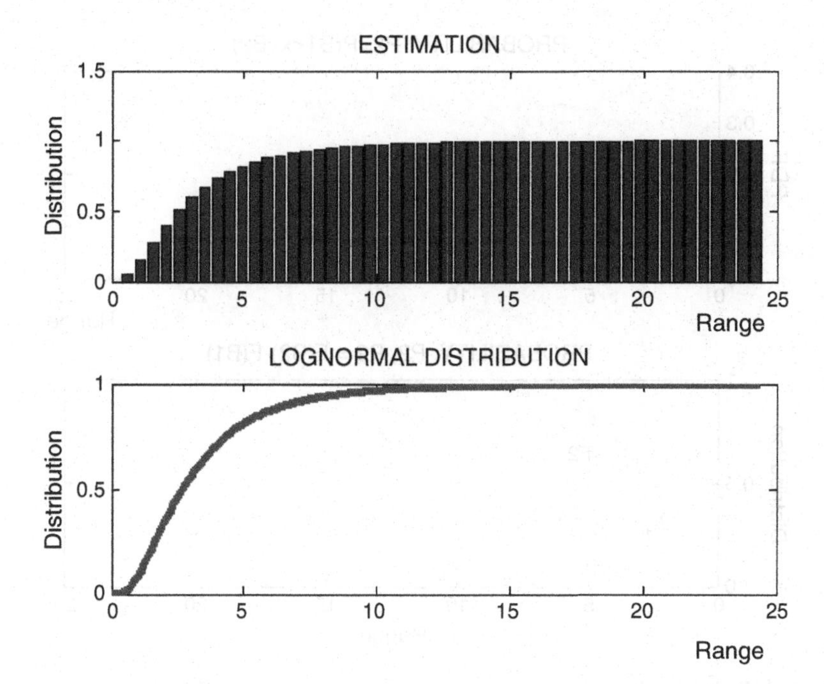

Fig. 5.21 Lognormal distribution

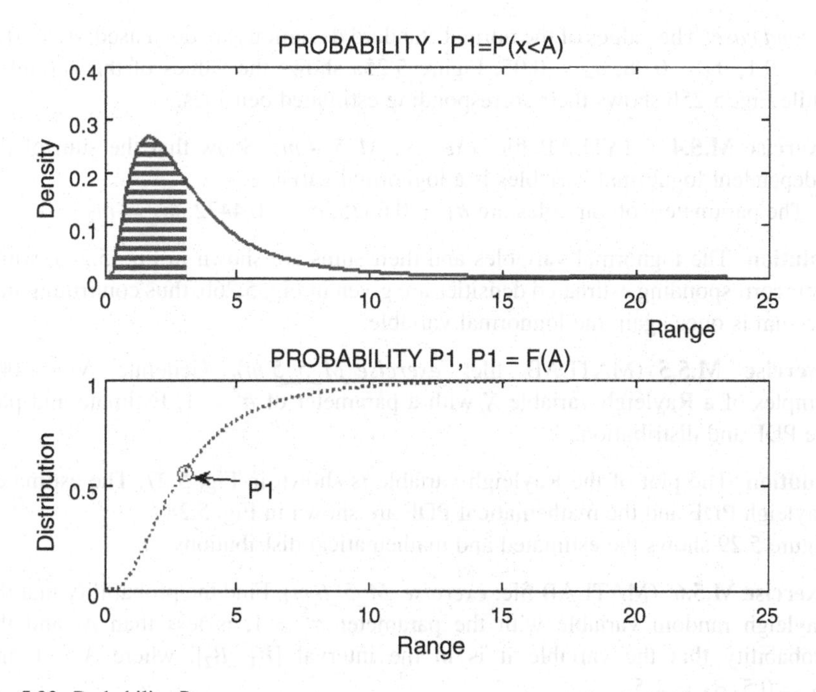

Fig. 5.22 Probability P_1

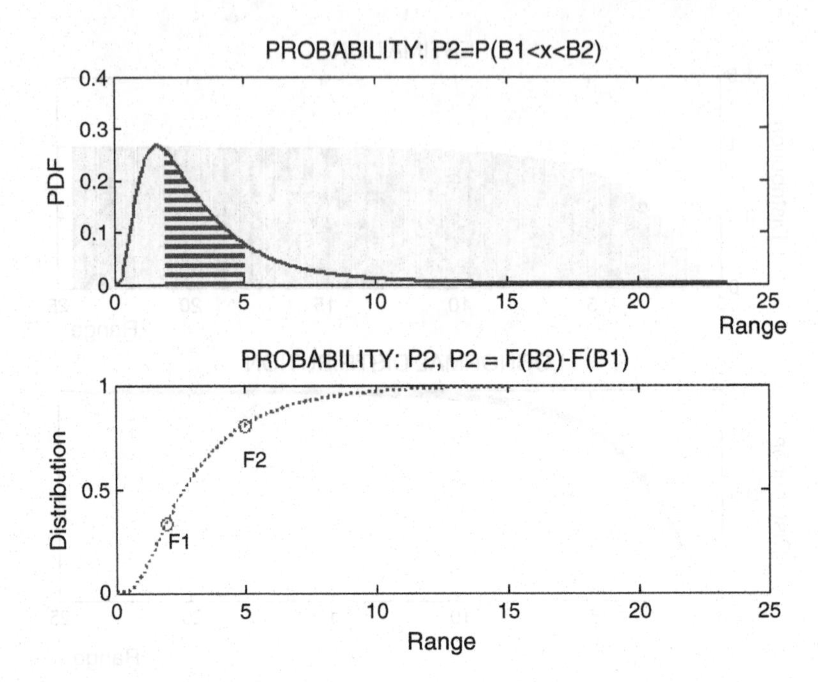

Fig. 5.23 Probability P_2

Second case: The values of the normal standard deviations are decreased: $\sigma_1 = 0.2$, $\sigma_2 = 0.1$, $\sigma_3 = 0.08$, $\sigma_4 = 0.05$. Figure 5.25a shows the values of the variables, while Fig. 5.25b shows their corresponding estimated densities.

Exercise M.5.4 (MATLAB file: *exercise_M_5_4.m*). Show that the sum of the independent lognormal variables is a lognormal variable.

The parameters of variables are $\sigma_1 = 0.6325$, $\sigma_2 = 0.4472$, $m_1 = m_2 = 1$.

Solution The lognormal variables and their sums are shown in Fig. 5.26a, while their corresponding estimated densities are given in Fig. 5.26b, thus confirming that the sum is once again the lognormal variable.

Exercise M.5.5 (MATLAB file: *exercise_M_5_5.m*). Generate $N = 1{,}000$ samples of a Rayleigh variable X with a parameter of $\sigma^2 = 1$. Estimate and plot the PDF and distribution.

Solution The plot of the Rayleigh variable is shown in Fig. 5.27. The estimated Rayleigh PDF and the mathematical PDF are shown in Fig. 5.28.
Figure 5.29 shows the estimated and mathematical distributions.

Exercise M.5.6 (MATLAB file: *exercise_M_5_6.m*). Find the probability that the Rayleigh random variable with the parameter $\sigma^2 = 1$, is less than A, and the probability that the variable it is in the interval $[B_1, B_2]$, where $A = 1$ and $B_1 = 0.5$, $B_2 = 1.5$.

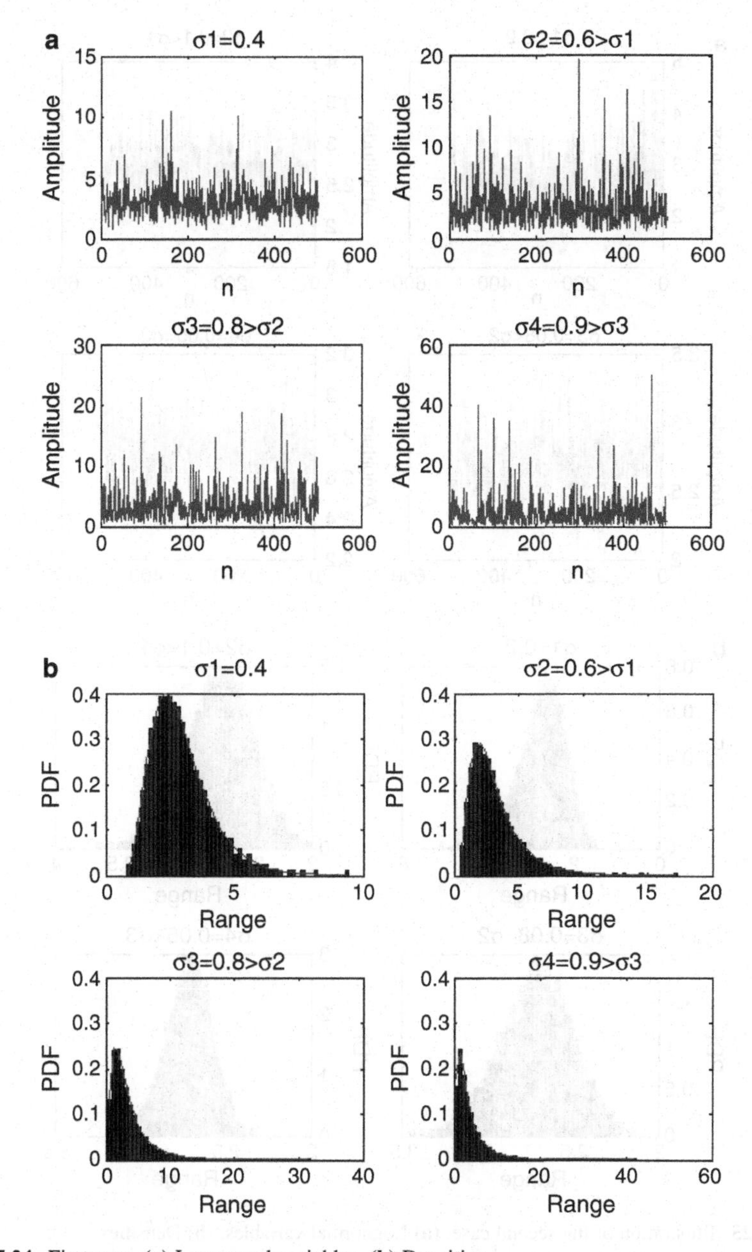

Fig. 5.24 First case. (**a**) Lognormal variables. (**b**) Densities

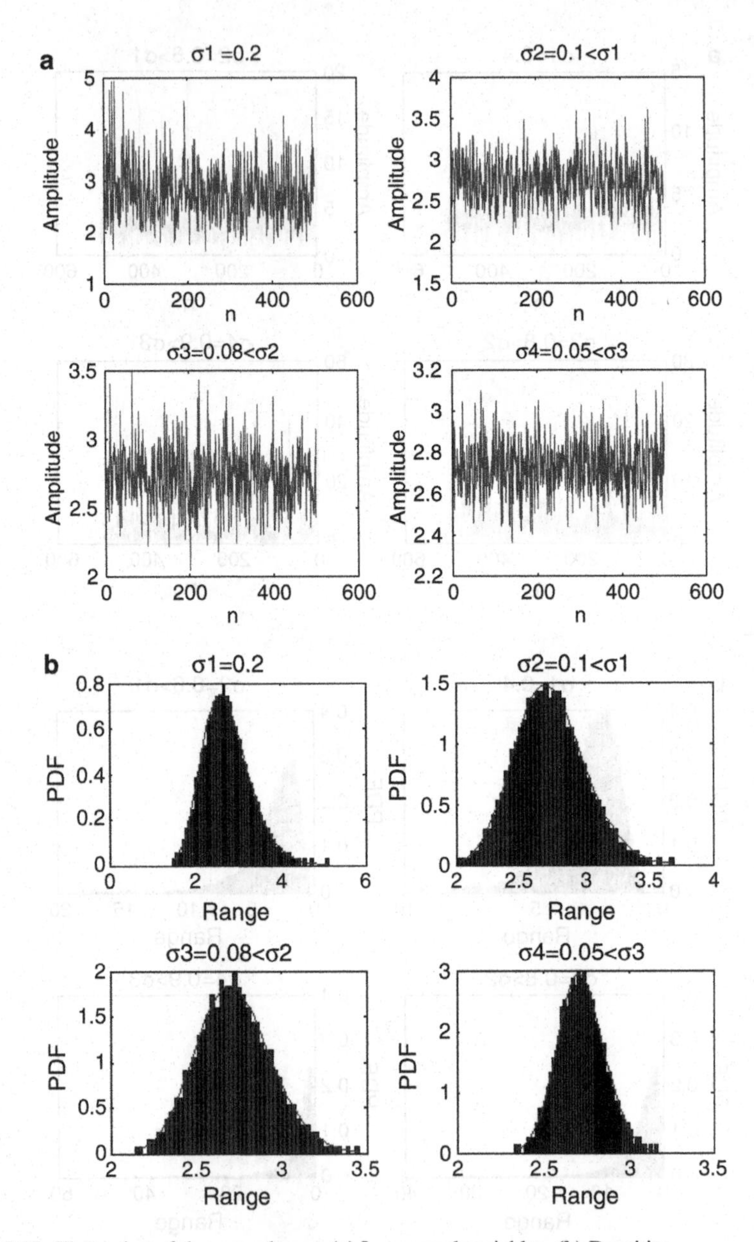

Fig. 5.25 Illustration of the second case. (**a**) Lognormal variables. (**b**) Densities

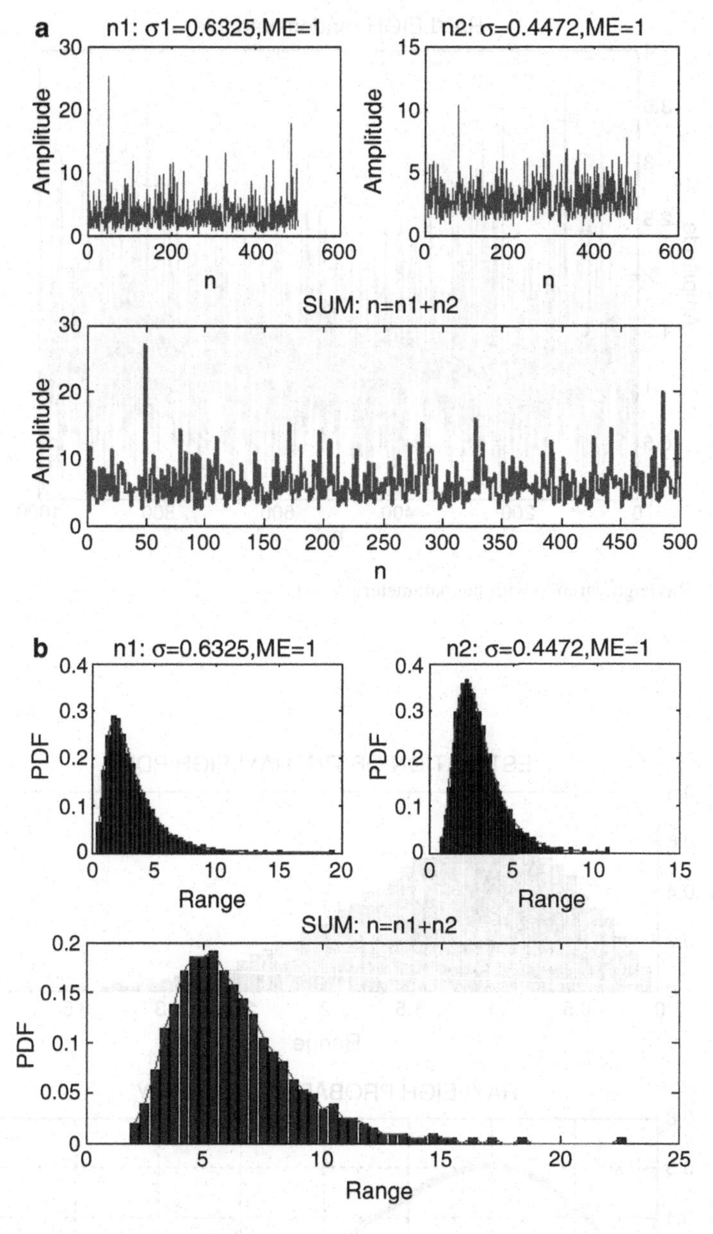

Fig. 5.26 Sum of lognormal variables. (**a**) Lognormal random variables. (**b**) Densities

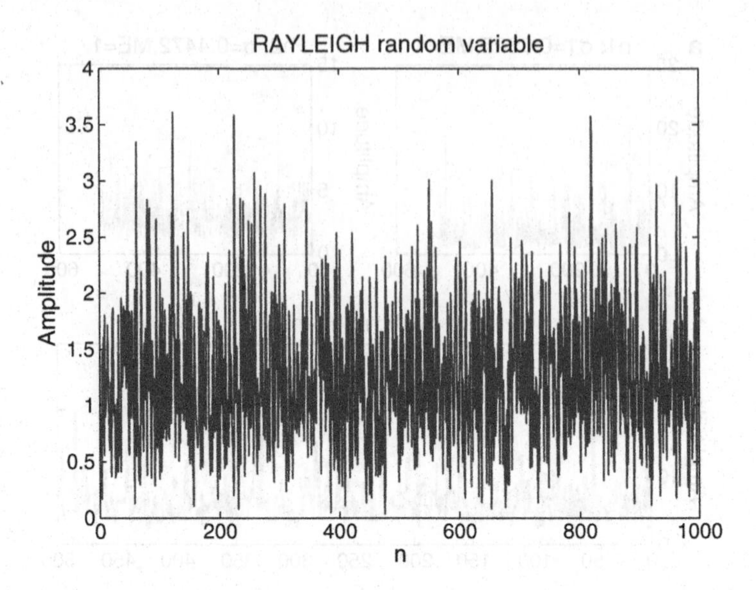

Fig. 5.27 Rayleigh variable with the parameter $\sigma^2 = 1$

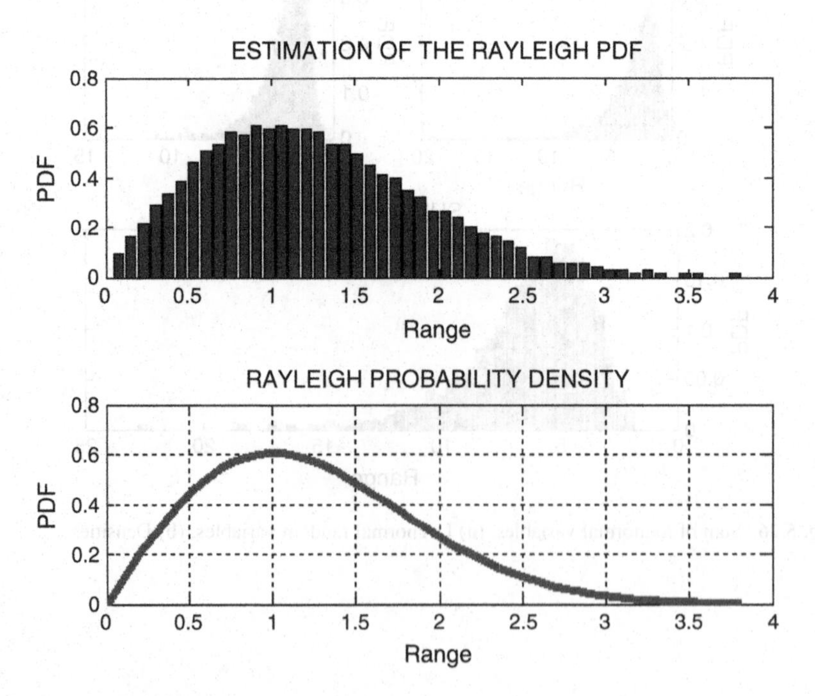

Fig. 5.28 Rayleigh PDFs

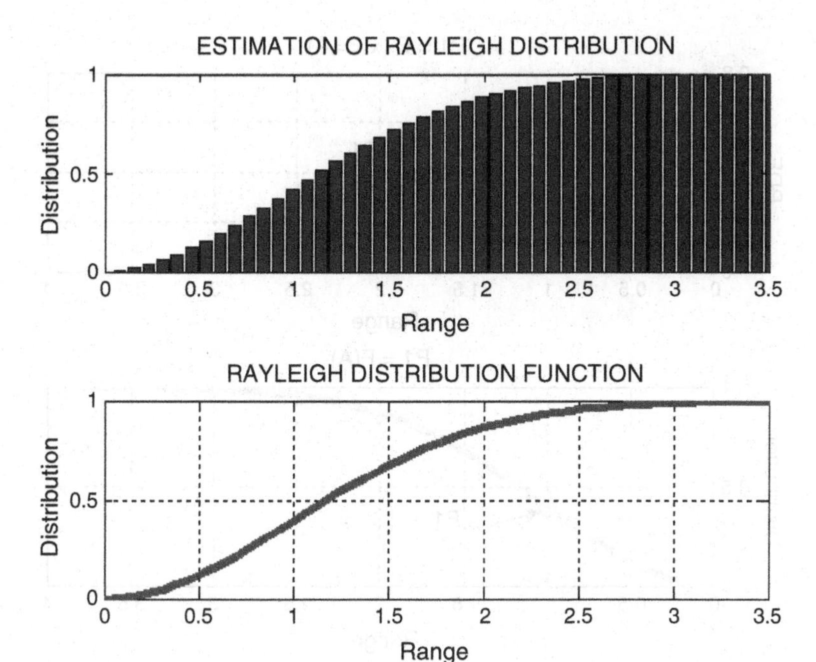

Fig. 5.29 Rayleigh distributions

Solution The probability

$$P\{X \leq 1\} = 0.3935, \tag{5.281}$$

and it is shown in Fig. 5.30.

Figure 5.31 shows the probability:

$$P\{0.5 < X \leq 1.5\} = F_X(1.5) - F_X(0.5) = 0.6753 - 0.1175 = 0.5578. \tag{5.282}$$

Exercise M.5.7 (MATLAB file: *exercise_M_5_7.m*). Generate Rician random variables with parameter $\sigma^2 = 1$, and the following values for A: 0.01, 0.5, 1, and 4. Estimate and plot the corresponding densities and distributions.

Solution The generated variables are shown in Fig. 5.32.

Figures 5.33 and 5.34 show the estimated and mathematical PDFs.

The estimated and mathematical distributions are shown in Figs. 5.35 and 5.36.

Exercise M.5.8 (MATLAB file: *exercise_M_5_8.m*). Generate the Rician random variables with the chosen parameter σ^2, and the different values of A:

(a) $A = A_1 \ll \sigma$
(b) $A = A_2 \gg \sigma$
(c) $A = A_3, A_1 < A_3 < A_2$

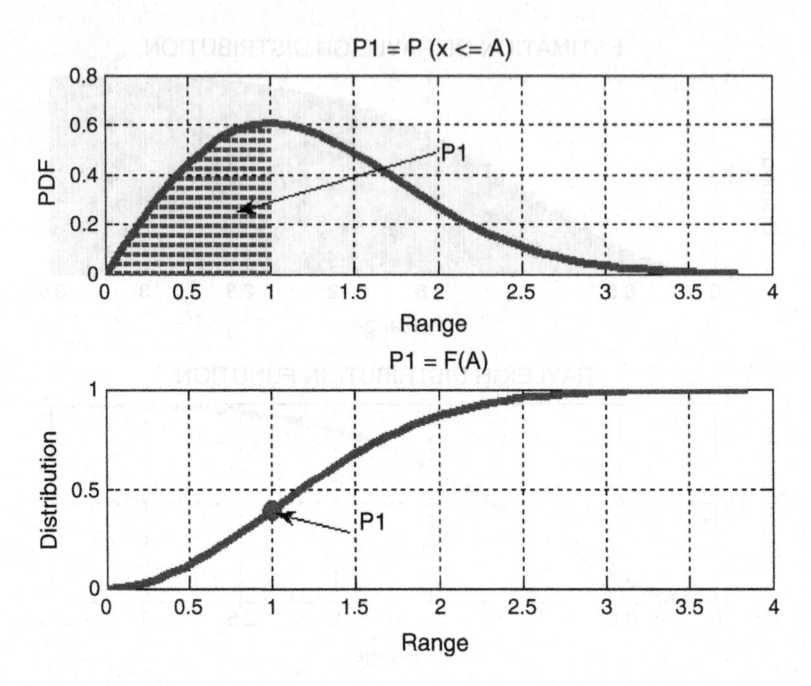

Fig. 5.30 Probability of the Rayleigh variable

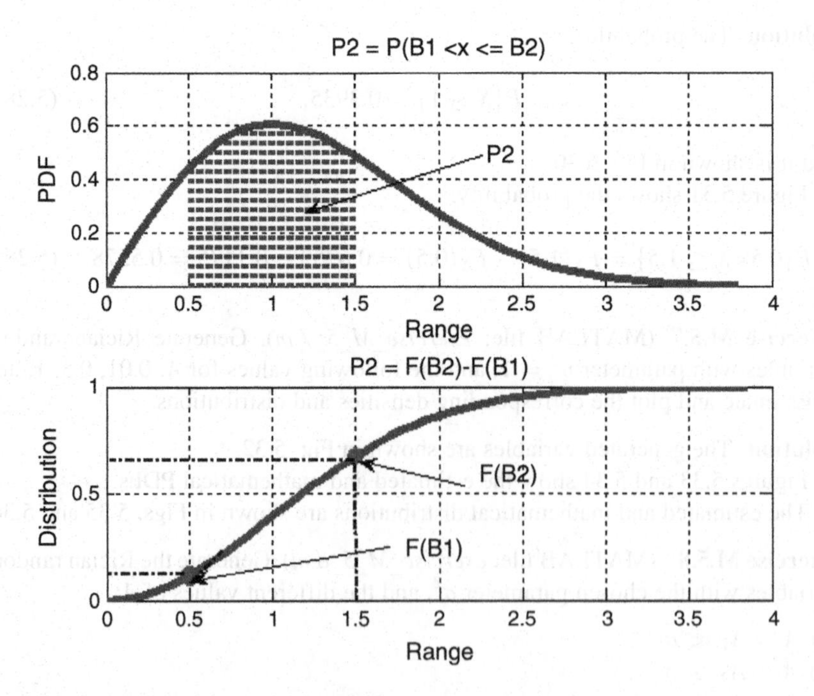

Fig. 5.31 Probability of the Rayleigh variable

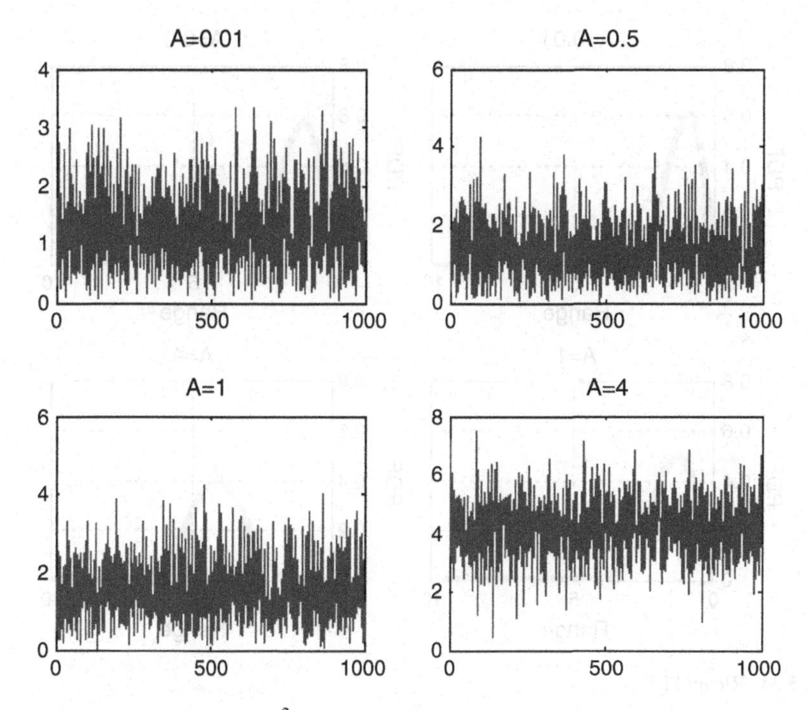

Fig. 5.32 Rician variables with $\sigma^2 = 1$, and different values of A

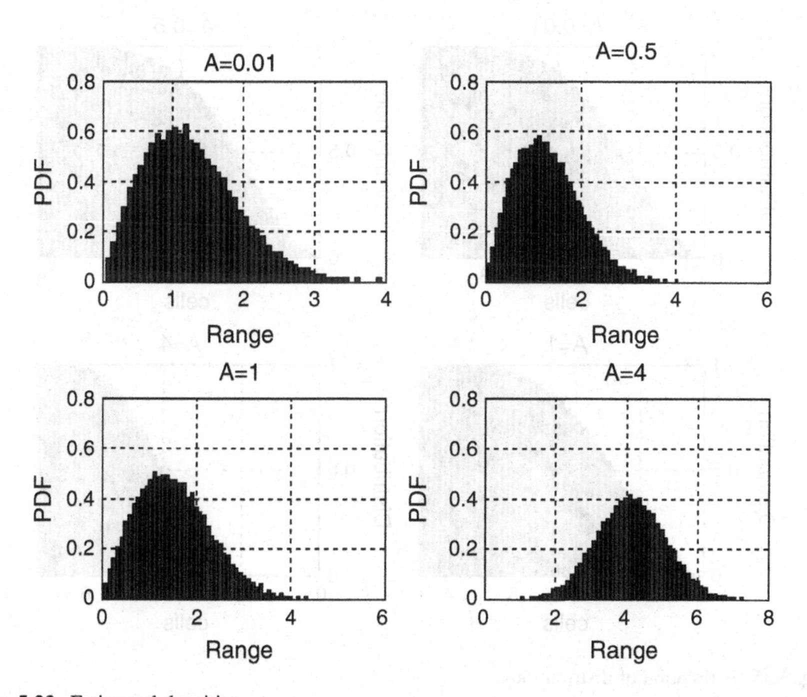

Fig. 5.33 Estimated densities

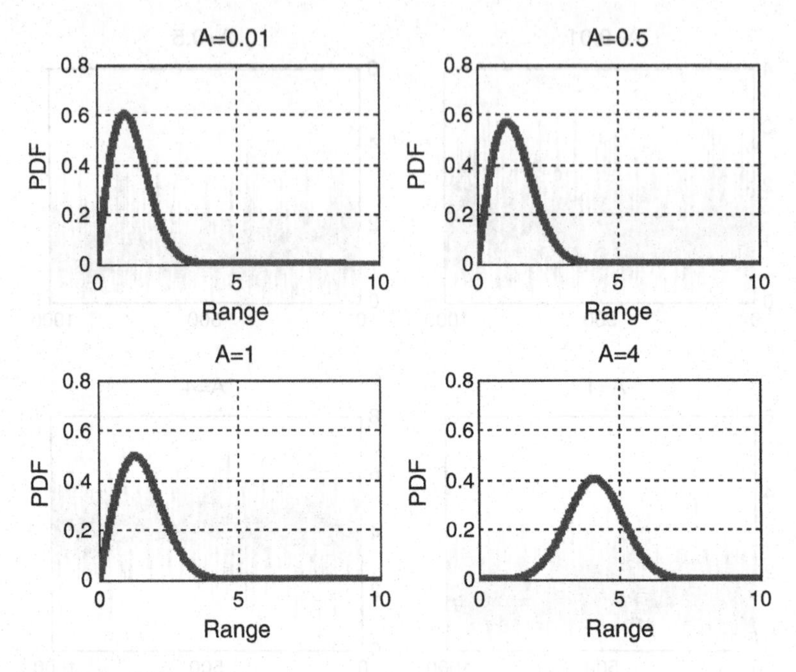

Fig. 5.34 Rician PDFs

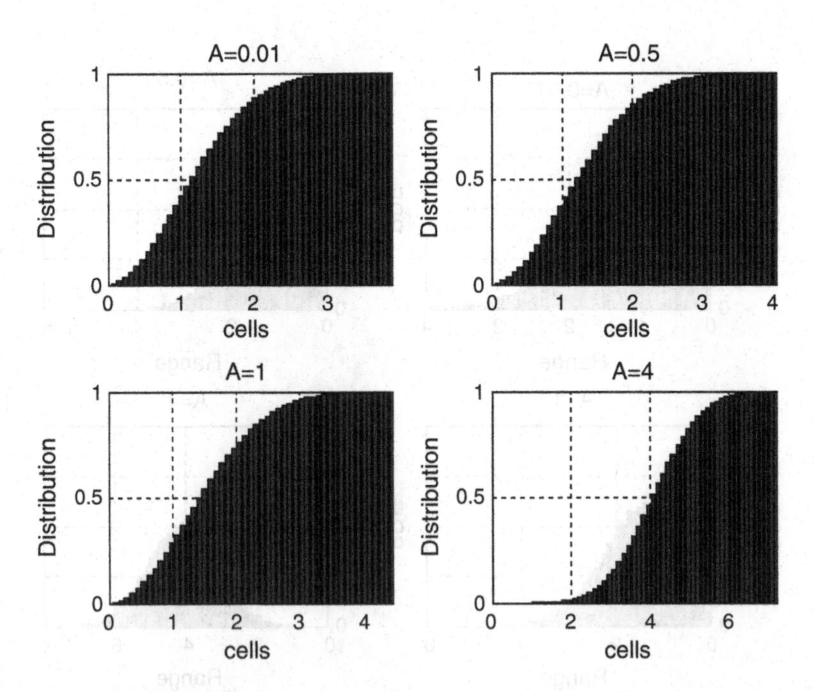

Fig. 5.35 Estimation of distributions

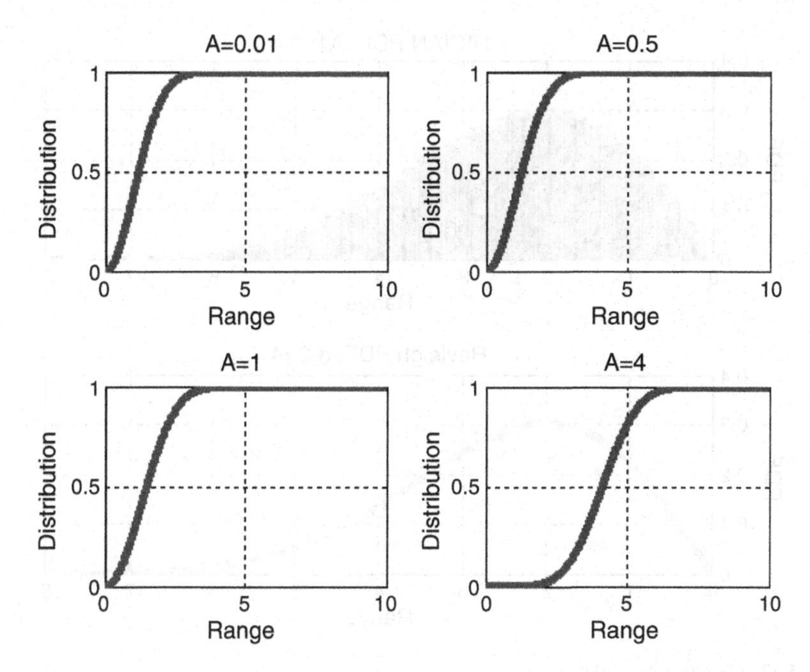

Fig. 5.36 Rician distributions

Generate the corresponding variables and estimate the corresponding densities and distributions. Compare the results of cases (a) and (b) with Rayleigh and normal random variables, respectively. Plot the corresponding mathematical densities and distributions.

Solution Figures 5.37–5.39 show results for the chosen value $\sigma^2 = 4$, $A_1 = 0.4$, $A_2 = 10$, $A_3 = 2$.

All cases including, $A = 0$, are compared in Figs. 5.40–5.44.

Exercise M.5.9 (MATLAB file: *exercise_M_5_9.m*). Generate $N = 1{,}000$ samples of an exponential variable X with parameter $\lambda = 0.8$. Estimate and plot PDF and distribution.

Solution The exponential variable is shown in Fig. 5.45a. The estimated and mathematical PDFs, along with the corresponding distribution functions are shown in Fig. 5.45b.

Exercise M.5.10 (MATLAB file: *exercise_M_5_10.m*). For the exponential variable from Exercise M.5.9 find the probability that the random variable is less than $A = 1$, and the probability that the random variable is in the interval [0.5, 2]. Show both probabilities in the PDF and distribution plots.

Solution Figure 5.46 illustrates the probability P_1.

$$P_1 = P\{X \leq 1\} = 0.5507.$$

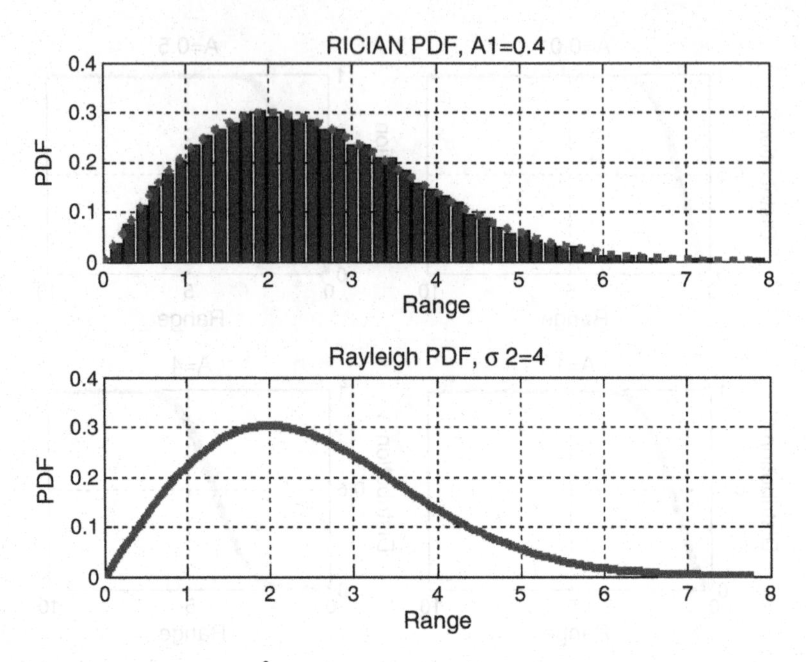

Fig. 5.37 Case (a): $A_1 = 0.4$, $\sigma^2 = 4$

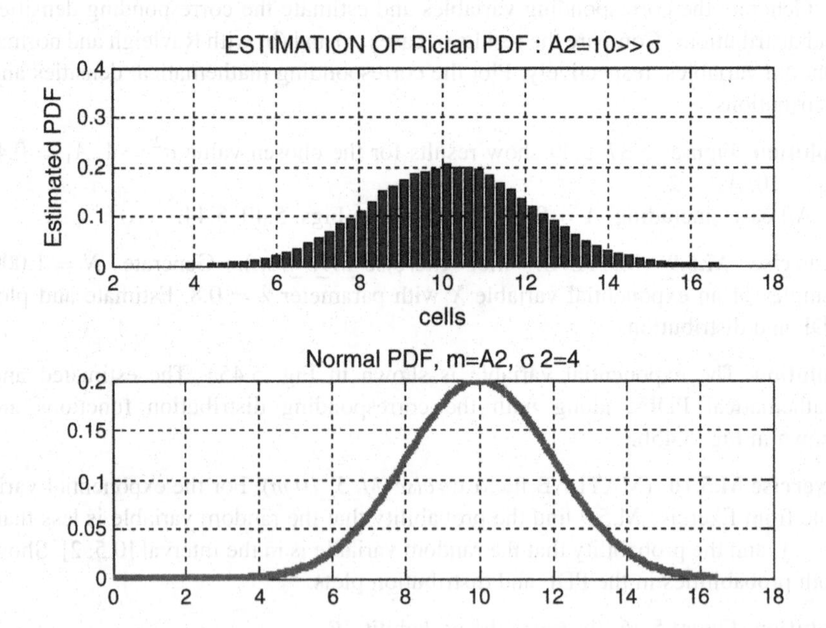

Fig. 5.38 Case (b): $A_2 = 10$, $\sigma^2 = 4$

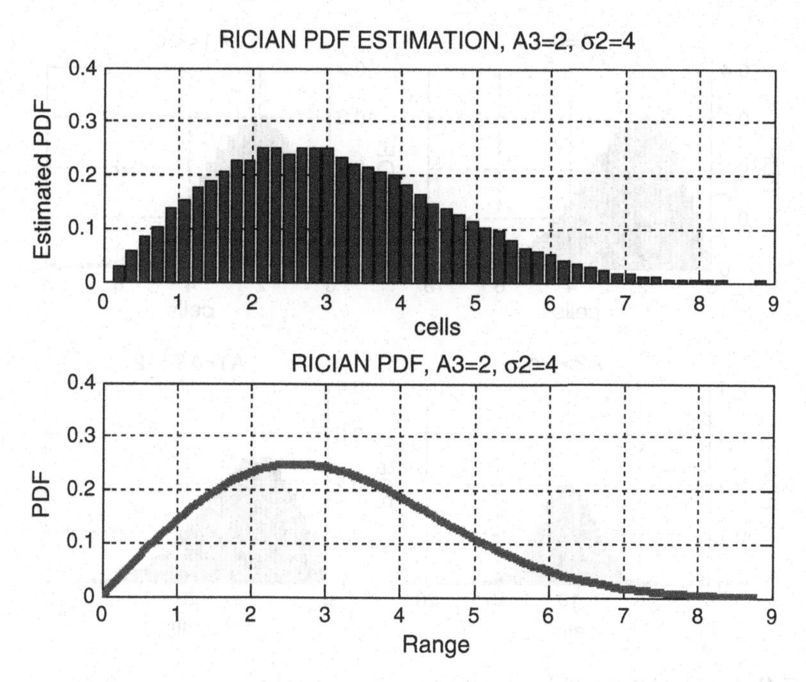

Fig. 5.39 Case (c): $A_3 = 2$, $\sigma^2 = 4$

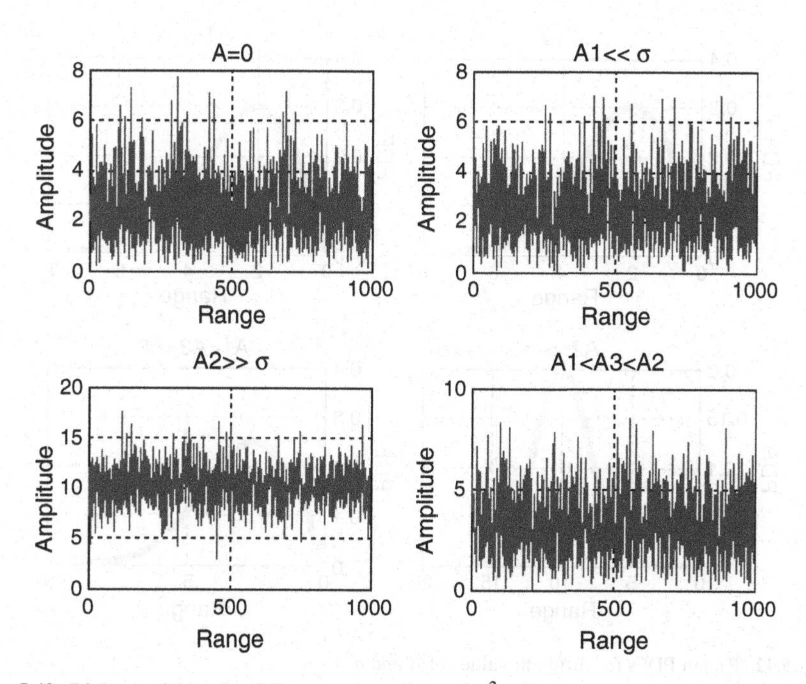

Fig. 5.40 Rician variables for different values of A and $\sigma^2 = 4$

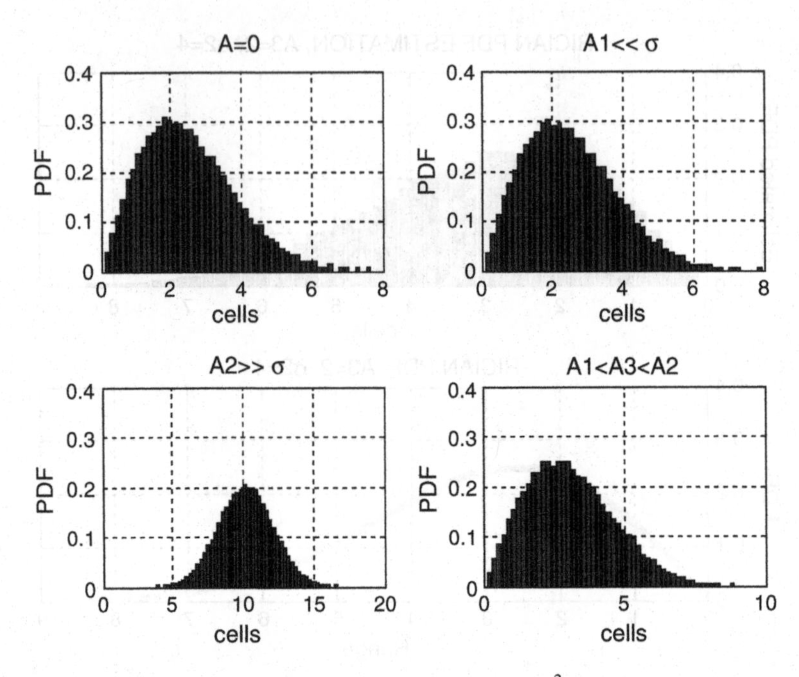

Fig. 5.41 Estimated Rician PDFs for different values of A and $\sigma^2 = 4$

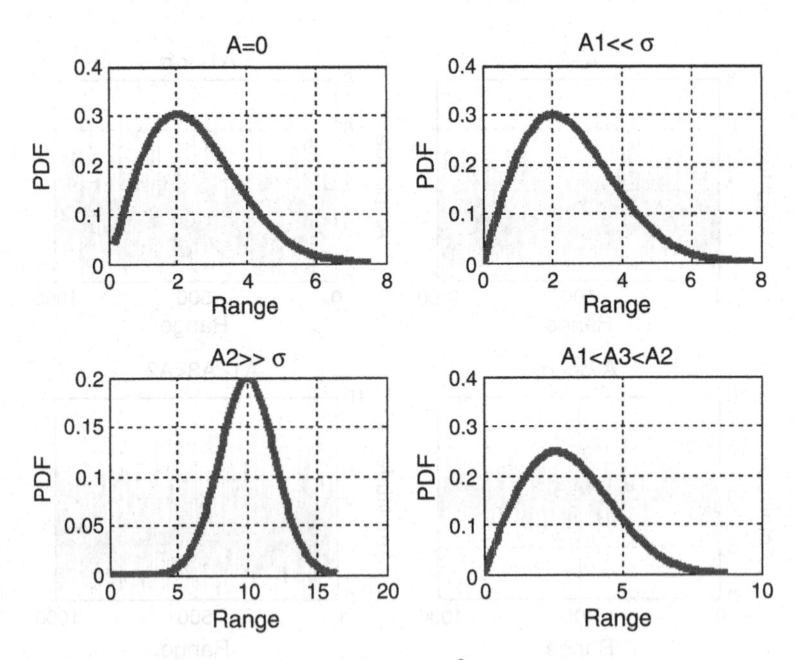

Fig. 5.42 Rician PDFs for different values of A and $\sigma^2 = 4$

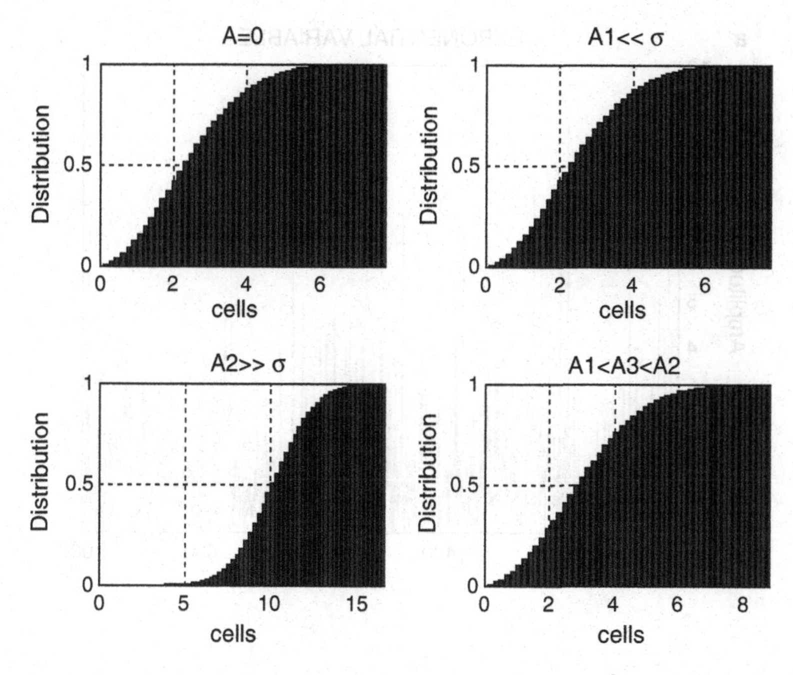

Fig. 5.43 Estimated Rician distributions for different values of A and $\sigma^2 = 4$

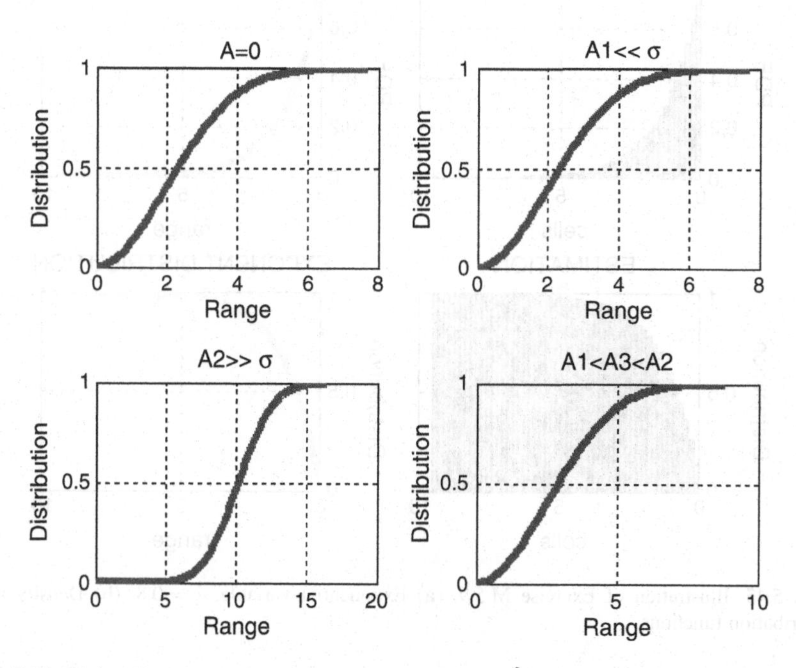

Fig. 5.44 Rician distributions for different values of A and $\sigma^2 = 4$

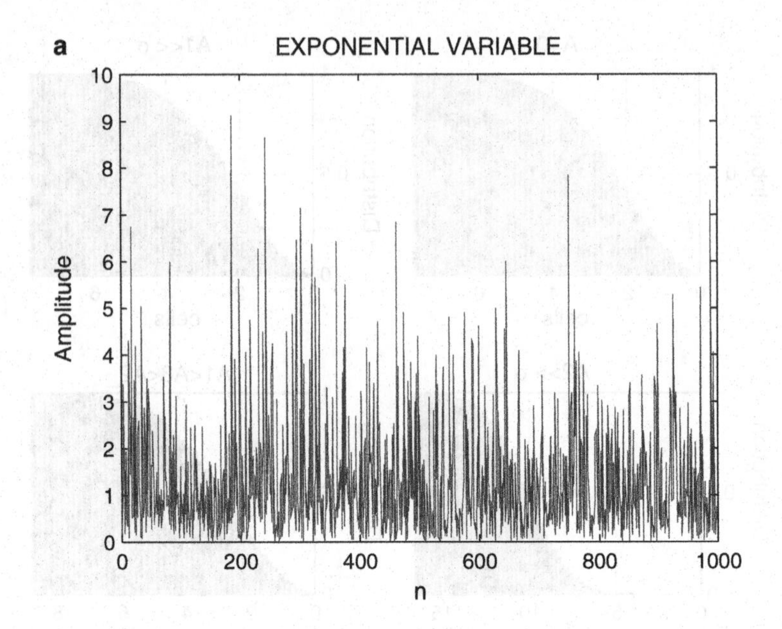

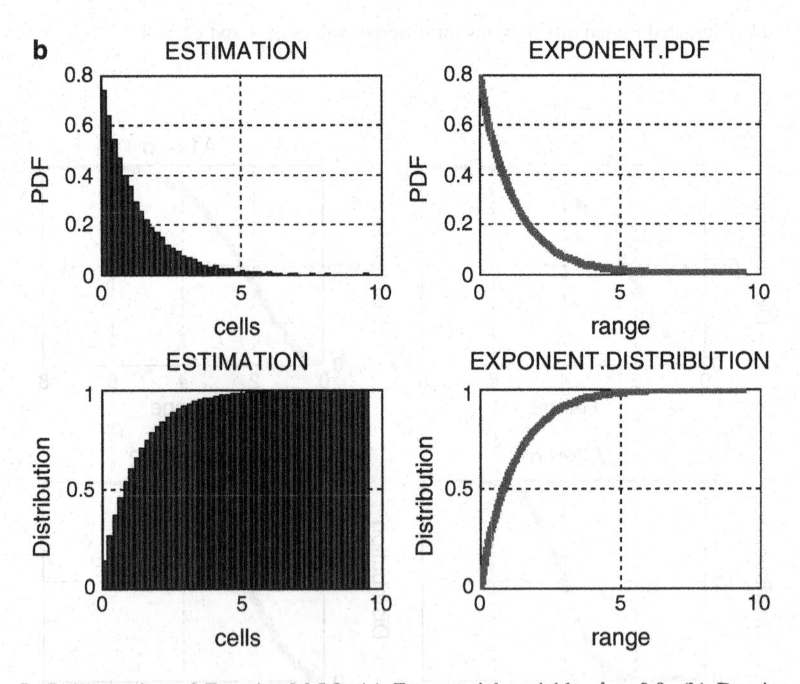

Fig. 5.45 Illustration of Exercise M.5.9. (**a**) Exponential variable, $\lambda = 0.8$. (**b**) Density and distribution functions

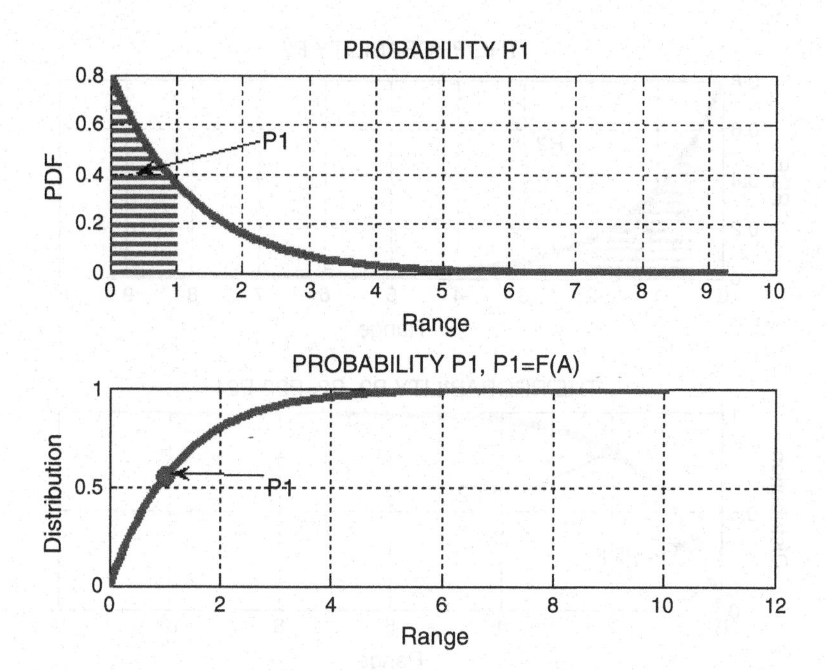

Fig. 5.46 Illustration of the probability P_1

The probability P_2 is illustrated in Fig. 5.47.

$$P_2 = P\{0.5 < X \leq 2\} = 0.4684.$$

5.11 Questions

Q.5.1. Explain why a Rician variable approaches to a Rayleigh variable in the case of a small values of A related to a deviation σ $(A \ll \sigma)$?

Q.5.2. Explain why a Rician variable approaches to a Gaussian variable for high values of A related to a deviation σ $(A \gg \sigma)$?

Q.5.3. When does a lognormal variable approach a normal variable?

Q.5.4. There are k events in a Poisson flow of events in the random time interval τ. Which random variable can describe the random interval τ?

Q.5.5. Why is a Poisson random variable said to be a "rare events variable?"

Q.5.6. (a) Which random variable has an equal mean value and standard deviation? (b) Which random variable has an equal mean value and variance?

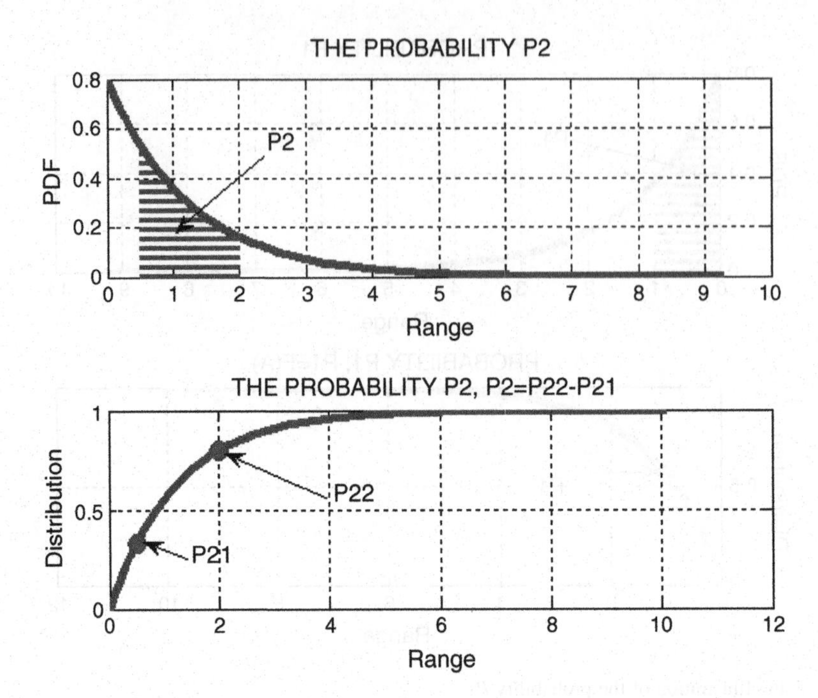

Fig. 5.47 Illustration of the probability P_2

5.12 Answers

A.5.1. The envelope of a narrowband Gaussian noise is described with a Rayleigh random variable.

A sinusoidal signal with an amplitude of A is added to a narrowband Gaussian noise. The envelope of the resulting noise is described with a Rician random variable. Therefore, if A is very small then the sum of a sinusoidal signal and a narrowband Gaussian noise is approximately the noise itself and, as a consequence, envelope is approximately described with a Rayleigh variable.

The Rician PDF is:

$$f_X(x) = \frac{x}{\sigma^2} e^{-\frac{x^2+A^2}{2\sigma^2}} I_0\left(\frac{xA}{\sigma^2}\right). \tag{5.283}$$

For small values of A compared to σ ($A \ll \sigma$), we have:

$$I_0(z) \approx 1 + \frac{z^2}{4} + \cdots = e^{z^2/4}. \tag{5.284}$$

Therefore, the Bessel function (5.284) approaches 1, and (5.283) becomes:

$$f_X(x) \approx \frac{x}{\sigma^2} e^{-\frac{x^2}{2\sigma^2}}, \tag{5.285}$$

which is a Rayleigh random variable.

A.5.2. In this case the signal is dominant and the sum of the signal and noise is highly concentrated around the value A, which is a characteristic of a normal random variable.

A.5.3. For small values of σ.

A.5.4. In a Poisson flow of events, a random time between events is an exponential random variable. Therefore, the random time interval τ is the sum of exponential random variables which is itself an Erlang's random variable.

A.5.5. Remember that the Poisson random variable can be obtained as a limiting case in Bernoulli trials where the number of trial n is very high and the probability of success p is very low.

A.5.6. (a) An exponential random variable has an equal mean value and standard deviation.

(b) A Poisson random variable has an equal mean value and variance.

Chapter 6
Random Processes

6.1 What Is a Random Process?

As in the case of a random variable, a random process can be defined using a random experiment. Remember that a random variable is obtained by assigning numbers on x-axis to all possible outcomes of s_i. However, if instead of numbers, time functions $x(t, s_i)$ are assigned to the outcomes of s_i from a sample space S, we obtain a random process $X(t, s)$, as shown in Fig. 6.1.

In this way, the *random process* $X(t, s)$ is a function that assigns family of time functions $x_i(t)$ to a space of events S. Generally, this function may be real or complex. However, we consider here only real processes.

The set of all time functions is called the *ensemble*, and a particular time function is called a *realization* or a *member of an ensemble*. Keeping in mind that mapping space S to an ensemble results of a random process, we will use a short denotation $X(t)$ for a process and $x_i(t)$ for a particular ith realization.

Example 6.1.1 Let a random experiment be a measurement of temperature in a town, and s_i be the temperature measurement of the ith day. The temperature during day is not constant. It is different, for example, at 6 a.m. than at 1 p.m. Therefore, the measured temperature is a function that changes over time. On the other hand, if we, for example, observe the temperatures at 1 p.m. every day, it is obvious that the temperature is a random variable that has all possible values in the range, say, from −20°C to 35°C.

Therefore, the random process is a family of time functions but in the given instant t_i, it represents a random variable. The random variable of the process $X(t)$ in the time instant t_i is denoted as a variable X_i.

G.J. Dolecek, *Random Signals and Processes Primer with MATLAB*,
DOI 10.1007/978-1-4614-2386-7_6, © Springer Science+Business Media New York 2013

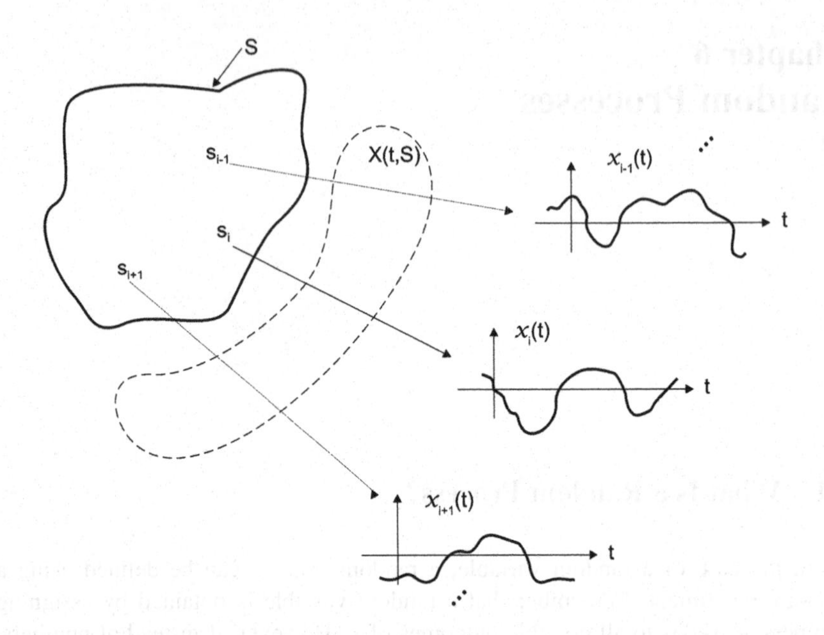

Fig. 6.1 Mapping a space of events S to a (x, y) space

6.1.1 Deterministic and Nondeterministic Random Processes

A random process is said to be *deterministic* if all future values of any realization can be determined from its past values [PEE93, p. 166]. As an example, consider sinusoidal realizations where the amplitude varies for different realizations, as shown in Fig. 6.2a. The future values of each realization are shown with dotted lines and are completely known. In the opposite case, shown in Fig. 6.2.b, the process is a *nondeterministic*. The waveforms of realizations are irregular and cannot be described using a mathematical expression. Consequently, future values–shown with dotted lines–cannot be precisely determined.

From the previous discussion, note that the term "random" in a random process is not related with a wave shape of its realizations, but with an uncertainty in which the time function is assigned to a particular outcome. Similarly, as in a die rolling experiment, one knows with certainty that a number from 1 to 6 will appear, and the uncertainty is which of those numbers will appear in each roll.

6.1.2 Continuous and Discrete Random Processes

The classification of random processes may be performed depending on whether amplitudes and time are continuous or discrete values. As both amplitude and time can be either continuous or discrete, all possible combinations are shown in Fig. 6.3.

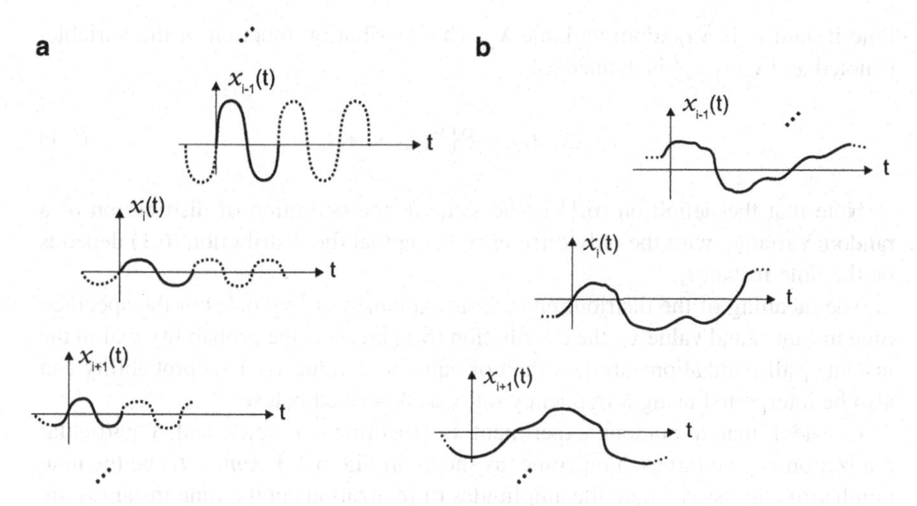

Fig. 6.2 Deterministic and nondeterministic random processes

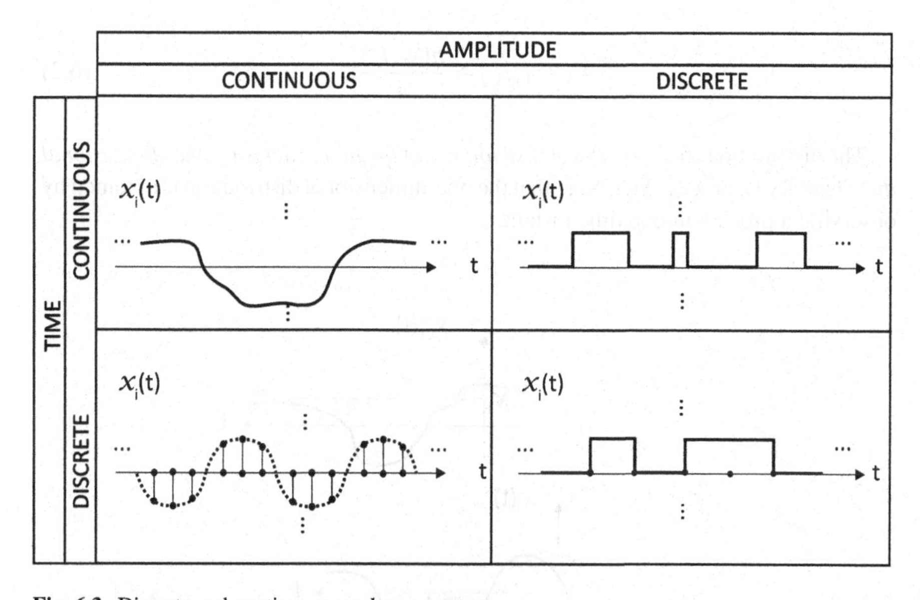

Fig. 6.3 Discrete and continuous random processes

6.2 Statistics of Random Processes

6.2.1 Description of a Process in One Point

We have already mentioned that a random process in a specified time t becomes a random variable. Let us consider the time instant t_1. A random variable defined in a time instant t_1 is a random variable X_1. The distribution function of the variable, denoted as $F_{X_1}(x_1, t_1)$ is defined as:

$$F_{X_1}(x_1, t_1) = P\{X_1 \leq x_1; t_1\}. \tag{6.1}$$

Note that the definition (6.1) is the same as the definition of distribution of a random variable, with the only difference being that the distribution (6.1) depends on the time instant t_1.

The meaning of the distribution (6.1) is explained in Fig. 6.4. For the specified time instant t_1 and value x_1, the distribution (6.1) presents the probability that in the instant t_1 all realizations are less than or equal to a value x_1. This probability can also be interpreted using a frequency ratio, as described below.

Consider that a random experiment is performed n times and a particular realization is given to each outcome, as shown in Fig. 6.4. Let $n(x_1, t_1)$ be the total number of successes where the amplitudes of realizations in the time instant t_1 are not more than x_1.

Taking n to be very large, the desired probability can be approximated as:

$$F_{X_1}(x_1, t_{1|}) \approx \frac{n(x_1, t_1)}{n}. \tag{6.2}$$

The distribution (6.2) is called a *distribution of the first order*, or a *one-dimensional distribution* of a process $X(t)$. Note that the one-dimensional distribution is obtained by observing a process in one time instant.

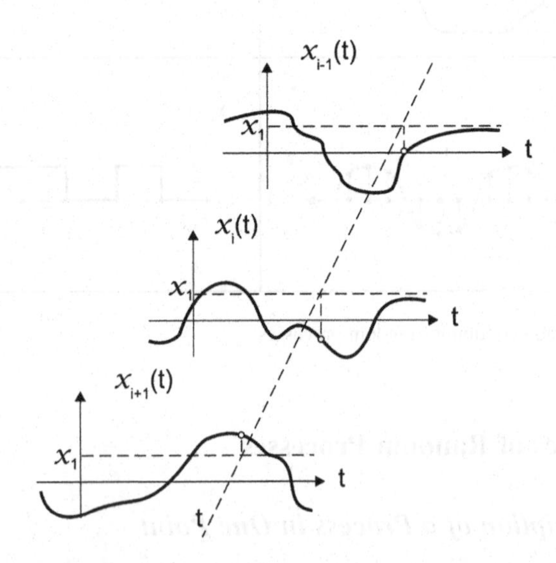

Fig. 6.4 Description of a process in a particular time instant

A density function is obtained as the derivation of the corresponding distribution function,

$$f_{X_1}(x_1, t_1) = \frac{\partial F_{X_1}(x_1; t_1)}{\partial x_1} = \frac{P\{x_1 < X_1 \le x_1 + \mathrm{d}x_1; t_1\}}{\mathrm{d}x_1}. \tag{6.3}$$

The density function (6.3) is called a *density function of the first order*, or a *one-dimensional density*.

6.2.2 Description of a Process in Two Points

Next we observe a process in two points denoted as t_1 and t_2. The corresponding random variables are X_1 and X_2. A *joint distribution function*, a *second order distribution*, or a *two-dimensional distribution* is defined as:

$$\begin{aligned} F_{X_1X_2}(x_1, x_2; t_1, t_2) &= P\{X(t_1) \le x_1, X(t_2) \le x_2\} \\ &= P\{X_1 \le x_1, X_2 \le x_2; t_1, t_2\}. \end{aligned} \tag{6.4}$$

Note again that this is the same definition as that of a joint distribution of two random variables with only exception being that now the joint distribution depends on time instants t_1 and t_2.

The meaning of this distribution can also be interpreted using a frequency ratio, as shown in Fig. 6.5. Consider a total number of outcomes for which the corresponding realizations are not more than x_1 in a given time instant t_1 and also not more than x_2 in a given time instant t_2. The obtained number is denoted as $n(x_1, x_2, t_1, t_2)$. Then, assuming a large n, a distribution function can be expressed as:

$$F_{X_1X_2}(x_1, x_2; t_1, t_2) = P\{X_1 \le x_1, X_2 \le x_2; t_1, t_2\} \approx \frac{n(x_1, x_2; t_1, t_2)}{n}. \tag{6.5}$$

The corresponding density function is given as:

$$\begin{aligned} f_{X_1X_2}(x_1, x_2; t_1, t_2) &= \frac{\partial^2 F_{X_1X_2}(x_1, x_2; t_1, t_2)}{\partial x_1 \partial x_2} \\ &= \frac{P\{x_1 < X(t_1) \le x_1 + \mathrm{d}x_1, x_2 < X(t_2) \le x_2 + \mathrm{d}x_2; t_1, t_2\}}{\mathrm{d}x_1\,\mathrm{d}x_2} \\ &= \frac{P\{x_1 < X_1 \le x_1 + \mathrm{d}x_1, x_2 < X_2 \le x_2 + \mathrm{d}x_2; t_1, t_2\}}{\mathrm{d}x_1\,\mathrm{d}x_2}. \end{aligned} \tag{6.6}$$

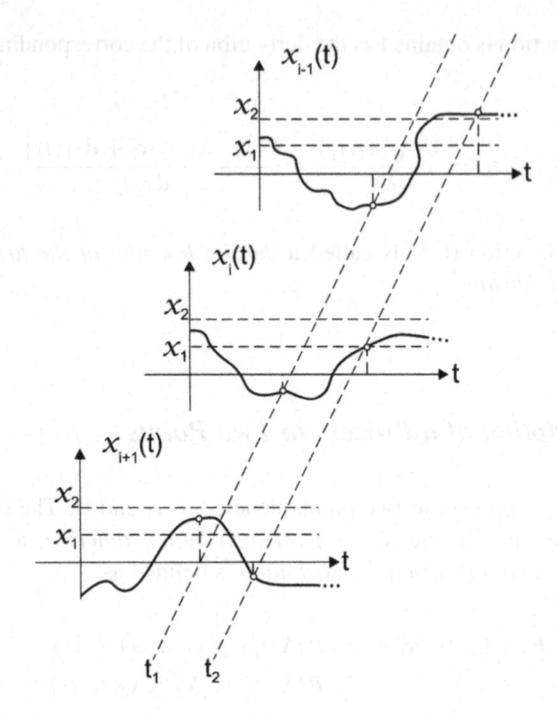

Fig. 6.5 Description of process in two time instants

6.2.3 *Description of Process in* n *Points*

The previous discussion can be generalized by observing a process in n points. Obviously, in this way we can obtain a more detailed description of a process. However, the resulting description will be highly complex.

By observing a process in n points, $t_1, \ldots, t_n$, we get n random variables:

$$X_1 = X(t_1), X_2 = X(t_2), \ldots, X_n = X(t_n). \tag{6.7}$$

We define a *joint distribution function* of nth order, as:

$$\begin{aligned} &F_{X_1X_2\ldots X_n}(x_1, x_2, \ldots, x_n; t_1, t_2, \ldots, t_n) \\ &= P\{X(t_1) \leq x_1, X(t_2) \leq x_2, \ldots, X(t_n) \leq x_n\} \\ &= P\{X_1 \leq x_1, X_2 \leq x_2, \ldots, X_n \leq x_n; t_1, t_2, \ldots, t_n\}. \end{aligned} \tag{6.8}$$

The corresponding *joint density function* of nth order is given as:

$$\begin{aligned} &f_{X_1X_2,\ldots,X_n}(x_1, x_2, \ldots, x_n; t_1, t_2, \ldots . t_n) = \frac{\partial^n F_{X_1X_2}(x_1, x_2, \ldots, x_n; t_1, t_2, \ldots, t_n)}{\partial x_1 \partial x_2 \ldots \partial x_n} \\ &= \frac{P\{x_1 < X(t_1) \leq x_1 + \mathrm{d}x_1, x_2 < X(t_2) \leq x_2 + \mathrm{d}x_2, \ldots, x_n < X(t_n) \leq x_n + \mathrm{d}x_n; t_1, t_2, \ldots, t_n\}}{\mathrm{d}x_1 \mathrm{d}x_2, \ldots, \mathrm{d}x_n}. \\ &= \frac{P\{x_1 < X_1 \leq x_1 + \mathrm{d}x_1, x_2 < X_2 \leq x_2 + \mathrm{d}x_2, \ldots . x_n < X_n \leq x_n + \mathrm{d}x_n; t_1, t_2, \ldots, t_n\}}{\mathrm{d}x_1 \mathrm{d}x_2, \ldots, \mathrm{d}x_n} \end{aligned} \tag{6.9}$$

Note that the only difference in the previously considered n-dimensional variable is that in this case the distribution as well as the density function are functions of the time instants, $t_1, \ldots, t_n$.

If we know an n-dimensional density then all densities of a lower order are also known, as demonstrated below:

$$f_{X_1X_2\ldots X_{n-1}}(x_1,\ldots,x_{n-1};t_1,\ldots t_{n-1}) = \int_{-\infty}^{\infty} f_{X_1X_2\ldots X_n}(x_1,\ldots,x_n;t_1,\ldots t_n)\mathrm{d}x_n.$$

$$f_{X_1X_2\ldots X_{n-2}}(x_1,\ldots,x_{n-2};t_1,\ldots,t_{n-2}) = \int_{-\infty}^{\infty} f_{X_1X_2\ldots X_{n-1}}(x_1,\ldots,x_{n-1};t_1,\ldots t_{n-1})\mathrm{d}x_{n-1}$$

$$\ldots \quad \ldots \quad \ldots$$

$$f_{X_1}(x_1;t_1) = \int_{-\infty}^{\infty} f_{X_1X_2}(x_1,x_2;t_1,t_2)\mathrm{d}x_2 \tag{6.10}$$

The calculation of n-dimensional distribution and density functions is a very complex task. Fortunately, in many practical situations, it is enough to observe a process in one or two time instants and then deal with a one-dimensional PDF and a two-dimensional PDFs.

6.3 Stationary Random Processes

A process is called *stationary* if none of its statistical characteristics change with time [PEE93, p. 168].

We can define different types of stationary processes according to observed statistical characteristics [PEE93, pp. 169–170].

A process is said to be a *first-order stationary process* if its first density function is independent of a time shifting Δ, that is:

$$f_X(x,t) = f_X(x,t+\Delta), \tag{6.11}$$

for any t and Δ.

Therefore,

$$f_X(x,t) = f_X(x). \tag{6.12}$$

Similarly, a process is called a *second-order stationary process* if its second order density has the following property for any value of t_1, t_2, and Δ:

$$f_{X_1X_2}(x_1,x_2;t_1,t_2) = f_{X_1X_2}(x_1,x_2;t_1+\Delta,t_2+\Delta). \tag{6.13}$$

Note that, in this case, the joint density function (6.13) is dependent on a time difference between time instants t_2 and t_1,

$$\tau = t_2 - t_1 \tag{6.14}$$

and can thus be expressed as:

$$f_{X_1X_2}(x_1, x_2; t_1, t_2) = f_{X_1X_2}(x_1, x_2; t_1 + \Delta, t_2 + \Delta) = f_{X_1X_2}(x_1, x_2; \tau). \tag{6.15}$$

By generalizing the previous discussion to a case of n random variables, X_i, $i = 1, \ldots, n$, we can define an *n-order stationary process.* In this case, an n-dimensional joint density function has the following property, for any values of $t_1, \ldots, t_n$ and Δ:

$$f_{X_1X_2,\ldots,X_n}(x_1, x_2, \ldots, x_n; t_1, t_2, \ldots, t_n) = f_{X_1X_2}(x_1, x_2, \ldots, x_n; t_1 + \Delta, t_2 + \Delta, \ldots, t_n + \Delta). \tag{6.16}$$

An n-order stationary process is also a k-order stationary process, where $k = 1, \ldots, n - 1$.

Example 6.3.1 Consider a random process

$$Y(t) = X_1 \cos \omega_0 t + X_2 \sin \omega_0 t, \tag{6.17}$$

where ω_0 is a constant frequency, and where X_1 and X_2 are normal independent random variables with parameters:

$$\begin{aligned} \overline{X_1} = \overline{X_2} = 0 \\ \sigma_1^2 = \sigma_2^2 = \sigma^2. \end{aligned} \tag{6.18}$$

Determine if this process is a first-order stationary process.

Solution The process $Y(t)$ is a linear combination of normal variables X_1 and X_2 in a time instant t and, consequently, is itself a normal random variable.

We can find the mean value and variance of a normal variable Y in a time instant t:

$$\begin{aligned} \overline{Y(t)} = \overline{X_1} \cos \omega_0 t + \overline{X_2} \sin \omega_0 t = 0, \\ \sigma_Y^2 = \sigma_1^2 \cos^2 \omega_0 t + \sigma_2^2 \sin^2 \omega_0 t = \sigma^2. \end{aligned} \tag{6.19}$$

Knowing that the normal PDF is completely determined by the parameters given in (6.19) which are both constants, it follows that the PDF of the variable Y does not depend on time:

$$f_Y(y, t) = f_Y(y) = \frac{1}{\sqrt{2\pi}\sigma} e^{-\frac{y^2}{2\sigma^2}}. \tag{6.20}$$

Therefore, the process $Y(t)$ is a first-order stationary process.

In many practical situations, it is necessary to consider only the first or second-order stationary processes. This can be further simplified by introducing a term of a wide-sense stationary process, considering a single process, or jointly wide-sense stationary processes. Wide-sense and jointly wide-sense processes are defined in Sect. 6.5.

6.4 Mean Value

A mean value considered in Chap. 2 for a single variable can also be introduced for a process observed in a particular time instant, as shown in Fig. 6.6. The process in a time instant t_1 is a random variable $X_1(t_1) = X_1$. Let the continuous range of variable X_1 be divided into k elements Δx, such that they are so small that if the

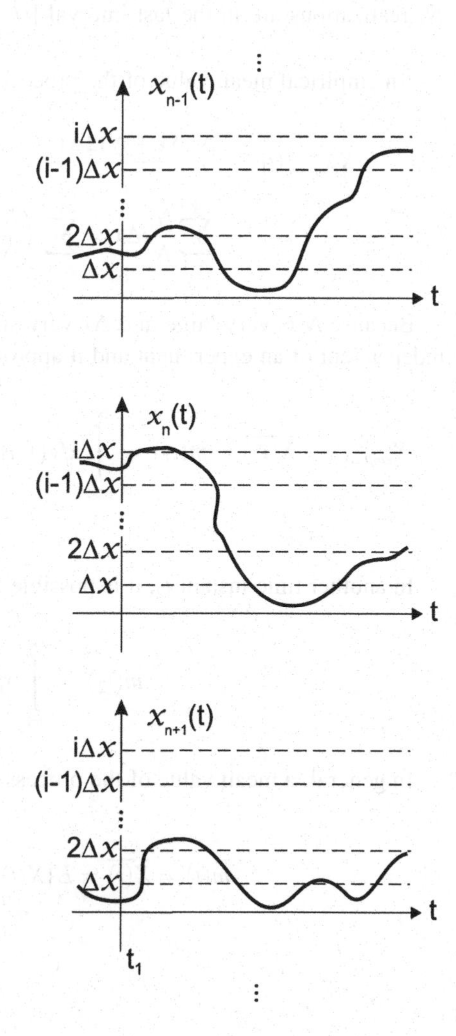

Fig. 6.6 Explanation of mean value of process

variable is in the interval Δx we say that it is equal to Δx. Similarly, if the variable is in the ith interval $i\Delta x$ we say that it is equal to $i\Delta x$.

Similarly, as in the case of a single variable, we first find an empirical mean value, as a result of a performed experiment.

The experiment is repeated N times under the same conditions. Process $X(t)$ is obtained by assigning N realizations $x(t)$ to each experiment's outcome. Consider all realizations in the time instant t_1, and suppose that we know the following:

N_1 realizations are in the interval $[0, \Delta x]$(like realization $x_{n+1}(t)$ in Fig.6.6),
N_2 realizations are in the interval $[\Delta x, 2\Delta x]$(like realization $x_{n-1}(t)$ in Fig.6.6),
...
N_i realizations are in the interval $[(i-1)\Delta x, i\Delta x]$ (like realization $x_n(t)$ in Fig.6.6),
...
N_k realizations are in the last interval $[(k-1\Delta x, k\Delta x)]$.

An empirical mean value of the process in the time instant t_1 is equal to:

$$X_{emp.av}(t_1) = \frac{N_1\Delta x + N_2 2\Delta x + \cdots + N_i i\Delta x + \cdots + N_k k\Delta x}{N}$$
$$= \sum_{i=1}^{k} \frac{N_i}{N} i\Delta x = \sum_{i=1}^{k} \left(\frac{N_i}{N}\frac{1}{\Delta x}\right) i\Delta x\Delta x. \tag{6.21}$$

Because N is very large and Δx very small, an empirical mean value becomes independent of an experiment and it approaches the mean value of a process:

$$X_{emp.av.} \to \overline{X(t_1)} = m(t_1) = \int_{-\infty}^{\infty} x f_X(x; t_1)\mathrm{d}x, \text{ as } N \to \infty,\ \Delta x \to \mathrm{d}x, i\Delta x \to x \tag{6.22}$$

In another time instant t_2, it is possible for us to obtain a different mean value:

$$m(t_2) = \int_{-\infty}^{\infty} x f_X(x; t_2)\mathrm{d}x. \tag{6.23}$$

In general, a mean value of the process depends on time:

$$m(t) = \overline{X(t)} = E\{X(t)\} = \int_{-\infty}^{\infty} x f_X(x; t)\mathrm{d}x. \tag{6.24}$$

If a process is a first-order stationary process then,

$$f_X(x,t) = f_X(x), \tag{6.25}$$

and, consequently, the mean value is a constant,

$$m(t) = \text{const.} \tag{6.26}$$

Therefore, an *expected or mean value of a first-order stationary process is a constant.*

Example 6.4.1 For a given process $X(t)$,

$$X(t) = \cos(\omega_0 t + \theta), \tag{6.27}$$

where ω_0 is a constant and θ is a uniform random variable,

$$f_\theta(\theta) = \begin{cases} \frac{1}{2\pi} & \text{for} \quad -\pi \le \theta \le \pi \\ 0 & \text{otherwise} \end{cases} \tag{6.28}$$

determine whether or not the process is a first-order stationary process.

Solution The mean value of the process is

$$\overline{X(t)} = \frac{1}{2\pi} \int_{-\pi}^{\pi} \cos(\omega_0 t + \theta) \mathrm{d}\theta = 0. \tag{6.29}$$

Therefore, the mean value does not depend on time. Consequently, the process is a first-order stationary process.

6.5 Autocorrelation Function

6.5.1 Definition

Statistical analysis of a random process is performed using characteristics, which characterize all of its process. One of the most important characteristic is the autocorrelation function.

An autocorrelation function (ACF) is defined as an expected value of the random variables of the process defined in the time instants t_1 and t_2,

$$R_{XX}(t_1,t_2) = \overline{X(t_1)X(t_2)} = \overline{X_1 X_2}, \tag{6.30}$$

or equivalently:

$$R_{XX}(t_1,t_2)=\int_{-\infty}^{\infty} x_1x_2f_{X_1X_2}(x_1,x_2;t_1,t_2)\mathrm{d}x_1\mathrm{d}x_2. \tag{6.31}$$

Note that an autocorrelation function is generally dependent on time instants t_1 and t_2. In many practical problems, solutions can be simplified if we can assume that an autocorrelation function does not depend on time instants t_1 and t_2 but rather on their difference $\tau = t_2 - t_1$. As mentioned in Sect. 6.3, a second-order stationary process will satisfy this condition. However, knowing that a mean value and autocorrelation function are the most important characteristics of a random process it is useful to define a less restrictive form of stationarity involving only those characteristics in stationary conditions of a process.

6.5.2 *WS Stationary Processes*

A process is said to be *wide sense (WS) stationary* if the two following conditions are satisfied: the mean value is constant and the autocorrelation function depends only on the difference of the time instants, $\tau = t_2 - t_1$:

$$\text{(a)}\quad \overline{X(t)}=\text{const.} \tag{6.32}$$

$$\text{(b)}\quad R_{XX}(\tau)=\overline{X(t)X(t+\tau)}=\int_{-\infty}^{\infty} x_1x_2f_{X_1X_2}(x_1,x_2;\tau)\mathrm{d}x_1\,\mathrm{d}x_2. \tag{6.33}$$

Example 6.5.1 Find whether a random process given by

$$X(t)=X_1\cos\omega_0 t+X_2\sin\omega_0 t \tag{6.34}$$

is stationary in the wide sense, if ω_0 is constant, the variables X_1 and X_2 are uncorrelated and have the mean values 0, and equal variances:

$$\begin{aligned}\overline{X_1}&=\overline{X_2}=0,\\ \sigma_1^2&=\sigma_2^2=\sigma^2.\end{aligned} \tag{6.35}$$

Solution In order to determine if the first condition (6.32) is satisfied, we must find the mean value of the process:

$$\overline{X(t)}=\overline{X_1\cos\omega_0 t+X_2\sin\omega_0 t}=\overline{X_1}\cos\omega_0 t+\overline{X_2}\sin\omega_0 t=0. \tag{6.36}$$

The mean value is zero and, thus constant. Therefore, the first condition (6.32) is satisfied.

Next we must verify whether or not the autocorrelation function depends only on τ,

$$\begin{aligned}
R_{XX}(t,t+\tau) &= \overline{X(t)X(t+\tau)} \\
&= \overline{(X_1\cos\omega_0 t + X_2\sin\omega_0 t)(X_1\cos\omega_0(t+\tau) + X_2\sin\omega_0(t+\tau))} \\
&= \overline{X_1^2\cos\omega_0 t\cos\omega_0(t+\tau)} + \overline{X_2^2\sin\omega_0 t\sin\omega_0(t+\tau)} \\
&\quad + \overline{X_1X_2\cos\omega_0 t\sin\omega_0(t+\tau)} + \overline{X_1X_2\sin\omega_0 t\cos\omega_0(t+\tau)}.
\end{aligned} \tag{6.37}$$

Knowing that variables X_1 and X_2 are uncorrelated, and using (6.35) we get:

$$\overline{X_1X_2} = \overline{X_1}\ \overline{X_2} = 0. \tag{6.38}$$

From (6.35), we have:

$$\begin{aligned}
\overline{X_1^2} &= \sigma_1^2 = \sigma^2, \\
\overline{X_2^2} &= \sigma_2^2 = \sigma^2.
\end{aligned} \tag{6.39}$$

Finally, from (6.37) to (6.39), we arrive at:

$$\begin{aligned}
R_{XX}(t,t+\tau) &= \sigma^2[\cos\omega_0 t\cos\omega_0(t+\tau) + \sin\omega_0 t\sin\omega_0(t+\tau)] \\
&= \sigma^2\cos\omega_0\tau.
\end{aligned} \tag{6.40}$$

The obtained result shows that the second condition (6.33) is satisfied as well, indicating that the process is a WS stationary process.

6.5.3 *What Does Autocorrelation Function Tell Us and Why Do We Need It?*

From the definition of an autocorrelation function, we know that an autocorrelation function is a statistical characteristic of a process that is obtained by observing the process in two points, the same as a joint density function of a second order. The question which arises is: "why do we need an autocorrelation function if we already have the second order joint density function?"

In order to answer this question, the following discussion is in order.

Consider two processes $X(t)$ and $Y(t)$, shown in Fig. 6.7.

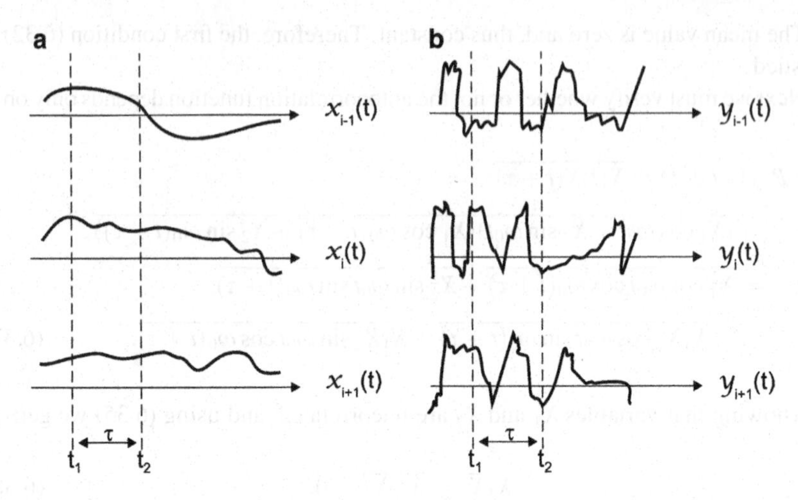

Fig. 6.7 Slow and fast processes *X* and *Y*

The process *X*(*t*) is assumed to be a WS stationary and, as a such, has the first order density function of:

$$f_X(x,t) = f_X(x). \tag{6.41}$$

The process *Y*(*t*) is obtained from the process *X*(*t*) by accelerating it. Therefore, both processes will have the same amplitude characteristics, i.e., both will have the same density function:

$$f_X(x) = f_Y(y). \tag{6.42}$$

However, even though both processes have the same statistical characteristic, that of a first order PDF, they are obviously different. The process *X*(*t*) changes in amplitude slowly over time, while the process *Y*(*t*) changes in amplitude very quickly. As a result, amplitudes of the process *X*(*t*) will be more similar to each other in the time instants t_1 and t_2, than will the amplitudes of the process *Y*(*t*) in the same time instants. In order to measure a similarity of amplitudes in time instants t_1 and t_2, thus obtaining a measure of a change of the process in time, we need an autocorrelation function.

The autocorrelation functions are given in the following equations and are shown in Fig. 6.8.

$$R_{XX}(\tau) = \overline{X(t)X(t+\tau)} = \int_{-\infty}^{\infty} x_1 x_2 f_{X_1X_2}(x_1, x_2; \tau)\mathrm{d}x_1\,\mathrm{d}x_2, \tag{6.43}$$

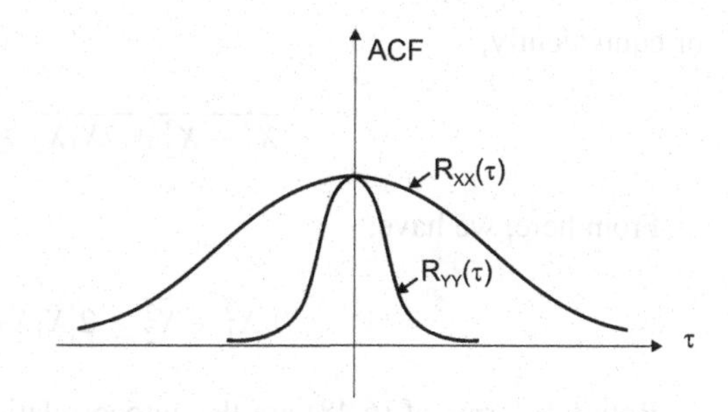

Fig. 6.8 Autocorrelation functions for slow ($R_{XX}(\tau)$) and fast ($R_{YY}(\tau)$) processes

$$R_{YY}(\tau) = \overline{Y(t)Y(t+\tau)} = \int_{-\infty}^{\infty} y_1 y_2 f_{X_1 X_2}(y_1, y_2; \tau) \mathrm{d}y_1 \, \mathrm{d}y_2. \tag{6.44}$$

From (6.43) and Fig. 6.7, we can note that the product $x_1 x_2$ will be positive in the majority of realizations, because the amplitudes of x_1 and x_2 will have the same sign in the majority of cases.

On the contrary, amplitudes of y_1 and y_2 for the process *Y*(*t*) will sometimes be with the same and sometimes with a contrary sign.

As a consequence, the average value $\overline{X_1 X_2}$ will be larger than that of $\overline{Y_1 Y_2}$, for the same value of τ.

Additionally, the random variables X_1 and X_2 still have a high correlation for higher values of τ in contrast with the variables Y_1 and Y_2 where the correlation is lost for small values of τ.

In this way, an autocorrelation function gives us information about whether process changes slowly or quickly, i.e., it gives us information about the rate of change of a process.

6.5.4 *Properties of Autocorrelation Function for WS Stationary Processes*

An autocorrelation function of a WS stationary process has the following properties:

P.1 The maximum value of the autocorrelation function is at $\tau = 0$,

$$|R_{XX}(\tau)| \leq R_{XX}(0). \tag{6.45}$$

Consider the random variables X_1 and X_2 of the process *X*(*t*) in the time instants t_1 and t_2, respectively. We have:

$$\overline{(X_1 \pm X_2)^2} \geq 0, \tag{6.46}$$

or equivalently,

$$\overline{X_1^2 + X_2^2 \pm 2X_1X_2} \geq 0. \tag{6.47}$$

From here, we have:

$$\overline{X_1^2} + \overline{X_2^2} \geq 2\left|\overline{X_1X_2}\right|. \tag{6.48}$$

Both left terms of (6.48) are the autocorrelation functions for $\tau = 0$, as shown below:

$$\begin{aligned} \overline{X_1(t_1)X_1(t_1+0)} &= \overline{X_1^2} = R_{XX}(0), \\ \overline{X_2(t_2)X_2(t_2+0)} &= \overline{X_2^2} = R_{XX}(0). \end{aligned} \tag{6.49}$$

Knowing that the right-side term from (6.48) is equal to $|R_{XX}(\tau)|$, where $\tau = t_2 - t_1$, and using (6.49), we obtain the desired result:

$$2R_{XX}(0) \geq 2|R_{XX}(\tau)|. \tag{6.50}$$

P.2 An autocorrelation function is an even function,

$$R_{XX}(-\tau) = R_{XX}(\tau). \tag{6.51}$$

For a WS process, we can write:

$$R_{XX}(\tau) = E\{X(t_1)X(t_2)\} = E\{X(t_3)X(t_4)\}, \tag{6.52}$$

where

$$t_1 = t, \qquad t_2 = t + \tau. \tag{6.53a}$$

$$t_3 = t, \qquad t_4 = t - \tau. \tag{6.53b}$$

Placing (6.53b) into (6.52), we get:

$$R_{XX}(\tau) = E\{X(t_3)X(t_4)\} = E\{X(t)X(t-\tau)\} = R_{XX}(-\tau), \tag{6.54}$$

P.3 The value of an autocorrelation function in $\tau = 0$ is a mean squared value of a process, that is a power of the process,

$$R_{XX}(0) = \overline{X^2(t)}. \tag{6.55}$$

If the mean value of the process is zero, $\overline{X(t)} = 0$, then

$$R_{XX}(0) = \overline{X^2(t)} = \sigma_X^2, \tag{6.56}$$

where σ_X^2 is the variance of the process.

For $\tau = 0$, from (6.33), we can write,

$$R_{XX}(0) = \overline{X(t)X(t+0)} = \overline{X^2(t)} = \sigma_X^2. \tag{6.57}$$

P.4 If the mean value of the process is not zero, then the autocorrelation function has a constant term which is equal to the squared mean value:

$$R_{XX}(\tau) = R_{X'X'}(\tau) + \overline{X(t)}^2 \tag{6.58}$$

where $X'(t)$ is a zero-mean process obtained from the process $X(t)$ by subtracting the mean value c.

Consider the WS process $X(t)$, where a mean value is a constant c, $E\{X(t)\} = c$, We can write:

$$X(t) = c + X'(t), \tag{6.59}$$

The autocorrelation function of the process $X(t)$ is

$$\begin{aligned} R_{XX}(\tau) &= E\{X(t)X(t+\tau)\} = E\{[c + X'(t)][c + X'(t+\tau)]\} \\ &= E\{c^2 + X'(t)X'(t+\tau) + cX'(t) + cX'(t+\tau)\} \\ &= c^2 + E\{X'(t)X'(t+\tau)\} = c^2 + R_{X'X'}(\tau) \end{aligned} \tag{6.60}$$

P.5 If a random process is a periodic process, then its autocorrelation function also has a periodic component of the same period as the process itself.

(a) Consider a periodic process $X(t) = A\cos(\omega_0 t + \theta)$ where A, and $\omega_0 = 2\pi/T$, where T is a period, are constants, and θ is a uniform random variable in the range $[0, 2\pi]$.

The autocorrelation function is:

$$\begin{aligned} R_{XX}(\tau) &= \overline{X(t)X(t+\tau)} = \overline{A\cos(\omega_0 t + \theta)A\cos(\omega_0(t+\tau) + \theta)} \\ &= \overline{\frac{A^2}{2}\cos(\omega_0 t - \omega_0 t - \omega_0 \tau) + \frac{A^2}{2}\cos(\omega_0(2t+\tau) + \theta)} \end{aligned} \tag{6.61}$$

We can easily find that the second term is equal to zero, resulting in:

$$R_{XX}(\tau) = \frac{A^2}{2}\cos\omega_0\tau. \tag{6.62}$$

The obtained result shows that the autocorrelation function is also a periodic function with the same period T.

(b) Next, we consider a process which has the periodic component $A\cos(\omega_0 t + \theta)$ (the same as in (a)) and a zero-mean, nonperiodic component $X'(t)$,

$$X(t) = A\cos(\omega_0 t + \theta) + X'(t). \quad (6.63)$$

Both components are independent.
The autocorrelation function is:

$$\begin{aligned}R_{XX}(\tau) &= \overline{X(t)X(t+\tau)} = \overline{[A\cos(\omega_0 t + \theta) + X'(t)][A\cos(\omega_0(t+\tau) + \theta) + X'(t+\tau)]}\\ &= \overline{\frac{A^2}{2}\cos(\omega_0 t - \omega_0 t - \omega_0\tau)} + \overline{\frac{A^2}{2}\cos(\omega_0(2t+\tau) + \theta)} + \overline{X'(t)X'(t+\tau)}\\ &+ \overline{X'(t)A(\cos\omega_0(t+\tau) + \theta)} + \overline{A\cos(\omega_0 t + \theta)X'(t+\tau)}.\end{aligned} \quad (6.64)$$

Knowing that the components of the process are independent, and using the results from (6.62), we arrive at:

$$R_{XX}(\tau) = \frac{A^2}{2}\cos\omega_0\tau + R_{X'X'}(\tau). \quad (6.65)$$

The obtained result again shows that the autocorrelation function has a periodic component with the same period as does a periodic component of the process.

(c) Finally, we consider a case in which the process has two independent periodic components with periods T_1 and T_2,

$$X(t) = A_1\cos(\omega_1 t + \theta) + A_2\cos(\omega_2 t + \theta), \quad (6.66)$$

where $A_1, A_2, \omega_1, \omega_2$ are constants, θ is a uniform random variable in the range $[0, 2\pi]$, $\omega_1 = 2\pi/T_1$, and $\omega_2 = 2\pi/T_2$.
The autocorrelation function is:

$$\begin{aligned}R_{XX}(\tau) = E\{&[A_1\cos(\omega_1 t + \theta) + A_2\cos(\omega_2 t + \theta)]\\ &\times [A_1\cos(\omega_1(t+\tau) + \theta) + A_2\cos(\omega_2(t+\tau) + \theta)]\}.\end{aligned} \quad (6.67)$$

Using the result described in the case (a) we can easily obtain:

$$R_{XX}(\tau) = \frac{A_1^2}{2}\cos\omega_1\tau + \frac{A_2{}^2}{2}\cos\omega_2\tau. \quad (6.68)$$

The obtained result shows that the autocorrelation function of the sum of independent periodic components is equal to the sum of the autocorrelation functions of the same periodic components.

P.6 If a process does not have a periodic component and its mean value is zero, then its autocorrelation function approaches to zero for high values of lag τ,

$$\lim_{|\tau| \to \infty} R_{XX}(\tau) = 0, \tag{6.69}$$

However, if a process has a nonzero mean value, then its autocorrelation function approaches a squared mean value, for high values of τ,

$$\lim_{|\tau| \to \infty} R_{XX}(\tau) = \overline{X}^2, \tag{6.70}$$

The proof is a self-explanatory knowing that the similarity of the amplitudes decreases as the distance of the time instants τ increases.

Example 6.5.2 A random process $X(t)$ consists of rectangular pulses with amplitudes U and $-U$ and random lengths, as shown in Fig. 6.9. The process has the following characteristics:

- The number of zero crossing in any interval is independent of the zero crossing in any other interval if the intervals are not overlapped either totally or partially.

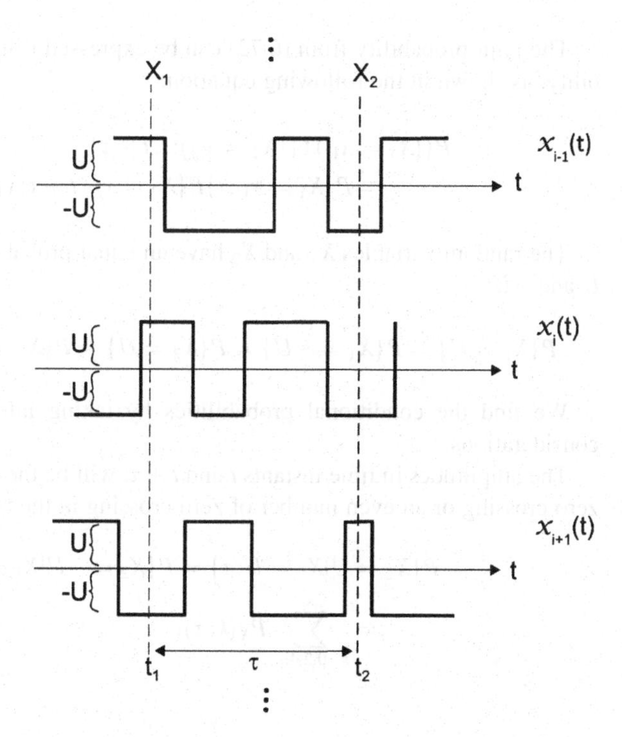

Fig. 6.9 Random process in Example 6.5.2

- The probability that more than one zero crossing in an infinitesimal interval appear is infinitely small compared to the probability of only one zero crossing.
- If a zero crossing occurs in the interval τ, then there is an equal probability of its occurring in any of the infinitesimal intervals $\mathrm{d}\tau$, in which the interval τ is divided.

Find the autocorrelation function and demonstrate its properties.

Solution According to the aforementioned properties of the given process, we can conclude that the process is a Poisson process (see Sect. 5.7).

Therefore, the number of the zero crossing in a given interval τ is described by the Poisson formula:

$$P\{k \text{ zero crossing in } \tau\} = \frac{(\lambda\tau)^k}{k!}\mathrm{e}^{-\lambda\tau}, \tag{6.71}$$

where λ is an average number of zero crossing in a unit of time.

Random variables in time instants t_1=t and t_2=$t + \tau$, are denoted as X_1 and X_2, respectively, as shown in Fig. 6.9. The autocorrelation function is:

$$R_{XX}(\tau) = \overline{X_1X_2} = \sum_{i=1}^{2}\sum_{j=1}^{2} x_{1i}x_{2j}P\{(X_1 = x_{1i}) \cap (X_2 = x_{2j}); \tau\}. \tag{6.72}$$

The joint probability from (6.72) can be expressed using the conditional probability, as shown in the following equation:

$$\begin{aligned} &P\{(X_1 = x_{1i}) \cap (X_2 = x_{2j}); t, t+\tau\} \\ &\quad = P\{X_1 = x_{1i}; t\}P\{X_2 = x_{2j}; t+\tau|X_1 = x_{1i}; t\}. \end{aligned} \tag{6.73}$$

The random variables X_1 and X_2 have an equal probability of taking the values U and $-U$:

$$P\{X_1 = U\} = P\{X_1 = -U\} = P\{X_2 = U\} = P\{X_2 = -U\} = 1/2. \tag{6.74}$$

We find the conditional probabilities by taking into account the following considerations.

The amplitudes in time instants t and $t + \tau$, will be the same if there is either no zero crossing or an even number of zero crossing in the time interval τ:

$$\begin{aligned} &P\{X_2 = U|X_1 = U; \tau\} = P\{X_2 = -U|X_1 = -U; \tau\} \\ &\quad = \sum_{k=0,2,4,\ldots}^{\infty} P_X(k; \tau), \end{aligned} \tag{6.75}$$

where $P_X(k;\tau)$ is the Poisson formula given in (6.71).

Similarly, the amplitudes of random variables X_1 and X_2 will have opposite signs, if there are an odd number of zero crossings in the interval τ:

$$P\{X_2 = U|X_1 = -U;\tau\} = P\{X_2 = -U|X_1 = U;\tau\} = \sum_{k=1,3,5,\ldots}^{\infty} P_X(k;\tau). \tag{6.76}$$

Finally, from (6.72) to (6.76), we get:

$$R_{XX}(\tau) = U^2 \sum_{k=0,2,4,\ldots}^{\infty} \frac{(\lambda|\tau|)^k}{k!} e^{-\lambda|\tau|} - U^2 \sum_{k=1,3,5,\ldots}^{\infty} \frac{(\lambda|\tau|)^k}{k!} e^{-\lambda|\tau|} = U^2 e^{-\lambda|\tau|}\left[\sum_{k=0,2,4,\ldots}^{\infty} \frac{(\lambda|\tau|)^k}{k!} - \sum_{k=1,3,5,\ldots}^{\infty} \frac{(\lambda|\tau|)^k}{k!}\right] = U^2 e^{-2\lambda|\tau|}. \tag{6.77}$$

Note that in the obtained result we have an absolute value of τ because the time interval τ in the Poisson formula (6.71) is always positive, while in an autocorrelation function time interval τ takes all values from $-\infty$ until $+\infty$.

The autocorrelation function is shown in Fig. 6.10 for $U = 1$ and $\lambda = 0.5$.

Next we verify the properties of the autocorrelation function for $U = 1$:

P.1 The autocorrelation function has its maximum value at $\tau = 0$, and it is equal to $U^2 = 1$.

P.2 From (6.77), it is clear that the autocorrelation function is an even function, i.e., $R_{XX}(\tau) = R_{XX}(-\tau)$.

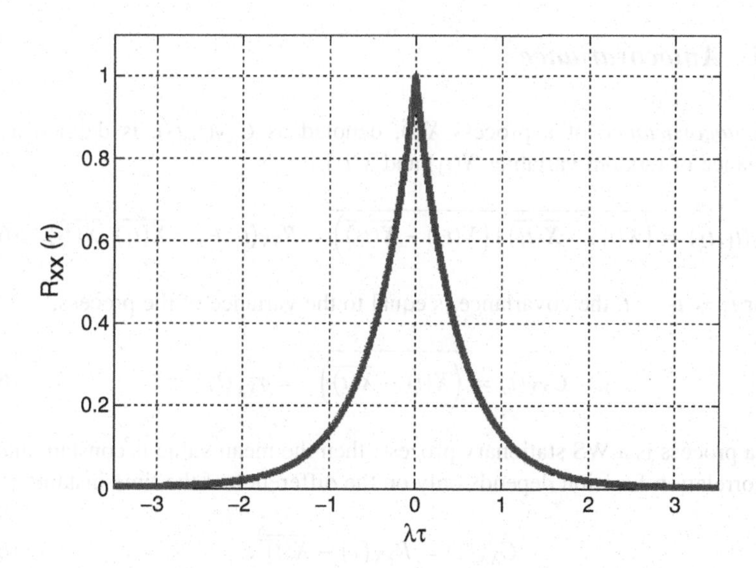

Fig. 6.10 Autocorrelation function in Example 6.5.2

P.3 The value of the autocorrelation function in $\tau = 0$ is the mean squared value of the process, i.e., the power of the process,

$$R_{XX}(0) = \overline{X^2(t)} = 1. \tag{6.78}$$

The mean squared value of the process is:

$$\overline{X^2(t)} = \frac{1}{2}U^2 + \frac{1}{2}(-U)^2 = U^2 = 1. \tag{6.79}$$

P.4 From (6.77), we note that the autocorrelation function does not have a constant term indicating that the mean value of the process is zero. The former is confirmed in the following.

$$\overline{X(t)} = \frac{1}{2}U + \frac{1}{2}(-U) = 0. \tag{6.80}$$

P.5 From (6.77), we can see that the autocorrelation function does not have a periodic component indicating that the process itself does not have a periodic component.

P.6 From (6.77), we have:

$$\lim_{|\tau|\to\infty} R_{XX}(\tau) = \lim_{|\tau|\to\infty} U^2 \mathrm{e}^{-2\lambda|\tau|} = 0. \tag{6.81}$$

6.5.5 *Autocovariance*

The *autocovariance* of a process $X(t)$, denoted as $C_{XX}(t_1,t_2)$, is defined as the covariance of random variables $X(t_1)$ and $X(t_2)$,

$$C_{XX}(t_1, t_2) = \overline{\left(X(t_1) - \overline{X(t_1)}\right)\left(X(t_2) - \overline{X(t_2)}\right)} = R_{XX}(t_1, t_2) - \overline{X(t_1)}\ \overline{X(t_2)}. \tag{6.82}$$

For $t_1 = t_2 = t$, the covariance is equal to the variance of the process,

$$C_{XX}(t) = \overline{\left(X(t) - \overline{X(t)}\right)^2} = \sigma_{XX}^2(t). \tag{6.83}$$

If a process is a WS stationary process, then the mean value is constant and the autocorrelation function depends only on the difference of the time instants τ,

$$C_{XX}(\tau) = R_{XX}(\tau) - \overline{X(t)}^2. \tag{6.84}$$

For $\tau = 0$, from (6.84), we have:

$$C_{XX}(0) = R_{XX}(0) - \overline{X(t)}^2 = \sigma_{XX}{}^2 = \text{const.} \tag{6.85}$$

Example 6.5.3 Find autocovariance and variance of a process with the following autocorrelation function,

$$R_{XX}(\tau) = 9 + \frac{1}{1+\tau^2}. \tag{6.86}$$

Solution According to (6.60), the constant term in (6.86) corresponds to the squared mean value:

$$9 = \overline{X(t)}^2 \tag{6.87}$$

resulting in:

$$E\{X(t)\} = \pm 3. \tag{6.88}$$

Additionally, the autocorrelation function (6.86) depends on the time distance τ, thus indicating that a process is a WS stationary process.

According to (6.55) and (6.86), we get:

$$R_{XX}(0) = \overline{X^2(t)} = 9 + \frac{1}{1+0} = 10. \tag{6.89}$$

From (6.85), we have:

$$\sigma_{XX}^2 = R_{XX}(0) - \overline{X(t)}^2 = 10 - 9 = 1. \tag{6.90}$$

The obtained result confirms that for a WS stationary process, variance is constant.

From (6.82), the autocovariance is equal to:

$$C_{XX}(\tau) = R_{XX}(\tau) - \overline{X(t)}^2 = 9 + \frac{1}{1+\tau^2} - 9 = \frac{1}{1+\tau^2}. \tag{6.91}$$

6.6 Cross-Correlation Function

6.6.1 Definition

When we are concerned with two random processes $X(t)$ and $Y(t)$ we define a *cross-correlation function* as:

$$R_{XY}(t_1, t_2) = E\{X(t_1)Y(t_2)\}, \tag{6.92}$$

where $X(t_1)$ and $Y(t_2)$ are the random variables of the processes $X(t)$ and $Y(t)$ in time instants t_1 and t_2, respectively.

We may also write a cross-correlation function as:

$$R_{YX}(t_1,t_2)=E\{Y(t_1)X(t_2)\}. \tag{6.93}$$

The cross-correlation functions in (6.92) and (6.93) are generally not equal.

Setting $t_1=t$ and $\tau=t_2-t_1$, we may write (6.92) and (6.93) in the following forms:

$$R_{XY}(t,t+\tau)=E\{X(t)Y(t+\tau)\}=\int_{-\infty}^{\infty}\int_{-\infty}^{\infty}xyf_{XY}(x,y;t,t+\tau)\mathrm{d}x\,\mathrm{d}y, \tag{6.94}$$

$$R_{YX}(t,t+\tau)=E\{Y(t)X(t+\tau)\}=\int_{-\infty}^{\infty}\int_{-\infty}^{\infty}xyf_{YX}(y,x;t,t+\tau)\mathrm{d}x\,\mathrm{d}y. \tag{6.95}$$

The processes are called *orthogonal* if their cross-correlation function is zero:

$$R_{XY}(\tau)=0. \tag{6.96}$$

If two processes are *statistically independent*, the cross-correlation function becomes:

$$R_{XY}(t,t+\tau)=E\{X(t)\}E\{Y(t+\tau)\}. \tag{6.97}$$

If, the processes (6.97) are additionally WS stationary then (6.97) can be rewritten as:

$$R_{XY}(t,t+\tau)=E\{X\}E\{Y\}=\text{const}. \tag{6.98}$$

6.6.2 Jointly WS Processes

Similarly, as in working with one process we may introduce more relaxed conditions for a second order stationarity of two processes, called a jointly wide-sense stationarity.

Two processes $X(t)$ and $Y(t)$ are said to be *jointly wide-sense stationary* if the two following conditions are satisfied:

(a) Both $X(t)$ and $Y(t)$ processes are wide stationary processes:

$$m_X=E\{X(t)\}=\text{const};\qquad m_Y=E\{Y(t)\}=\text{const}, \tag{6.99}$$

$$R_{XX}(\tau) = E\{X(t)X(t+\tau)\}, \qquad R_{YY}(\tau) = E\{Y(t)Y(t+\tau)\}. \tag{6.100}$$

(b) The cross-correlation functions (6.92) and (6.93) depend only on the time difference τ:

$$R_{XY}(\tau) = E\{X(t)Y(t+\tau)\} \tag{6.101}$$

or

$$R_{YX}(\tau) = E\{Y(t)X(t+\tau)\}. \tag{6.102}$$

Example 6.6.1 Consider two random processes

$$\begin{aligned} X(t) &= \xi \cos \omega_0 t + \rho \sin \omega_0 t, \\ Y(t) &= \rho \cos \omega_0 t - \xi \sin \omega_0 t, \end{aligned} \tag{6.103}$$

where ω_0 is a constant and ξ and ρ are uncorrelated random variables,

$$E\{\xi\rho\} = E\{\xi\}E\{\rho\}, \tag{6.104}$$

with zero mean values;

$$E\{\xi\} = E\{\rho\} = 0, \tag{6.105}$$

and equal variances

$$\sigma_\xi^2 = \sigma_\rho^2 = \sigma^2. \tag{6.106}$$

Determine whether the processes are jointly WS stationary.

Solution We first determine whether both processes are WS stationary.

Both processes have mean values of zero and, therefore, the first condition (6.99) is satisfied.

Next we will investigate whether both autocorrelation functions are only dependent on a time difference τ.

To this end we find:

$$\begin{aligned} R_{XX}(t, t+\tau) &= E\{X(t)X(t+\tau)\} \\ &= E\{[\xi \cos \omega_0 t + \rho \sin \omega_0 t][\xi \cos \omega_0(t+\tau) + \rho \sin \omega_0(t+\tau)]\} \\ &= E\{\xi^2\} \cos \omega_0 t \cos \omega_0(t+\tau) + E\{\rho^2\} \sin \omega_0 t \sin \omega_0(t+\tau) \\ &\quad + E\{\xi\rho\} \sin \omega_0 t \cos \omega_0(t+\tau) + E\{\xi\rho\} \cos \omega_0 t \sin \omega_0(t+\tau). \end{aligned} \tag{6.107}$$

From (6.104) to (6.106), we have

$$E\{\xi^2\} = E\{\rho^2\} = \sigma^2. \tag{6.108}$$

From the conditions (6.104) and (6.105), we can write:

$$E\{\xi\rho\} = 0 \tag{6.109}$$

resulting in the last two terms being zero in (6.107).

Finally, from (6.107), we arrive at:

$$R_{XX}(t, t+\tau) = \sigma^2 \cos(\omega_0 t - \omega_0 t - \omega_0 \tau) = \sigma^2 \cos(\omega_0 \tau) = R_{XX}(\tau). \tag{6.110}$$

In a similar way, we find

$$R_{YY}(t, t+\tau) = \sigma^2 \cos(\omega_0 t - \omega_0 t - \omega_0 \tau) = \sigma^2 \cos(\omega_0 \tau) = R_{YY}(\tau). \tag{6.111}$$

The obtained results indicate that both processes are WS stationary. Thus, the first condition (6.99)–(6.100) is satisfied.

Next we examine the second condition (6.101):

$$\begin{aligned} R_{XY}(t, t+\tau) &= E\{X(t)Y(t+\tau)\} \\ &= E\{[\xi \cos \omega_0 t + \rho \sin \omega_0 t][\rho \cos \omega_0 (t+\tau) - \xi \sin \omega_0 (t+\tau)]\} \\ &= -E\{\xi^2\} \cos \omega_0 t \sin \omega_0 (t+\tau) + E\{\rho^2\} \sin \omega_0 t \cos \omega_0 (t+\tau) \\ &\quad + E\{\xi\rho\} \cos \omega_0 t \cos \omega_0 (t+\tau) - E\{\xi\rho\} \sin \omega_0 t \sin \omega_0 (t+\tau) \\ &= \sigma^2 [\sin \omega_0 t \cos \omega_0 (t+\tau) - \cos \omega_0 t \sin \omega_0 (t+\tau)] = \sigma^2 \sin(\omega_0 t - \omega_0 t - \omega_0 \tau) \\ &= -\sigma^2 \sin(\omega_0 \tau) \end{aligned} \tag{6.112}$$

The obtained result shows that the cross-correlation function depends only on τ and therefore, the second condition is also satisfied and processes $X(t)$ and $Y(t)$ are jointly WS stationary.

6.6.3 *Properties of Cross-Correlation Function for Jointly WS Stationary Processes*

We will now discuss the properties of a cross-correlation function.

P.1 As opposed to an autocorrelation function, *there is no way to physically interpret $R_{XY}(0)$ and $R_{YX}(0)$. Additionally, the values $R_{XY}(0)$ and $R_{YX}(0)$ may not be the maximum values of corresponding cross-correlation functions.*

P.2 In general, a *cross-correlation function is neither even nor odd.* The following relation holds:

$$R_{XY}(-\tau) = R_{YX}(\tau). \qquad (6.113)$$

From the definition of a cross-correlation function (6.94)–(6.95), we write

$$R_{XY}(-\tau) = E\{X(t)Y(t-\tau)\}. \qquad (6.114)$$

Introducing

$$t' = t - \tau, \qquad (6.115)$$

into (6.114), we can write:

$$R_{XY}(-\tau) = E\{Y(t')X(t'+\tau)\} = R_{YX}(\tau). \qquad (6.116)$$

P.3 This property establishes the upper limit for a cross correlation function as a geometric mean of the corresponding autocorrelation functions at the origin:

$$|R_{XY}(\tau)| \leq \sqrt{R_{XX}(0)R_{YY}(0)}. \qquad (6.117)$$

For any two random variables, we can write:

$$[E\{XY\}]^2 \leq E\{X^2\}E\{Y^2\}. \qquad (6.118)$$

Then we have:

$$[E\{X(t)Y(t+\tau)\}]^2 \leq E\{X^2(t)\}E\{Y^2(t+\tau)\} \qquad (6.119)$$

resulting in the desired result (6.117).

P.4 This property establishes another upper limit for a cross-correlation function as an arithmetic mean of the corresponding autocorrelation functions in the origin.

This is a stronger limit than that of (6.117) because a geometric mean of two positive numbers cannot exceed an arithmetic mean of the same numbers:

$$|R_{XY}(\tau)| \leq \frac{1}{2}[R_{XX}(0) + R_{YY}(0)]. \qquad (6.120)$$

To establish this property, we write:

$$[X(t) \pm Y(t+\tau)]^2 = X^2(t) + Y^2(t) \pm 2X(t)Y(t+\tau) \geq 0. \qquad (6.121)$$

Taking the expectation of (6.121), we arrive at:

$$R_{XX}(0) + R_{YY}(0) \pm 2R_{XY}(\tau) \geq 0. \tag{6.122}$$

From here, we easily obtain the desired result (6.120).

Example 6.6.2 Verify the properties of the cross-correlation function $R_{XY}(\tau) = -\sigma^2 \sin\omega_0\tau$ from Example 6.6.1 for the variables X and Y.

Solution

P.1 For $\tau = 0$ the cross-correlation function becomes zero and thus, it does not take its maximum value for $\tau = 0$.

P.2 Next we investigate the property (6.116). To this end, we find $R_{XY}(-\tau)$ from the cross-correlation function (6.112):

$$R_{XY}(-\tau) = \sigma^2 \sin \omega_0 \tau. \tag{6.123}$$

We also find $R_{YX}(\tau)$,

$$\begin{aligned}
R_{YX}(\tau) &= E\{Y(t)X(t+\tau)\} \\
&= E\{[\rho \cos \omega_0 t - \xi \sin \omega_0 t][\xi \cos \omega_0 (t+\tau) + \rho \sin \omega_0 (t+\tau)]\} \\
&= E\{\rho^2\} \cos \omega_0 t \sin \omega_0 (t+\tau) - E\{\xi^2\} \sin \omega_0 t \cos \omega_0 (t+\tau) \\
&\quad + E\{\xi\rho\} \cos \omega_0 t \cos \omega_0 (t+\tau) - E\{\xi\rho\} \sin \omega_0 t \sin \omega_0 (t+\tau) \\
&= \sigma^2 [\cos \omega_0 t \sin \omega_0 (t+\tau) - \sin \omega_0 t \cos \omega_0 (t+\tau)] = \sigma^2 \sin(\omega_0 t + \omega_0 \tau - \omega_0 t) \\
&= \sigma^2 \sin(\omega_0 \tau)
\end{aligned} \tag{6.124}$$

Comparing (6.123) and (6.124), we see that the following is satisfied:

$$R_{XY}(-\tau) = R_{YX}(\tau). \tag{6.125}$$

P.3 In order to verify the property (6.117) we use the results (6.112) and (6.108):

$$R_{XX}(0) = R_{YY}(0) = \sigma^2 \tag{6.126}$$

resulting in the following:

$$|R_{XY}(\tau)| = \sigma^2 |\sin \omega_0 \tau| \leq \sqrt{R_{XX}(0) R_{YY}(0)} = \sigma^2, \tag{6.127}$$

As

$$|\sin \omega_0 \tau| \leq 1, \tag{6.128}$$

the inequality (6.117) is satisfied.

P.4 In order to verify the property (6.120), we use the results (6.126) and (6.112).

From (6.120), we write:

$$|R_{XY}(\tau)| = \sigma^2|\sin\omega_0\tau| \leq \frac{1}{2}[R_{XX}(0) + R_{YY}(0)] = \frac{1}{2}2\sigma^2 = \sigma^2 \tag{6.129}$$

resulting in (6.128), which is always true. Therefore, (6.120) is satisfied.

6.6.4 Cross-Covariance

A cross-covariance function for two processes $X(t)$ and $Y(t)$ is defined as:

$$C_{XY}(t, t+\tau) = E\{[X(t) - E\{X(t)\}][Y(t+\tau) - E\{Y(t+\tau)\}]\}. \tag{6.130}$$

This expression can be rewritten in the following form:

$$\begin{aligned} C_{XY}(t, t+\tau) &= E\{X(t)Y(t+\tau)\} - E\{X(t)\}E\{Y(t+\tau)\} \\ &\quad - E\{X(t)\}E\{Y(t+\tau)\} + E\{X(t)\}E\{Y(t+\tau)\} \\ &= E\{X(t)Y(t+\tau)\} - E\{X(t)\}E\{Y(t+\tau)\} \\ &= R_{XY}(t, t+\tau) - E\{X(t)\}E\{Y(t+\tau)\} \end{aligned} \tag{6.131}$$

If processes are at least jointly WS stationary, then (6.131) reduces to:

$$C_{XY}(\tau) = R_{XY}(\tau) - E\{X\}E\{Y\}. \tag{6.132}$$

If a cross-covariance is equal to zero,

$$C_{XY}(\tau) = 0, \tag{6.133}$$

then processes $X(t)$ and $Y(t)$ are called *uncorrelated*. Using (6.131), the following relation represents uncorrelated processes:

$$R_{XY}(t, t+\tau) = E\{X(t)\}E\{Y(t+\tau)\}. \tag{6.134}$$

Similarly, if uncorrelated processes are at least jointly WS stationary, from (6.132) we have:

$$R_{XY}(\tau) = E\{X\}E\{Y\}. \tag{6.135}$$

The same results (6.134) and (6.135) also stand for independent process. In other words, *independent processes are also uncorrelated*. However, the opposite is not necessarily true (except in the case of jointly Gaussian processes).

6.7 Ergodic Processes

6.7.1 Time Averaging

In the previous discussion, we observed processes (all realizations) in one or more points. The obtained average values, like mean value and autocorrelation function are *ensemble average values*.

Next we consider time average values obtained by averaging particular realizations of the process in time. Similarly, as we introduced the notation $E\{\}$ for the ensemble averaging, we introduce the notation $A\{\}$ for the averaging of time.

A *time average* of a function $g(x(t))$ is defined as:

$$A\{g(x(t))\} = \lim_{T\to\infty} \frac{1}{T} \int_{-T/2}^{T/2} g(x(t))\mathrm{d}t. \tag{6.136}$$

Two time average values: mean time value and time autocorrelation function have a special importance. Those mean values are obtained observing a realization $x(t)$ of the process in time.

A *mean time value* is defined as:

$$A\{x(t)\} = \lim_{T\to\infty} \frac{1}{T} \int_{-T/2}^{T/2} x(t)\mathrm{d}t = \overline{x(t)}^{t}. \tag{6.137}$$

Similarly, we can define time autocorrelation function for a realization $x(t)$ as:

$$R_{xx}(\tau) = A\{x(t)x(t+\tau)\} = \overline{x(t)x(t+\tau)}^{t} = \lim_{T\to\infty} \frac{1}{T} \int_{-T/2}^{T/2} x(t)x(t+\tau)\mathrm{d}t, \tag{6.138}$$

where $\overline{x(t)}^{t}$ presents the averaging of a realization $x(t)$ of the process $X(t)$ in time.

6.7.2 What Is Ergodicity?

A process is said to be *ergodic* if all of its statistical averages are equal to its corresponding time averages. From (6.137), it is clear that for an ergodic process, any time average cannot be a function of time. In other words, in ergodic processes the corresponding statistical averages cannot be functions of time. Knowing that only stationary processes possess such characteristics, we can conclude that an ergodic process must be stationary. However, the opposite is not true (i.e., there are stationary processes which are not ergodic processes). As an example of this, consider the process of DC voltages, shown in Fig. 6.11. This process is obviously stationary because the statistical characteristics are independent of the time instants in which the process is observed. However, because the amplitudes of the realizations are different, the time averages will vary from one realization to another. Therefore, the process is stationary but not ergodic.

The mutual relations between stationarity and ergodicity can be illustrated, as shown in Fig.6.12.

The ergodicity makes it possible to describe a process knowing only one realization. This is of practical interest because usually one has knowledge of only one time function of a process and, based on ergodicity, can find a statistical average of the process by finding the time average of just one realization of the process.

6.7.3 *Explanation of Ergodicity*

Ergodicity can be explained using the example of die rolling. If we consider an experiment in which one die is rolled many times, then we intuitively can expect the same result as in an experiment in which a large number of dice are rolled only once.

Similarly, we can expect the same result if the measurement of one single source of random voltage is repeated many times. This can be seen as the simultaneous measurements of the random voltages of a large number of independent, identical random voltage sources. The set of all voltages generated by the independent sources is an ensemble or process while the particular measured random voltages are the realizations of the process.

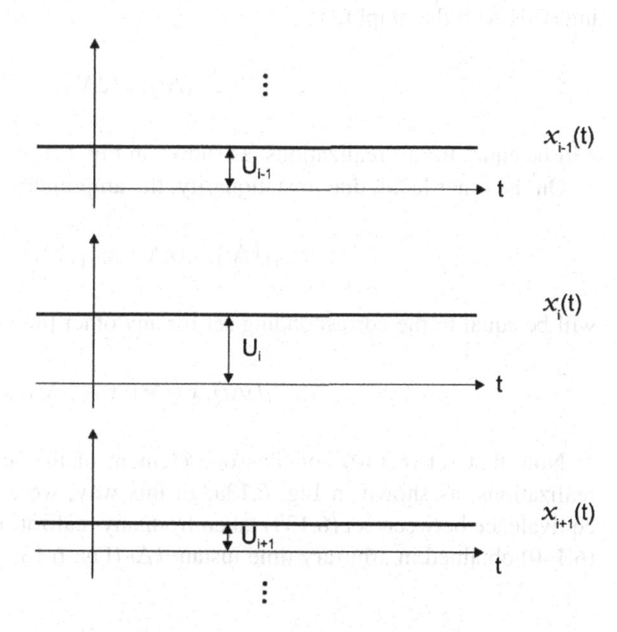

Fig. 6.11 Example of a stationary process that is not ergodic

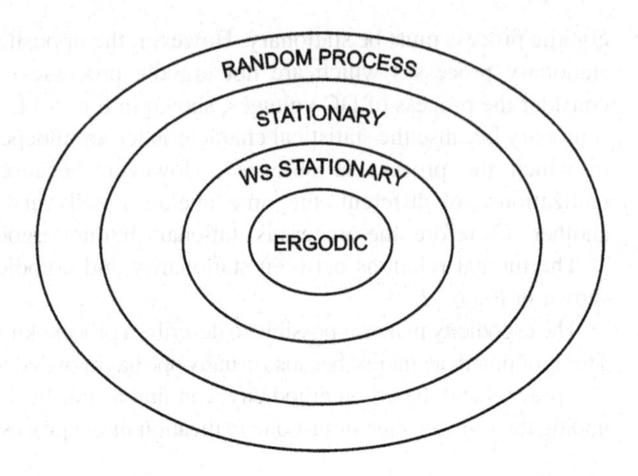

Fig. 6.12 Stationarity and ergodicity

Since the particular realizations of ergodic process are generated by equal independent sources, we can expect that, with enough time, all time functions (realizations) will pass the same number of times the same amplitude values, but with a different order of appearance. This is shown in Fig.6.13a, where time realizations are divided into small intervals [LEE 60].

If all amplitudes from Fig. 6.13a are ordered using a criterion of decreasing amplitude and not as they appear, we will obtain the same realizations as shown in Fig. 6.13b.

In another words, if we imagine that each realization is divided into small intervals of infinitesimal duration Δt, then we can expect a very large set of those intervals with the amplitudes

$$\ldots, x_i(0), x_i(\Delta t), x_i(2\Delta t), \ldots \tag{6.139}$$

will be equal for all realizations, as shown in Fig. 6.13b.

On the other hand, due to stationarity, the amplitudes of the intervals in $k\Delta t$

$$\ldots, x_{i-1}(k\Delta t), x_i(k\Delta t), x_{i+1}(k\Delta t), \ldots \tag{6.140}$$

will be equal to the corresponding set for any other time interval $j\Delta t$,

$$\ldots, x_{i-1}(j\Delta t), x_i(j\Delta t), x_{i+1}(j\Delta t), \ldots \tag{6.141}$$

Note that set (6.140) contains one element of the set (6.139), taken from all realizations, as shown in Fig. 6.13a. In this way, we can conclude that there is equivalence between set (6.139) taken from any realization (Fig. 6.13b) and the set (6.140) obtained in arbitrary time instant $k\Delta t$ (Fig. 6.13c).

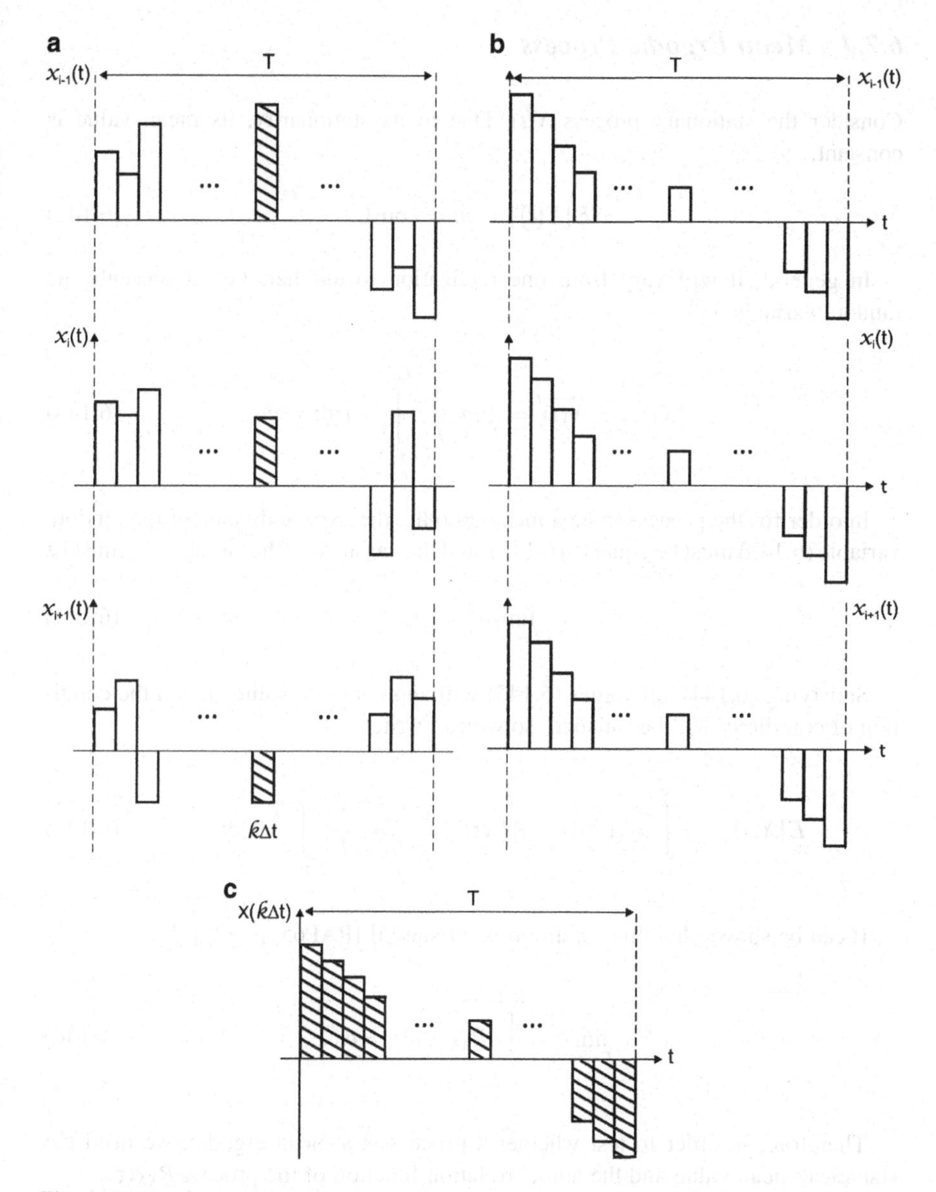

Fig. 6.13 Illustration of ergodicity

Consequently, there is an equivalence of the average value taken in the set (6.139) (time average) and an average value taken in the set (6.140) (statistical average).

In many practical cases, it is not necessary to consider all statistical characteristics but only some of them. Consequently, it is necessary to define ergodicity as related to a particular statistical characteristic. The special importance has the mean value and the autocorrelation function.

6.7.4 Mean Ergodic Process

Consider the stationary process $X(t)$. Due to its stationarity, its mean value is constant,

$$E\{X(t)\} = m = \text{const.} \tag{6.142}$$

In general, it will vary from one realization to another, i.e., it presents the random variable m_X,

$$A\{X(t)\} = \overline{X(t)}^t = \lim_{T\to\infty} \frac{1}{T} \int_{-T/2}^{T/2} x(t)\mathrm{d}t = m_X. \tag{6.143}$$

In order for the process to be a mean ergodic, the expected value of the random variable (6.143) must be equal to (6.142), and the variance of the variable m_X must be

$$\lim_{T\to\infty} \sigma_{m_X}^2 = 0. \tag{6.144}$$

Satisfying (6.144), all values (6.143) will approach the value m, and the condition of ergodicity will be satisfied, so we can write:

$$E\{X(t)\} = \int_{-\infty}^{\infty} x f_X(x)\mathrm{d}x = A\{X(t)\} = \lim_{T\to\infty} \frac{1}{T} \int_{-T/2}^{T/2} x(t)\mathrm{d}t. \tag{6.145}$$

It can be shown that this condition is satisfied if [PAP65, p. 328]:

$$\lim_{T\to\infty} \frac{1}{T} \int_{-T/2}^{T/2} R_{XX}(\tau)\mathrm{d}\tau = m^2. \tag{6.146}$$

Therefore, in order to test whether a process is a mean ergodic, we need the statistical mean value and the autocorrelation function of the process $R_{XX}(\tau)$.

Example 6.7.1 Find whether or not the process $X(t)$ is a mean ergodic, if the autocorrelation of the process is given as:

$$R_{XX}(\tau) = \mathrm{e}^{-2\lambda|\tau|}. \tag{6.147}$$

Solution The expression for autocorrelation function (6.147) indicates that the process has a zero mean value,

$$E\{X(t)\} = m = 0, \tag{6.148}$$

and that the process is a WS stationary.

The condition necessary for the process to be a mean ergodic (6.146) is:

$$\begin{aligned}\lim_{T\to\infty}\frac{1}{T}\int_{-T/2}^{T/2}R_{XX}(\tau)\mathrm{d}\tau &= \lim_{T\to\infty}\frac{1}{T}\int_{-T/2}^{T/2}\mathrm{e}^{-2\lambda|\tau|}\,\mathrm{d}\tau\\ &= \lim_{T\to\infty}\frac{1}{T}\left[\int_{-T/2}^{0}\mathrm{e}^{2\lambda\tau}\,\mathrm{d}\tau+\int_{0}^{T/2}\mathrm{e}^{-2\lambda\tau}\,\mathrm{d}\tau\right]\\ &= \lim_{T\to\infty}\left[\frac{1-\mathrm{e}^{-\lambda T}}{2\lambda T}+\frac{1-\mathrm{e}^{-\lambda T}}{2\lambda T}\right]=\lim_{T\to\infty}\left[-\frac{\mathrm{e}^{-\lambda T}}{\lambda T}\right]\\ &= \lim_{T\to\infty}\frac{\lambda\mathrm{e}^{-\lambda T}}{\lambda}=0.\end{aligned} \tag{6.149}$$

From (6.148) to (6.149), we conclude that the condition for a process to be a mean ergodic is satisfied. Therefore, we have:

$$\lim_{T\to\infty}\frac{1}{T}\int_{-T/2}^{T/2}x(t)\,\mathrm{d}t = \int_{-\infty}^{\infty}xf_X(x)\,\mathrm{d}x = m_X = 0. \tag{6.150}$$

6.7.5 Autocorrelation and Cross-Correlation Ergodic Processes

A time-averaged autocorrelation function is obtained by applying a time average on a particular realization $x(t)$ of a stationary process $X(t)$:

$$A\{X(t)X(t+\tau)\} = R_{xx}(\tau) = \lim_{T\to\infty}\frac{1}{T}\int_{-T/2}^{T/2}x(t)x(t+\tau)\mathrm{d}t. \tag{6.151}$$

For a particular realization of the process, the obtained result is a deterministic function of τ which varies from one realization to another and presents a random variable.

The condition for ergodicity is that those variations approach zero, i.e., the variance of the random variable $R_{XX}(\tau)$ must approach zero.

$$\lim_{T\to\infty}\sigma^2_{R_{XX}} = 0. \tag{6.152}$$

If (6.152) is satisfied, we can write

$$\lim_{T\to\infty}\frac{1}{T}\int_{-T/2}^{T/2}x(t)x(t+\tau)\mathrm{d}t=\int_{-\infty}^{\infty}\int_{-\infty}^{\infty}x_1x_2f_{X_1X_2}(x_1,x_2;\tau)\mathrm{d}x_1\,\mathrm{d}x_2, \tag{6.153}$$

where X_1 and X_2 are random variables of the process $X(t)$ in time instants t and $t+\tau$, respectively.

Example 6.7.2 Determine whether the process

$$X(t)=Y\cos(\omega_0t+\theta) \tag{6.154}$$

is a mean and autocorrelation ergodic, if Y is uniform over [0, 1], and if ω_0 and θ are constants.

Solution The mean value is:

$$E\{X(t)\}=\int_0^1 y\cos(\omega_0t+\theta)f_Y(y)\mathrm{d}y=\cos(\omega_0t+\theta)/2. \tag{6.155}$$

The mean value is not a constant and thus the process is not a first-order stationary and consequently, it is not a mean ergodic.

The autocorrelation function is:

$$\begin{aligned}E\{X(t)X(t+\tau)\}&=E\{Y\cos(\omega_0t+\theta)Y\cos(\omega_0(t+\tau)+\theta)\}\\&=E\{Y^2\}[\cos\omega_0\tau+\cos(2\omega_0t+\omega_0\tau+2\theta)]/2\\&=\frac{1}{6}[\cos\omega_0\tau+\cos(2\omega_0t+\omega_0\tau+2\theta)].\end{aligned} \tag{6.156}$$

From (6.156), the autocorrelation function is a function of time t and thus cannot be equal to a time-averaged autocorrelation function which does not depend on time t. Therefore, the process is not autocorrelation ergodic.

Similarly, for mutually ergodic processes $X(t)$ and $Y(t)$, we can make a time and a statistical cross-correlation function equal as:

$$\begin{aligned}\overline{X(t)Y(t+\tau)}&=\overline{X(t)Y(t+\tau)}^t\\&=\int_{-\infty}^{\infty}\int_{-\infty}^{\infty}x_1y_2f_{X_1Y_2}(x_1,y_2;\tau)\mathrm{d}x_1\,\mathrm{d}y_2=\lim_{T\to\infty}\frac{1}{T}\int_{-T/2}^{T/2}x(t)y(t+\tau)\mathrm{d}t\end{aligned} \tag{6.157}$$

6.8 Numerical Exercises

Exercise 6.1 Find whether or not the process *X*(*t*) is a WS stationary process

$$X(t) = A\cos(\omega_0 t + \theta), \tag{6.158}$$

where A and ω_0 are constants and θ is a uniform variable over $[0, 2\pi]$.

Answer The first condition for a WS stationarity says that the mean value of the process must be constant.

$$E\{X(t)\} = \int_0^{2\pi} A\cos(\omega_0 t + \theta) f_\theta(\theta)\mathrm{d}\theta = \int_0^{2\pi} A\cos(\omega_0 t + \theta)\frac{1}{2\pi}\mathrm{d}\theta = 0. \tag{6.159}$$

Therefore, the first condition is satisfied.

The second condition says that the autocorrelation function has to be a function of a time difference τ.

$$\begin{aligned} E\{X(t)X(t+\tau)\} &= E\{A\cos(\omega_0 t + \theta)A\cos(\omega_0(t+\tau) + \theta)\} \\ &= \frac{A^2}{2}E\{\cos\omega_0\tau + \cos(2\omega_0 t + \omega_0\tau + 2\theta)\} \\ &= \frac{A^2}{2}\cos\omega_0\tau + \frac{A^2}{2}E\{\cos(2\omega_0 t + \omega_0\tau + 2\theta)\} \\ &= \frac{A^2}{2}\cos\omega_0\tau. \end{aligned} \tag{6.160}$$

The second condition is also satisfied and, thus, the process is a WS stationary.

Exercise 6.2 Given the process

$$Y(t) = \cos\,(\omega_0 t + X), \tag{6.161}$$

where ω_0 is a constant and X is a random variable with the characteristic function $\phi_X(\omega)$. Find which condition must be imposed to the characteristic function $\phi_X(\omega)$ in order for the process *Y*(*t*) to be a WS stationary.

Answer The mean value of the process *Y*(*t*) is:

$$\begin{aligned} E\{\cos(\omega_0 t + X)\} &= E\{\cos(\omega_0 t)\cos X - \sin(\omega_0 t)\sin X\} \\ &= \cos(\omega_0 t)E\{\cos X\} - \sin(\omega_0 t)E\{\sin X\}. \end{aligned} \tag{6.162}$$

The mean value (6.162) is constant if the following condition is satisfied:

$$E\{\cos(X)\} = E\{\sin(X)\} = 0. \tag{6.163}$$

The autocorrelation function is:

$$\begin{aligned}
E\{Y(t)Y(t+\tau)\} &= E\{\cos(\omega_0 t + X)\cos(\omega_0(t+\tau)+X)\} \\
&= \frac{1}{2}\cos(\omega_0\tau) + \frac{1}{2}E\{\cos((2\omega_0 t + \omega_0\tau) + 2X)\} \\
&= \frac{1}{2}\cos(\omega_0\tau) + \frac{1}{2}E\{\cos(2\omega_0 t + \omega_0\tau)\cos(2X) - \sin(2\omega_0 t + \omega_0\tau)\sin(2X)\} \\
&= \frac{1}{2}\cos(\omega_0\tau) + \frac{1}{2}\cos(2\omega_0 t + \omega_0\tau)E\{\cos(2X)\} - \frac{1}{2}\sin(2\omega_0 t + \omega_0\tau)E\{\sin(2X)\}
\end{aligned} \tag{6.164}$$

The expression (6.164) will be dependent only on τ if

$$E\{\cos(2X)\} = E\{\sin(2X)\} = 0. \tag{6.165}$$

Next we relate the conditions (6.163) and (6.165) with the characteristic function $\phi_X(\omega)$.

$$\phi_X(\omega) = E\{e^{j\omega X}\} = E\{\cos(\omega X) + jE\{\sin(\omega X)\}. \tag{6.166}$$

From here, we have:

$$\phi_X(1) = E\{\cos(X) + jE\{\sin(X)\}, \tag{6.167}$$

$$\phi_X(2) = E\{\cos(2X) + jE\{\sin(2X)\}. \tag{6.168}$$

Placing (6.167) and (6.165), we get the following conditions for WS stationarity of the process $Y(t)$:

$$\begin{aligned}
\phi_X(1) &= 0, \\
\phi_X(2) &= 0.
\end{aligned} \tag{6.169}$$

Exercise 6.3 Express the autocorrelation function of the process

$$Y(t) = X(t+b) - X(t) \tag{6.170}$$

in terms of the autocorrelation function of the process $X(t)$, $R_{XX}(t_1, t_2)$

Answer

$$\begin{aligned}
R_{YY}(t_1,t_2) &= E\{[X(t_1+b) - X(t_1)][X(t_2+b) - X(t_2)]\} \\
&= E\{X(t_1+b)X(t_2+b)\} - E\{X(t_1)X(t_2+b)\} \\
&\quad - E\{X(t_1+b)X(t_2) + E\{X(t_1)X(t_2)\} \\
&= R_{XX}(t_1+b, t_2+b) - R_{XX}(t_1, t_2+b) - R_{XX}(t_1+b, t_2) + R_{XX}(t_1, t_2).
\end{aligned} \tag{6.171}$$

Exercise 6.4 Find the mean value and variance of the process $X(t)$ if its autocorrelation function is given as

$$R_{XX}(\tau) = \frac{2}{1+\tau^2}. \tag{6.172}$$

Answer The autocorrelation function does not have a constant term. This indicates that the mean value of the process is zero:

$$E\{X(t) = 0. \tag{6.173}$$

For $\tau = 0$, we have:

$$R_{XX}(0) = E\{X(t)X(t+0)\} = E\{X^2(t)\} = \frac{2}{1+0} = 2. \tag{6.174}$$

Using (6.172) and (6.174), we get the variance of the process:

$$\sigma^2 = E\{X^2(t)\} - E\{X(t)\}^2 = E\{X^2(t)\} = 2 \tag{6.175}$$

Exercise 6.5 Determine whether or not the process $X(t)$ is a WS stationary and find its variance if it is known that the autocorrelation function is:

$$R_{XX}(\tau) = \frac{4}{1+\tau^2} + 16. \tag{6.176}$$

Answer The autocorrelation function has a constant term which is equal to a squared mean value:

$$16 = E\{X(t)\}^2. \tag{6.177}$$

From here, it follows that the mean value of the process $X(t)$ is a constant:

$$E\{X(t)\} = \pm\, 4. \tag{6.178}$$

From (6.176), we can calculate the mean squared value:

$$R_{XX}(0) = E\{X^2(t)\} = \frac{4}{1+0} + 16 = 20. \tag{6.179}$$

Using (6.178) and (6.179), we get:

$$\sigma^2 = E\{X^2(t)\} - E\{X(t)\}^2 = 20 - 16 = 4. \tag{6.180}$$

The process $X(t)$ is WS stationary because both conditions (6.32) and (6.33) are satisfied.

Exercise 6.6 The autocorrelation function of the process $X(t)$ is given as:

$$R_{XX}(\tau) = p^2 + p(1-p)\mathrm{e}^{-c|\tau|}, \tag{6.181}$$

where p and c are constants. Find the variance of the process.

Answer The constant term in (6.181) is equal to a squared mean value, resulting in:

$$E\{X(t)\}^2 = p^2. \tag{6.182}$$

Similarly, from (6.181), we have:

$$R_{XX}(0) = E\{X^2(t)\} = p^2 + p(1-p). \tag{6.183}$$

From (6.182) and (6.183), we get:

$$\sigma^2 = E\{X^2(t)\} - E\{X(t)\}^2 = p^2 + p(1-p) - p^2 = p(1-p). \tag{6.184}$$

Exercise 6.7 Determine whether or not a process (6.185) is a WS stationary process.

$$Y(t) = X\cos(\omega_0 t + \theta), \tag{6.185}$$

where ω_0 and θ are constants and X is a random variable.

Answer Let us verify that the first condition (6.32) is satisfied:

$$E\{Y(t)\} = E\{X\cos(\omega_0 t + \theta)\} = \cos(\omega_0 t + \theta)E\{X\}. \tag{6.186}$$

This result indicates that the mean value $E\{Y(t)\}$ is not constant but dependent on time, unless $E\{X\} = 0$. In this case, $E\{Y(t)\} = 0$ and the first condition (6.32) is satisfied.

To verify the second condition (6.33), we must find the autocorrelation function:

$$\begin{aligned} R_{XX}(t, t+\tau) &= E\{Y(t)Y(t+\tau)\} = E\{X\cos(\omega_0 t + \theta)X\cos(\omega_0(t+\tau) + \theta)\} \\ &= E\{X^2\}\cos(\omega_0 t + \theta)\cos(\omega_0(t+\tau) + \theta) \end{aligned}, \tag{6.187}$$

or equivalently

$$R_{XX}(t,t+\tau) = \frac{1}{2}E\{X^2\} \times [\cos(\omega_0 t + \theta - \omega_0(t+\tau) - \theta) + \cos(\omega_0 t + \theta + \omega_0(t+\tau) + \theta)]$$
$$= \frac{1}{2}E\{X^2\}[\cos(-\omega_0\tau) + \cos(2\omega_0 t + 2\theta + \omega_0\tau)]$$
$$= \frac{1}{2}E\{X^2\}[\cos(\omega_0\tau) + \cos(2(\omega_0 t + \theta) + \omega_0\tau)]. \quad (6.188)$$

The received result indicates that the autocorrelation function depends not only on τ but also on t. Therefore, the second condition is not satisfied and the process is not a WS stationary.

Exercise 6.8 Verify and explain whether or not each of the following can be an autocorrelation function.

(a) $$g(\tau) = \frac{2\tau}{1+\tau^2}. \quad (6.189)$$

(b) $$g(\tau) = \frac{2\tau^2}{1+\tau^2}. \quad (6.190)$$

(c) $$g(\tau) = \sin\omega_0\tau, \quad \omega_0 = \text{constant}. \quad (6.191)$$

(d) $$g(\tau) = \cos\omega_0\tau, \quad \omega_0 = \text{constant}. \quad (6.192)$$

Answer An autocorrelation function which is a function of a time difference τ must have the properties listed in Sect. 6.5.4.

(a) This function cannot be an autocorrelation function because it does not have the property (6.45)

$$|g(\tau)| \leq g(0). \quad (6.193)$$

For example, $g(0) = 0$, $g(2) = 4/5$; and thus $g(2) > g(0)$.

Additionally, this function does not have the property (6.51) (i.e., it is not an even function).

(b) This function satisfies the property (6.51), i.e., it is an even function. However, it cannot be an autocorrelation function because it does not have the property (6.45). For example, $g(0) = 0$, $g(1) = 1$, and $g(1) > g(0)$.

(c) This function cannot be an autocorrelation function because it is not an even function.

(d) Yes, this function can be an autocorrelation function because it is an even function. This function is a periodic autocorrelation function indicating that the corresponding process is also a periodic process.

Exercise 6.9 Verify that there is no process $X(t)$ which has the following autocorrelation function:

$$R(\tau) = \begin{cases} 1 & \text{for} \quad |\tau| \leq 1 \\ 0 & \quad \text{otherwise} \end{cases}. \tag{6.194}$$

Answer From (6.49) and (6.194), we have:

$$R(0) = E\{X^2\} = 1. \tag{6.195}$$

The property (6.60) yields

$$E\{X\}^2 = 1. \tag{6.196}$$

Therefore, the variance of the process X is

$$\sigma^2 = E\{X^2\} - E\{X\}^2 = 1 - 1 = 0. \tag{6.197}$$

There is no such random process with a variance of zero.

Exercise 6.10 The autocorrelation function is given in Fig. 6.14. Find the variance of the process $X(t)$.

Answer We observe in Fig. 6.14:

$$E\{X^2(t)\} = 9; \qquad E\{X(t)\}^2 = 4. \tag{6.198}$$

The variance is

$$\sigma_X^2 = E\{X^2(t)\} - E\{X(t)\}^2 = 9 - 4 = 5. \tag{6.199}$$

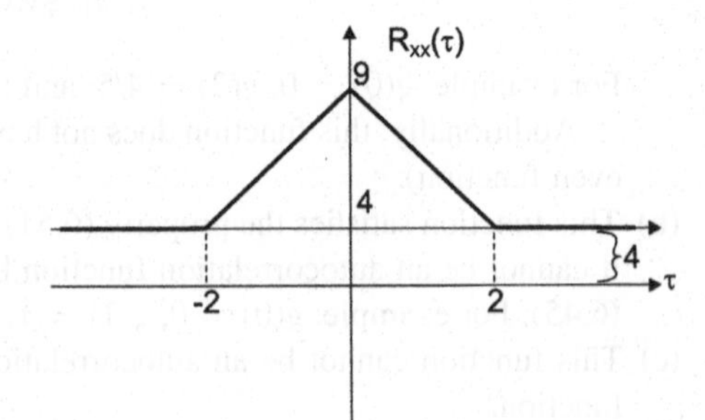

Fig. 6.14 Autocorrelation function $R_{XX}(\tau)$ in Exercise 6.10.

Exercise 6.11 Find the autocorrelation function of the process

$$Y(t) = X_1 X(t), \tag{6.200}$$

where the process $X(t)$ is described in Example 6.5.2, and X_1 is a discrete random variable which has only values -1 and 1 and

$$P\{-1\} = P\{1\} = 1/2. \tag{6.201}$$

The random variable X_1 and the process $X(t)$ are independent for any value of t.

Answer The autocorrelation function of the process $Y(t)$ is given as:

$$R_{YY}(t_1, t_2) = E\{X_1 X(t_1) X_1 X(t_2)\} = E\{X_1^2 X(t_1) X(t_2)\}. \tag{6.202}$$

From the independence of the variable X_1 and the process $X(t)$, we get:

$$R_{YY}(t_1, t_2) = E\{X_1^2\} E\{X(t_1) X(t_2)\}. \tag{6.203}$$

The mean squared value of the variable X_1 is found to be

$$E\{X_1^2\} = 1^2 P\{X_1 = 1\} + (-1)^2 P\{X_1 = -1\} = \frac{1}{2} + \frac{1}{2} = 1. \tag{6.204}$$

The second factor in (6.203) is the autocorrelation function determined in (6.77):

$$E\{X(t_1) X(t_2)\} = R_{XX}(\tau) = U^2 \, \mathrm{e}^{-2\lambda|\tau|}, \tag{6.205}$$

where

$$|\tau| = |t_2 - t_1|. \tag{6.206}$$

Finally, from (6.203) to (6.206), we arrive at:

$$R_{YY}(t_1, t_2) = R_{YY}(\tau) = U^2 \, \mathrm{e}^{-2\lambda|\tau|}. \tag{6.207}$$

Exercise 6.12 Consider the random process $Y(t)$ given by

$$Y(t) = X \, \mathrm{e}^{-2t}, \tag{6.208}$$

where X is a binomial random variable with parameters given in (5.116)–(5.118).
Find the mean value and variance of the process $Y(t)$.

Answer The mean value and mean squared values of the random variable X are:

$$E\{X\} = np; \qquad E\{X^2\} = n^2p^2 + np(1-p). \tag{6.209}$$

The mean value of the process $Y(t)$ is:

$$E\{Y(t)\} = \mathrm{e}^{-2t}E\{X\} = np\,\mathrm{e}^{-2t}. \tag{6.210}$$

Similarly, the mean squared value of the process $Y(t)$ is:

$$E\{Y^2(t)\} = \mathrm{e}^{-4t}E\{X^2\} = \left[n^2p^2 + np(1-p)\right]\mathrm{e}^{-4t}. \tag{6.211}$$

From (6.210) and (6.211), we get:

$$\begin{aligned}\sigma_Y^2 &= E\{Y^2(t)\} - E\{Y(t)\}^2 = \left[n^2p^2 + np(1-p)\right]\mathrm{e}^{-4t} - n^2p^2\,\mathrm{e}^{-4t} \\ &= np(1-p)\mathrm{e}^{-4t}.\end{aligned} \tag{6.212}$$

Exercise 6.13 Determine whether the process defined in (6.158) is mean and autocorrelation ergodic.

Answer A process is said to be a mean ergodic if its statistical mean value is equal to its time-averaged mean.

The statistical mean value in (6.159) is found to be zero.
Its time-averaged mean is:

$$\begin{aligned}A\{X(t)\} &= \lim_{T\to\infty}\frac{1}{T}\int_{-T/2}^{T/2} X(t)\mathrm{d}t = \lim_{T\to\infty}\frac{1}{T}\int_{-T/2}^{T/2} A\cos(\omega_0 t + \theta)\mathrm{d}t \\ &= \frac{A}{T}\int_{-T/2}^{T/2}\cos(\omega_0 t + \theta)\mathrm{d}t = 0,\end{aligned} \tag{6.213}$$

where

$$T = 2\pi/\omega_0. \tag{6.214}$$

Thus, the process is mean ergodic because

$$A\{X(t)\} = E\{X(t)\}. \tag{6.215}$$

The statistical autocorrelation function in (6.160) is found to be:

$$E\{X(t)X(t+\tau)\} = \frac{A^2}{2}\cos\omega_0\tau. \tag{6.216}$$

The time-averaged autocorrelation function from (6.151), using (6.158), is found to be:

$$A\{X(t)X(t+\tau)\} = \lim_{T\to\infty} \frac{1}{T} \int_{-T/2}^{T/2} x(t)x(t+\tau)\mathrm{d}t$$

$$= \lim_{T\to\infty} \frac{1}{T} \int_{-T/2}^{T/2} A^2 \cos(\omega_0 t+\theta)\cos(\omega_0(t+\tau)+\theta)\mathrm{d}t$$

$$= \lim_{T\to\infty} \frac{A^2}{2T} \int_{-T/2}^{T/2} [\cos\omega_0\tau + \cos(2\omega_0 t+\omega_0\tau+2\theta)]\mathrm{d}t = \frac{A^2}{2}\cos(\omega_0\tau). \quad (6.217)$$

Thus, we have:

$$A\{X(t)X(t+\tau)\} = E\{X(t)X(t+\tau)\}. \quad (6.218)$$

Hence, the process $X(t)$ is autocorrelation ergodic as well.

Exercise 6.14 A stationary random process $X(t)$ has rectangular pulses of amplitude 1. The process has zero amplitude values between pulses. A realization of this process is shown in Fig. 6.15. Transitions from amplitude 1–0 and vice versa are described by a Poisson formula with a parameter of λ. The probabilities of amplitudes 1 and 0 are equal. Find the autocorrelation function of this process.

Answer This process is a discrete process which only takes values of amplitudes 0 and 1. Consequently, in time instants t and $t+\tau$, we have discrete random variables X_1 and X_2.

Then, the autocorrelation function is:

$$R_{XX}(t,t+\tau) = E\{X(t)X(t+\tau)\} = E\{X_1X_2;\tau\}$$

$$= \sum_{i=1}^{2}\sum_{j=1}^{2} x_{1i}x_{2j}P\{X_1=x_{1i}, X_2=x_{2j};\tau\}, \quad (6.219)$$

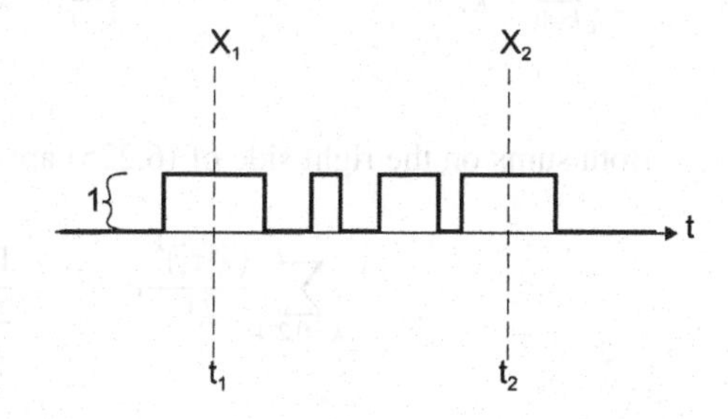

Fig. 6.15 A realization of the process given in Exercise 6.14

where

$$\begin{aligned} x_{11} &= x_{21} = 0, \quad x_{12} = x_{22} = 1, \\ P\{X_1 = 1\} &= P\{X_2 = 1\} = 1/2. \end{aligned} \tag{6.220}$$

Using (6.219) and (6.220), we get:

$$R_{XX}(t, t+\tau) = 1 \times 1 \times P\{X_1 = 1, X_2 = 1; \tau\} = P\{X_1 = 1, X_2 = 1; \tau\}. \tag{6.221}$$

The joint probability from (6.221) can be expressed in terms of a conditional probability and the probability from (6.220):

$$P\{X_1 = 1, X_2 = 1; \tau\} = P\{X_1 = 1\} P\{X_2 = 1 | X_1 = 1; \tau\}. \tag{6.222}$$

The conditional probability in (6.222) corresponds to the probability that there are even numbers of transitions in an interval τ. Using the Poisson formula (5.149), we get:

$$P\{X_2 = 1 | X_1 = 1; \tau\} = \sum_{k=0,2,4,\ldots} \frac{(\lambda|\tau|)^k}{k!} e^{-\lambda|\tau|}. \tag{6.223}$$

The absolute value of τ is used in (6.223) because the value of τ in (5.149) is a positive value, while in a probability (6.222) it can be both, positive or negative.
The conditional probability (6.223) can be calculated as described below:

$$\sum_{k=0}^{\infty} \frac{(\lambda|\tau|)^k}{k!} e^{-\lambda|\tau|} + \sum_{k=0}^{\infty} \frac{(\lambda|\tau|)^k}{k!} e^{-\lambda|\tau|} (-1)^k = 2 \sum_{k=0,2,4,\ldots} \frac{(\lambda|\tau|)^k}{k!} e^{-\lambda|\tau|}. \tag{6.224}$$

From here, we have:

$$\begin{aligned} \sum_{k=0,2,4,\ldots} \frac{(\lambda|\tau|)^k}{k!} e^{-\lambda|\tau|} &= \frac{1}{2} \left[\sum_{k=0}^{\infty} \frac{(\lambda|\tau|)^k}{k!} e^{-\lambda|\tau|} + e^{-\lambda|\tau|} e^{-(-\lambda|\tau|)} \sum_{k=0}^{\infty} \frac{(-\lambda|\tau|)^k}{k!} e^{-\lambda|\tau|} \right] \\ &= \frac{1}{2} \left[\sum_{k=0}^{\infty} \frac{(\lambda|\tau|)^k}{k!} e^{-\lambda|\tau|} + e^{-\lambda|\tau|} e^{-\lambda|\tau|} \sum_{k=0}^{\infty} \frac{(-\lambda|\tau|)^k}{k!} e^{-(-\lambda|\tau|)} \right]. \end{aligned} \tag{6.225}$$

Both sums on the right side of (6.225) are equal to 1, resulting in:

$$\sum_{k=0,2,4} \frac{(\lambda|\tau|)^k}{k!} e^{-\lambda|\tau|} = \frac{1}{2} \left[1 + e^{-2\lambda|\tau|} \right]. \tag{6.226}$$

Finally, placing (6.226) into (6.223), and using (6.221) and (6.222), we arrive at:

$$R_{XX}(\tau) = \frac{1}{4}\left[1 + e^{2\lambda|\tau|}\right]. \tag{6.227}$$

Exercise 6.15 Consider two random processes:

$$\begin{aligned} X(t) &= \xi \sin \omega_0 t + \rho \cos \omega_0 t \\ Y(t) &= \xi \sin \omega_0 t - \rho \cos \omega_0 t \end{aligned}, \tag{6.228}$$

where ω_0 is a constant, and ξ and ρ are uncorrelated random variables:

$$E\{\xi\rho\} = E\{\xi\}E\{\rho\}. \tag{6.229}$$

The random variables ξ and ρ have means of zero,

$$E\{\xi\} = E\{\rho\} = 0, \tag{6.230}$$

and equal variances

$$\sigma_\xi^2 = \sigma_\rho^2 = \sigma^2. \tag{6.231}$$

Determine whether the processes are jointly WS stationary.

Answer We must first determine whether both processes are WS stationary.

Both processes have a mean value of 0 and, therefore, the first condition for WS is satisfied.

Next we investigate if both autocorrelation functions depend only on the time difference τ.

To this end we find:

$$\begin{aligned} R_{XX}(t, t+\tau) &= E\{X(t)X(t+\tau)\} \\ &= E\{[\xi \sin \omega_0 t + \rho \cos \omega_0 t][\xi \sin \omega_0 (t+\tau) + \rho \cos \omega_0 (t+\tau)]\} \\ &= E\{\xi^2\} \sin \omega_0 t \sin \omega_0 (t+\tau) + E\{\rho^2\} \cos \omega_0 t \cos \omega_0 (t+\tau) \\ &\quad + E\{\xi\rho\} \sin \omega_0 t \cos \omega_0 (t+\tau) + E\{\xi\rho\} \cos \omega_0 t \sin \omega_0 (t+\tau). \end{aligned} \tag{6.232}$$

From (6.230) and (6.231), we have:

$$E\{\xi^2\} = E\{\rho^2\} = \sigma^2. \tag{6.233}$$

Using the conditions (6.229) and (6.230), we can write:

$$E\{\xi\rho\} = 0. \tag{6.234}$$

Using (6.234), last two terms in (6.232) become zero.

Finally, from (6.232), we arrive at:

$$R_{XX}(t, t+\tau) = \sigma^2 \cos(\omega_0 t - \omega_0 t - \omega_0 \tau) = \sigma^2 \cos \omega_0 \tau = R_{XX}(\tau). \tag{6.235}$$

In a similar way, we find:

$$R_{YY}(t, t+\tau) = \sigma^2 \cos(\omega_0 t - \omega_0 t - \omega_0 \tau) = \sigma^2 \cos \omega_0 \tau = R_{YY}(\tau). \tag{6.236}$$

The obtained results (6.235) and (6.236) indicate that both processes are WS stationary. Thus, the conditions (6.99) and (6.100) are satisfied.

Next we determine whether the conditions (6.101) and (6.102) are satisfied.

The cross-correlation function is:

$$R_{XY}(t, t+\tau) = E\{[\xi \sin \omega_0 t + \rho \cos \omega_0 t][\xi \sin \omega_0 (t+\tau) - \rho \cos \omega_0 (t+\tau)]\}. \tag{6.237}$$

Using (6.233) and (6.234), we finally arrive at:

$$\begin{aligned} R_{XY}(t, t+\tau) &= -\sigma^2 [\cos \omega_0 t \cos \omega_0 (t+\tau) - \sin \omega_0 t \sin \omega_0 (t+\tau)] \\ &= -\sigma^2 \cos(2\omega_0 t + \omega_0 \tau). \end{aligned} \tag{6.238}$$

The cross-correlation function depends on t and τ and the processes are not jointly WS stationary.

6.9 MATLAB Exercises

Exercise M.6.1 (MATLAB file *exercise_M_6_1.m*). Generate a uniform random process with $N = 2{,}000$ realizations where each realization has $M = 2{,}000$ elements. The process is uniform in the interval $[-1, 1]$.

(a) Observe the process at three different points:

$$M_1 = 1, \quad M_2 = 530, \text{ and } M_3 = 800. \tag{6.239}$$

(b) Denote the corresponding random variables as X_1, X_2, and X_3 and draw their plots.

(c) Estimate the mean values of variables X_1, X_2, and X_3.

(d) Estimate the variances of variables X_1, X_2, and X_3.

(e) Plot the histograms of variables X_1, X_2, and X_3.

(f) Estimate the PDFs of variables X_1, X_2, and X_3.

(g) Is this process stationary?

Solution The process is generated using the MATLAB code $X = (R2 - R1)*rand(M,N) + R1$; where $R1 = -1$ and $R2 = 1$.

(a) The random variables are shown in Fig. 6.16 for the first $n = 1{,}000$ samples.
(b) The mean values are estimated using the MATLAB file *mean.m*. The following mean values are obtained:

$$m_1 = 0.0065; \qquad m_2 = 0.0028; \qquad m_3 = 0.0055. \tag{6.240}$$

All estimated mean values are approximately equal to the mathematical mean value which is equal to zero, $m = 0$.
(c) The variances are estimated using the MATLAB file *var.m* and the following values for the variances are obtained:

$$\sigma_1^2 = 0.3333; \qquad \sigma_2^2 = 0.3357; \qquad \sigma_3^2 = 0.3336. \tag{6.241}$$

The estimated variances are approximately equal to the mathematical value of the variance which is $\sigma^2 = 0.3333$.
(d) The histograms are shown in Fig. 6.17.
The histograms from Fig. 6.17 indicate that all three variables are uniform in $[-1, 1]$.

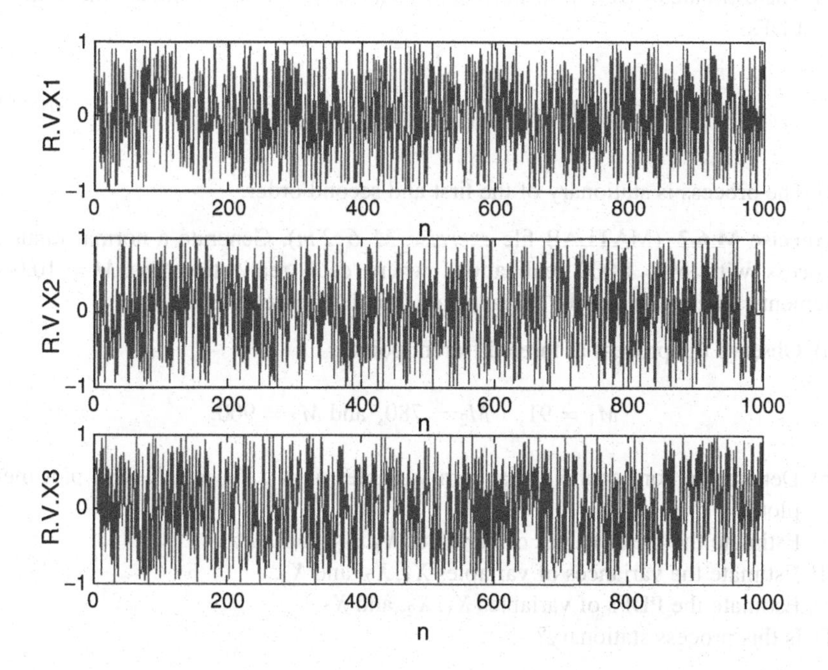

Fig. 6.16 Random variables X_1, X_2, and X_3 in Exercise M.6.1

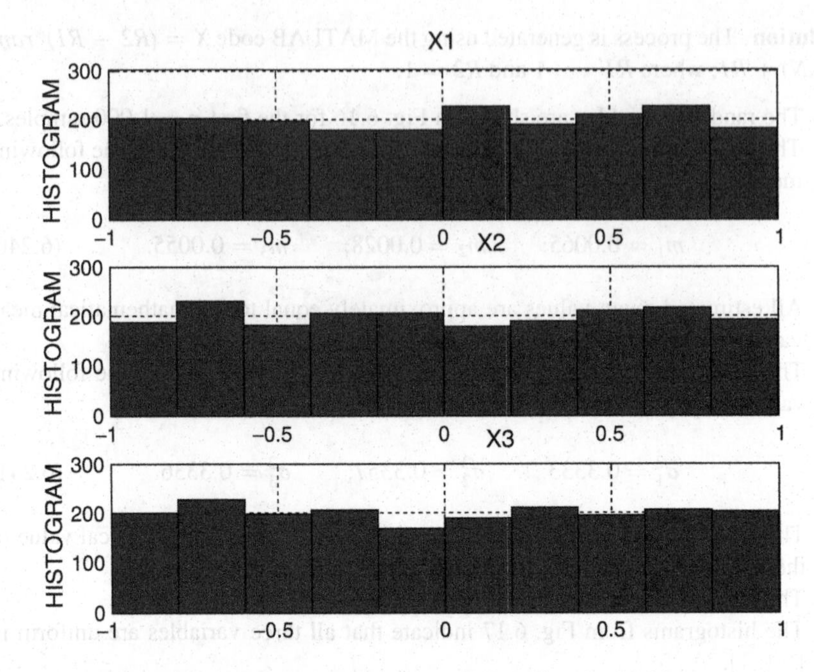

Fig. 6.17 Histograms for random variables X_1, X_2, and X_3 in Exercise M.6.1

(e) The estimated PDFs shown in Fig. 6.18 indicate that they estimate the uniform PDFs:

$$f_X(x) = \begin{cases} 1/2 & \text{for} \quad -1 \le x \le 1 \\ 0 & \text{otherwise} \end{cases}. \tag{6.242}$$

(f) The process is stationary of the first and second order.

Exercise M.6.2 (MATLAB file *exercise_M_6_2.m*). Generate a normal random process with $N = 1{,}000$ realizations where each realization has $M = 10{,}000$ elements. The mean and the variance are 0 and 1, respectively.

(a) Observe the process at three different points:

$$M_1 = 91, \quad M_2 = 780, \text{ and } M_3 = 900. \tag{6.243}$$

(b) Denote the corresponding random variables as X_1, X_2, and X_3 and plot their plots.
(c) Estimate the mean values of variables X_1, X_2, and X_3.
(d) Estimate the variances of variables X_1, X_2, and X_3.
(e) Estimate the PDFs of variables X_1, X_2, and X_3.
(f) Is this process stationary?

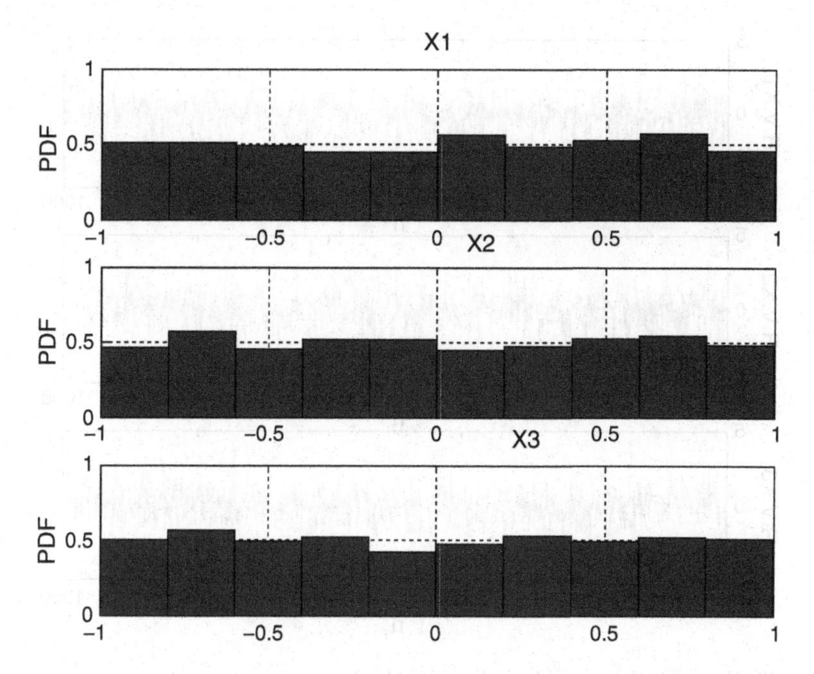

Fig. 6.18 Estimated PDFs for random variables X_1, X_2, and X_3 in Exercise M.6.1

Solution The process is generated using the MATLAB code $X = randn(M,N)$.

(a) The random variables are shown in Fig. 6.19 for the first $n = 1{,}000$ samples.
(b) The mean values are estimated using the MATLAB file *mean.m*. The following mean values are obtained:

$$m_1 = -0.0001; \qquad m_2 = -0.0002; \qquad m_3 = -0.0048. \tag{6.244}$$

All estimated mean values are approximately equal to the theoretical mean value which is equal to zero, $m = 0$.
(c) The variances are estimated using the MATLAB file *var.m* and the following values for the variances are obtained:

$$\sigma_1^2 = 0.9918; \qquad \sigma_2^2 = 0.9849; \qquad \sigma_3^2 = 0.9702. \tag{6.245}$$

The estimated variances are approximately equal to the theoretical value of the variance, which is $\sigma^2 = 1$.
(d) The estimated PDFs are shown in Fig. 6.20. The estimated PDFs are very similar indicating that the corresponding PDFs are independent of the time in which the random variable is observed.
(e) The process is at least the first and second-order stationary.

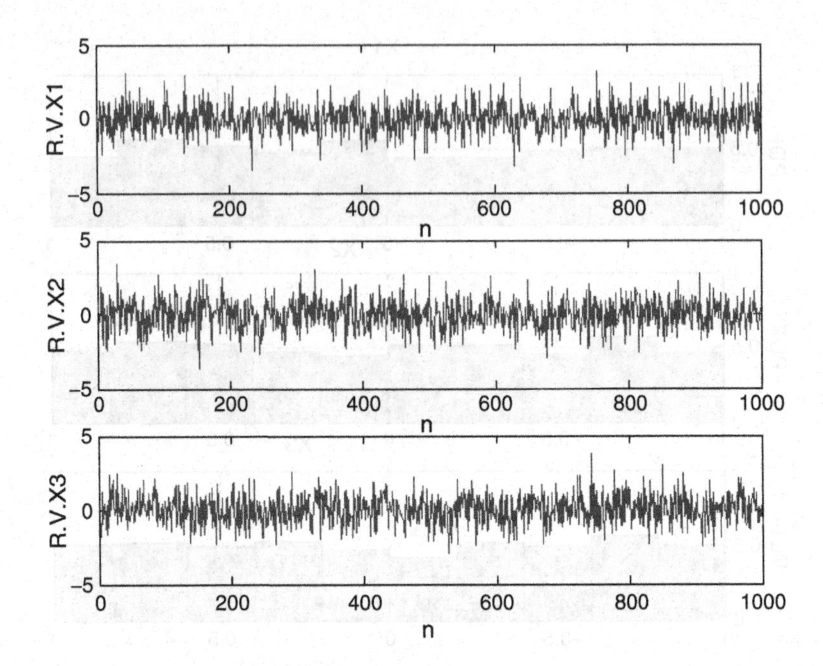

Fig. 6.19 Normal random variables in Exercise M.6.2

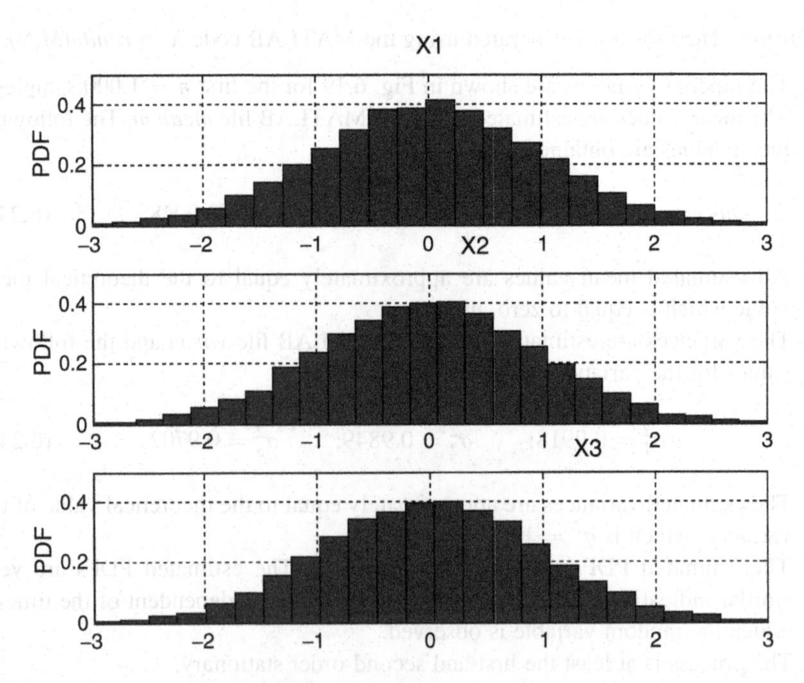

Fig. 6.20 Estimated PDFs for random variables X_1, X_2, and X_3 in Exercise M.6.2

Exercise M.6.3 (MATLAB file *exercise_M_6_3.m*). Generate a random process $X(t)$ with $N = 2{,}000$ realizations where each realization has $M = 2{,}000$ elements. Observe the process at three different points:

$$M_1 = 100, \quad M_2 = 450, \text{ and } M_3 = 900. \tag{6.246}$$

(a) Denote the corresponding random variables as X_1, X_2, and X_3 and plot their plots.
(b) Estimate the mean values of variables X_1, X_2, and X_3.
(c) Estimate the variances of variables X_1, X_2, and X_3.
(d) Estimate the PDFs of variables X_1, X_2, and X_3.
(e) Is this process stationary?

Solution The process is generated using the MATLAB file: *X_E_6_3.m*

(a) The random variables are shown in Fig. 6.21 for the first $n = 1{,}000$ samples.
(b) The mean values are estimated using the MATLAB file *mean.m*. The following mean values are obtained:

$$m_1 = 100.0067; \qquad m_2 = 450.0204; \qquad m_3 = 900.0040. \tag{6.247}$$

All estimated mean values are different.
(c) The variances are estimated using the MATLAB file *var.m*. The following values for the variances are obtained:

$$\sigma_1^2 = 0.0331; \qquad \sigma_2^2 = 0.3245; \qquad \sigma_3^2 = 0.0321. \tag{6.248}$$

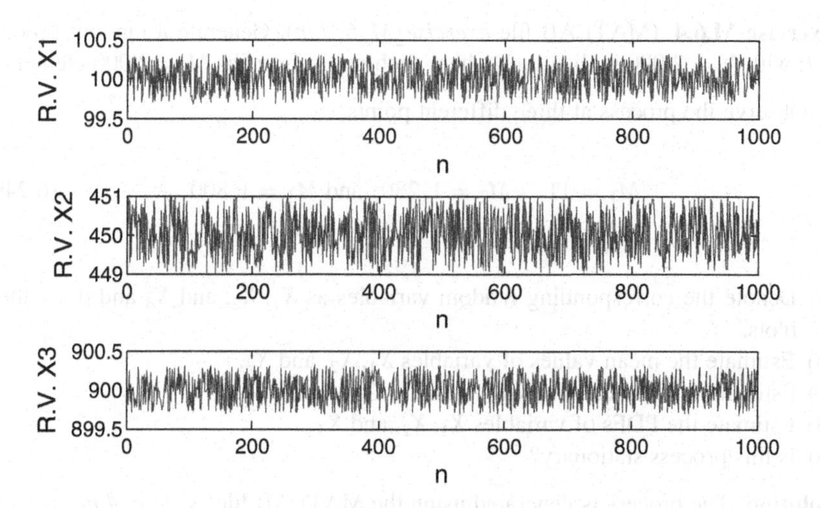

Fig. 6.21 Random variables X_1, X_2, and X_3 in Exercise M.6.3

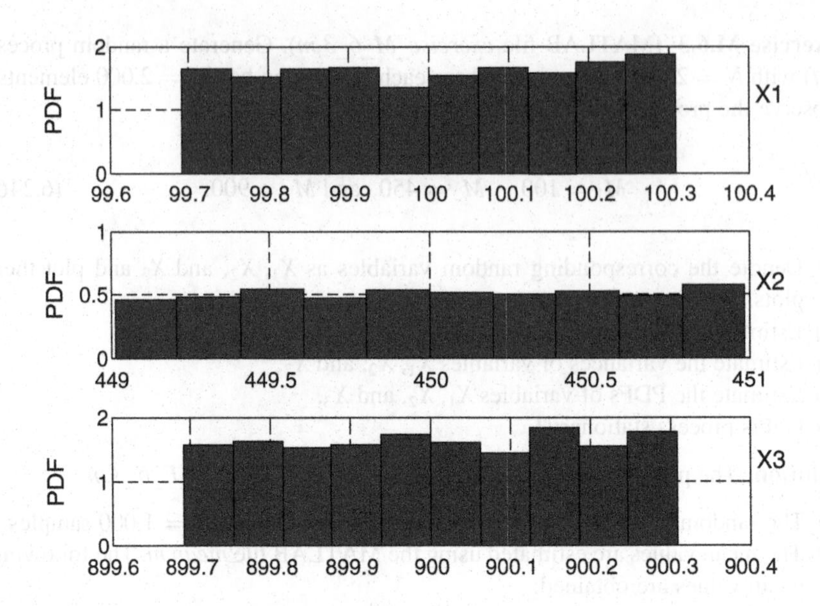

Fig. 6.22 Estimated PDFs in Exercise M.6.3

The estimated variances are different in different time instants.

(d) The estimated PDFs are shown in Fig. 6.22. Estimated PDFs indicate that all three variables are uniform but in different intervals.
(e) The process is not stationary because its mean value, variance, and PDF depend on the time instants in which the variables are defined.

Exercise M.6.4 (MATLAB file *exercise_M_6_4.m*). Generate a random process $X(t)$ with $N = 2{,}000$ realizations where each realization has $M = 2{,}000$ elements.

Observe the process at three different points,

$$M_1 = 11, \quad M_2 = 1,230, \text{ and } M_3 = 1,800. \tag{6.249}$$

(a) Denote the corresponding random variables as X_1, X_2, and X_3 and draw their plots.
(b) Estimate the mean values of variables X_1, X_2, and X_3.
(c) Estimate the variances of variables X_1, X_2, and X_3.
(d) Estimate the PDFs of variables X_1, X_2, and X_3.
(e) Is this process stationary?

Solution The process is generated using the MATLAB file: *X_E_6_4.m*

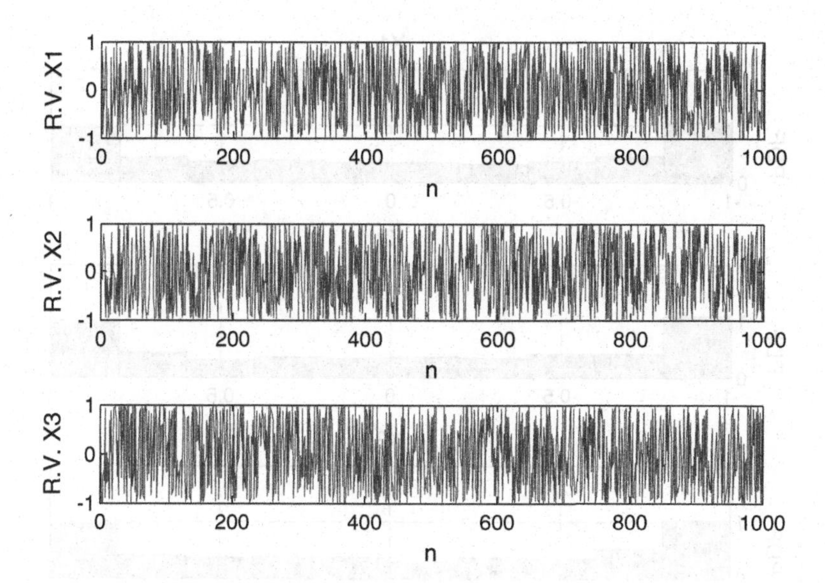

Fig. 6.23 Random variables in Exercise M.6.4

(a) The random variables are shown in Fig. 6.23 for the first $n = 1{,}000$ samples.
(b) The mean values are estimated using the MATLAB file *mean.m*. The following mean values are obtained:

$$m_1 = -0.0147; \qquad m_2 = -0.0241; \qquad m_3 = -0.0188. \tag{6.250}$$

All estimated mean values are similar (the error is a result of the estimation procedure).
(c) The variances are estimated using the MATLAB file *var.m*. The following values for the variances are obtained:

$$\sigma_1^2 = 0.4894; \qquad \sigma_2^2 = 0.4919; \qquad \sigma_3^2 = 0.4863. \tag{6.251}$$

All estimated variances are similar (the error is a result of the estimation procedure).
(d) The estimated PDFs are shown in Fig. 6.24. The estimated PDFs indicate that all three variables have similar PDFs (the error is a result of the estimation procedure).
(e) The process is at least the first and second-order stationary.

Exercise M.6.5 (MATLAB file *exercise_M_6_5.m*). Generate a uniform random process with $N = 3{,}000$ realizations where each realization has $M = 3{,}000$ elements. The process is uniform in the interval [1, 5].

Observe the process at one point $M_1 = 1{,}203$. The random variable X_1 is obtained.

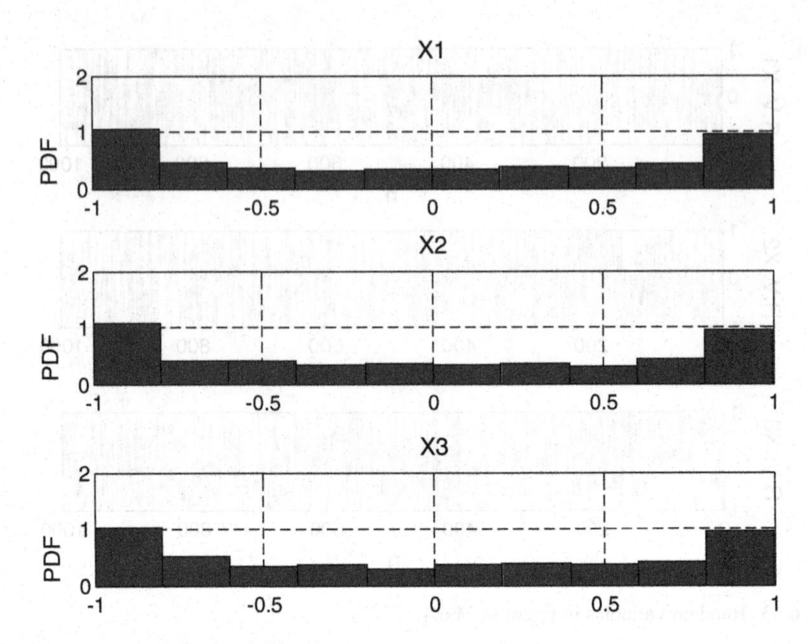

Fig. 6.24 Estimated PDFs in Exercise M.6.4

Choose a realization of the process $N_1 = 547$.
The amplitudes of the realization are random variable Y_1.

$$M_1 = 1,203, \qquad N_1 = 547. \tag{6.252}$$

(a) Plot the process at the point M_1.
Plot the N_1th realization.
(b) Estimate the mean values of X_1, and Y_1.
(c) Estimate the variances of variables of X_1, and Y_1.
(d) Estimate the PDFs of variables X_1, and Y_1 .
(e) Is this process ergodic?

Solution The process is generated using the MATLAB code $X = (R2 - R1)*rand(M,N) + R1$; where $R1 = 1$ and $R2 = 5$.

(a) The random variables are shown in Fig. 6.25 for the first $n = 1{,}000$ samples.
(b) The estimated mean values are approximately equal:

$$m_X = 3.002, \qquad m_Y = 3.019. \tag{6.253}$$

(c) The estimated variances are approximately equal:

$$\sigma_X^2 = 1.3608, \qquad \sigma_Y^2 = 1.3237. \tag{6.254}$$

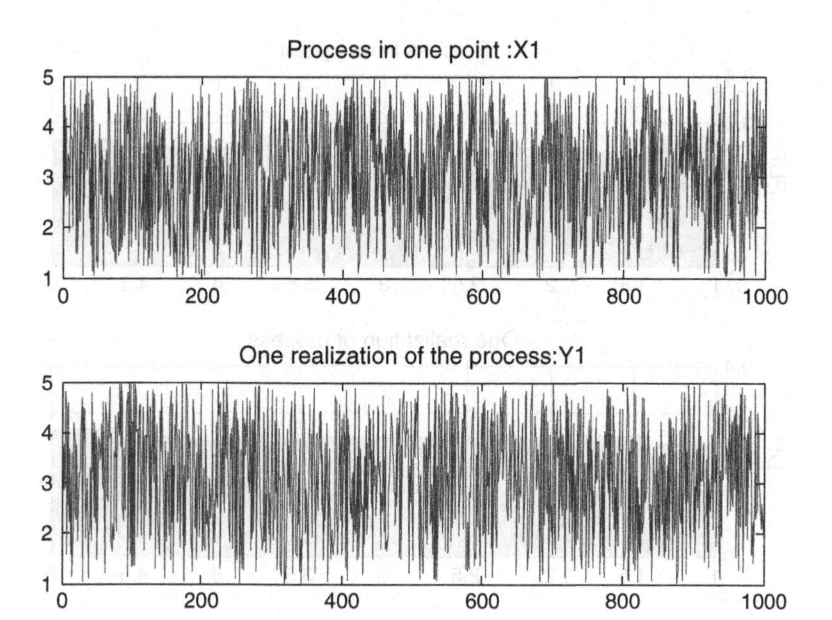

Fig. 6.25 Process in one point and one realization of the process

(d) The estimated PDFs are approximately equal as shown in Fig. 6.26.
(e) The process is a mean, mean squared, and PDF ergodic.

Exercise M.6.6 (MATLAB file *exercise_M_6_6.m*). Generate a stationary random process $X(t)$ with $N = 3{,}000$ realizations where each realization has $M = 3{,}000$ elements. The process is generated using the MATLAB file X_E_6_6.m.

Case 1 Observe the process in one point $M_1 = 411$. The random variable X_1 is obtained.

Choose a realization of the process $N_1 = 411$. The amplitudes of the realization are random variable Y_1.

$$N_1 = 411, \qquad M_1 = 411. \tag{6.255}$$

(a) Plot the process at the point M_1.
Plot N_1th realization.
(b) Estimate the mean values of X_1, and Y_1.
(c) Estimate the variances of variables of X_1, and Y_1.
(d) Estimate the PDFs of variables X_1, and Y_1.

Case 2 Do the same as for *Case 1* but with the following values:

$$M_1 = 2{,}800, \qquad N_1 = 1{,}790. \tag{6.256}$$

(e) Determine whether this process is ergodic?

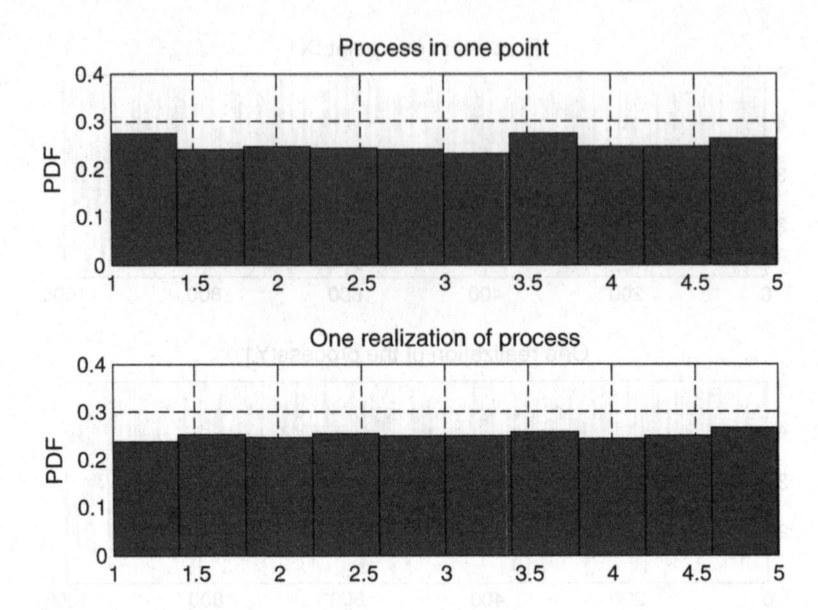

Fig. 6.26 Estimated PDFs in Exercise M.6.5

Solution

(a) The first $n = 1{,}000$ samples of the random variables are shown in Figs. 6.27 and 6.28.

(b) The estimated mean values are approximately equal:

$$\begin{aligned} m_X &= -0.0508, \qquad m_Y = -0.0271 \quad \text{(Case 1)} \\ m_X &= -0.0180, \qquad m_Y = -0.0394 \quad \text{(Case 2)}. \end{aligned} \tag{6.257}$$

(c) The estimated variances are approximately equal:

$$\begin{aligned} \sigma^2{}_X &= 1.9427, \qquad \sigma^2{}_Y = 1.9258, \quad \text{(Case 1)} \\ \sigma^2{}_X &= 1.9653, \qquad \sigma^2{}_Y = 1.9508. \quad \text{(Case 2)} \end{aligned} \tag{6.258}$$

(d) The estimated PDFs are very similar as shown in Figs. 6.29 and 6.30.

Exercise M.6.7 (MATLAB file *exercise_M_6_7.m*). Generate two uncorrelated signals X_1 (normal) and X_2 (uniform) both with $m = 0$ and $\sigma^2 = 1$. Plot the autocorrelation functions.

Solution The uniform signal is generated using the transformation described in Exercise M.2.3.

The signals are shown in Fig. 6.31.

The estimated PDFs are shown in Fig. 6.32.

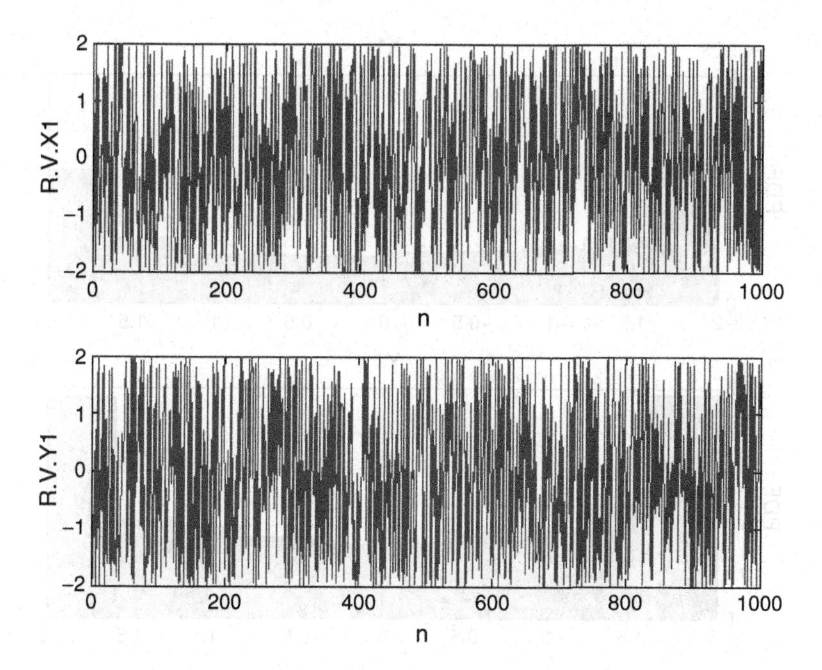

Fig. 6.27 Exercise M.6.6: Random variables (Case 1)

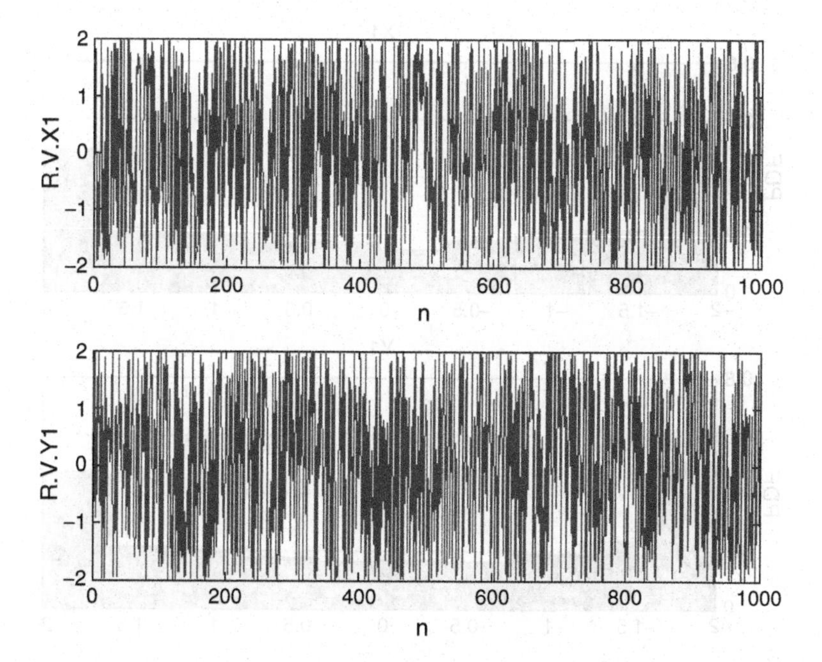

Fig. 6.28 Exercise M.6.6: Random variables (Case 2)

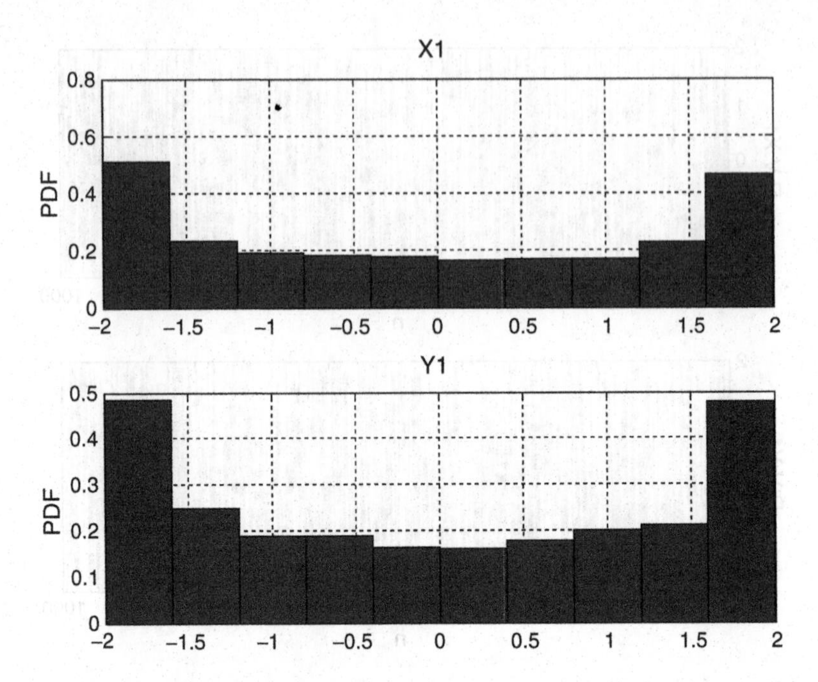

Fig. 6.29 Exercise M.6.6: Estimated PDFs (Case 1)

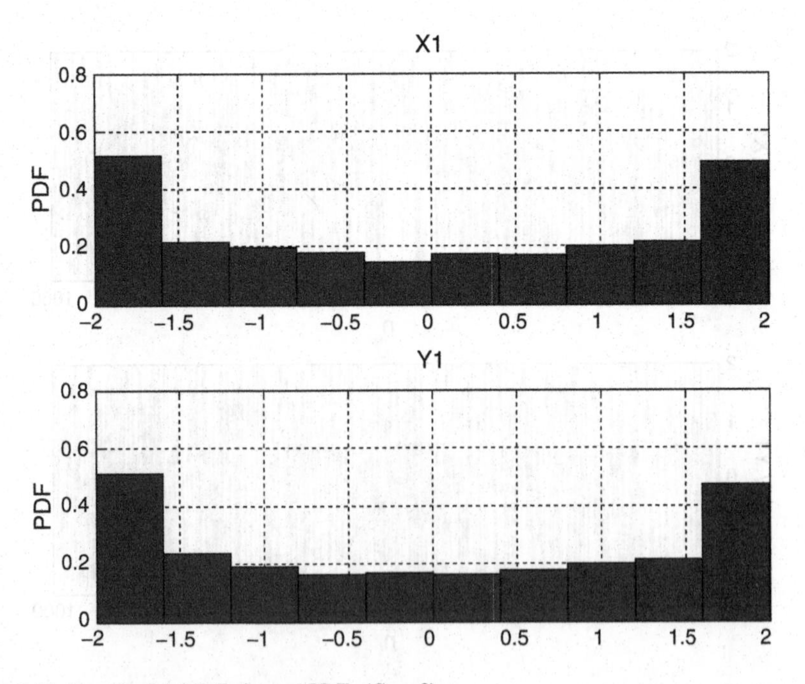

Fig. 6.30 Exercise M.6.6: Estimated PDFs (Case 2)

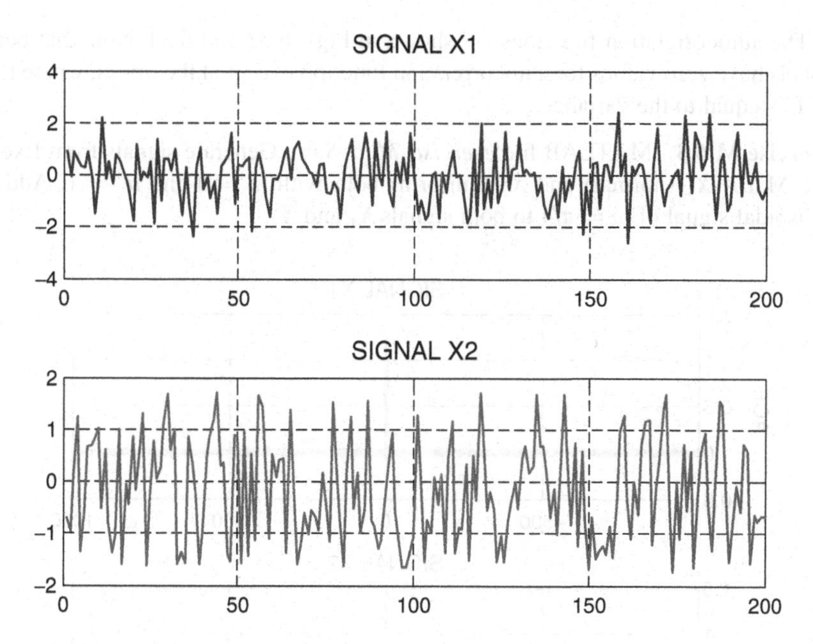

Fig. 6.31 Random signals in Exercise M.6.7

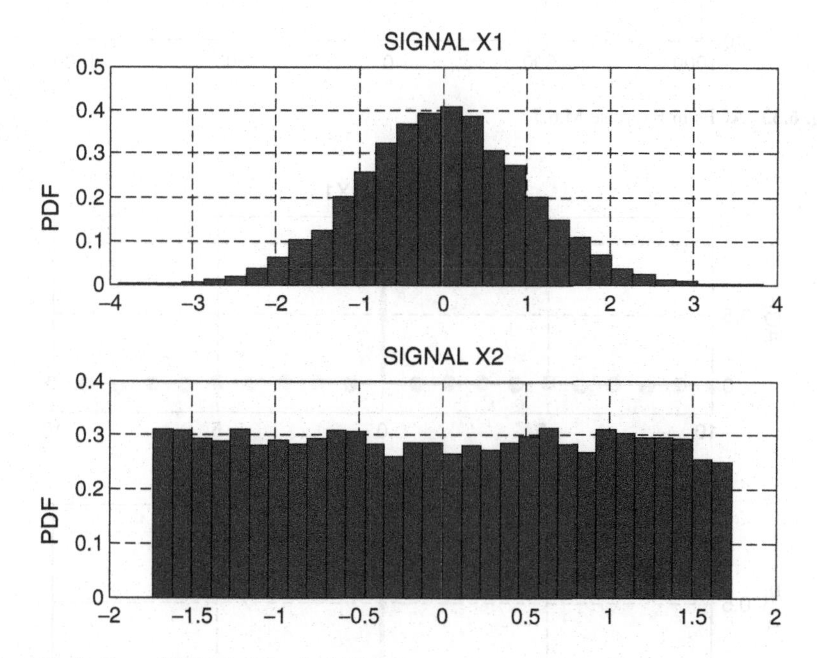

Fig. 6.32 Estimated PDFs in Exercise M.6.7

The autocorrelation functions are shown in Figs. 6.33 and 6.34. Note that both signals have zero values for autocorrelation function except at the origin, where the ACF is equal to the variance.

Exercise M.6.8 (MATLAB file *exercise_M_6_8.m*). Generate signals from Exercise M.6.7: X_1 (normal) and X_2 (uniform) both with $m = 0$ and $\sigma^2 = 1$. Add a sinusoidal signal of $\omega = \pi/4$ to both signals X_1 and X_2.

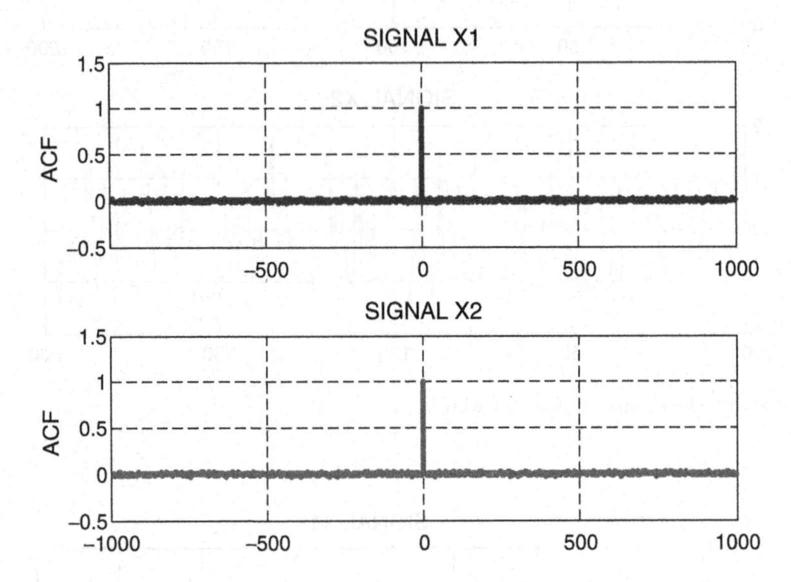

Fig. 6.33 ACFs in Exercise M.6.7

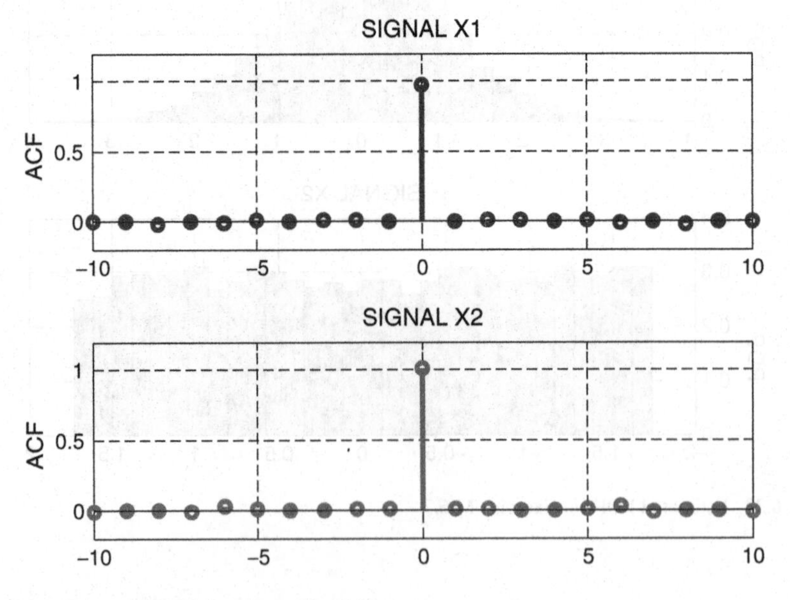

Fig. 6.34 Zooms of ACF plots from Fig. 6.33

Consider two cases: The amplitude of the sinusoidal signal is $A = 0.8$ and $A = 2$. Plot the signals, PDFs, and autocorrelation functions.

Solution
First case: $A = 0.8$;

The signals, PDFs, and ACFs are shown in Figs. 6.35–6.37.

Note that both autocorrelation functions have a periodical component.

Second case: $A = 2$;

The signals, PDFs, and ACFs are shown in Figs. 6.38–6.40.

Note that both autocorrelation functions have a periodical component. However, a sinusoidal signal is dominant compared with the random signals.

Exercise M.6.9 (MATLAB file *exercise_M_6_9.m*). Generate two uncorrelated signals X_1 (normal) and X_2 (uniform) both with $m = 2$ and $\sigma^2 = 4$. Plot the signals, PDFs, and autocorrelation functions.

Solution The uniform and normal signals are generated using the transformation method described in Sect. 2.6.

The signals, PDFs, and ACFs are shown in Figs. 6.41–6.43. Due to the nonzero mean, both autocorrelation functions have a constant term of 4 which is m^2. The autocorrelation functions at the origin are:

$$\begin{aligned}
E\{X_1^2\} &= E\{X_2^2\} = 8, \\
\sigma_1^2 &= E\{X_1^2\} - (E\{X_1\})^2 = 8 - 4 = 4, \\
\sigma_2^2 &= E\{X_2^2\} - (E\{X_2\})^2 = 8 - 4 = 4.
\end{aligned} \tag{6.259}$$

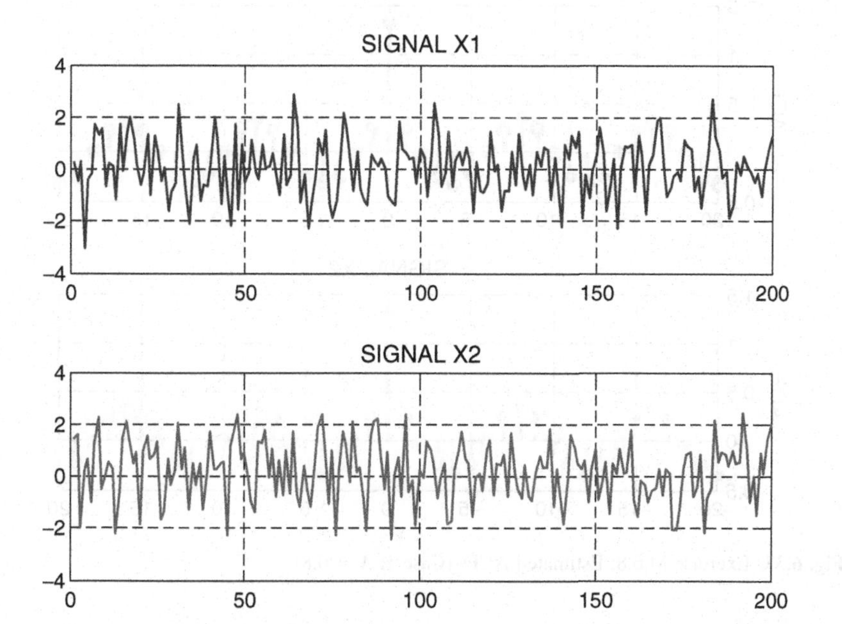

Fig. 6.35 Exercise M.6.8: Random signals (Case 1: $A = 0.8$)

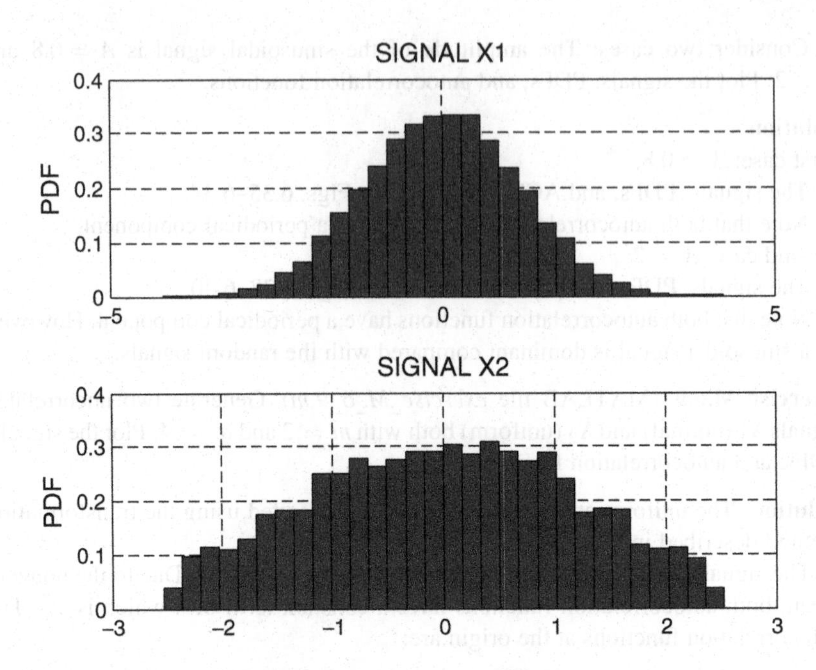

Fig. 6.36 Exercise M.6.8: Estimated PDFs (Case 1: $A = 0.8$)

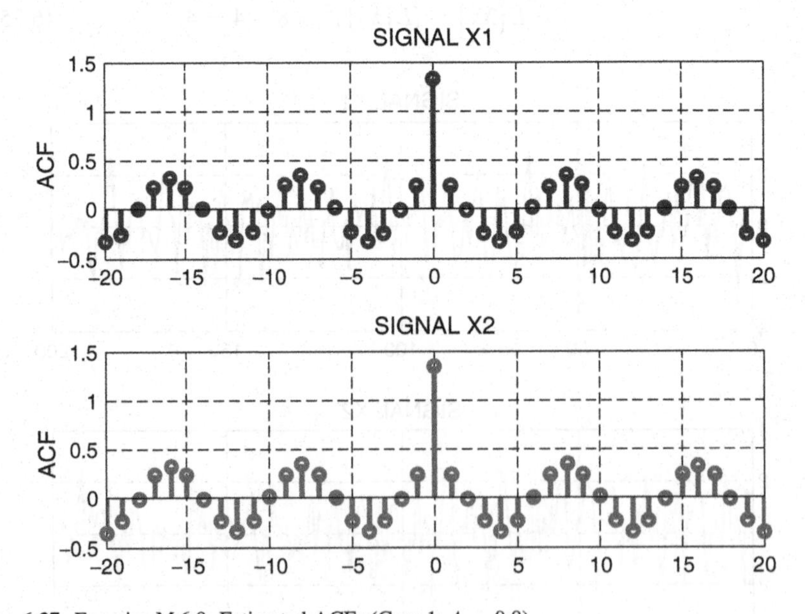

Fig. 6.37 Exercise M.6.8: Estimated ACFs (Case 1: $A = 0.8$)

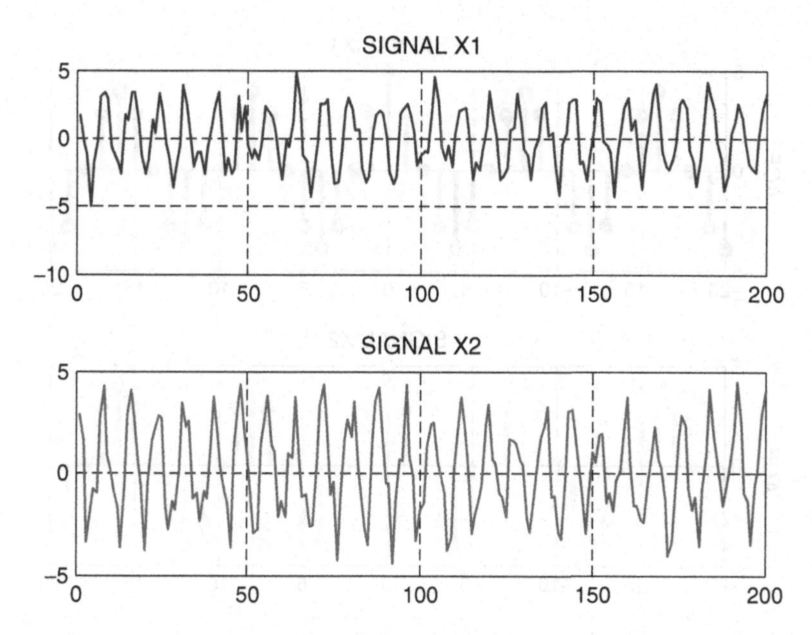

Fig. 6.38 Exercise M.6.8: Random signals (Case 2: $A = 2$)

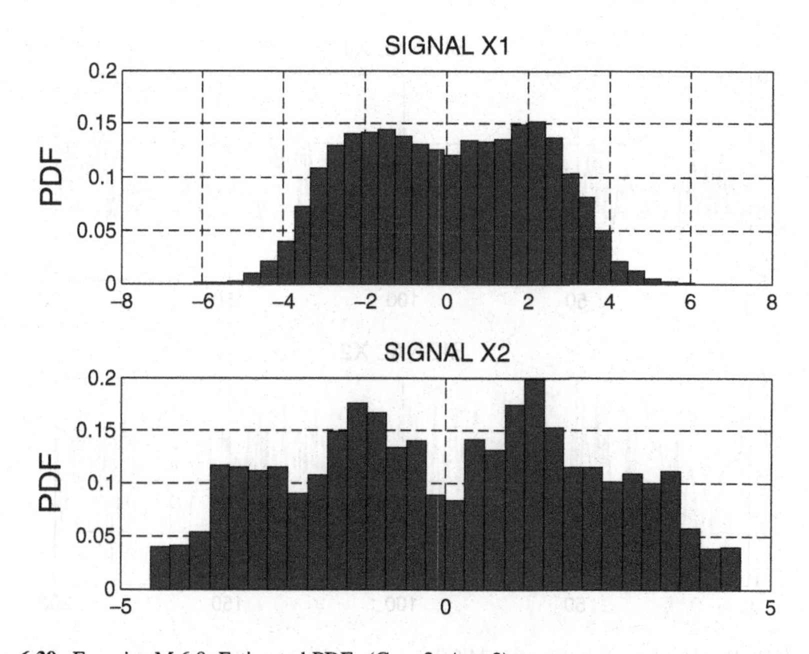

Fig. 6.39 Exercise M.6.8: Estimated PDFs (Case 2: $A = 2$)

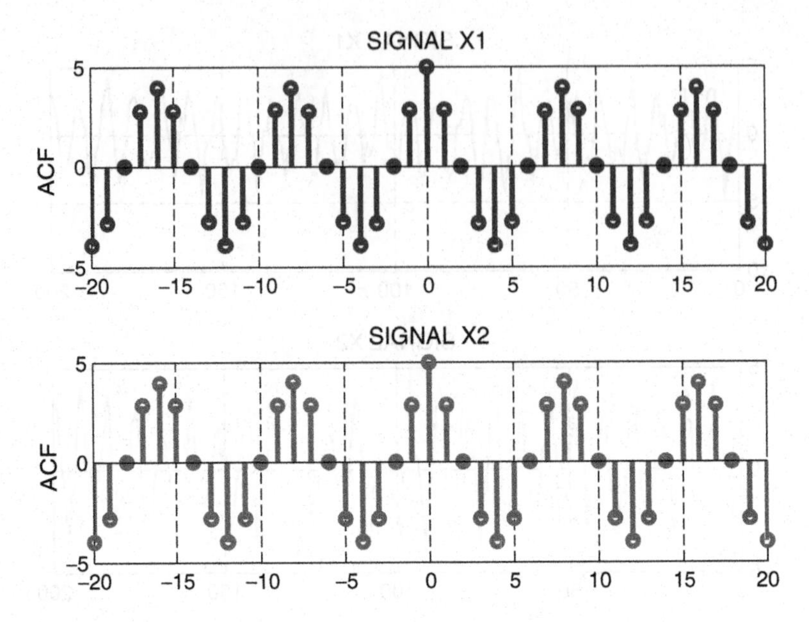

Fig. 6.40 Exercise M.6.8: Estimated ACFs (Case 2: $A = 2$)

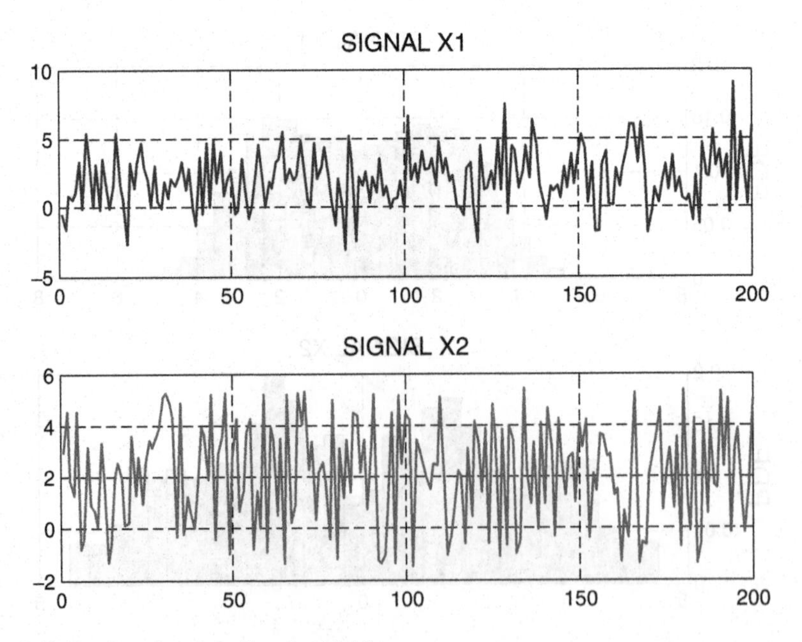

Fig. 6.41 Random signals in Exercise M.6.9

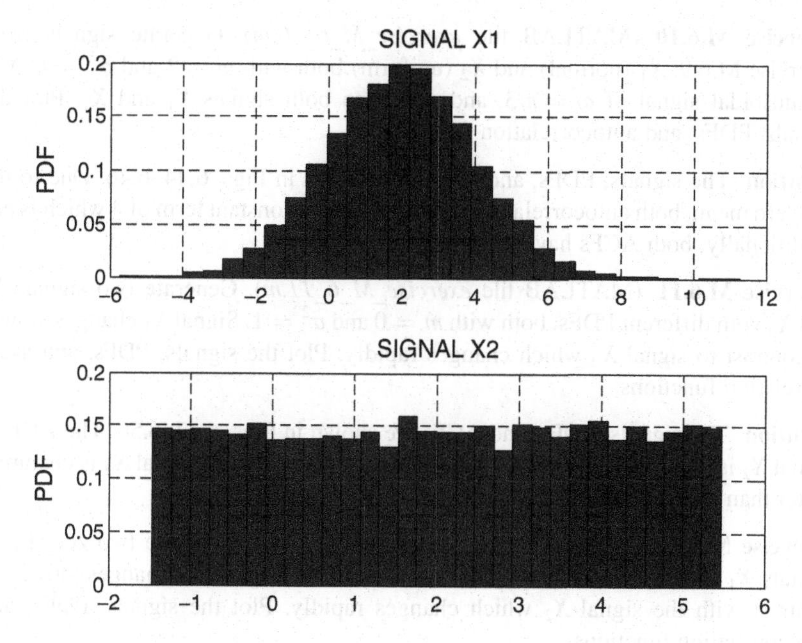

Fig. 6.42 Estimated PDFs in Exercise M.6.9

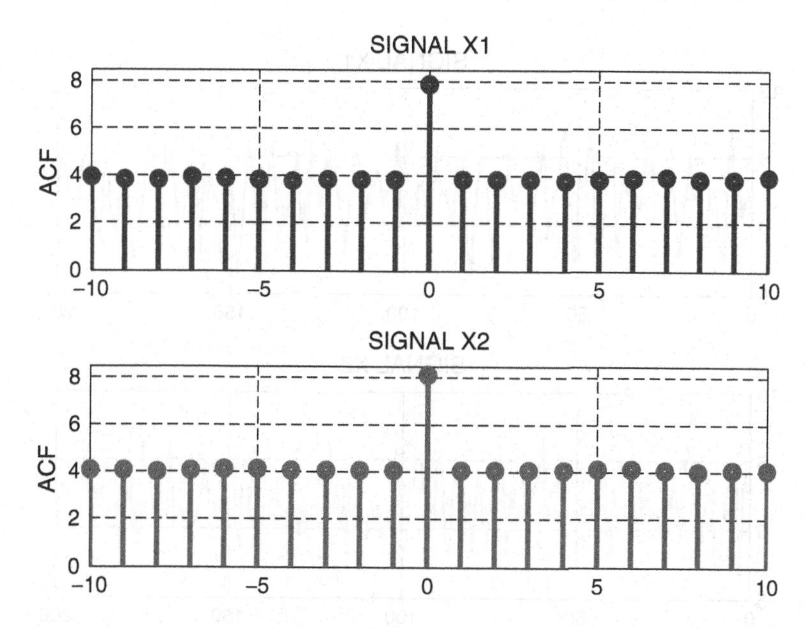

Fig. 6.43 Estimated ACFs in Exercise M.6.9

Exercise M.6.10 (MATLAB file *exercise_M_6_10.m*). Generate signals from Exercise M.6.9: X_1 (normal) and X_2 (uniform) both with $m = 2$ and $\sigma^2 = 4$. Add a sinusoidal signal of $\omega = \pi/3$, and $A = 2$ to both signals X_1 and X_2. Plot the signals, PDFs, and autocorrelation functions.

Solution The signals, PDFs, and ACFs are shown in Figs. 6.44–6.46. Due to the nonzero mean, both autocorrelation functions have a constant term of 4 which is m^2. Additionally, both ACFs have a periodical component.

Exercise M.6.11 (MATLAB file *exercise_M_6_11.m*). Generate two signals X_1 and X_2 with different PDFs, both with $m = 0$ and $\sigma^2 = 1$. Signal X_1 changes slowly in contrast to signal X_2 which changes rapidly. Plot the signals, PDFs, and autocorrelation functions.

Solution The signals, PDFs, and ACFs are shown in Figs. 6.47–6.50. The ACF of signal X_2 is decaying faster than that of the signal X_1 because signal X_2 is changing faster than is signal X_1.

Exercise M.6.12 (MATLAB file *exercise_M_6_12.m*). Generate two zero mean signals X_1 and X_2 with equal PDFs and variances. Signal X_1 changes slowly in contrast with the signal X_2 which changes rapidly. Plot the signals, PDFs, and autocorrelation functions.

Solution The signals, PDFs, and ACFs are shown in Figs. 6.51–6.54. The ACF of signal X_2 is decaying faster than that of signal X_1 because the signal X_2 is changing faster than the signal X_1.

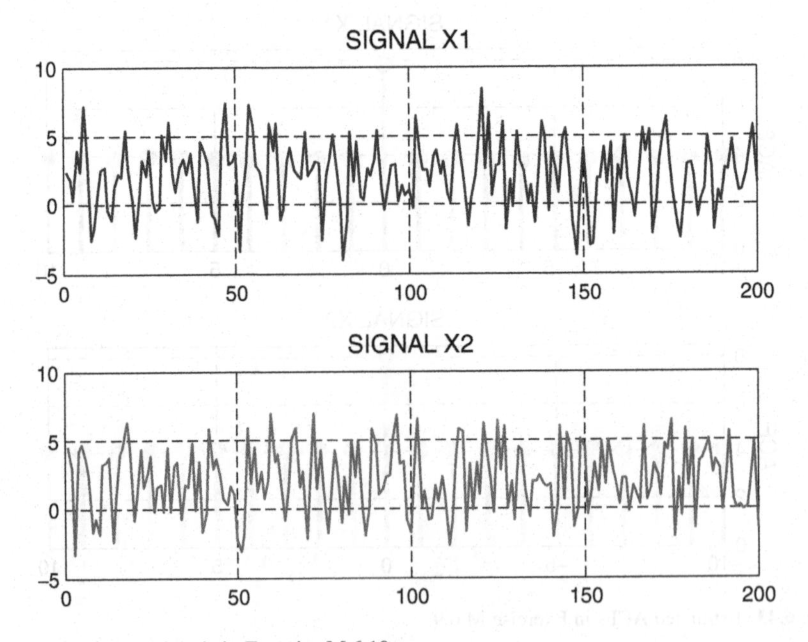

Fig. 6.44 Random signals in Exercise M.6.10

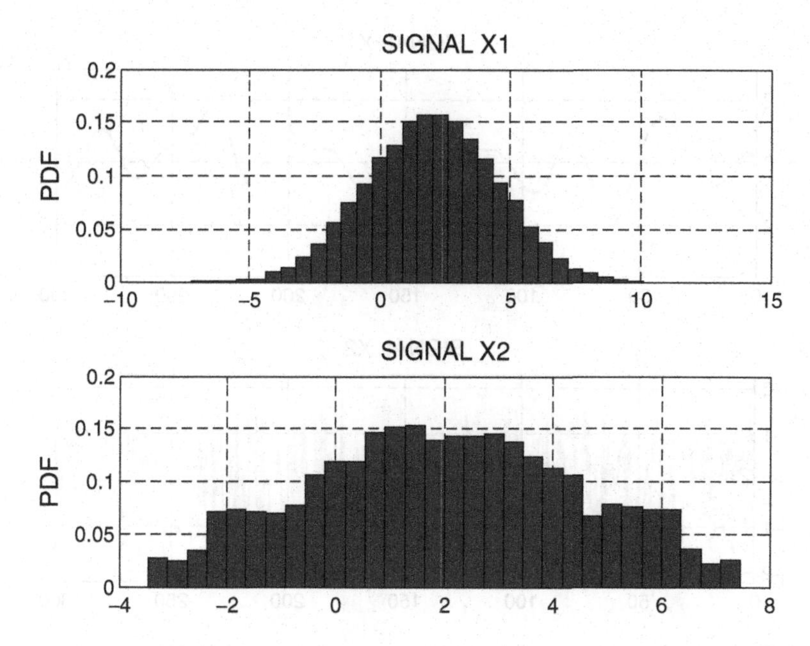

Fig. 6.45 Estimated PDFs in Exercise M.6.10

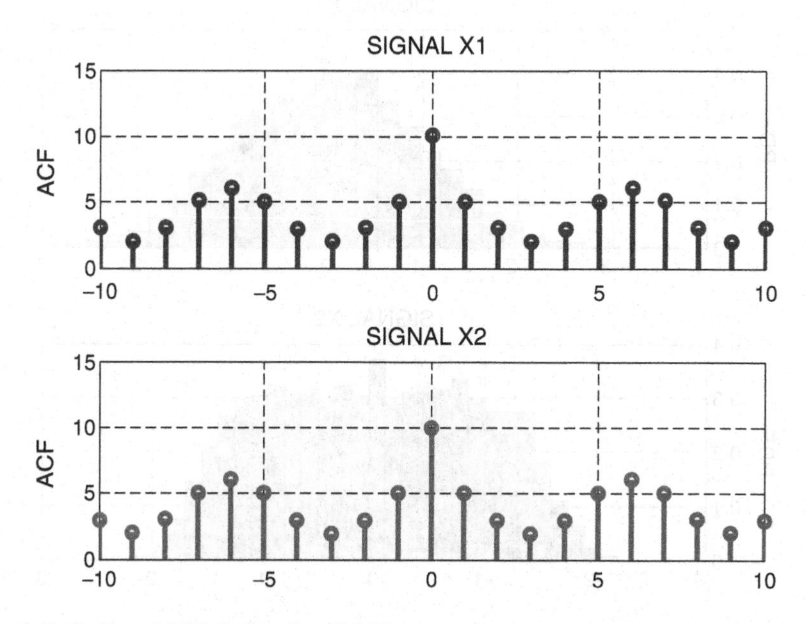

Fig. 6.46 Estimated ACFs in Exercise M.6.10

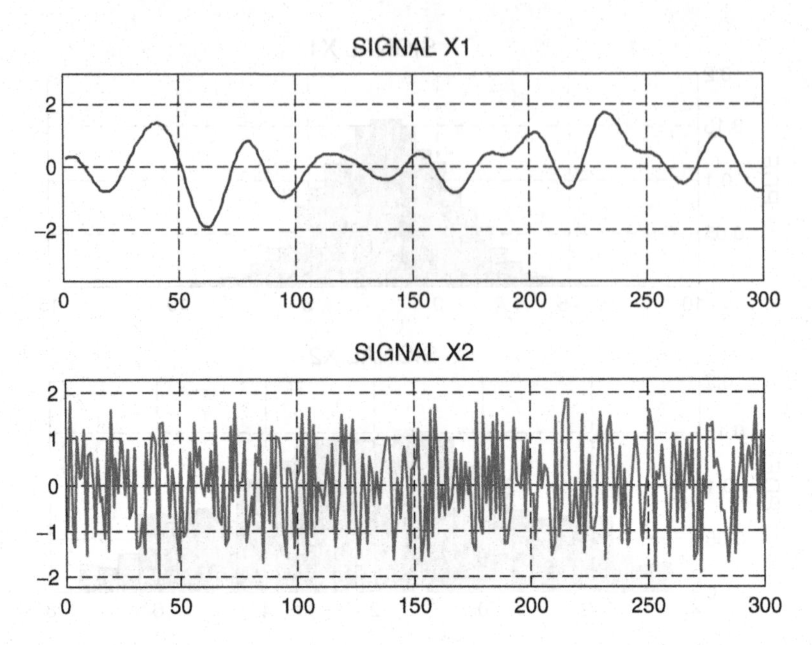

Fig. 6.47 Random signals in Exercise M.6.11

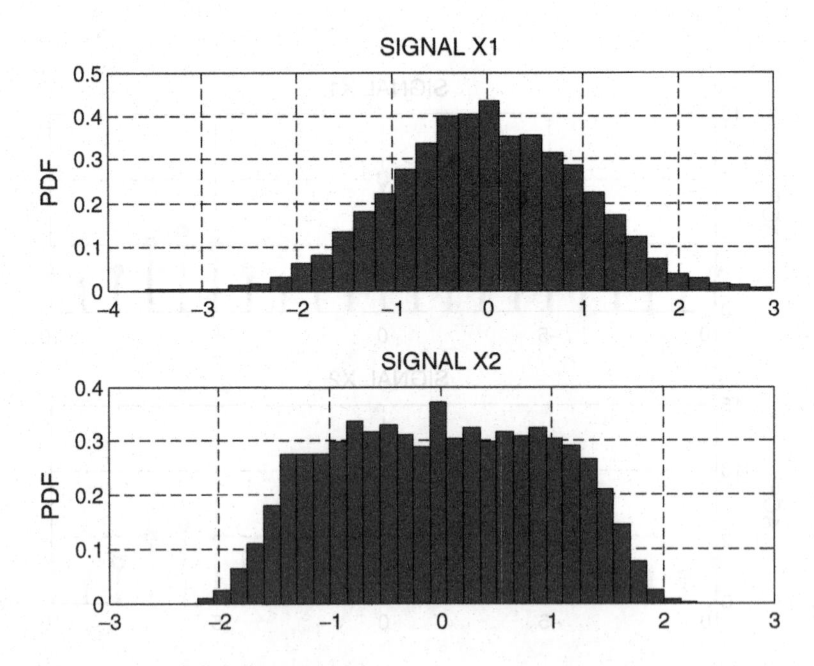

Fig. 6.48 Estimated PDFs in Exercise M.6.11

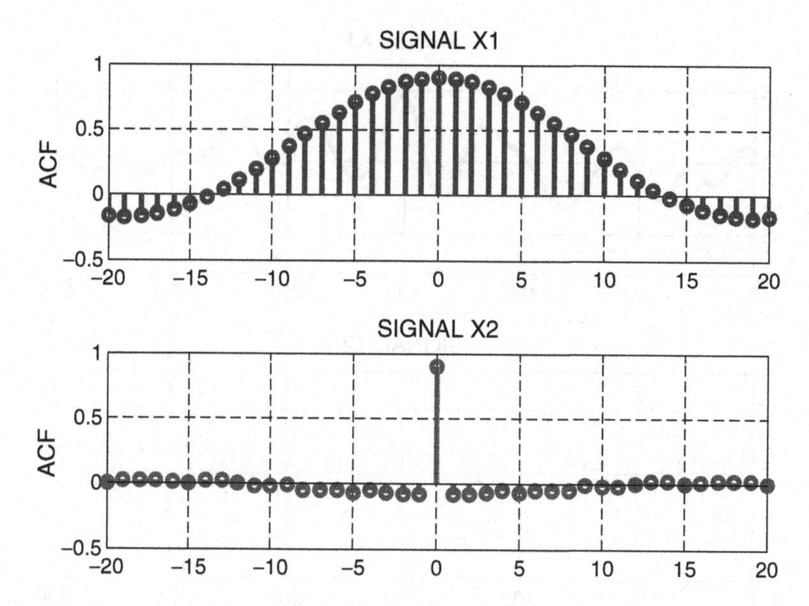

Fig. 6.49 Estimated ACFs in Exercise M.6.11

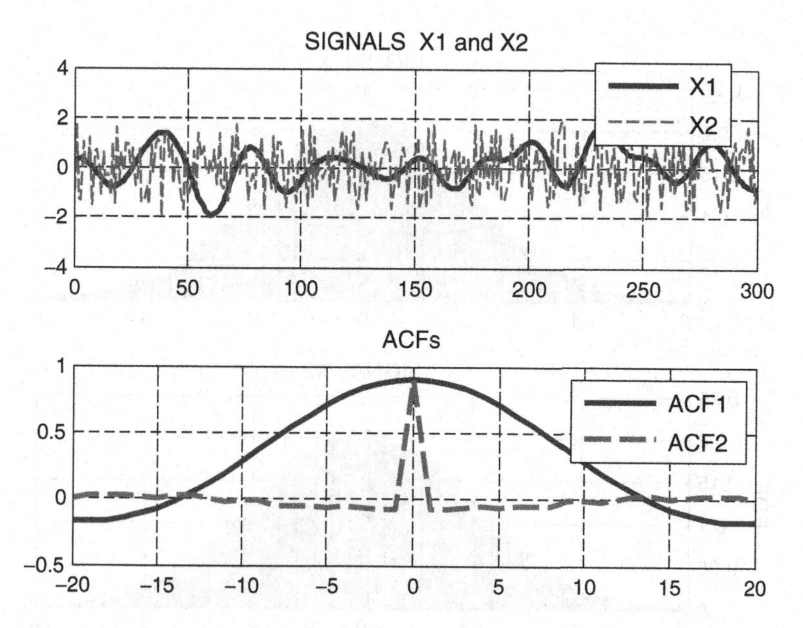

Fig. 6.50 Random signals and their ACFs in Exercise M.6.11

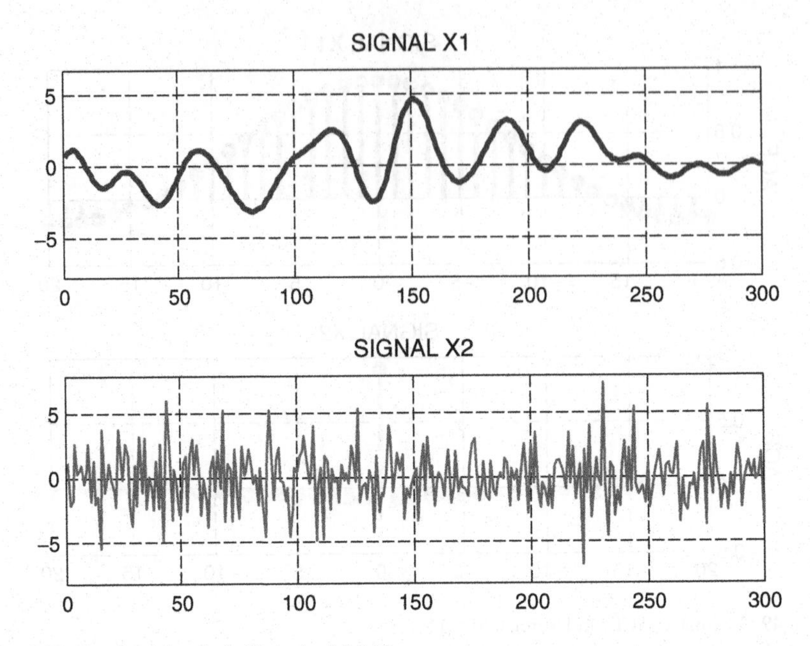

Fig. 6.51 Random signals in Exercise M.6.12

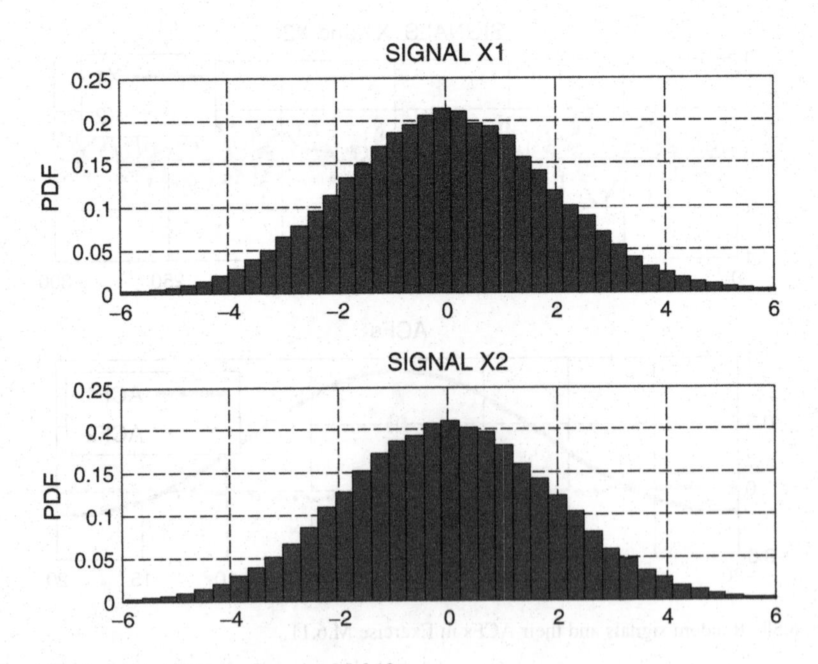

Fig. 6.52 Estimated PDFs in Exercise M.6.12

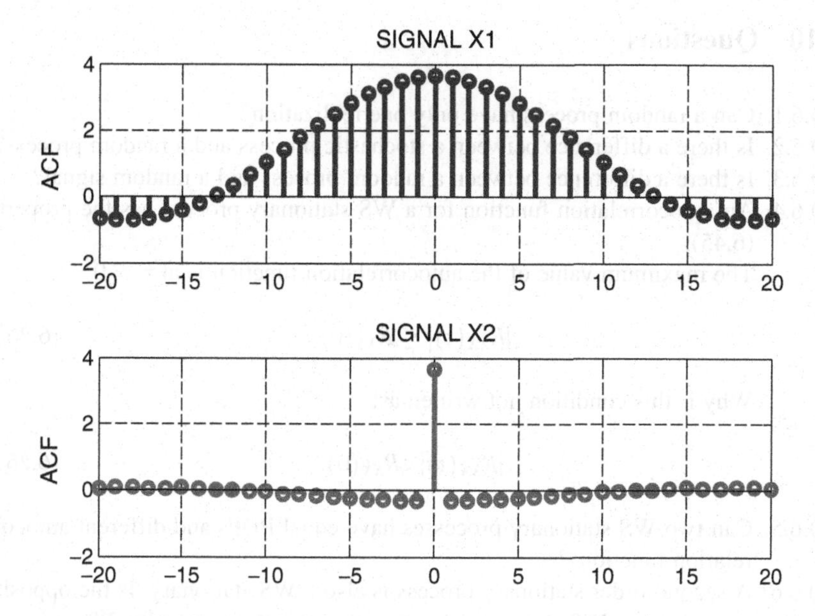

Fig. 6.53 Estimated ACFs in Exercise M.6.12

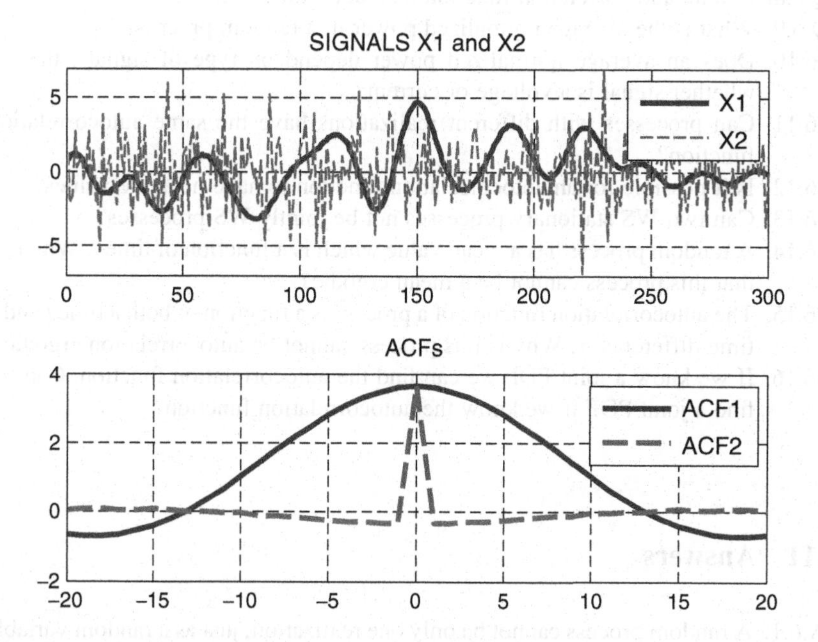

Fig. 6.54 Random signals and their ACFs in Exercise M.6.12

6.10 Questions

Q.6.1. Can a random process have only one realization?

Q.6.2. Is there a difference between a stochastic process and a random process?

Q.6.3. Is there a difference between a random process and a random signal?

Q.6.4. An autocorrelation function for a WS stationary process has the property (6.45):
The maximum value of the autocorrelation function is at $\tau = 0$

$$|R_{XX}(\tau)| \leq R_{XX}(0). \tag{6.260}$$

Why is this condition not written as:

$$|R_{XX}(\tau)| < R_{XX}(0). \tag{6.261}$$

Q.6.5. Can two WS stationary processes have equal PDFs and different autocorrelation functions?

Q.6.6. A second-order stationary process is also a WS stationary. Is the opposite true (i.e., is a WS stationary process also stationary to order 2)?

Q.6.7. Is it possible to obtain the mean value of a WS stationary process from its autocorrelation function?

Q.6.8. Can an autocorrelation function be a delta function?

Q.6.9. What is the average normalized power in a random process?

Q.6.10. Does an average normalized power depend on type of signals, that is, whether signal is a voltage or current?

Q.6.11. Can processes with different realizations have the same autocorrelation function?

Q.6.12. Is it possible for an autocorrelation function to have negative values?

Q.6.13. Can two WS stationary processes not be jointly WS processes?

Q.6.14. A random process has a mean value which is a function of time t. Why is it that this process cannot be a mean ergodic?

Q.6.15. The autocorrelation function of a process is a function of both a time t and a time difference τ. Why is this process cannot be autocorrelation ergodic?

Q.6.16. If we know a joint PDF we can find the autocorrelation function. Can we find a joint PDF if we know the autocorrelation function?

6.11 Answers

A.6.1. A random process cannot be only one realization, just as a random variable cannot be only one number. Remember that a random process is an abstraction based on the mapping of a sample space to a collection of functions. In order to apply a statistical description of a process, we need a theoretically infinite number of realizations (see Sect. 6.2).

A.6.2. The word "stochastic" is a synonym for the word "random" and thus the two can be used interchangeably.

A.6.3. The terms "random process" and "random signal" are sometimes used interchangeably. However, a random process is a collection of all realizations in ensemble. In contrast, a random signal is only one realization of a process.

If a process is ergodic then characteristics of the process can be obtained observing one realization.

For example, if a process is mean ergodic, then one can find the mean value by observing one realization which is equivalent to a statistical mean value obtained observing process (all realizations) in one point.

A.6.4. If a WS process does not have a periodic component, then

$$|R_{XX}(\tau)| < R_{XX}(0). \tag{6.262}$$

However, in order to include a periodic WS stationary process (see, for example, Exercise 6.1 where the autocorrelation function periodically takes the value $R_{XX}(0)$), in the inequality (6.262) is added sign of the equality.

A.6.5. Yes they can, as explained in Sect. 6.5.3.

A.6.6. The opposite is not necessarily true. A WS stationary process is not necessarily a second-order stationary. See, for example, [THO71, p. 211].

A.6.7. If an autocorrelation function does not have a constant term, it follows that the mean value of the process is zero. However, if an autocorrelation function has a constant term c, this term is equal

$$c = E\{X(t)\}^2. \tag{6.263}$$

From here,

$$E\{X(t)\} = \pm\sqrt{c}. \tag{6.264}$$

Therefore, in this case only the magnitude of $E\{X(t)\}$ is determined and the sign cannot be determined.

A.6.8. Yes it can. Such a process is called white noise and is explained in Chap. 7.

A.6.9. For a particular realization $x(t)$, we have an instantaneous normalized power measured using a $1 - \Omega$ impedance

$$p(t) = x^2(t). \tag{6.265}$$

This value varies from one realization to another. As such, it cannot be representative of the process. A power averaged for all realizations is called an *average power* and this quantity is thus representative of the process. An average *normalized power* is an average power measured in the $1 - \Omega$ impedance:

$$P_{\text{av}} = E\{X^2(t)\}. \tag{6.266}$$

Note that this value is related to an autocorrelation function for a WS stationary process:

$$P_{\text{av}} = E\{X^2(t)\} = R_{XX}(0). \tag{6.267}$$

A.6.10. Considering that this random process is a measured voltage, and r is the impedance in Ohms, then the average power is:

$$P_{\text{av,v}} = E\{X^2(t)\}/r. \tag{6.268}$$

Similarly, if the random process is a measured current, and r is the impedance in Ohms, then the average power is:

$$P_{\text{av,c}} = rE\{X^2(t)\}. \tag{6.269}$$

However, it is often desirable not to be concerned with whether a signal is a voltage or a current [MIL04, p. 300]. Consequently, average normalized power which is a power measured in a $1 - \Omega$ (6.267) is used rather than average powers (6.268) or (6.269).

A.6.11. An autocorrelation does not provide a complete description of the process. It describes a process only in two points. As such, it is possible that different realizations result in a same autocorrelation function.

A.6.12. There is no such restriction for an autocorrelation function to be a nonnegative (see, for example, the autocorrelation function in (6.160))

A.6.13. Two WS stationary processes are not necessarily jointly WS stationary processes because the additional conditions (6.101) and (6.102) must be satisfied (see Exercise 6.15).

A.6.14. From (6.143), a time-averaged mean value cannot be a function of time but only a constant. Therefore, if a statistical mean is not a constant cannot be equal to a time-averaged mean value. Consequently, the process cannot be mean ergodic.

A.6.15. From (6.151), a time-averaged autocorrelation function cannot be a function of time. Therefore, a statistical autocorrelation function which depends on time t cannot be equal to a time-averaged autocorrelation function and, consequently, a process cannot be autocorrelation ergodic.

A.6.16. It is not possible. There are different processes and thus different joint PDFs which have equal autocorrelation functions. Therefore, knowledge of an autocorrelation function is not equivalent to knowledge of a joint PDF and is less informative.

Chapter 7
Spectral Characteristics of Random Processes

7.1 Power Spectral Density

The previous discussion was focused on a description of a random process in time. However, as in the description of deterministic signals, it is of interest to also describe a random process in the frequency domain. We will consider here only real processes.

7.1.1 Fourier Transformation of Deterministic Signals

The spectral description of a deterministic signal $f(t)$ is obtained by taking the *Fourier transform* $F(\omega)$, also known as the *spectrum*, of a signal $f(t)$:

$$F(\omega) = \int_{-\infty}^{\infty} f(t)\mathrm{e}^{-j\omega t}\mathrm{d}t. \tag{7.1}$$

If $f(t)$ is voltage, then $F(\omega)$ represents voltage per Hertz and describes how the voltage is distributed with frequencies.

In order for the Fourier transformation (7.1) to exist, the following condition must be satisfied:

$$\int_{-\infty}^{\infty} |f(t)|\mathrm{d}t<\infty. \tag{7.2}$$

G.J. Dolecek, *Random Signals and Processes Primer with MATLAB*,
DOI 10.1007/978-1-4614-2386-7_7,

The *inverse Fourier transformation* finds the time domain signal $f(t)$ from its spectrum

$$f(t) = \frac{1}{2\pi} \int_{-\infty}^{\infty} F(\omega) e^{j\omega t} d\omega. \tag{7.3}$$

The question that one might naturally ask is: "Can we also apply the Fourier transformation to the random signals?"

The answer is provided in the following section.

7.1.2 How to Apply the Fourier Transformation to Random Signals?

The direct application of the Fourier transform to a random process is not possible for the reasons listed below:

- There is a problem of the existence of the Fourier transform (7.2), i.e., for a majority of the realizations of the process, the condition (7.2) is not satisfied and the Fourier transform may not exist.
- Even in a case in which the Fourier transform of a particular realization exists, it is obvious that the result cannot represent the whole process. The Fourier transform of the individual realizations would be different.
- In the majority of cases, realizations of a process have irregular forms and cannot be represented in analytical forms and, consequently, there is a problem of practical calculation of the Fourier transform.

Next we will describe how to overcome these difficulties by applying the Fourier transform to a random process.

7.1.2.1 Problem of the Existence of the Fourier Transformation

For a particular realization $x(t)$ of a random process $X(t)$, which exists in the interval $[-\infty, \infty]$, the condition

$$\int_{-\infty}^{\infty} |x(t)| dt < \infty \tag{7.4}$$

may not be satisfied and, thus, the Fourier transform does not exist. The mean energy W of a particular realization would also be infinity

$$W = \lim_{T \to \infty} \int_{-T/2}^{T/2} x^2(t) dt. \tag{7.5}$$

However, a mean power P

$$P = \lim_{T\to\infty} \frac{W}{T} = \lim_{T\to\infty} \frac{1}{T} \int_{-T/2}^{T/2} x^2(t)\mathrm{d}t \tag{7.6}$$

is finite. This conclusion suggests the application of the Fourier transform to a power rather than to amplitude of a realization. This idea is developed below.

Consider the part of a realization $x(t)$ of a process $X(t)$ in the interval $[-T/2, T/2]$, as shown in Fig. 7.1.

$$x_T(t) = \begin{cases} x(t) & \text{for} \quad -T/2 \le t \le T/2, \\ 0 & \text{otherwise.} \end{cases} \tag{7.7}$$

For the finite interval T, x_T will satisfy the condition (7.4) and, thus, x_T will have the Fourier transform:

$$X_T(\omega) = \int_{-T/2}^{T/2} x_T(t)\mathrm{e}^{-j\omega t}\mathrm{d}t = \int_{-\infty}^{\infty} x_T(t)\mathrm{e}^{-j\omega t}\mathrm{d}t = \int_{-T/2}^{T/2} x(t)\mathrm{e}^{-j\omega t}\mathrm{d}t. \tag{7.8}$$

The energy of the signal $X_T(t)$ is:

$$W_T = \int_{-T/2}^{T/2} x_T^2(t)\mathrm{d}t = \int_{-\infty}^{\infty} x_T^2(t)\mathrm{d}t. \tag{7.9}$$

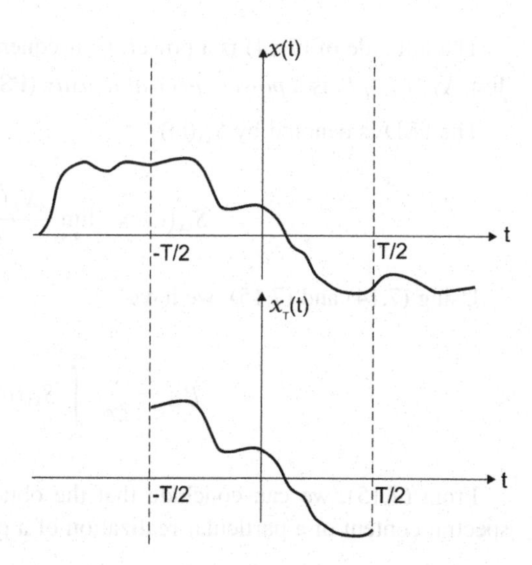

Fig. 7.1 Realization $x(t)$ in the interval $[-T/2,T/2]$

According to the Parseval's theorem (see Appendix E), the energy W_T can be expressed as:

$$W_T = \frac{1}{2\pi} \int_{-\infty}^{\infty} |X_T(\omega)|^2 d\omega. \tag{7.10}$$

The obtained expression is also the energy of a signal $x(t)$ in the interval $[-T/2, T/2]$. Consequently, the energy of the whole signal $x(t)$ is obtained from (7.10), when $T \to \infty$,

$$W = \lim_{T \to \infty} W_T. \tag{7.11}$$

From (7.11), the power of the signal $x(t)$ is:

$$P_{xx} = \lim_{T \to \infty} \frac{W_T}{T}. \tag{7.12}$$

Using (7.10), we finally arrive at:

$$P_{xx} = \lim_{T \to \infty} \frac{1}{T} \left[\frac{1}{2\pi} \int_{-\infty}^{\infty} |X_T(\omega)|^2 d\omega \right]. \tag{7.13}$$

This expression can be rewritten as:

$$P_{xx} = \frac{1}{2\pi} \int_{-\infty}^{\infty} \lim_{T \to \infty} \frac{|X_T(\omega)|^2}{T} d\omega. \tag{7.14}$$

The left side of (7.14) is a power. Consequently, the expression in the integral, $\lim_{T \to \infty} |X_T(\omega)|^2 / T$, is a *power spectral density* (PSD) with a unit [Power/Hertz]. The PSD is denoted by $S_{xx}(\omega)$

$$S_{xx}(\omega) = \lim_{T \to \infty} \frac{|X_T(\omega)|^2}{T}. \tag{7.15}$$

Using (7.14) and (7.15), we have:

$$P_{xx} = \frac{1}{2\pi} \int_{-\infty}^{\infty} S_{xx}(\omega) d\omega. \tag{7.16}$$

From (7.15), we can conclude that the obtained PSD exists and presents the spectral content of a particular realization of a process $X(t)$.

7.1.2.2 How Can We Obtain PSD of the Process from the PSD of a Particular Realization?

In order to obtain the power spectral density of the process $S_{XX}(\omega)$, we need to find the average value of power spectral densities over the whole ensemble. In other words, we have to find an average value of power spectral densities $S_{xx}(\omega)$ for all realizations:

$$S_{XX}(\omega) = E\{S_{xx}(\omega)\}. \tag{7.17}$$

Therefore, we have:

$$S_{XX}(\omega) = E\left\{\lim_{T\to\infty} \frac{1}{T}|X_T(\omega)|^2\right\} = \lim_{T\to\infty} \frac{E\left\{|X_T(\omega)|^2\right\}}{T}. \tag{7.18}$$

The mean power of the process is given below:

$$P_{XX} = \frac{1}{2\pi}\int_{-\infty}^{\infty} S_{XX}(\omega)\mathrm{d}\omega. \tag{7.19}$$

7.1.2.3 How Can We Find the Spectral Density (7.18) If the Realizations Have an Irregular Form and Cannot Be Represented in Analytical Form?

Obviously, in this case, there is a problem related to the practical calculation of the PSD. The solution is given in the following.

7.1.3 *PSD and Autocorrelation Function*

As mentioned before, the obtained expression (7.18) for the PSD is not convenient for a practical application, because it is necessary to calculate the Fourier transform of the squared amplitudes of a random process, which usually cannot be expressed in an analytical form. Consequently, it is of interest to express the PSD in a more convenient form, as is explained below.

From the Fourier transform of a signal $x_T(t)$, defined in (7.8):

$$X_T(\omega) = \int_{-\infty}^{\infty} x_T(t)\mathrm{e}^{-j\omega t}\mathrm{d}t = \int_{-T/2}^{T/2} x(t)\mathrm{e}^{-j\omega t}\mathrm{d}t, \tag{7.20}$$

we write:

$$
\begin{aligned}
|X_T(\omega)|^2 &= X_T(-\omega)X_T(\omega) \\
&= \int_{-T/2}^{T/2} x(t_1)\mathrm{e}^{j\omega t_1}\mathrm{d}t_1 \int_{-T/2}^{T/2} x(t_2)\mathrm{e}^{-j\omega t_2}\mathrm{d}t \\
&= \int_{-T/2}^{T/2}\int_{-T/2}^{T/2} x(t_1)x(t_2)\mathrm{e}^{-j\omega(t_2-t_1)}\mathrm{d}t_1\mathrm{d}t_2.
\end{aligned} \tag{7.21}
$$

Using (7.18) and (7.21) and knowing that expectation is applied to a process $X(t)$ we have [PEE93, p. 206]:

$$
S_{XX}(\omega) = \lim_{T\to\infty}\frac{1}{T}\overline{|X_T(\omega)|^2} = \lim_{T\to\infty}\frac{1}{T}\overline{\int_{-T/2}^{T/2}\int_{-T/2}^{T/2} X(t_1)X(t_2)\mathrm{e}^{-j\omega(t_2-t_1)}\mathrm{d}t_1\mathrm{d}t_2}. \tag{7.22}
$$

By interchanging the operations of expectation and integration, we have:

$$
S_{XX}(\omega) = \lim_{T\to\infty}\frac{1}{T}\int_{-T/2}^{T/2}\int_{-T/2}^{T/2} \overline{X(t_1)X(t_2)}\mathrm{e}^{-j\omega(t_2-t_1)}\mathrm{d}t_1\mathrm{d}t_2. \tag{7.23}
$$

The expected value in (7.23) presents an autocorrelation function:

$$
\overline{X(t_1)X(t_2)} = R_{XX}(t_1,t_2), \tag{7.24}
$$

resulting in:

$$
S_{XX}(\omega) = \lim_{T\to\infty}\frac{1}{T}\int_{-T/2}^{T/2}\int_{-T/2}^{T/2} R_{XX}(t_1,t_2)\mathrm{e}^{-j\omega(t_2-t_1)}\mathrm{d}t_1\mathrm{d}t_2. \tag{7.25}
$$

Expressing t_1 and t_2 as t and $t+\tau$, respectively (where $\tau = t_2 - t_1$), we get:

$$
S_{XX}(\omega) = \lim_{T\to\infty}\frac{1}{T}\int_{-T/2-t}^{T/2-t}\int_{-T/2}^{T/2} R_{XX}(t,t+\tau)\mathrm{e}^{-j\omega\tau}\mathrm{d}t\mathrm{d}\tau, \tag{7.26}
$$

or alternatively

$$S_{XX}(\omega) = \int_{-\infty}^{\infty} \left\{ \lim_{T\to\infty} \frac{1}{T} \int_{-T/2}^{T/2} R_{XX}(t, t+\tau)\mathrm{d}t \right\} \mathrm{e}^{-j\omega\tau}\mathrm{d}\tau. \tag{7.27}$$

The expression in the brackets presents a time average of the autocorrelation function of the process $X(t)$

$$A\{R_{XX}(t, t+\tau)\} = \lim_{T\to\infty} \frac{1}{T} \int_{-T/2}^{T/2} R_{XX}(t, t+\tau)\mathrm{d}t. \tag{7.28}$$

If we suppose that the process is at least WS stationary, its autocorrelation function does not depend on absolute time but only of a time difference τ, resulting in:

$$A\{R_{XX}(t, t+\tau)\} = \lim_{T\to\infty} \frac{1}{T} \int_{-T/2}^{T/2} R_{XX}(\tau)\mathrm{d}t = R_{XX}(\tau). \tag{7.29}$$

Placing (7.29) into (7.27), we finally get:

$$S_{XX}(\omega) = \int_{-\infty}^{\infty} R_{XX}(\tau)\mathrm{e}^{-j\omega\tau}\mathrm{d}\tau, \tag{7.30}$$

or equivalently

$$R_{XX}(\tau) = \frac{1}{2\pi} \int_{-\infty}^{\infty} S_{XX}(\omega)\mathrm{e}^{j\omega\tau}\mathrm{d}\tau. \tag{7.31}$$

The obtained expressions (7.30) and (7.31) are known as the *Wiener–Khinchin theorem*, which states that a *PSD and a autocorrelation function of a process, which is at least WS stationary, are the Fourier transform pair.*

The importance of this theorem is that one can obtain a spectral characteristic (i.e., the PSD of a random process as the Fourier transform of an autocorrelation function), which itself is a deterministic function that represents the process. Consequently, the Wiener–Khinchin theorem solves all problems related to the application of the Fourier transform to a random process, mentioned in Sect. 7.1.2.

Example 7.1.1 Find a PSD for the process considered in Example 6.5.2, where the autocorrelation is obtained in the expression (6.77).

Solution Using (7.30) and (6.77), we write:

$$\begin{aligned} S_{XX}(\omega) &= \int_{-\infty}^{\infty} U^2 \mathrm{e}^{-2\lambda|\tau|} \mathrm{e}^{-j\omega\tau} \mathrm{d}\tau, \\ &= U^2 \left[\int_{-\infty}^{0} \mathrm{e}^{(2\lambda - j\omega)\tau} \mathrm{d}\tau + \int_{0}^{\infty} \mathrm{e}^{-(2\lambda + j\omega)\tau} \mathrm{d}\tau \right], \\ &= 2U^2 \frac{2\lambda}{(2\lambda)^2 + \omega^2}. \end{aligned} \tag{7.32}$$

The spectral density is shown in Fig. 7.2 for $\lambda = 1$ and $U = 1$.

As expected, one may note from Fig. 7.2, that the spectrum band becomes wider for higher values of λ, i.e. average number of zero crossing in a unit of time.

7.1.4 Properties of PSD

The PSD has a number of useful properties. Some of them are presented below. We will consider only a real process which is at least WS stationary.

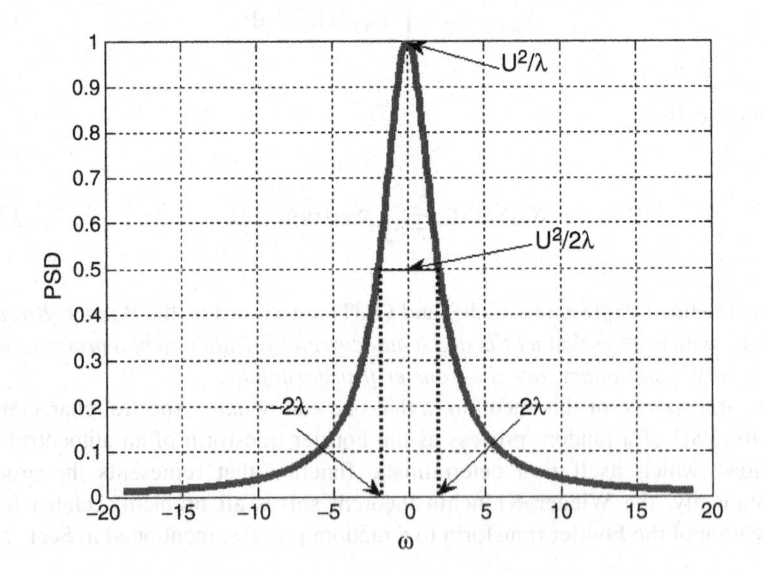

Fig. 7.2 Power spectral density (PSD) in Example 7.1.1

P.1 The power density is a real function.
From (7.30), we have:

$$\begin{aligned} S_{XX}(\omega) &= \int_{-\infty}^{\infty} R_{XX}(\tau)\mathrm{e}^{-j\omega\tau}\mathrm{d}\tau, \\ &= \int_{-\infty}^{\infty} R_{XX}(\tau)(\cos\omega\tau - j\sin\omega\tau)\mathrm{d}\tau, \\ &= \int_{-\infty}^{\infty} R_{XX}(\tau)\cos\omega\tau\mathrm{d}\tau - j\int_{-\infty}^{\infty} R_{XX}(\tau)\sin\omega\,\tau\mathrm{d}\tau. \end{aligned} \tag{7.33}$$

The second integral in (7.33) is zero, because the integral over a symmetric range of the product of $R_{XX}(\tau)$ (which is an even function) and $\sin\omega\tau$ (which is an odd function) is zero.
Therefore, we get:

$$S_{XX}(\omega) = \int_{-\infty}^{\infty} R_{XX}(\tau)\cos\omega\tau\mathrm{d}\tau, \tag{7.34}$$

which is a real function.

P.2 A PSD is an even function in a radian frequency ω:

$$S_{XX}(\omega) = S_{XX}(-\omega). \tag{7.35}$$

From (7.34), we have:

$$S_{XX}(-\omega) = \int_{-\infty}^{\infty} R_{XX}(\tau)\cos(-\omega\tau)\mathrm{d}\tau = \int_{-\infty}^{\infty} R_{XX}(\tau)\cos\omega\tau\mathrm{d}\tau = S_{XX}(\omega). \tag{7.36}$$

P.3 The average power of a random process is obtained as the area below the PSD, as shown in Fig. 7.3.

$$E\{X^2(t)\} = \overline{P} = \frac{1}{2\pi}\int_{-\infty}^{\infty} S_{XX}(\omega)\mathrm{d}\omega = \int_{-\infty}^{\infty} S_{XX}(f)\mathrm{d}f, \tag{7.37}$$

where $\omega = 2\pi f$ and f is a frequency in Hz.
From (7.31), we have:

$$R_{XX}(0) = E\{X(t)X(t+0)\} = \frac{1}{2\pi}\int_{-\infty}^{\infty} S_{XX}(\omega)\mathrm{e}^{j\omega 0}\mathrm{d}\omega = \int_{-\infty}^{\infty} S_{XX}(f)\mathrm{d}f. \tag{7.38}$$

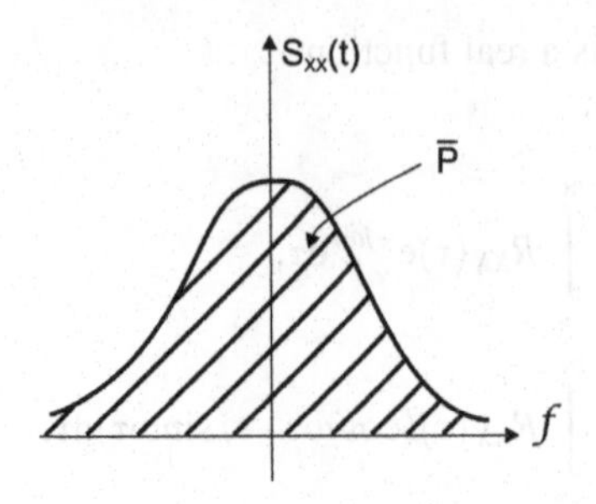

Fig. 7.3 Average power and spectral density

P.4 The PSD is a nonnegative function

$$S_{XX}(\omega) \geq 0. \tag{7.39}$$

This property follows from the definition (7.18)

$$S_{XX}(\omega) = \lim_{T\to\infty} \frac{1}{T}\overline{|X_T(\omega)|^2}, \tag{7.40}$$

where one can see that a PSD is defined by an expected value of a nonnegative function, which itself is also a nonnegative function.

Example 7.1.2 We prove the properties of the spectral density from Example 7.1.1.

Solution The spectral density is:

$$S_{XX}(\omega) = 2U^2 \frac{2\lambda}{(2\lambda)^2 + \omega^2}. \tag{7.41}$$

From (7.41), we can conclude:

P.1. The spectral density is real.

P.2. The spectral density is an even function.

$$S_{XX}(-\omega) = 2U^2 \frac{2\lambda}{(2\lambda)^2 + (-\omega)^2} = S_{XX}(\omega). \tag{7.42}$$

P.3. The average power is obtained as:

$$\begin{aligned} E\{X^2(t)\} &= \overline{P} = \frac{1}{2}U^2 + \frac{1}{2}(-U)^2 = U^2 \\ &= \frac{1}{2\pi}\int_{-\infty}^{\infty} 2U^2 \frac{2\lambda}{(2\lambda)^2 + \omega^2} \mathrm{d}\omega = \frac{U^2}{\pi}\tan^{-1}\left(\frac{\omega}{2\lambda}\right)\Big|_{-\infty}^{\infty} = U^2. \end{aligned} \tag{7.43}$$

The reader can find, as an exercise, that there is half of the power in the range $[-2\lambda, 2\lambda]$ (see Fig. 7.2).

P.4.

$$S_{XX}(\omega) = 2U^2 \frac{2\lambda}{(2\lambda)^2 + \omega^2} \geq 0. \tag{7.44}$$

7.1.5 Interpretation of PSD

So far, we have concluded that a mean power is an area under a PSD, as given in Fig. 7.3, where both negative and positive frequencies are shown. However, it is well known that negative frequencies do not exist because the frequency is the number of oscillations per second and, obviously, cannot be negative.

In order to explain this situation, let us imagine an experiment [LEE60] in which random voltage signal $x(t)$, which does contains neither periodic nor DC components, is filtered by a bank of ideal low-pass filters with the cutoff frequencies, $f_1, f_2, \ldots, f_k, \ldots, f_K$ (as shown in Fig. 7.4). The filters are ideal, and pass all frequency components less than and equal to the corresponding cutoff frequency and eliminate all frequencies which are greater than the corresponding cutoff frequency. For example, all frequency components of the signal $f_1, \ldots, f_k$ will be at the output of the kth filter.

Outputs of the filters are connected with the unit value resistors and the mean power at the resistors is measured. The results of measurements are shown in the diagram of the mean power $\bar{P}(f)$.

For example, the mean power P_k measured at the output of the k-filter represents the mean power of the signal in the frequency band $[0, f_k]$. Note that the power which belongs to the frequency $f = 0$ is zero because the signal $x(t)$ has a zero mean value. This is explained below.

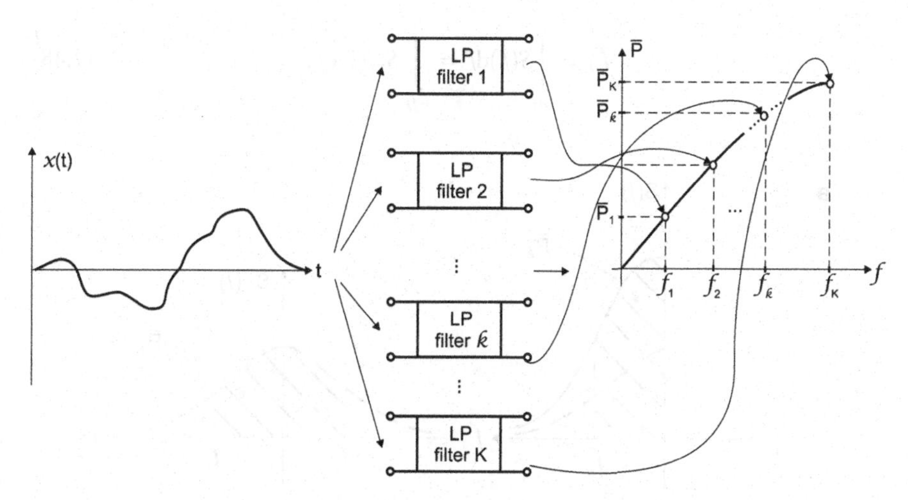

Fig. 7.4 Measurement of mean power $\bar{P}(f)$.

Denoting a power, which belongs to a infinitesimal frequency band d*f*, as *S*(*f*)d*f*, we get:

$$\bar{P}_k = \int_0^{f_k} S(f)\mathrm{d}f, \tag{7.45}$$

or generally

$$\bar{P}(f) = \int_0^{f} S(f)\mathrm{d}f. \tag{7.46}$$

From here, we have:

$$S(f) = \frac{\mathrm{d}\bar{P}(f)}{\mathrm{d}f}. \tag{7.47}$$

The obtained result, shown in Fig. 7.5a, is the spectral density of a random signal, and is, naturally, defined only for the positive frequencies. However, the negative frequencies are introduced in a frequency analysis for mathematical reasons in order to present the Fourier series in exponential form.

Therefore, in order to obtain the mathematical spectral density, all values of *S*(*f*) are halved, and the top part of the plot is mapped to the left part, resulting in the diagram shown in Fig. 7.5b.

Therefore,

$$\bar{P}_k = \int_0^{f_k} S(f)\mathrm{d}f = \int_{-f_k}^{f_k} S_{xx}(f)\mathrm{d}f. \tag{7.48}$$

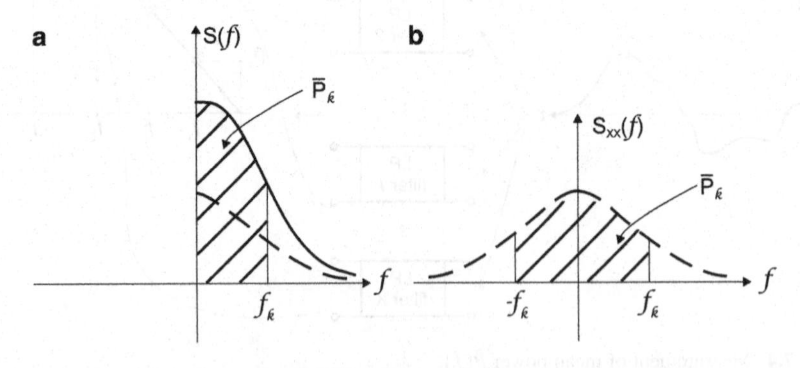

Fig. 7.5 Physical (**a**) and mathematical spectral density (**b**)

The obtained expression shows that a mean power of a physical spectral density $S(f)$ in the frequency band $[0, f_k]$ corresponds to the mean power in the frequency band $[-f_k, f_k]$ of a mathematical PSD $S_{xx}(f)$.

Denoting $\omega = 2\pi f$, from (7.48), we obtain:

$$\overline{P}_k = \frac{1}{2\pi}\int_{0}^{\omega_k} S(\omega)\mathrm{d}\omega = \frac{1}{2\pi}\int_{-\omega_k}^{\omega_k} S_{xx}(\omega)\mathrm{d}\omega. \tag{7.49}$$

7.2 Classification of Processes

7.2.1 *Low-Pass, Band-Pass, Band-Limited, and Narrow-Band Processes*

A process is called a *low-pass* process if the values of its PSD are clustered around the frequency $f = 0$ and vanish for high frequencies, as shown in Fig. 7.6a.

A random process is said to be a *band-pass* process if its power density spectrum is clustered around a frequency band B that does not include $f = 0$, as shown in Fig. 7.6b. Note that a PSD is not necessarily zero at $f = 0$, but that it has nonsignificant values at small frequencies including $f = 0$ compared to the frequencies in the band B.

A random process is a *band-limited* if its power density spectrum is zero outside of the band B, which does not include $f = 0$, as shown in Fig. 7.6c.

A band-limited process is said to be a *narrow-band* process if $B < < f_o$, where f_o is a characteristic frequency. This frequency f_o can be chosen as a frequency of the center of B (Fig. 7.7a) or as a frequency in which a PSD has a maximum value (Fig. 7.7b).

7.2.2 *White and Colored Processes*

A process is called a *white process* if its spectral density is constant for all frequencies from $-\infty$ to ∞, as shown in Fig. 7.8a

$$Sxx(f) = N_0/2 \quad \text{for} \quad -\infty < f < \infty. \tag{7.50}$$

The term "white" comes from optics as an analogy of a white light, which contains all visible frequencies in its spectrum [PEE93, p. 214]. As a white process is usually a noise, many authors use the term white noise instead of white process.

An autocorrelation function of a white noise is:

$$R_{XX}(\tau) = \int_{-\infty}^{\infty} S_{XX}(f)\mathrm{e}^{j2\pi f\tau}\mathrm{d}f = \frac{N_0}{2}\int_{-\infty}^{\infty} \mathrm{e}^{j2\pi f\tau}\mathrm{d}f = \frac{N_0}{2}\delta(\tau). \tag{7.51}$$

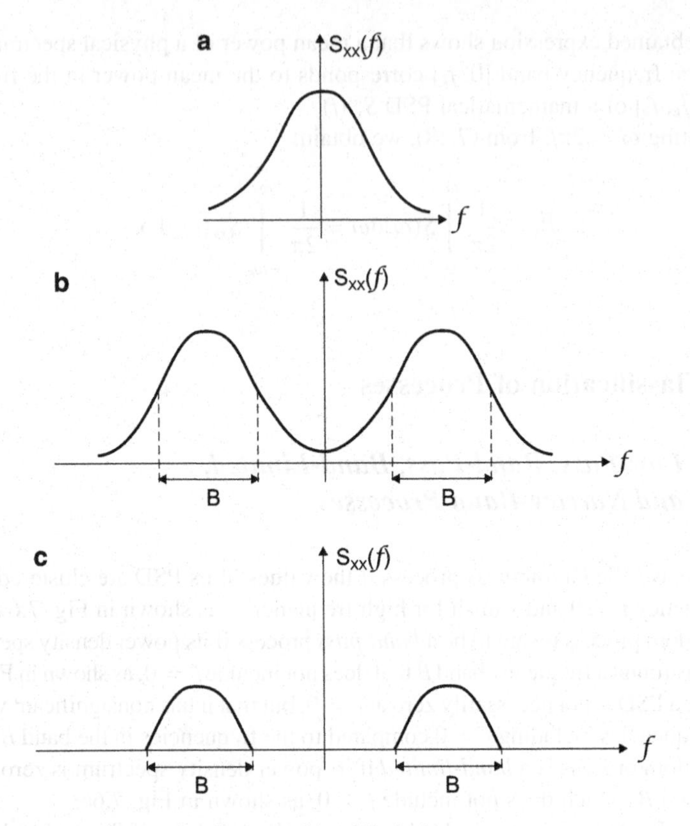

Fig. 7.6 PSDs for low-pass, band-pass, and band-limited processes

The obtained result shows that the autocorrelation function of a white noise is a delta function (Fig. 7.8b) in its origin. This means that there is no correlation between values of a process, for any τ, except $\tau = 0$. Consequently, this process is a maximally unpredictable process.

The area under a spectral density is infinite, indicating that a mean power of a white noise is infinite. As a process with infinite power cannot really exist, we can conclude that there is no white process in the real world.

Instead, consider a low-pass process, which has a flat PSD in the frequency band $[-B, B]$, as shown in Fig. 7.9. The corresponding autocorrelation function is:

$$R_{XX}(\tau) = \int_{-B}^{B} S_{XX}(f) e^{j2\pi f\tau} df = \frac{N_0}{2} \int_{-B}^{B} e^{j2\pi f\tau} df = N_0 B \frac{\sin 2\pi B\tau}{2\pi B\tau}. \quad (7.52)$$

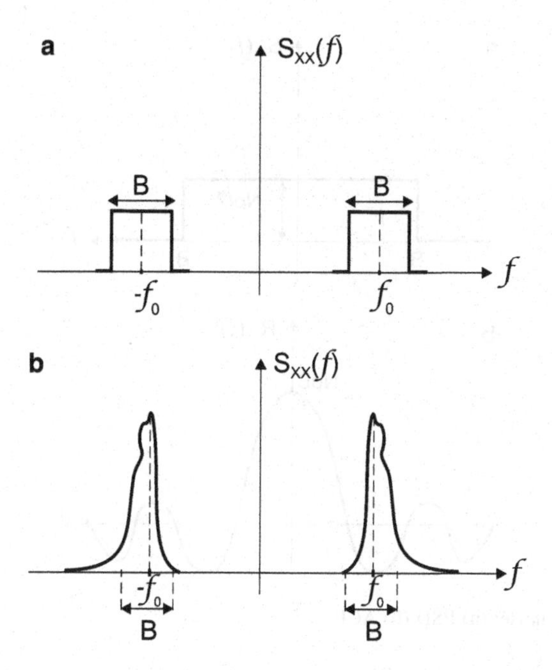

Fig. 7.7 Narrowband processes and f_0: (**a**) in the middle of B (**b**) corresponds to max PSD

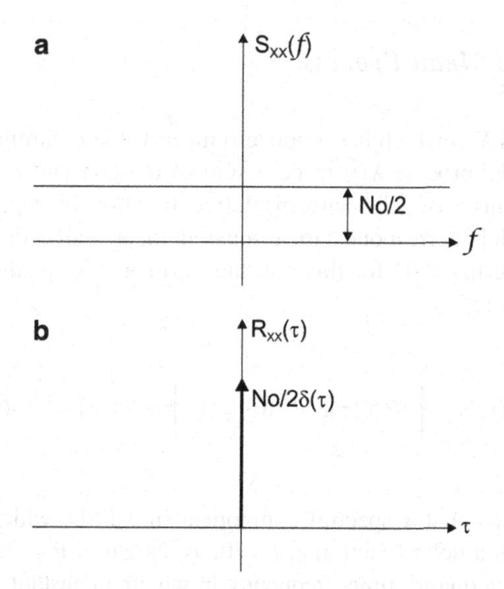

Fig. 7.8 White process: (**a**) PSD (**b**) ACF

This autocorrelation function is in the form of a sin x/x function and it is shown in Fig. 7.9b. As opposed to a white process, this process is called a *colored process* [PEE93, p. 216]. The average power of a colored process is a finite value, equal to

$$\overline{P} = 2BN_0/2 = BN_0. \tag{7.53}$$

A colored process is not necessarily a low-pass process. It can be also a band-pass process.

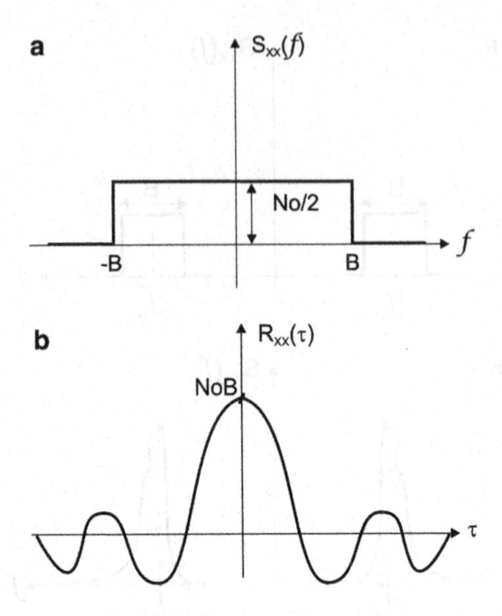

Fig. 7.9 Colored noise: (**a**) PSD (**b**) ACF

7.2.3 *Nonzero Mean Process*

Consider a process $X(t)$ which has a nonzero mean value c. Similarly, as in (6.59), we can consider the process $X(t)$ to be a sum of the constant c and a zero-mean process $X'(t)$. In this case, the autocorrelation function of a process $X(t)$ has a constant term which corresponds to a squared mean value denoted as c^2 (see (6.60)). Let us find the PSD for this constant term $A = c^2$ of the autocorrelation function, shown in Fig. 7.10.

$$S_{XX}(f) = \int_{-\infty}^{\infty} R_{XX}(\tau)\mathrm{e}^{-j\omega\tau}\mathrm{d}\tau = A\int_{-\infty}^{\infty} \mathrm{e}^{-j\omega\tau}\mathrm{d}\tau = A\delta(f). \tag{7.54}$$

This result shows that a spectral component in a PSD, which corresponds to a nonzero mean, is a delta function at $f = 0$, as shown in Fig. 7.10b. This is also in agreement with a duality time–frequency in which a constant PSD corresponds to an autocorrelation function which is a delta function and vice versa. A constant autocorrelation function has a corresponding PSD which is a delta function at $f = 0$.

An example of the autocorrelation function and the corresponding PSD is shown in Fig. 7.11.

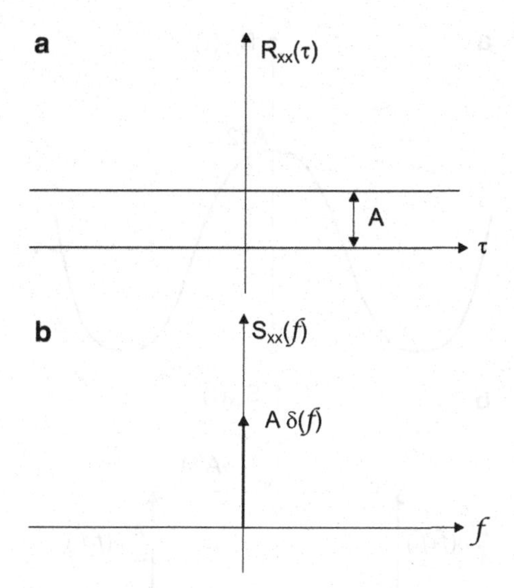

Fig. 7.10 ACF and PSD: (**a**) Constant term of ACF (**b**) PSD of constant term

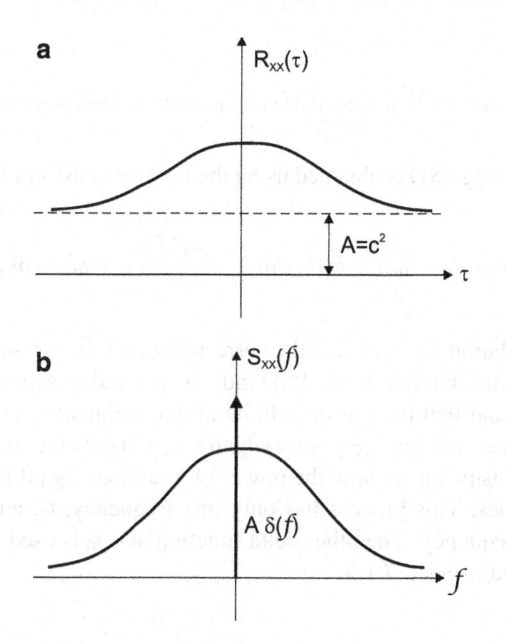

Fig. 7.11 Example of a nonzero mean process: (**a**) ACF (**b**) PSD

7.2.4 Random Sinusoidal Process

Consider a periodic process $X(t) = A\cos(\omega_0 t + \theta)$ where A and $\omega_0 = 2\pi/T$ (T is the period) are constants, while θ is a uniform random variable over $[0, 2\pi]$. The autocorrelation function has been found in (6.62) and is rewritten here:

$$R_{XX}(\tau) = \frac{A^2}{2}\cos\omega_0\tau. \tag{7.55}$$

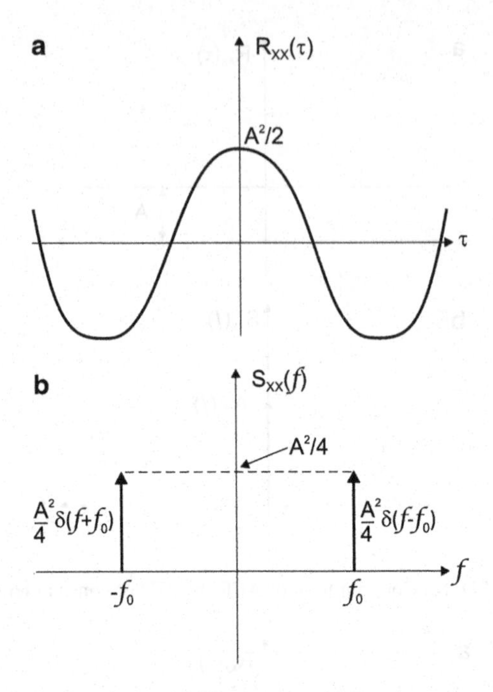

Fig. 7.12 Autocorrelation function and PSD of a sinusoidal random process: (**a**) ACF (**b**) PSD

The corresponding PSD is obtained using the Fourier transform from Appendix E:

$$S_{XX}(\omega) = \frac{A^2}{2}\pi[\delta(\omega - \omega_0) + \delta(\omega + \omega_0)] = \frac{A^2}{4}[\delta(f - f_0) + \delta(f + f_0)]. \quad (7.56)$$

The autocorrelation function and the corresponding PSD are shown in Fig. 7.12. As expected, a delta function in the PSD indicates that all power is concentrated at the frequency f_0 and that the power is finite at that frequency. This is because the random process has only one frequency which is f_0. On the other hand, this confirms that a spectral density shows how the power of a random signal is distributed over various frequencies. This process has only one frequency, f_0, and thus all power belongs to this frequency. The other delta function at $-f_0$ is used for mathematical reasons, explained in Sect. 7.1.5.

7.2.5 Bandwidth of Random Process

There are different definitions of a bandwidth of a random process that are given below. The definition, which can be used, mainly depends on the application. However, in all cases, a measured bandwidth is only measured for positive frequencies [MIL04, p. 380].

7.2.5.1 The Root-Mean Square (RMS) Bandwidth, B_{rms}

Consider a low-pass process $X(t)$ with a zero-mean value. We found that a PSD has certain important properties (e.g., it is a nonnegative and a real function). Now we will recall that a PDF has the same characteristics. Therefore, there is a similarity between a PSD and a PDF with exception that the area under a PSD is not necessarily unity. Next, if we normalize a PSD by dividing it with its area, the normalized PSD will have the unity area, and thus equivalent behavior to that of the PDF, which can be useful in different analyses.

$$S_{XX_{\text{norm}}}(f) = \frac{S_{XX}(f)}{\int_{-\infty}^{\infty} S_{XX}(f)\mathrm{d}f}. \tag{7.57}$$

Knowing that the standard deviation is a measure of the spread of a density function, we can define the analog quantities for the normalized PSD, just changing x for f, and PDF for PSD, as shown below ($m = 0$):

$$\text{STD} = \sigma = \sqrt{\sigma^2} = \sqrt{\int_{-\infty}^{\infty} x^2 f_X(x)\mathrm{d}x} \rightarrow \sqrt{\int_{-\infty}^{\infty} f^2 S_{XX_{\text{norm}}}(f)\mathrm{d}f}. \tag{7.58}$$

A parameter of a PSD which is equivalent to a standard deviation of a PDF is called a *root-mean squared bandwidth* and is denoted as B_{rms}

$$B_{\text{rms}} = \sqrt{\int_{-\infty}^{\infty} f^2 S_{XX_{\text{norm}}}\mathrm{d}f} = \sqrt{\frac{\int_{0}^{\infty} f^2 S_{XX}(f)\mathrm{d}f}{\int_{0}^{\infty} S_{XX}(f)\mathrm{d}f}}. \tag{7.59}$$

Note that, as mentioned before, the bandwidth B_{rms} is defined only for positive frequencies, so the limits of the integrals are between 0 and ∞ (i.e., only the right part of the PSD).

The above concept can also be extended to a band-pass process. For more details, see [PEE93, p. 205].

7.2.5.2 Half-Power Bandwidth

A *half-power bandwidth* is a bandwidth in which a PSD is greater than or equal to half of its peak value $S_{XX\max}$, as illustrated in Fig. 7.13a for a LP process. This quantity is also called a *3-dB bandwidth* because

$$-10\log(1/2) = 3\text{dB}. \tag{7.60}$$

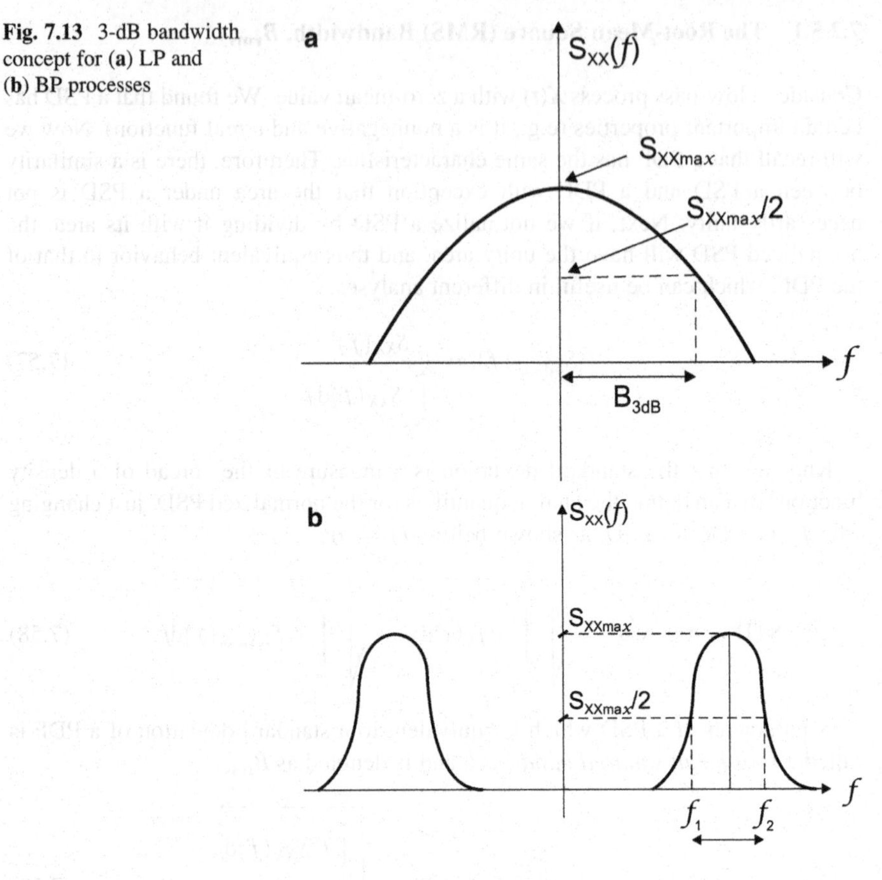

Fig. 7.13 3-dB bandwidth concept for (**a**) LP and (**b**) BP processes

So, that is why this denotation $B_{3\mathrm{dB}}$ becomes.

$$S_{XX}(\omega) \geq S_{XX\max}/2 \quad \text{for} \quad 0 \leq f \leq B_{3\mathrm{dB}}. \tag{7.61}$$

This concept can also be applied to a band-pass process, as shown in Fig. 7.13b

$$S_{XX}(\omega) \geq S_{XX\max}/2 \quad \text{for} \quad f_1 \leq f \leq f_2, \quad f_2 - f_1 = B_{3\mathrm{dB}}. \tag{7.62}$$

7.2.5.3 Absolute Bandwidth

An absolute bandwidth B_{abs} for a LP band-limited process is defined as a largest frequency for which the PSD is nonzero in value

$$S_{XX}(f) = \begin{cases} \neq 0 & \text{for} \quad 0<f<B_{\mathrm{abs}}, \\ 0 & \text{for} \quad f \geq B_{\mathrm{abs}}. \end{cases} \tag{7.63}$$

Similarly, for a BP band-limited process

$$S_{XX}(f) = \begin{cases} \neq 0 & \text{for} \quad f_1 < f < f_2, B_{\text{abs}} = f_2 - f_1, \\ 0 & \text{for} \quad f \geq f_2, f \leq f_1. \end{cases} \tag{7.64}$$

This concept is illustrated in Fig. 7.14.

Example 7.2.1 Find the different bandwidths defined in this section for the process from Example 7.1.1.

Solution The spectral density from Example 7.1.1 is rewritten here

$$S_{XX}(\omega) = 2U^2 \frac{2\lambda}{(2\lambda)^2 + \omega^2}. \tag{7.65}$$

From (7.65) and also from the plot in Fig. 7.2, we can see that the bandwidth is not limited. Therefore, absolute bandwidth is infinite

$$B_{\text{abs}} = \infty. \tag{7.66}$$

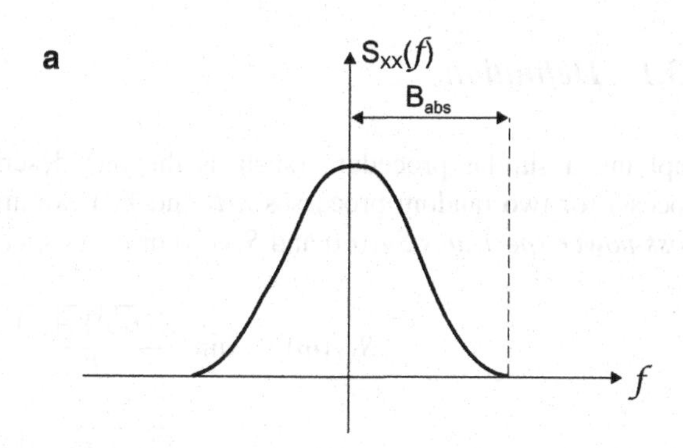

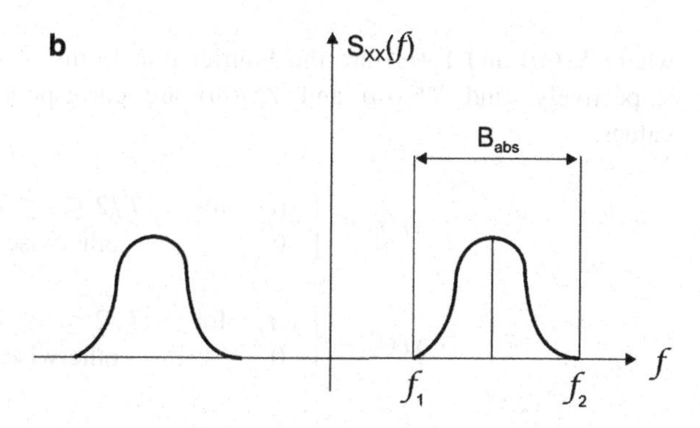

Fig. 7.14 Absolute bandwidth concept for (**a**) LP and (**b**) BP processes

The 3-dB bandwidth is shown in Fig. 7.2. and is calculated from

$$2\pi B_{3\mathrm{dB}} = 2\lambda, \tag{7.67}$$

or equivalently

$$B_{3\mathrm{dB}} = \lambda/\pi. \tag{7.68}$$

The RMS bandwidth B_{rms} is infinite

$$B_{\mathrm{rms}} = \infty, \tag{7.69}$$

because the integral in the nominator of (7.59) is infinite

$$\int_{0}^{\infty} f^2 S_{XX}(f)\mathrm{d}f = \infty \tag{7.70}$$

7.3 Cross-Power Spectrum

7.3.1 Definition

Applying a similar procedure (such as the one described in Sect. 7.1 for one process) for two random processes $X(t)$ and $Y(t)$, we arrive at the expressions for *cross-power spectrums* $S_{XY}(\omega)$ and $S_{YX}(\omega)$ or *cross-spectral density functions*

$$S_{XY}(\omega) = \lim_{T\to\infty} \frac{\overline{X_T^*(\omega)Y_T(\omega)}}{T}. \tag{7.71}$$

$$S_{YX}(\omega) = \lim_{T\to\infty} \frac{\overline{Y_T^*(\omega)X_T(\omega)}}{T}, \tag{7.72}$$

where $X_T(\omega)$ and $Y_T(\omega)$ are the Fourier transforms of processes $x_T(t)$ and $y_T(t)$, respectively, and $X^*_T(\omega)$ and $Y^*_T(\omega)$ are corresponding complex-conjugated values.

$$x_T(t) = \begin{cases} x(t) & \text{for } -T/2 \le t \le T/2, \\ 0 & \text{otherwise.} \end{cases} \tag{7.73}$$

$$y_T(t) = \begin{cases} y(t) & \text{for } -T/2 \le t \le T/2, \\ 0 & \text{otherwise.} \end{cases} \tag{7.74}$$

If $X(t)$ and $Y(t)$ are jointly wide sense stationary random processes with the cross-correlation functions $R_{XY}(\tau)$ and $R_{YX}(\tau)$, then–as in one random variables–the cross-spectral density and the cross-correlation function are Fourier transform pairs.

$$\begin{aligned} S_{XY}(\omega) &= \int_{-\infty}^{\infty} R_{XY}(\tau)\mathrm{e}^{-j\omega\tau}\mathrm{d}\tau, \\ S_{XY}(f) &= \int_{-\infty}^{\infty} R_{XY}(\tau)\mathrm{e}^{-j2\pi f\tau}\mathrm{d}\tau. \end{aligned} \tag{7.75}$$

$$R_{XY}(\tau) = \frac{1}{2\pi}\int_{-\infty}^{\infty} S_{XY}(\omega)\mathrm{e}^{j\omega\tau}\mathrm{d}\omega = \int_{-\infty}^{\infty} S_{XY}(f)\mathrm{e}^{j2\pi f\tau}\mathrm{d}f. \tag{7.76}$$

Similarly, we have:

$$\begin{aligned} S_{YX}(\omega) &= \int_{-\infty}^{\infty} R_{YX}(\tau)\mathrm{e}^{-j\omega\tau}\mathrm{d}\tau, \\ S_{YX}(f) &= \int_{-\infty}^{\infty} R_{YX}(\tau)\mathrm{e}^{-j2\pi f\tau}\mathrm{d}\tau \end{aligned} \tag{7.77}$$

and

$$R_{YX}(\tau) = \frac{1}{2\pi}\int_{-\infty}^{\infty} S_{YX}(\omega)\mathrm{e}^{j\omega\tau}\mathrm{d}\omega = \int_{-\infty}^{\infty} S_{YX}(f)\mathrm{e}^{j2\pi f\tau}\mathrm{d}f. \tag{7.78}$$

The properties of cross-spectral functions are related with the properties of cross-correlation functions. These are discussed in the next section.

7.3.2 Properties of Cross-Spectral Density Function

P.1 Cross-spectral density functions $S_{XY}(\omega)$ and $S_{YX}(\omega)$ are not necessarily real.

As opposed to an autocorrelation function, which is an even function, a cross-correlation function is not necessarily even and consequently a cross-spectral density is not necessarily real.

P.2 The following relation between $S_{XY}(\omega)$ and $S_{YX}(\omega)$ holds:

$$S_{XY}(\omega) = S_{YX}(-\omega) \quad \text{or} \quad S_{XY}(f) = S_{YX}(-f). \tag{7.79}$$

We rewrite here the relation (6.113) for a cross-correlation function

$$R_{XY}(\tau) = R_{YX}(-\tau). \tag{7.80}$$

Placing (7.80) into (7.75), we get:

$$\begin{aligned} S_{XY}(\omega) &= \int_{-\infty}^{\infty} R_{YX}(-\tau) e^{-j\omega\tau} d\tau = \int_{-\infty}^{\infty} R_{YX}(-\tau) e^{j\omega(-\tau)} d\tau \\ &= \int_{-\infty}^{\infty} R_{YX}(\tau) e^{j\omega(\tau)} d\tau = S_{YX}(-\omega). \end{aligned} \tag{7.81}$$

P.3 The following relations hold for $S_{XY}(\omega)$ and $S_{YX}(\omega)$:

$$\begin{aligned} S_{XY}(-\omega) &= S_{XY}^*(\omega), \\ S_{XY}(\omega) &= S_{XY}^*(-\omega). \end{aligned} \tag{7.82}$$

Similarly, we have:

$$\begin{aligned} S_{YX}(-\omega) &= S_{YX}^*(\omega), \\ S_{YX}(\omega) &= S_{YX}^*(-\omega). \end{aligned} \tag{7.83}$$

From (7.75), we write:

$$\begin{aligned} S_{XY}(-\omega) &= \int_{-\infty}^{\infty} R_{XY}(\tau) e^{-j(-\omega)\tau} d\tau = \int_{-\infty}^{\infty} R_{XY}(\tau) e^{-(-j)\omega\tau} d\tau = S_{XY}^*(\omega), \\ S_{YX}(-\omega) &= \int_{-\infty}^{\infty} R_{YX}(\tau) e^{-j(-\omega)\tau} d\tau = \int_{-\infty}^{\infty} R_{YX}(\tau) e^{-(-j)\omega\tau} d\tau = S_{YX}^*(\omega). \end{aligned} \tag{7.84}$$

P.4 If two processes (that are at least WS stationary, both with zero means) are uncorrelated, then their cross-spectral densities are zero

$$S_{XY}(\omega) = S_{YX}(\omega) = 0. \tag{7.85}$$

For uncorrelated processes from (6.135), we have:

$$R_{XY}(\tau) = \overline{XY} = 0 \tag{7.86}$$

and (7.85) follows from the definition (7.75).

Example 7.3.1 Consider the random processes from Example 6.6.1, where the cross-correlation function was given as:

$$R_{XY} = -\sigma^2 \sin \omega_0 \tau, \tag{7.87}$$

where ω_0 is a particular deterministic frequency. Find the cross-spectral density and demonstrate its properties.

Solution From (7.75), the cross-PSD is:

$$\begin{aligned}
S_{XY}(\omega) &= -\sigma^2 \int_{-\infty}^{\infty} \sin \omega_0 \tau \mathrm{e}^{-j\omega\tau} \mathrm{d}\tau = \frac{-\sigma^2}{2j} \int_{-\infty}^{\infty} \left(\mathrm{e}^{j\omega_0\tau} - \mathrm{e}^{-j\omega_0\tau}\right)\mathrm{e}^{-j\omega\tau} \mathrm{d}\tau, \\
&= \frac{\sigma^2}{2j} \int_{-\infty}^{\infty} \left(\mathrm{e}^{-(j\omega_0 + j\omega)\tau} - \mathrm{e}^{-(-j\omega_0 + j\omega)\tau}\right) \mathrm{d}\tau \\
&= j\frac{\sigma^2}{2} \int_{-\infty}^{\infty} \left(-\mathrm{e}^{-(j\omega_0 + j\omega)\tau} + \mathrm{e}^{-(-j\omega_0 + j\omega)\tau}\right) \mathrm{d}\tau, \\
&= j\frac{\sigma^2}{2}\delta(\omega - \omega_0) - j\frac{\sigma^2}{2}\delta(\omega + \omega_0).
\end{aligned} \tag{7.88}$$

P.1. From (7.88), the cross-spectral density is imaginary, i.e., it is not a real function.

P.2. To verify the property in (7.79), we find a cross-spectral power density $S_{YX}(\omega)$.

Using (7.80) and (7.87), we get the cross-correlation function:

$$R_{YX}(\tau) = \sigma^2 \sin \omega_0 \tau. \tag{7.89}$$

From (7.77), we get:

$$\begin{aligned}
S_{YX}(\omega) &= \sigma^2 \int_{-\infty}^{\infty} \sin \omega_0 \tau \mathrm{e}^{-j\omega\tau} \mathrm{d}\tau = \frac{\sigma^2}{2j} \int_{-\infty}^{\infty} \left(\mathrm{e}^{j\omega_0\tau} - \mathrm{e}^{-j\omega_0\tau}\right)\mathrm{e}^{-j\omega\tau} \mathrm{d}\tau, \\
&= \frac{\sigma^2}{2j} \int_{-\infty}^{\infty} \left(-\mathrm{e}^{-(j\omega_0 + j\omega)\tau} + \mathrm{e}^{-(-j\omega_0 + j\omega)\tau}\right) \mathrm{d}\tau \\
&= j\frac{\sigma^2}{2} \int_{-\infty}^{\infty} \left(+\mathrm{e}^{-(j\omega_0 + j\omega)\tau} - \mathrm{e}^{-(-j\omega_0 + j\omega)\tau}\right) \mathrm{d}\tau, \\
&= -j\frac{\sigma^2}{2}\delta(\omega - \omega_0) + j\frac{\sigma^2}{2}\delta(\omega + \omega_0).
\end{aligned} \tag{7.90}$$

Comparing (7.89) and (7.90), we see that:

$$S_{YX}(\omega) = S_{XY}(-\omega). \tag{7.91}$$

P.3. From (7.88). we have:

$$\begin{aligned} S_{XY}(-\omega) &= j\frac{\sigma^2}{2}\delta(-\omega-\omega_0) - j\frac{\sigma^2}{2}\delta(-\omega+\omega_0), \\ &= -j\frac{\sigma^2}{2}\delta(\omega+\omega_0) + j\frac{\sigma^2}{2}\delta(\omega-\omega_0) = S_{XY}^*(\omega). \end{aligned} \tag{7.92}$$

7.4 Operation of Processes

7.4.1 Sum of Random Processes

In many problems, we need to find the spectral characteristics of a process $Y(t)$ which is obtained as a sum of two processes $X_1(t)$ and $X_2(t)$, as shown in Fig. 7.15.

$$Y(t) = X_1(t) + X_2(t). \tag{7.93}$$

Consider that processes $X_1(t)$ and $X_2(t)$ are both at least WS stationary processes. Consequently, their sum will also be WS stationary (i.e., the process $Y(t)$ will also be at least WS stationary).

The autocorrelation function of the process $Y(t)$ is:

$$\begin{aligned} R_{YY}(\tau) = E\{Y(t)Y(t+\tau)\} &= E\{(X_1(t)+X_2(t))(X_1(t+\tau)+X_2(t+\tau))\}, \\ &= E\{X_1(t)X_1(t+\tau)\} + E\{X_2(t)X_2(t+\tau)\} \\ &+ E\{X_1(t)X_2(t+\tau)\} + E\{X_2(t)X_1(t+\tau)\}. \end{aligned} \tag{7.94}$$

We can recognize the autocorrelation and cross-correlation functions in the expression (7.94). Therefore, from (7.94), it follows:

$$R_{YY}(\tau) = R_{X_1X_1}(\tau) + R_{X_2X_2}(\tau) + R_{X_1X_2}(\tau) + R_{X_2X_1}(\tau). \tag{7.95}$$

The PSD of the process $Y(t)$ is obtained using the Fourier transform of (7.95)

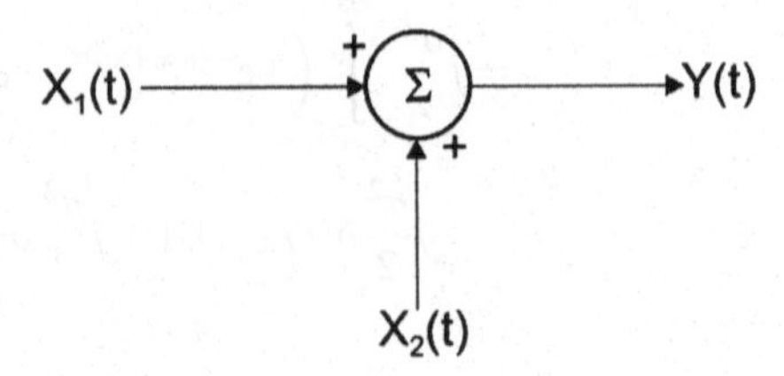

Fig. 7.15 Sum of two processes

$$S_{YY}(\omega) = S_{X_1X_1}(\omega) + S_{X_2X_2}(\omega) + S_{X_1X_2}(\omega) + S_{X_2X_1}(\omega). \tag{7.96}$$

Consider that the processes $X_1(t)$ and $X_2(t)$ are uncorrelated. We have:

$$R_{X_1X_2}(\tau) = R_{X_2X_1}(\tau) = E\{X_1\}E\{X_2\}. \tag{7.97}$$

Therefore, for uncorrelated processes, (7.95) becomes:

$$R_{YY}(\tau) = R_{X_1X_1}(\tau) + R_{X_2X_2}(\tau) + 2E\{X_1\}E\{X_2\}. \tag{7.98}$$

It is useful to have the expression of (7.98) for a case where

$$E\{X_1\} = E\{X_2\} = 0. \tag{7.99}$$

From (7.98) and (7.99), we get:

$$R_{YY}(\tau) = R_{X_1X_1}(\tau) + R_{X_2X_2}(\tau). \tag{7.100}$$

An autocorrelation function of a sum of two at least WS stationary, uncorrelated, zero-mean processes is equal to the sum of the autocorrelation functions of the particular processes.

Using the Fourier transform of (7.100), we get a PSD of the sum:

$$S_{YY}(\tau) = S_{X_1X_1}(\tau) + S_{X_2X_2}(\tau). \tag{7.101}$$

A PSD of a sum of two at least WS stationary, uncorrelated, zero-mean processes is equal to a sum of power spectral densities of particular processes.

The mean power of the sum is:

$$\begin{aligned} P_{YY} &= \overline{Y^2} = \overline{X_1^2} + \overline{X_2^2} = \frac{1}{2\pi}\int_{-\infty}^{\infty} S_{YY}(\omega)\mathrm{d}\omega, \\ &= \frac{1}{2\pi}\int_{-\infty}^{\infty} S_{X_1X_1}(\omega)\mathrm{d}\omega + \frac{1}{2\pi}\int_{-\infty}^{\infty} S_{X_2X_2}(\omega)\mathrm{d}\omega, \\ &= P_{X_1X_1} + P_{X_2X_2}. \end{aligned} \tag{7.102}$$

The mean power of a sum of two at least WS uncorrelated zero-mean processes is equal to a sum of mean powers of marginal processes.

7.4.2 Multiplication of Random Process with a Sinusoidal Signal

There are many practical applications where one may need to multiply a random process with a sinusoidal function, as shown in Fig. 7.16.

We can describe a time and spectral characteristics of the output process $Y(t)$

$$Y(t) = X(t)U_0 \cos \omega_0 t. \tag{7.103}$$

The autocorrelation function is:

$$\begin{aligned} R_{YY}(t, t+\tau) &= E\{Y(t)Y(t+\tau)\} \\ &= E\{[X(t)U_0 \cos \omega_0 t][X(t+\tau)U_0 \cos \omega_0 (t+\tau)]\}, \\ &= U_0^2 \cos \omega_0 t \cos \omega_0 (t+\tau) E\{X(t)X(t+\tau)\}. \end{aligned} \tag{7.104}$$

Using the definition for the autocorrelation function (6.32) and the trigonometric relation from Appendix D, we arrive at:

$$R_{YY}(t, t+\tau) = \frac{U_0^2}{2} R_{XX}(t, t+\tau)[\cos \omega_0 \tau + \cos(2\omega_0 t + \omega_0 \tau)]. \tag{7.105}$$

If the process $X(t)$ is at least WS stationary, then

$$R_{XX}(t, t+\tau) = R_{XX}(\tau), \tag{7.106}$$

and (7.105) reduces to:

$$R_{YY}(t, t+\tau) = \frac{U_0^2}{2} R_{XX}(\tau)[\cos \omega_0 \tau + \cos(2\omega_0 t + \omega_0 \tau)]. \tag{7.107}$$

Note that the autocorrelation function (7.107) is a function of time and it is necessary to find the time average for it:

$$\begin{aligned} A\{R_{YY}(t, t+\tau)\} &= \lim_{T\to\infty} \frac{1}{T} \int_{-T/2}^{T/2} R_{YY}(t, t+\tau) \mathrm{d}t, \\ &= \frac{U_0^2}{2} R_{XX}(\tau) \cos \omega_0 \tau + \frac{U_0^2}{2} R_{XX}(\tau) \lim_{T\to\infty} \frac{1}{T} \int_{-T/2}^{T/2} \cos(2\omega_0 t + \omega_0 \tau) \mathrm{d}t, \\ &= \frac{U_0^2}{2} R_{XX}(\tau) \cos \omega_0 \tau, \\ &= R_{YY}(\tau). \end{aligned} \tag{7.108}$$

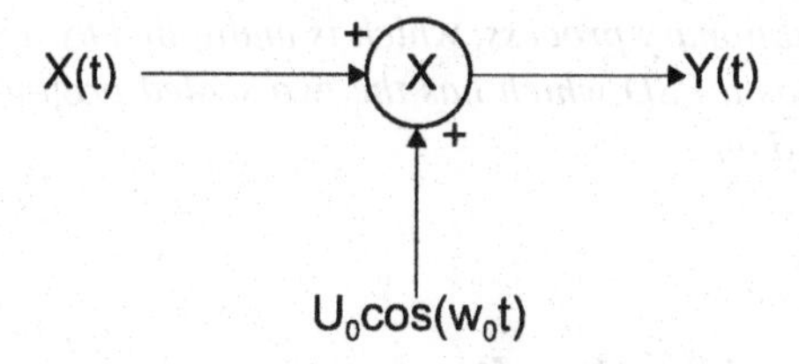

Fig. 7.16 Multiplication of random process with sinusoidal signal

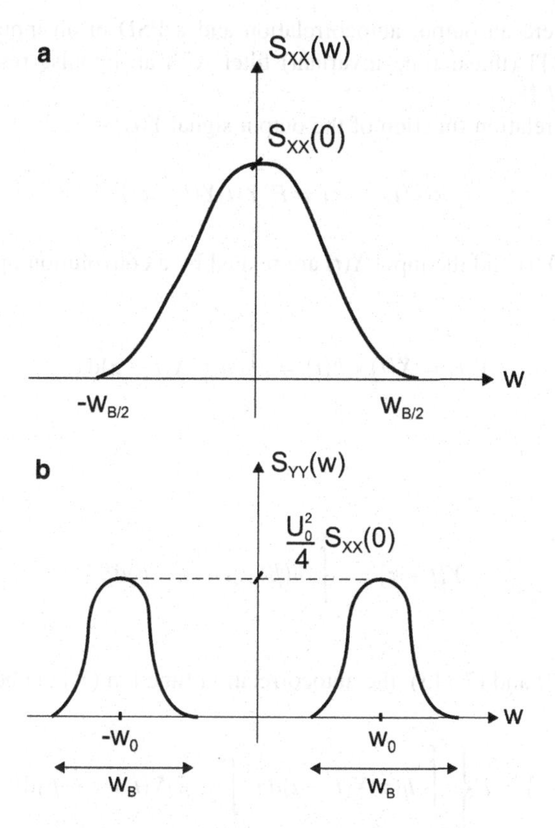

Fig. 7.17 Power spectral densities: (a) Input (b) Output

The obtained expression can be rewritten as (see Appendix D):

$$R_{YY}(\tau) = \frac{U_0^2}{4} R_{XX}(\tau)\left(\mathrm{e}^{j\omega_0\tau} + \mathrm{e}^{-j\omega_0\tau}\right). \tag{7.109}$$

Using the shifting property of the Fourier transform (see Appendix E), we arrive at:

$$S_{YY}(\omega) = \frac{U_0^2}{4}\left[S_{XX}(\omega - \omega_0) + S_{XX}(\omega + \omega_0)\right]. \tag{7.110}$$

As a result, the output PSD has two symmetrical power spectral components around the frequencies ω_0 and $-\omega_0$, which are scaled by $U_0{}^2/4$, as shown in Fig. 7.17b.

An at least WS stationary process, which is multiplied by a sinusoidal signal of the frequency ω_0, has a PSD which has the two scaled original PSDs around the frequencies $-\omega_0$ and ω_0.

7.4.3 Filtering of Random Process

We consider here an output autocorrelation and a PSD of an input process $X(t)$ filtered by a LTI (linear time invariant) filter with an impulse response $h(t)$, as shown in Fig. 7.18.

The autocorrelation function of the output signal $Y(t)$ is:

$$R_{YY}(t, t+\tau) = E\{Y(t)Y(t+\tau)\}. \tag{7.111}$$

The output $Y(t)$ and the input $X(t)$ are related by a convolution operation

$$Y(t) = X(t) * h(t) = \int_{-\infty}^{\infty} h(\alpha)X(t-\alpha)\mathrm{d}\alpha. \tag{7.112}$$

Similarly,

$$Y(t+\tau) = \int_{-\infty}^{\infty} h(\beta)X(t+\tau-\beta)\mathrm{d}\beta. \tag{7.113}$$

From (7.112) and (7.113), the autocorrelation function (7.111) becomes:

$$\begin{aligned} R_{YY}(t, t+\tau) &= E\left\{ \int_{-\infty}^{\infty} h(\alpha)X(t-\alpha)\mathrm{d}\alpha \int_{-\infty}^{\infty} h(\beta)X(t+\tau-\beta)\mathrm{d}\beta \right\} \\ &= \int_{-\infty}^{\infty}\int_{-\infty}^{\infty} E\{X(t-\alpha)X(t+\tau-\beta)\}h(\alpha)h(\beta)\mathrm{d}\alpha\mathrm{d}\beta. \end{aligned} \tag{7.114}$$

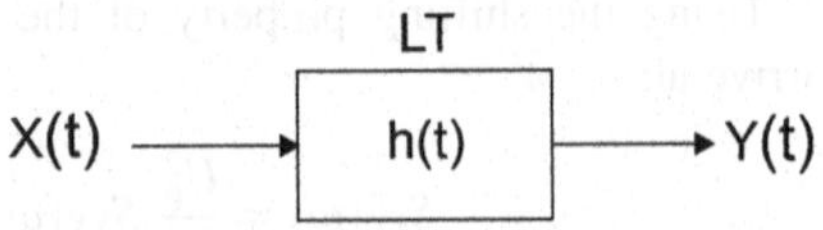

Fig. 7.18 LTI filtering of process $X(t)$

Denoting

$$t_1 = t - \alpha, \\ t_2 = t + \tau - \beta, \tag{7.115}$$

the expected value in the integrals of (7.114) is the autocorrelation function:

$$E\{X(t-\alpha)X(t+\tau-\beta)\} = E\{X(t_1)X(t_2)\} = R_{XX}(t_1, t_2). \tag{7.116}$$

If a process $X(t)$ is at least WS stationary, then the autocorrelation function (7.116) depends only on the time difference,

$$t_2 - t_1 = \tau - \beta + \alpha, \tag{7.117}$$

and the autocorrelation function (7.116) becomes

$$E\{X(t-\alpha)X(t+\tau-\beta)\} = R_{XX}(\tau - \beta + \alpha). \tag{7.118}$$

Using (7.118), the output autocorrelation function (7.114) becomes:

$$R_{YY}(t, t+\tau) = R_{YY}(\tau) = \int_{-\infty}^{\infty}\int_{-\infty}^{\infty} R_{XX}(\tau - \beta + \alpha)h(\alpha)h(\beta)\mathrm{d}\alpha\mathrm{d}\beta. \tag{7.119}$$

The output PSD is equal to the Fourier transform of the output autocorrelation function (7.119).

$$S_{YY}(\omega) = \int_{-\infty}^{\infty} R_{YY}(\tau)\mathrm{e}^{-j\omega\tau}\mathrm{d}\tau. \tag{7.120}$$

Placing (7.119) into (7.94), we get:

$$S_{YY}(\omega) = \int_{-\infty}^{\infty} h(\alpha) \int_{-\infty}^{\infty} h(\beta) \int_{-\infty}^{\infty} R_{XX}(\tau - \beta + \alpha)\mathrm{e}^{-j\omega\tau} \mathrm{d}\tau\mathrm{d}\alpha\mathrm{d}\beta. \tag{7.121}$$

We introduce the following auxiliary variable into (7.121):

$$\gamma = \tau - \beta + \alpha, \tag{7.122}$$

Thus, the expression (7.121) becomes:

$$S_{YY}(\omega) = \int_{-\infty}^{\infty} h(\alpha)\mathrm{e}^{j\omega\alpha}\mathrm{d}\alpha \int_{-\infty}^{\infty} h(\beta)\mathrm{e}^{-j\omega\beta}\mathrm{d}\beta \int_{-\infty}^{\infty} R_{XX}(\gamma)\mathrm{e}^{-j\omega\gamma}\mathrm{d}\gamma. \tag{7.123}$$

In the expression (7.123), we can recognize the PSD $S_{XX}(\omega)$:

$$S_{XX}(\omega) = \int_{-\infty}^{\infty} R_{XX}(\gamma) e^{-j\omega\gamma} d\gamma. \tag{7.124}$$

Similarly, the second integral in (7.123) is a Fourier transform of the impulse response $h(t)$ of the filter, which is a transfer function of a filter $H(\omega)$:

$$H(\omega) = \int_{-\infty}^{\infty} h(\beta) e^{-j\omega\beta} d\beta. \tag{7.125}$$

Similarly, the first integral in (7.123) is:

$$H(-\omega) = \int_{-\infty}^{\infty} h(\alpha) e^{j\omega\alpha} d\alpha. \tag{7.126}$$

Finally, placing (7.124)–(7.126) into (7.123), we arrive at:

$$S_{YY}(\omega) = H(-\omega)H(\omega)S_{XX}(\omega). \tag{7.127}$$

Knowing that a magnitude characteristic of the filter can be expressed as:

$$|H(\omega)|^2 = H(\omega)H(-\omega), \tag{7.128}$$

we get the final result for the output PSD

$$S_{YY}(\omega) = |H(\omega)|^2 S_{XX}(\omega). \tag{7.129}$$

Note that the final result (7.129) is real because the PSD is a real function. From (7.129), the average power of the output process is:

$$P_{YY} = \frac{1}{2\pi} \int_{-\infty}^{\infty} S_{YY}(\omega) d\omega = \frac{1}{2\pi} \int_{-\infty}^{\infty} |H(\omega)|^2 S_{XX}(\omega) d\omega. \tag{7.130}$$

Example 7.4.1 Find the mean power at the output of the filter shown in Fig. 7.19, if in the input is a white noise with the PSD equal to $N_0/2$.

Solution The transfer function of the filter is:

$$H(\omega) = \frac{1}{1 + j\omega\frac{L}{R}}. \tag{7.131}$$

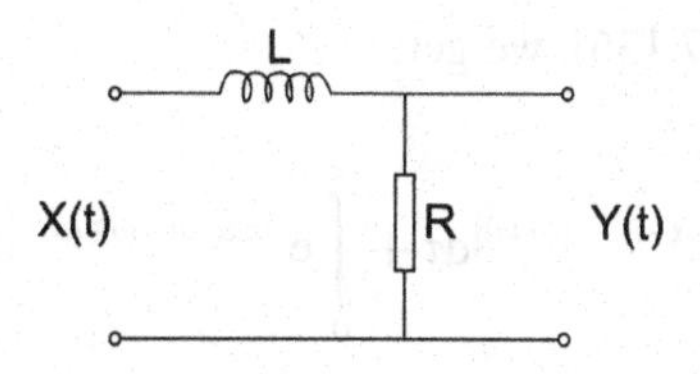

Fig. 7.19 Filter with the input noise

Using (7.129), we find the PSD at the output of the filter

$$S_{YY}(\omega) = \left| \frac{1}{1 + j\omega\frac{L}{R}} \right|^2 \frac{N_0}{2} = \frac{1}{1 + \left(\omega\frac{L}{R}\right)^2} \frac{N_0}{2}. \tag{7.132}$$

The average power of the output process $Y(t)$ is obtained by placing (7.132) into (7.130):

$$P_{YY} = \frac{1}{2\pi} \int_{-\infty}^{\infty} \frac{N_0}{2} \frac{1}{1 + \left(\omega\frac{L}{R}\right)^2} d\omega = \frac{N_0}{4\pi} \frac{R^2}{L^2} \int_{-\infty}^{\infty} \frac{1}{\left(\frac{R}{L}\right)^2 + \omega^2} d\omega. \tag{7.133}$$

Finally, using the integral (7) from Appendix A, we arrive at:

$$P_{YY} = \frac{N_0 R}{4L}. \tag{7.134}$$

7.5 Numerical Exercises

Exercise E.7.1 The autocorrelation function of a process $X(t)$ is given as:

$$R_{XX}(\tau) = R(0)e^{-\alpha|\tau|} \cos \omega_0 \tau \tag{7.135}$$

where α and ω_0 are constants. Find the PSD of the process.

Answer According to the Wiener–Khinchin theorem, a PSD is a Fourier transform of an autocorrelation function:

$$S_{XX}(\omega) = F\{R_{XX}(\tau)\} = \int_{-\infty}^{\infty} R(0)e^{-\alpha|\tau|} \cos(\omega_0 \tau) e^{-j\omega\tau} d\tau. \tag{7.136}$$

We express $\cos(\omega_0 \tau)$ in a complex form:

$$\cos(\omega_0 \tau) = \frac{e^{j\omega_0 \tau} + e^{-j\omega_0 \tau}}{2}. \tag{7.137}$$

Placing (7.137) into (7.136), we get:

$$\begin{aligned}
S_{XX}(\tau) &= \frac{R_{XX}(0)}{2}\left[\int_{-\infty}^{0} e^{\tau[\alpha+j(-\omega+\omega_0)]}d\tau + \int_{0}^{\infty} e^{-\tau[\alpha-j(-\omega+\omega_0)]}d\tau\right] \\
&+ \frac{R_{XX}(0)}{2}\left[\int_{-\infty}^{0} e^{\tau[\alpha-j(\omega+\omega_0)]}d\tau + \int_{0}^{\infty} e^{-\tau[\alpha+j(\omega+\omega_0)]}d\tau\right], \\
&= \frac{R_{XX}(0)}{2}\left[\frac{1}{\alpha+j(\omega_0-\omega)} + \frac{1}{\alpha-j(\omega_0-\omega)} + \frac{1}{\alpha-j(\omega+\omega_0)} + \frac{1}{\alpha+j(\omega+\omega_0)}\right], \\
&= \alpha R_{XX}(0)\left[\frac{1}{\alpha^2+(\omega_0-\omega)^2} + \frac{1}{\alpha^2+(\omega_0+\omega)^2}\right].
\end{aligned} \tag{7.138}$$

The autocorrelation function and the PSD are shown in Fig. 7.20 for $\alpha = 0.5$, $R(0) = 2$, and $\omega_0 = \pi/8$.

Exercise E.7.2 The spectral density of a process $X(t)$ is given as

$$S_{XX}(\omega) = \frac{1}{\alpha\sqrt{\pi}} e^{-\omega^2/4\alpha^2}. \tag{7.139}$$

where α is a constant. Find the autocorrelation function of the process.

Answer The autocorrelation function is an inverse Fourier transformation of a PSD:

$$\begin{aligned}
R_{XX}(\tau) &= \frac{1}{2\pi}\int_{-\infty}^{\infty} S_{XX}(\omega)e^{j\omega\tau}d\omega = \frac{1}{2\pi}\int_{-\infty}^{\infty} \frac{1}{\alpha\sqrt{\pi}} e^{-\omega^2/4\alpha^2} e^{j\omega\tau}d\omega, \\
&= \frac{1}{2\pi}\int_{-\infty}^{\infty} \frac{1}{\alpha\sqrt{\pi}} e^{-\omega^2/4\alpha^2}\cos(\omega\tau)d\omega \\
&= \frac{1}{\alpha\pi\sqrt{\pi}}\int_{0}^{\infty} e^{-\omega^2/4\alpha^2}\cos(\omega\tau)d\omega.
\end{aligned} \tag{7.140}$$

Using integral 3 from Appendix A, we arrive at:

$$R_{XX}(\tau) = \frac{1}{\pi} e^{-(\alpha\tau)^2}. \tag{7.141}$$

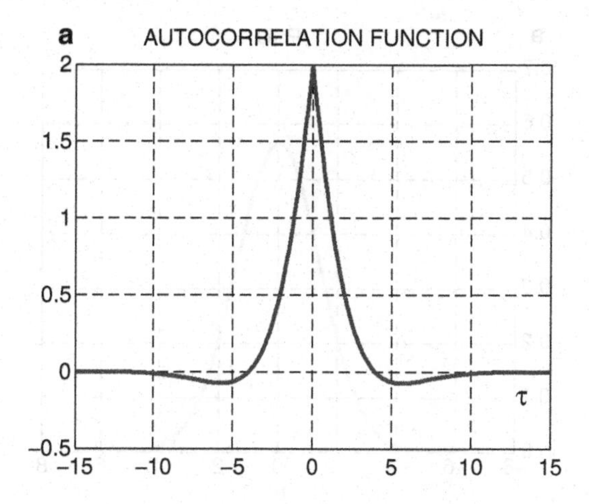

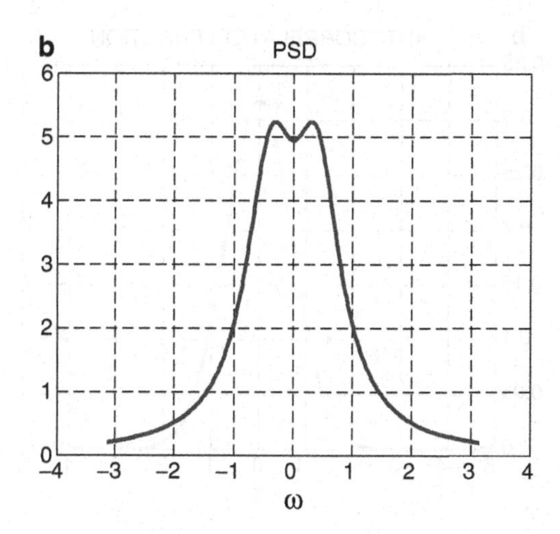

Fig. 7.20 (**a**) ACF and (**b**) PSD for $\alpha = 1$, $R(0) = 2$, and $\omega_0 = \pi/8$

The PSD and autocorrelation function are shown in Fig. 7.21 for $\alpha = 1$.

Exercise E.7.3 Show that a process $X(t)$ that is at least WS stationary which has an autocorrelation function given as:

$$R_{XX}(\tau) = \begin{cases} c & \text{for} \quad -\tau_0 < \tau < \tau_0, \\ 0 & \quad\quad \text{otherwise.} \end{cases} \tag{7.142}$$

cannot exist.

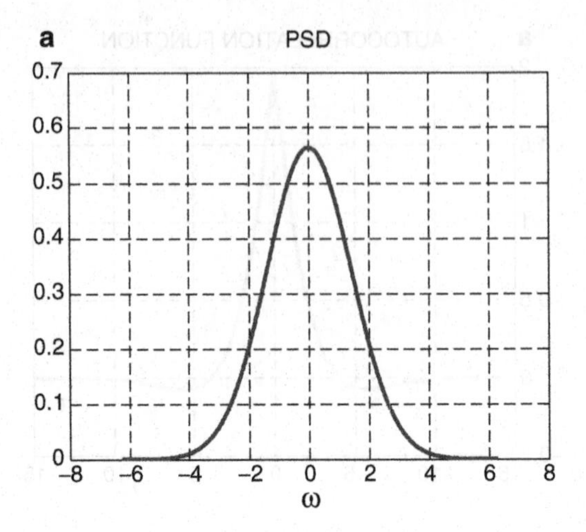

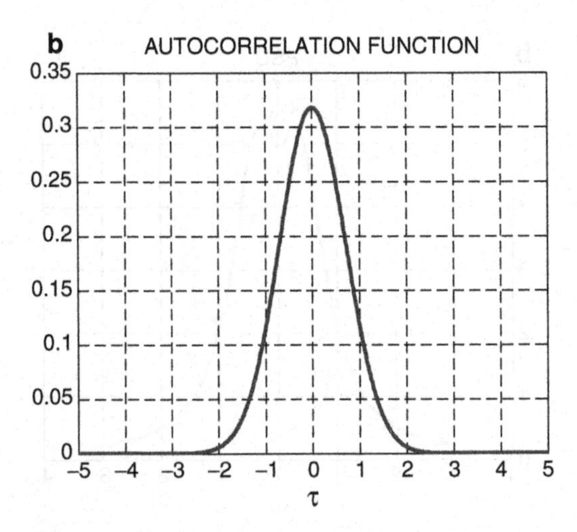

Fig. 7.21 (**a**) PSD and (**b**) ACF for $\alpha = 1$

Answer Let us suppose the contrary that there is a process $X(t)$ which has the autocorrelation function given in (7.142). As such, we find the PSD of the process as a Fourier transform of its autocorrelation function:

$$S_{XX}(\omega) = \int_{-\tau_0}^{\tau_0} c\mathrm{e}^{-j\omega\tau}\mathrm{d}\tau = 2c\left[\int_{0}^{\tau_0} \cos(\omega\tau)\mathrm{d}\tau + j\int_{0}^{\tau_0} \sin(-\omega\tau)\mathrm{d}\tau\right]. \tag{7.143}$$

Knowing that the PSD is a real function, we write:

$$S_{XX}(\omega) = \int_{-\tau_0}^{\tau_0} c\mathrm{e}^{-j\omega\tau}\mathrm{d}\tau = 2c\int_{0}^{\tau_0} \cos(\omega\tau)\mathrm{d}\tau = 2c\sin(\omega\tau_0). \tag{7.144}$$

The obtained result shows that the PSD for certain values of ω may be a negative. However, according to the property (7.39), the PSD cannot be negative. Therefore, our assumption was wrong (i.e., there is no process that is at least WS stationary, which has the autocorrelation function (7.142)).

Exercise E.7.4 The PSD $S_{XX}(\omega)$ of a process $X(t)$ that is at least WS stationary is given as:

$$S_{XX}(\omega) = \begin{cases} c & \text{for} \quad -\omega_1<\omega<\omega_1, \\ 0 & \quad \text{otherwise.} \end{cases} \tag{7.145}$$

Find the autocorrelation function, mean value, and variance of the process.

Answer

$$R_{XX}(\tau) = \frac{1}{2\pi}\int_{-\omega_1}^{\omega_1} c e^{j\omega\tau} d\omega = \frac{1}{\pi}\int_{0}^{\omega_1} c\cos(\omega\tau) d\omega = \frac{c}{\pi\tau}\sin(\omega_1\tau). \tag{7.146}$$

The autocorrelation function (Fig. 7.22) does not have a constant term. Consequently, its mean value is zero. Therefore, the variance is equal to the mean square value:

$$\begin{aligned} \sigma_{XX}^2 = E\{X^2(t)\} = R_{XX}(0) &= \frac{c}{\pi}\lim_{\tau\to 0}\frac{\sin(\omega_1\tau)}{\tau} \\ &= \frac{c\omega_1}{\pi}\lim_{\tau\to 0}\frac{\sin(\omega_1\tau)}{\omega_1\tau} = \frac{c\omega_1}{\pi}. \end{aligned} \tag{7.147}$$

Exercise E.7.5 Find the PSD of the process $X(t)$ which has the following autocorrelation function (Fig. 7.23a):

$$R_{XX}(\tau) = \frac{1 + e^{-2\lambda|\tau|}}{4}, \tag{7.148}$$

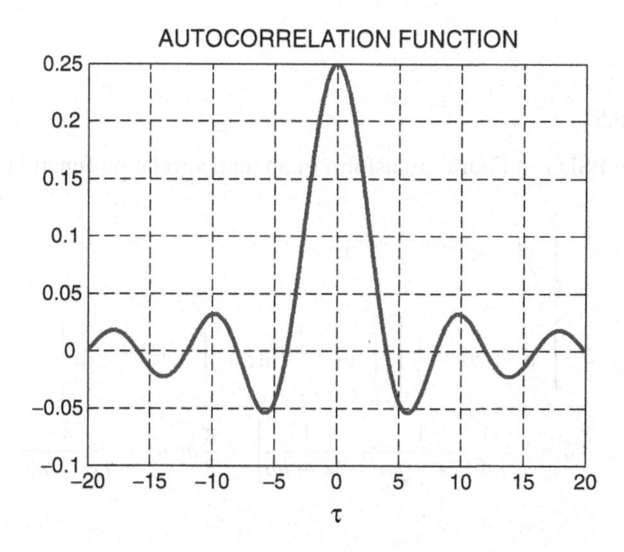

Fig. 7.22 ACF in Exercise E.7.4

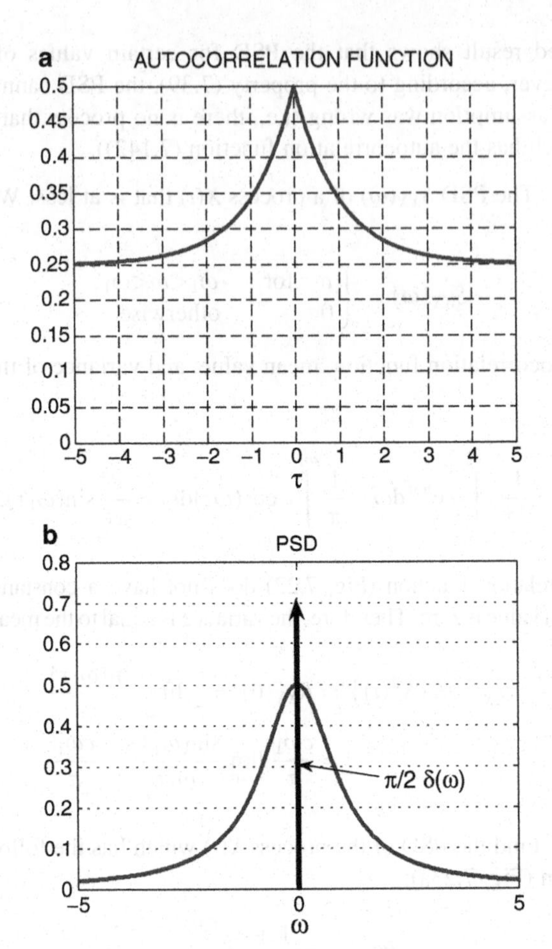

Fig. 7.23 (**a**) ACF and (**b**) PSD for a nonzero mean process

where $\lambda = 0.5$.

Answer The PSD is a Fourier transform of an autocorrelation function:

$$
\begin{aligned}
S_{XX}(\omega) &= \frac{1}{4}\int_{-\infty}^{\infty}\left(1 + \mathrm{e}^{-2\lambda|\tau|}\right)\mathrm{e}^{-j\omega\tau}\mathrm{d}\tau, \\
&= \frac{1}{4}\int_{-\infty}^{\infty}\mathrm{e}^{-j\omega\tau}\mathrm{d}\tau + \frac{1}{4}\left[\int_{-\infty}^{0}\mathrm{e}^{(2\lambda - j\omega)\tau}\mathrm{d}\tau + \int_{0}^{\infty}\mathrm{e}^{-(2\lambda + j\omega)\tau}\mathrm{d}\tau\right], \\
&= \frac{\pi}{2}\delta(\omega) + \frac{1}{4}\left[\frac{1}{2\lambda - j\omega} + \frac{1}{2\lambda + j\omega}\right] = \frac{\pi}{2}\delta(\omega) + \frac{\lambda}{4\lambda^2 + \omega^2}.
\end{aligned}
\tag{7.149}
$$

The PSD is shown in Fig. 7.23b.

Note that the delta function at $\omega = 0$ indicates that the process has a DC component. The same can be observed from the autocorrelation function (7.148) which has a constant term of 0.25 (see Sect. 6.5 and Fig. 7.23a).

Exercise E.7.6 Find the PSD of a process $X(t)$ that is at least WS stationary, and show that the process has neither a DC component nor a periodic component, if the autocorrelation function is given as (Fig. 7.24a):

$$R_{XX}(\tau) = 4e^{-3|\tau|}. \tag{7.150}$$

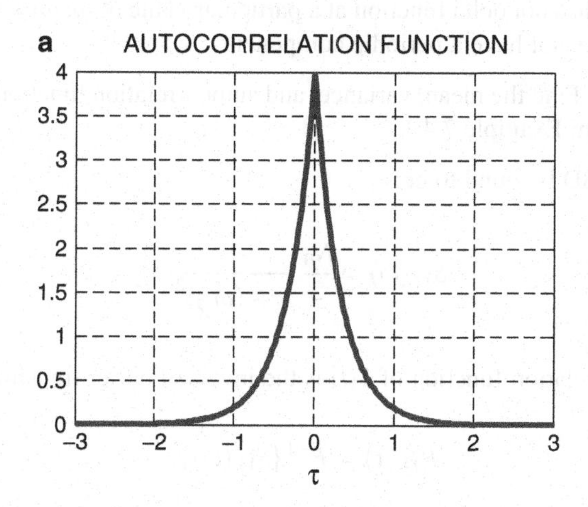

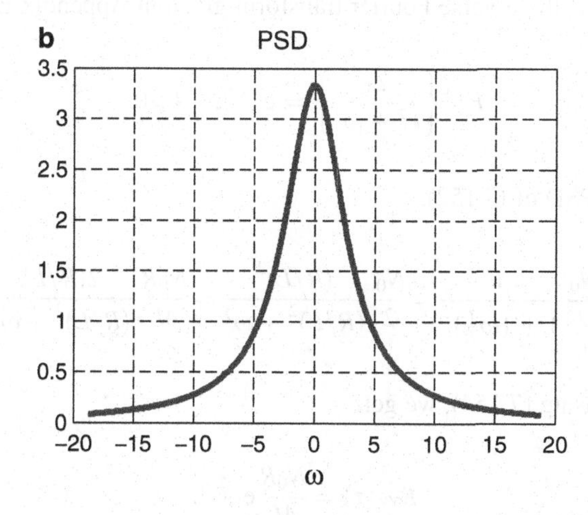

Fig. 7.24 Exercise E.7.6: (**a**) ACF and (**b**) PSD

Answer The PSD is:

$$S_{XX}(\omega) = 5\int_{-\infty}^{\infty} e^{-3|\tau|}e^{-j\omega\tau}d\tau = 5\left[\int_{-\infty}^{0} e^{\tau(3-j\omega)}d\tau + \int_{0}^{\infty} e^{-\tau(3+j\omega)}d\tau\right],$$

$$= 5\left[\frac{1}{3-j\omega} + \frac{1}{3+j\omega}\right] = \frac{30}{9+\omega^2}. \tag{7.151}$$

The PSD (Fig. 7.24b) does not have a delta function for $\omega = 0$, which indicates that the process does not have a finite power at the frequency for $\omega = 0$, i.e., it does not have a DC component.

The process has not delta function at a particular value of ω, $\omega \neq 0$. This shows that process does not have a periodic component.

Exercise E.7.7 Find the mean, variance, and autocorrelation function of the output process $Y(t)$ from Example 7.4.1.

Answer The PSD is found to be:

$$S_{YY}(\omega) = \frac{N_0}{2}\frac{1}{1+\left(\omega\frac{L}{R}\right)^2}. \tag{7.152}$$

The autocorrelation function of $Y(t)$ is the inverse Fourier transform of $S_{YY}(\omega)$

$$R_{YY}(\tau) = F^{-1}\{S_{YY}(\omega)\}. \tag{7.153}$$

In order to use the inverse Fourier transform given in Appendix E,

$$F^{-1}\left\{\frac{2k}{k^2+\omega^2}\right\} = e^{-k|t|}, \quad k>0. \tag{7.154}$$

we rewrite the PSD of (7.152):

$$S_{YY}(\omega) = \frac{N_0}{2}\frac{1}{1+\left(\omega\frac{L}{R}\right)^2} = \frac{N_0}{2}\frac{(R/L)^2}{(R/L)^2+\omega^2} = \frac{N_0 R}{4L}\frac{2(R/L)}{(R/L)^2+\omega^2}. \tag{7.155}$$

From here, using (7.154), we get:

$$R_{YY}(\tau) = \frac{N_0 R}{4L}e^{-\frac{R}{L}|\tau|}. \tag{7.156}$$

The variance of the process is:

$$\sigma_{YY}^2 = E\{Y^2(t)\} = R_{YY}(0) = \frac{N_0 R}{4L}. \quad (7.157)$$

Exercise E.7.8 Show that the PSD of the process $X(t)$ is like a Gaussian PDF curve, if the autocorrelation of the process, shown in Fig. 7.25a, is:

$$R_{XX}(\tau) = e^{-\tau^2}. \quad (7.158)$$

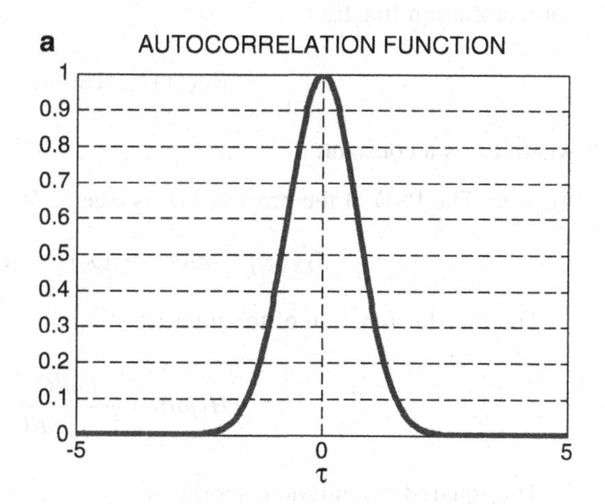

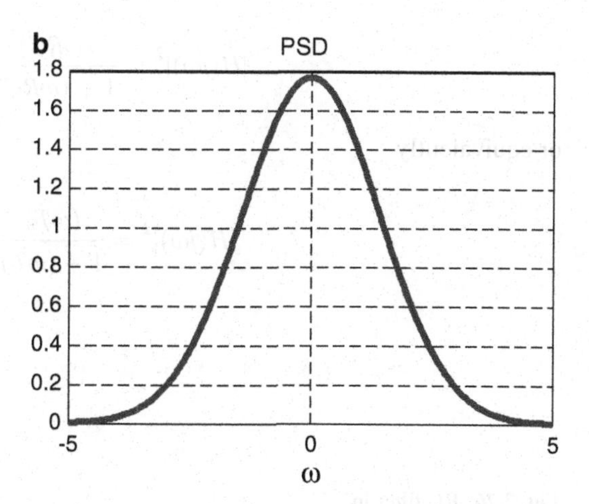

Fig. 7.25 Exercise E.7.8: (**a**) ACF and (**b**) PSD

Answer We have:

$$S_{XX}(\omega) = \int_{-\infty}^{\infty} e^{-\tau^2} e^{-j\omega\tau} d\tau = 2\int_{0}^{\infty} e^{-\tau^2} \cos(\omega\tau) d\tau. \tag{7.159}$$

Using integral 3 from Appendix A to (7.159), we arrive at:

$$S_{XX}(\omega) = 2\frac{\sqrt{\pi}}{2} e^{-\omega^2/4} = \sqrt{\pi} e^{-\omega^2/4}. \tag{7.160}$$

Exercise E.7.9 Find the output PSD of the process *Y*(*t*) at the output of the RC filter shown in Fig. 7.26. Find PSD, mean value, and variance of the process *Y*(*t*).

Consider that a WS stationary process *X*(*t*) in the input of the filter has the autocorrelation function

$$R_{XX}(\tau) = \cos(\omega_0 \tau), \tag{7.161}$$

where ω_0 is a constant.

Answer The PSD of the process *X*(*t*) is (see (7.56)):

$$S_{XX}(\omega) = \pi[\delta(\omega + \omega_0) + \delta(\omega - \omega_0)]. \tag{7.162}$$

The transfer function of the filter is:

$$H(j\omega) = \frac{j\omega RC}{1 + j\omega RC}. \tag{7.163}$$

The squared magnitude response is:

$$|H(j\omega)|^2 = \frac{(\omega RC)^2}{1 + (\omega RC)^2}, \tag{7.164}$$

or equivalently

$$|H(j\omega)|^2 = \frac{(\omega T)^2}{1 + (\omega T)^2}. \tag{7.165}$$

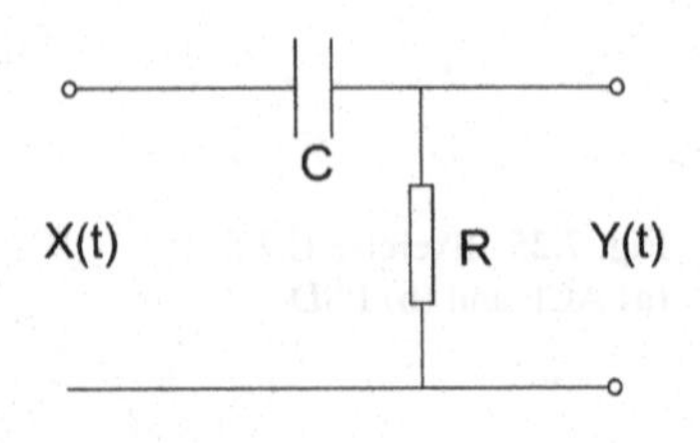

Fig. 7.26 RC filter in Exercise E.7.9

where T is a time constant

$$T = RC. \tag{7.166}$$

Using (7.129), (7.162), and (7.165), we arrive at:

$$S_{YY}(\omega) = \frac{(\omega T)^2 \pi}{1 + (\omega T)^2} [\delta(\omega + \omega_0) + \delta(\omega - \omega_0)]. \tag{7.167}$$

There is no delta function for $\omega = 0$ and consequently the output process $Y(t)$ has no DC component (the same result comes from the fact that the capacitor does not pass a DC component).

$$E\{Y(t)\} = 0. \tag{7.168}$$

The variance is obtained from

$$\sigma_{YY}^2 = E\{Y^2(t)\} = R_{YY}(0) = \frac{1}{2\pi} \int_{-\infty}^{\infty} S_{YY}(\omega) \mathrm{d}\omega$$

$$= \frac{\pi}{2\pi} \left[\int_{-\infty}^{\infty} \frac{(\omega T)^2}{1 + (\omega T)^2} \delta(\omega + \omega_0) \mathrm{d}\omega + \int_{-\infty}^{\infty} \frac{(\omega T)^2}{1 + (\omega T)^2} \delta(\omega - \omega_0) \mathrm{d}\omega \right]. \tag{7.169}$$

Using the characteristics of delta function from (2.72), we arrive at:

$$\sigma_{YY}^2 = \frac{1}{2} \left[\frac{(\omega_0 T)^2}{1 + (\omega_0 T)^2} + \frac{(\omega_0 T)^2}{1 + (\omega_0 T)^2} \right] = \frac{(\omega_0 T)^2}{1 + (\omega_0 T)^2}. \tag{7.170}$$

Exercise E.7.10 Find the autocorrelation function of the output process $Y(t)$ from Exercise E.7.9 and confirm that the mean and variance of the process $Y(t)$ are given as in (7.168) and (7.170), respectively.

Answer The autocorrelation function is the inverse Fourier transform of the PSD (7.167)

$$R_{YY}(\tau) = F^{-1}\{S_{YY}(\omega)\} = \frac{1}{2\pi} \int_{-\infty}^{\infty} \frac{(\omega T)^2 \pi}{1 + (\omega T)^2} [\delta(\omega + \omega_0) + \delta(\omega - \omega_0)] e^{j\omega\tau} \mathrm{d}\omega,$$

$$= \frac{1}{2} \left[\int_{-\infty}^{\infty} \frac{(\omega T)^2 e^{j\omega\tau}}{1 + (\omega T)^2} \delta(\omega + \omega_0) \mathrm{d}\omega + \int_{-\infty}^{\infty} \frac{(\omega T)^2 e^{j\omega\tau}}{1 + (\omega T)^2} \delta(\omega - \omega_0) \mathrm{d}\omega \right]. \tag{7.171}$$

Using the characteristics of the delta function from (2.72), we arrive at:

$$R_{YY}(\tau) = \frac{1}{2}\left[\frac{(\omega_0 T)^2 \mathrm{e}^{-j\omega_0\tau}}{1+(\omega_0 T)^2} + \frac{(\omega_0 T)^2 \mathrm{e}^{j\omega_0\tau}}{1+(\omega_0 T)^2}\right] = \frac{(\omega_0 T)^2}{1+(\omega_0 T)^2}\left[\frac{\mathrm{e}^{j\omega_0\tau}+\mathrm{e}^{-j\omega_0\tau}}{2}\right],$$
$$= \frac{(\omega_0 T)^2}{1+(\omega_0 T)^2}\cos(\omega_0\tau). \quad (7.172)$$

The autocorrelation function (7.171) does not have a constant term and consequently

$$E\{Y(t)\} = 0. \quad (7.173)$$

The variance is:

$$\sigma_{YY}^2 = E\{Y^2(t)\} - E\{Y(t)\} = E\{Y^2(t)\} = R_{YY}(0) = \frac{(\omega_0 T)^2}{1+(\omega_0 T)^2}. \quad (7.174)$$

The obtained results (7.173) and (7.174) are equal to the results (7.168) and (7.170), respectively.

Exercise E.7.11 If the input process $X(t)$ from Exercise E.7.9 has a Gaussian PDF what will be the PDF of the output process $Y(t)$?

Answer Since the input process is subjected to a LTI filtering, it follows that the input Gaussian process is subjected to a linear transformation. As a consequence, the output process is also a Gaussian process (see Sect. 4.3).

The process $Y(t)$ is a WS stationary (with a constant mean and autocorrelation function that depends only on τ) and, consequently, its PDF is independent of time. This means that the PDF of the output process is:

$$f_Y(y) = N(0, \sigma_{YY}^2). \quad (7.175)$$

where the variance is given in (7.174).

Exercise E.7.12 Consider that the input to the RC filter shown in Fig. 7.27 is a white noise with the PSD $S_{XX}(\omega) = N_0/2$. Find the PSD and the autocorrelation function of the process at the output of the filter.

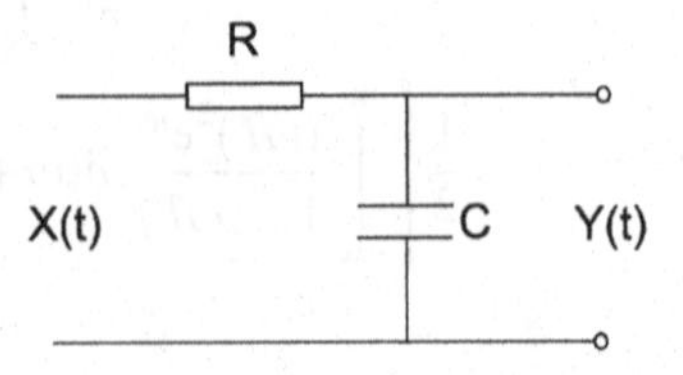

Fig. 7.27 RC filter in Exercise E.7.12

Answer Introducing the time constant of the filter

$$T = RC, \tag{7.176}$$

we get the transfer function of the filter:

$$H(j\omega) = \frac{1}{1 + jT\omega}. \tag{7.177}$$

From (7.177), the squared magnitude response is

$$|H(j\omega)|^2 = \frac{1}{1 + (\omega T)^2}. \tag{7.178}$$

The output PSD is:

$$S_{YY}(\omega) = |H(j\omega)|^2 S_{XX}(\omega) == \frac{1}{1 + (\omega T)^2} \frac{N_0}{2}. \tag{7.179}$$

The corresponding autocorrelation function is obtained as

$$R_{YY}(\tau) = F^{-1}\{S_{YY}(\omega)\} = F^{-1}\left\{\frac{1}{1 + (\omega T)^2} \frac{N_0}{2}\right\}$$

$$= \frac{N_0}{4T} F^{-1}\left\{2 \frac{1/T}{(1/T)^2 + \omega^2}\right\}. \tag{7.180}$$

From Appendix E, we find the inverse Fourier transform

$$F^{-1}\left\{2 \frac{1/T}{(1/T)^2 + \omega^2}\right\} = \mathrm{e}^{-|\tau|/T}. \tag{7.181}$$

Placing (7.181) into (7.180), we get:

$$R_{YY}(\tau) = \frac{N_0}{4T} \mathrm{e}^{-|\tau|/T}. \tag{7.182}$$

Exercise E.7.13 Considering that the input process into the filter from Exercise E.7.12 was a Gaussian noise, find the probability that the absolute value of the output signal is less than σ_{YY}.

Answer The output process is also a Gaussian process, as discussed in Exercise E.7.11. Since the process $Y(t)$ is a WS stationary, it follows that its PDF is independent of time.

Consequently, the PDF of the process $Y(t)$ is a normal PDF with parameters:

$$E\{Y(t))\} = 0. \tag{7.183}$$

$$\sigma^2{}_{YY} = R_{YY}(0) = N_0/4T. \tag{7.184}$$

The probability is calculated using (4.33):

$$P\left\{-\frac{1}{2}\sqrt{\frac{N_0}{T}} \le Y \le \frac{1}{2}\sqrt{\frac{N_0}{T}}\right\} = \operatorname{erf}\left(\frac{1}{\sqrt{2}}\right). \tag{7.185}$$

Exercise E.7.14 Find the minimum value of the filter constant T for which the probability that the output signal in absolute value is less than its STD σ_{YY}.

Answer From (7.185), we found:

$$\frac{1}{2}\sqrt{\frac{N_0}{T}} = \operatorname{erf}\left(\frac{1}{\sqrt{2}}\right). \tag{7.186}$$

From here, we get:

$$T = \frac{N_0}{4\left(\operatorname{erf}(1/\sqrt{2})\right)^2}. \tag{7.187}$$

Exercise E.7.15 The process $X(t)$ at the input of the filter shown in Fig. 7.28a has the autocorrelation function

$$R_{XX}(\tau) = \cos(\omega_0\tau). \tag{7.188}$$

Find the mean value and the mean power at the output of the filter.

Answer The PSD of the input process is (see Exercise E.7.9):

$$S_{XX}(\omega) = \pi[\delta(\omega + \omega_0) + \delta(\omega - \omega_0)]. \tag{7.189}$$

From Fig. 7.28a, we have:

$$Y(t) = X(t) - X(t - T). \tag{7.190}$$

If the input signal is a delta function $\delta(t)$, then the output of the filter is denoted as $h(t)$ and is called an impulse response (Fig. 7.28b):

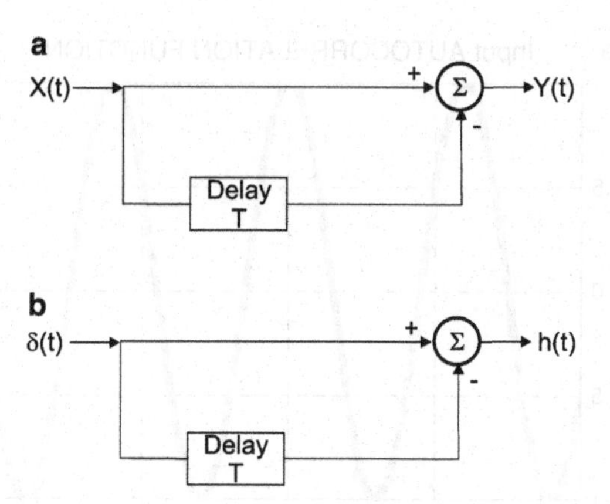

Fig. 7.28 Filter in Exercise E.7.15

$$h(t) = \delta(t) - \delta(t - T). \tag{7.191}$$

The transfer function of the filter is obtained by taking the Fourier transform of both sides of (7.191):

$$H(j\omega) = 1 - \mathrm{e}^{-j\omega T}, \tag{7.192}$$

where

$$F\{h(t)\} = H(j\omega). \tag{7.193}$$

The squared magnitude response of the filter is:

$$\begin{aligned} |H(j\omega)|^2 &= \left|1 - \mathrm{e}^{-j\omega T}\right|^2 = \left|1 - [\cos(\omega T) - j\sin(\omega T)]\right|^2 \\ &= \left|[1 - \cos(\omega T)]^2 + \sin^2(\omega T)\right| = 2[1 - \cos(\omega T)]. \end{aligned} \tag{7.194}$$

The output power density spectrum is:

$$S_{YY}(\omega) = |H(j\omega)|^2 S_{XX}(\omega) = 2\pi[1 - \cos(\omega T)][\delta(\omega + \omega_0) + \delta(\omega + \omega_0)]. \tag{7.195}$$

From (7.195), we conclude that

$$E\{Y(t)\} = 0. \tag{7.196}$$

The variance (mean power) is obtained in the following:

$$\begin{aligned} \sigma_{YY}^2 = R_{YY}(0) &= \frac{1}{2\pi}\int_{-\infty}^{\infty} 2\pi[1 - \cos(\omega T)][\delta(\omega + \omega_0) + \delta(\omega - \omega_0)]\mathrm{d}\omega \\ &= 1 - \cos(\omega_0 T) + 1 - \cos(\omega_0 T) = 2[1 - \cos(\omega_0 T)]. \end{aligned} \tag{7.197}$$

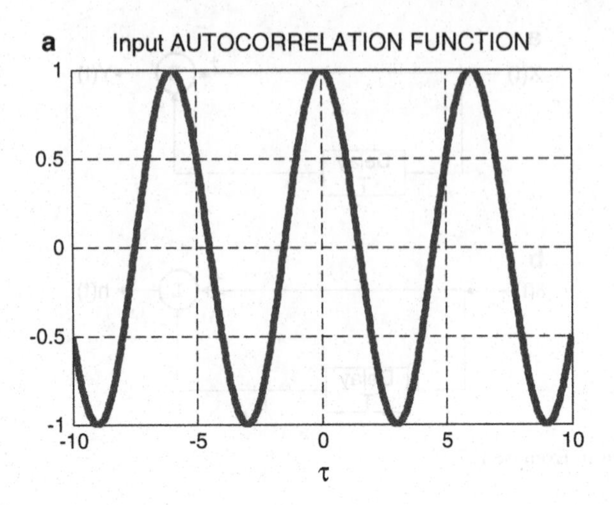

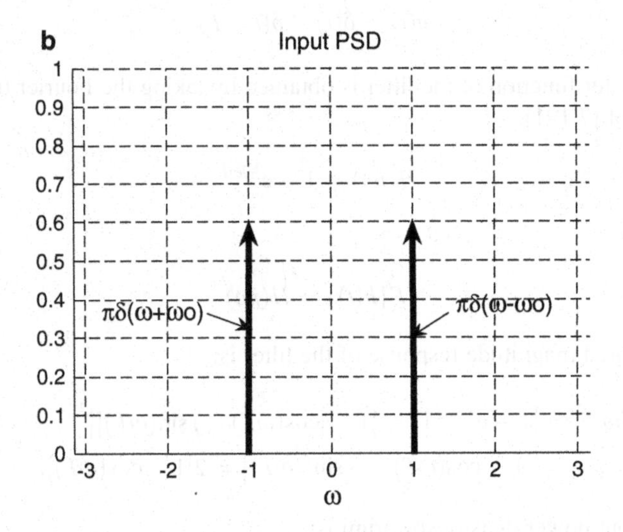

Fig. 7.29 (**a**) Input ACF. (**b**) Input PSD

7.6 MATLAB Exercises

Exercise M.7.1 (MATLAB file *exercise_M_7_1.m*) Using the periodogram method, estimate the PSDs and show PSD plots for two white signals X_1 (normal) and X_2 (uniform), both with $m = 0$ and $\sigma^2 = 1$.

Solution The PSD is estimated using the MATLAB file *periodogram.m*

The estimated PSDs are shown in Fig. 7.30 in a normal scale and in a logarithmic scale in Fig. 7.31.

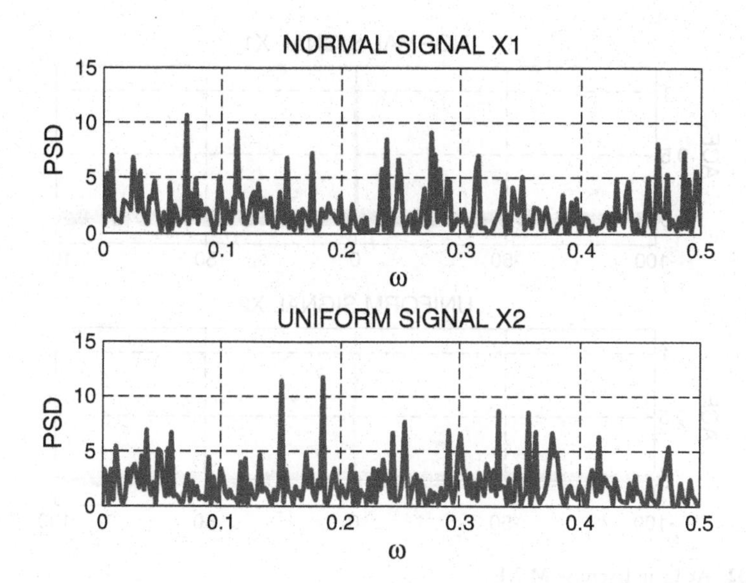

Fig. 7.30 PSDs in Exercise M.7.1

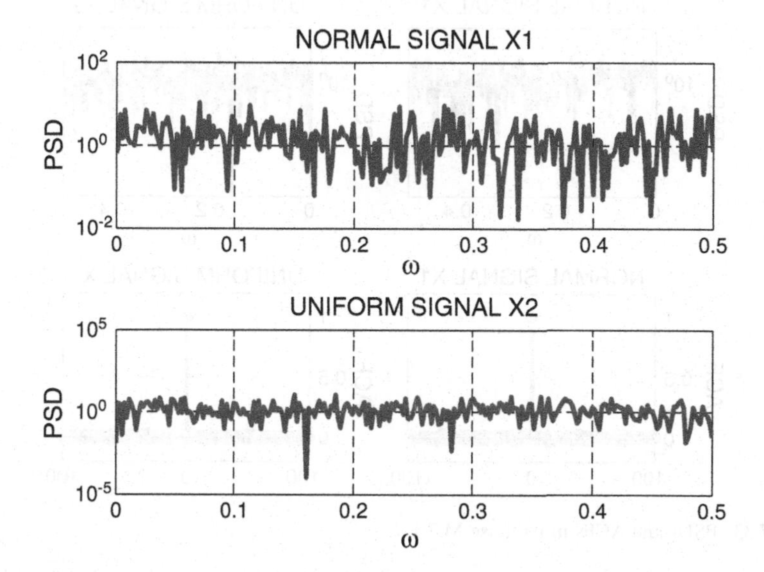

Fig. 7.31 PSDs (logarithmic scale) in Exercise M.7.1

Figure 7.32 presents the ACFs. For the sake of comparison, both ACFs and PSDs are shown together in Fig. 7.33.

Exercise M.7.2 (MATLAB file *exercise_M_7_2.m*) Sinusoidal signals of amplitude A and frequencies $f_1 = 150$ Hz and $f_2 = 250$ Hz are added to the random

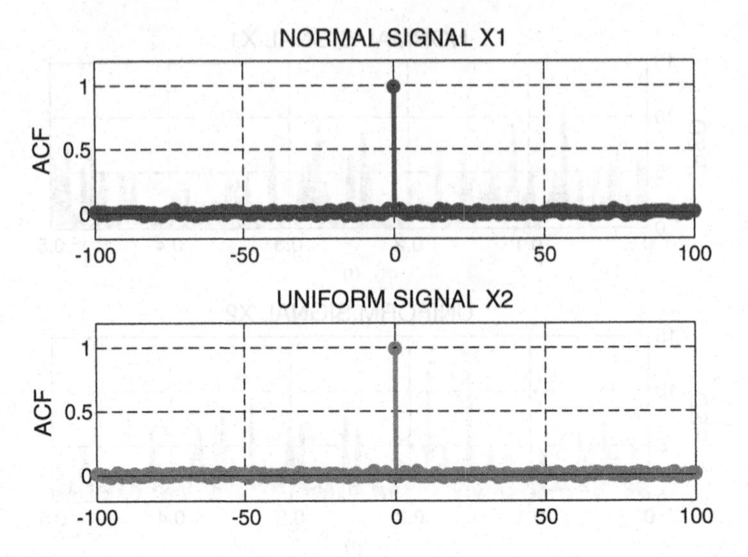

Fig. 7.32 ACFs in Exercise M.7.1

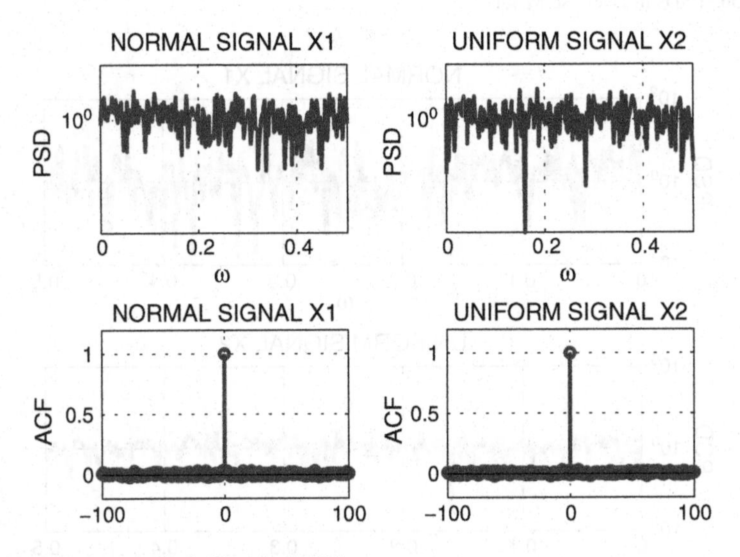

Fig. 7.33 PSDs and ACFs in Exercise M.7.1

signals X_1 and X_2 from Exercise M.6.7, respectively. Using the periodogram method, estimate their PSDs and show the PSD plots for $A = 1$ and $A = 2$.

Solution The PSD is estimated using the MATLAB file *periodogram.m*

The estimated PSDs for $A = 1$ on a linear and a logarithmic scale are shown in Figs. 7.34. and 7.35. Figures 7.36 and 7.37 show the estimated PSDs for $A = 2$. Note the peaks in the PSDs at the frequency of sinusoidal signal.

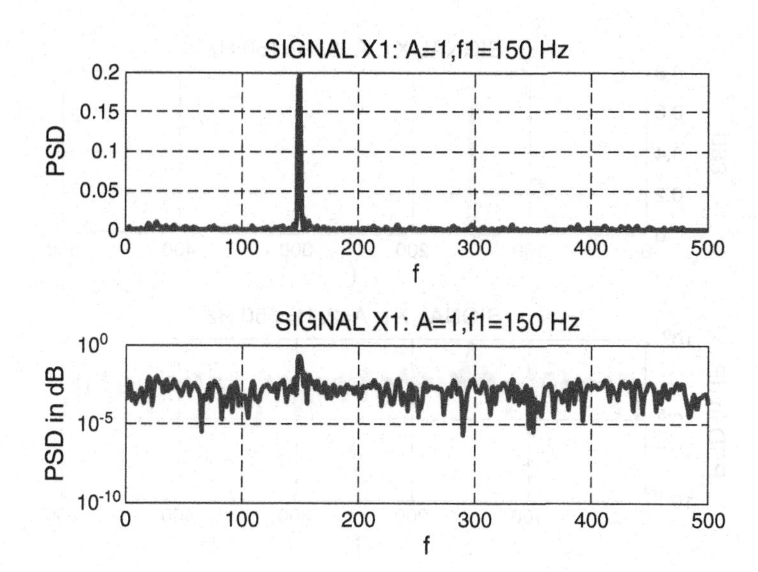

Fig. 7.34 Signal $X_1(A = 1)$: PSDs

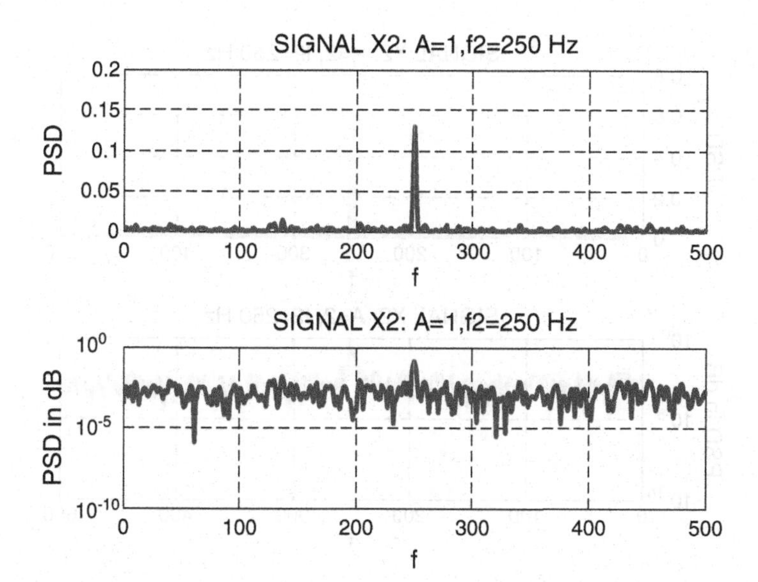

Fig. 7.35 Signal $X_2(A = 1)$: PSDs

Exercise M.7.3 (MATLAB file *exercise_M_7_3.m*) Using the periodogram method, estimate the PSDs and show the PSD plots for normal white signals with:

1. $m = 4$ and $\sigma^2 = 9$.
2. $m = 5$ and $\sigma^2 = 16$.

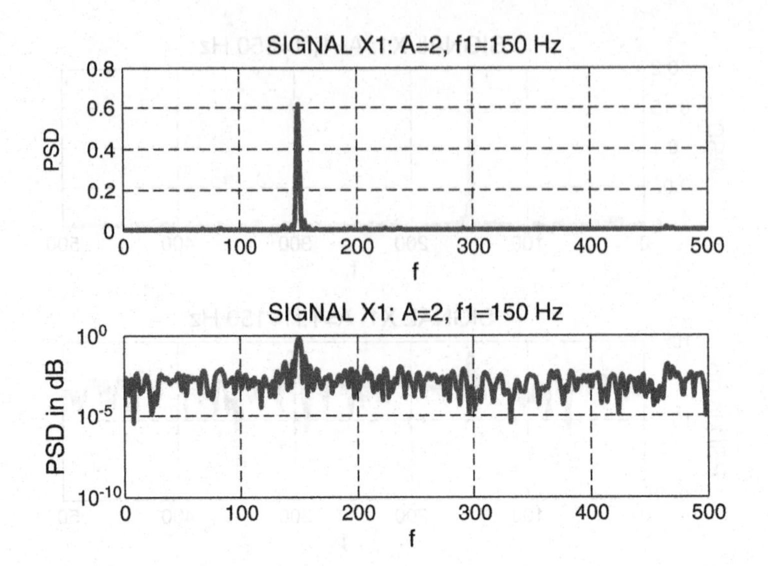

Fig. 7.36 Signal $X_1(A = 2)$: PSDs

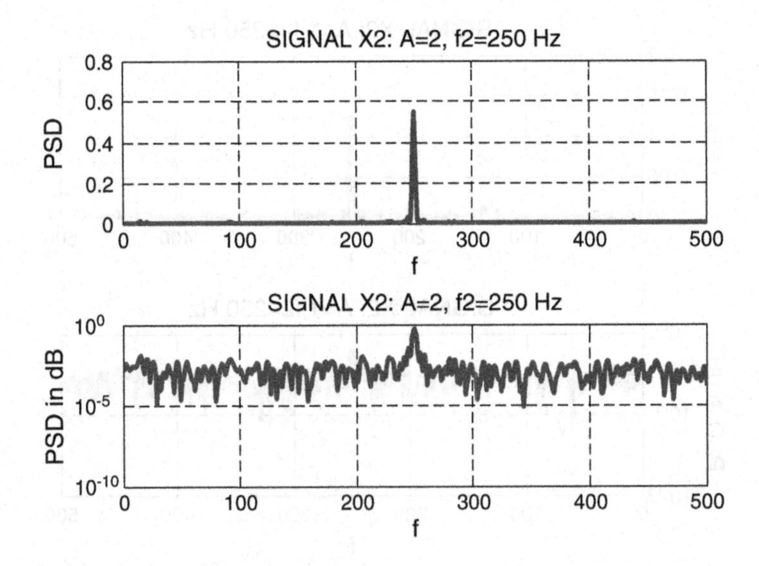

Fig. 7.37 Signal $X_2(A = 2)$: PSDs

Solution The PSD is estimated using the MATLAB file *periodogram.m*

The estimated PSDs in a linear and a logarithmic scale are shown in Figs. 7.38 and 7.39.

The PSD has the peak at zero frequency because signal has a nonzero mean.

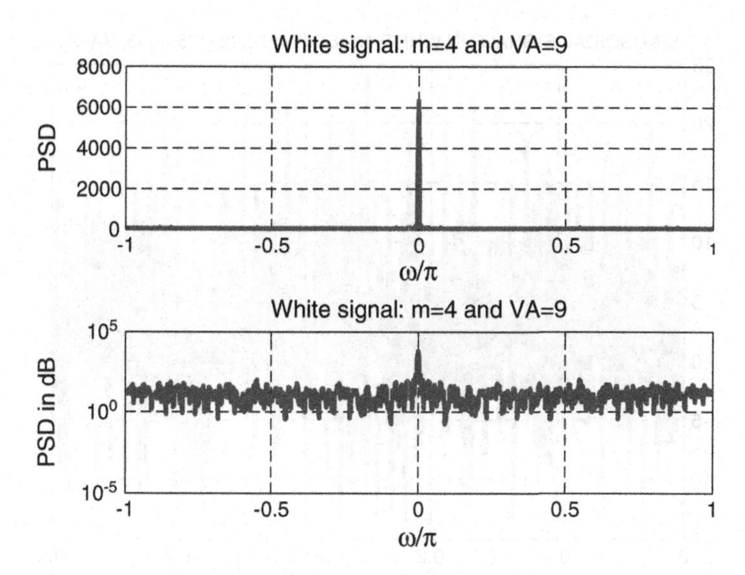

Fig. 7.38 White noise (PSDs) with a mean of $m = 4$ and $\sigma^2 = 9$

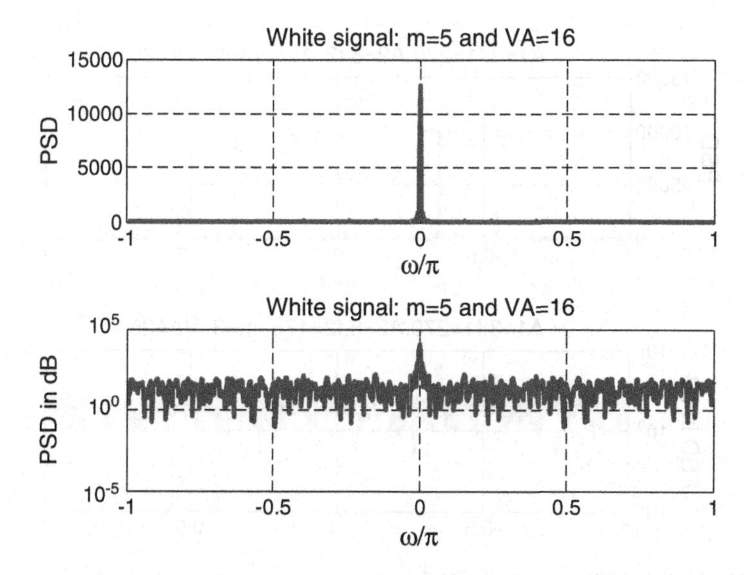

Fig. 7.39 White noise (PSDs) with a mean of $m = 5$ and $\sigma^2 = 16$

Exercise M.7.4 (MATLAB file *exercise_M_7_4.m*) Using the periodogram method, estimate the PSD and show the PSD plots for sinusoidal signals with $A_1 = 3$, $f_1 = 270$ Hz as well as $A_2 = 8$, $f_2 = 175$ Hz in a normal white noise with $m = 3$ and $\sigma^2 = 25$, as shown in Fig. 7.40.

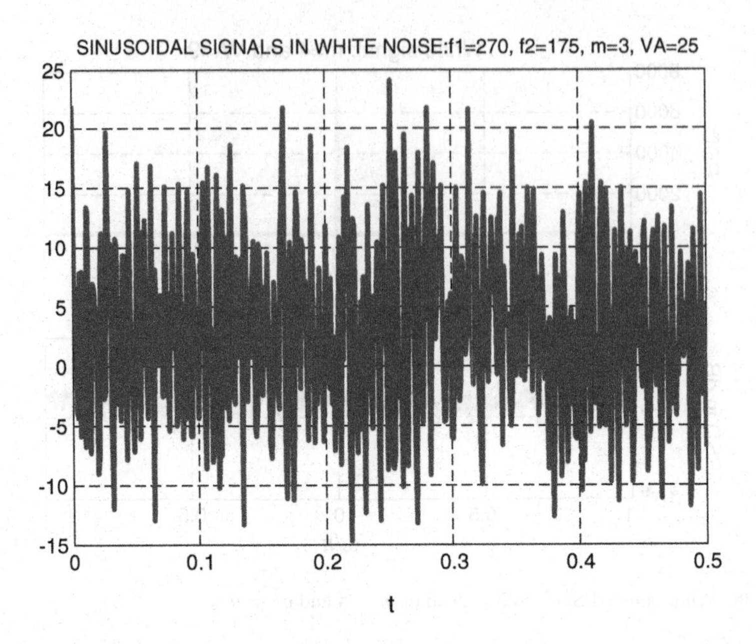

Fig. 7.40 Sinusoidal signals in nonzero mean white noise

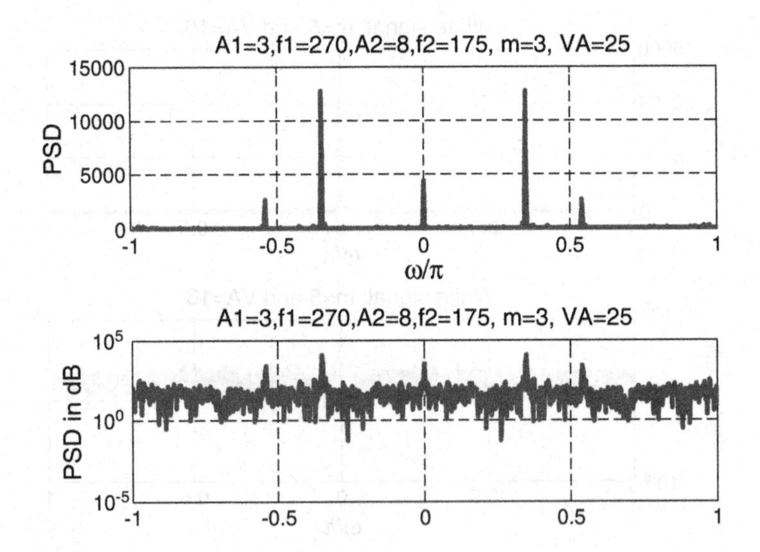

Fig. 7.41 Estimated PSDs in Exercise M.7.4

Solution The PSD is estimated using the MATLAB file *periodogram.m*

The estimated PSDs on a normal and a logarithmic scale are shown in Fig. 7.41. The PSD has the peak at zero frequency because signal has a nonzero mean.

Additionally, the PSD has peaks at the frequencies of the sinusoidal signal, 175 and 270 Hz.

Exercise M.7.5 (MATLAB file *exercise_M_7_5.m*) Given a normal noise with $m = 4$ and $\sigma^2 = 4$.

1. If the noise is filtered with a low-pass filter $H_1(z)$, using the periodogram method, estimate the PSD and plot the PSD at the filter output.
2. If the noise is filtered with a high-pass filter $H_2(z)$, using the periodogram method, estimate the PSD and plot the PSD at the filter output.

Solution
1. The estimated PSDs of the input and output signals are shown in Fig. 7.42. The filter attenuates the high frequencies of the input signal.
2. The estimated PSDs of the input and output signals are shown in Fig. 7.43. The filter attenuates the low frequencies of the input signal including a DC component.

Exercise M.7.6 (MATLAB file *exercise_M_7_6.m*) A normal noise with $m = 0$ and $\sigma^2 = 16$ is LP filtered and then multiplied with a sinusoidal signal of amplitude $A = 1$ and $f = 300$ Hz.
Estimate and plot the PSDs at the filter output and at the product device output.

Solution The estimated PSDs are shown in Fig. 7.44. Note that the PSD at the output of the product device has replicas of the input spectrum around the sinusoidal frequency $f = 300$ Hz or $\omega/\pi = 300/500 = 0.6$.

Exercise M.7.7 (MATLAB file *exercise_M_7_7.m*) Given two independent low-pass and high-pass normal signals $y_1(t)$ and $y_2(t)$, respectively, both with $m = 0$ and ${\sigma_1}^2 = 16$, ${\sigma_2}^2 = 1$. A sinusoidal signal $x(t)$ of amplitude $A = 1$ and $f = 250$ Hz is added to the signal $y_2(t)$ yielding

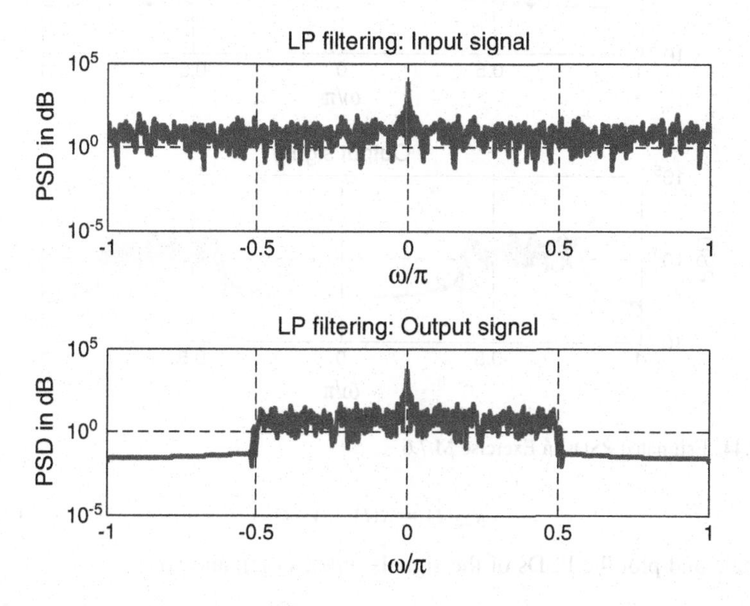

Fig. 7.42 LP filtering in Exercise M.7.5

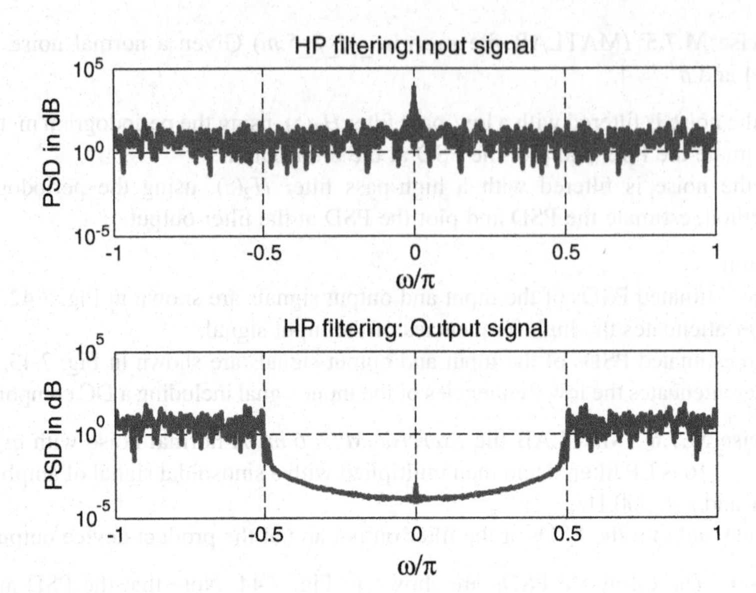

Fig. 7.43 HP filtering in Exercise M.7.5

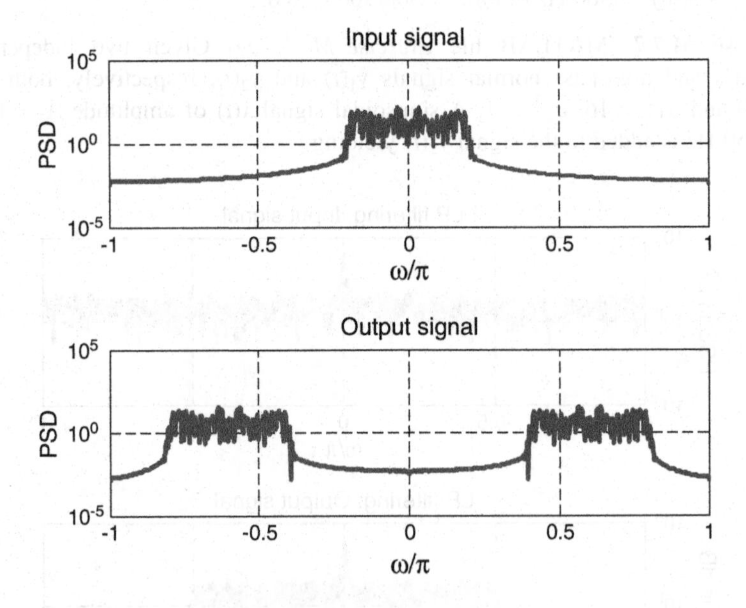

Fig. 7.44 Estimated PSDs in Exercise M.7.6

$$y_{22}(t) = x(t) + y_2(t). \tag{7.198}$$

Estimate and plot the PSDs of the signals, $y_1(t)$, $y_{22}(t)$, and $y(t)$.

$$y(t) = y_1(t) + y_{22}(t). \tag{7.199}$$

Solution The PSD of the signal Y is

$$S_{YY}(\omega) = S_{Y_1Y_1}(\omega) + S_{Y_{22}Y_{22}}(\omega). \tag{7.200}$$

The estimated PSDs are shown in Figs. 7.45–7.47.
Figure 7.48 shows all estimated PSDs in one plot.

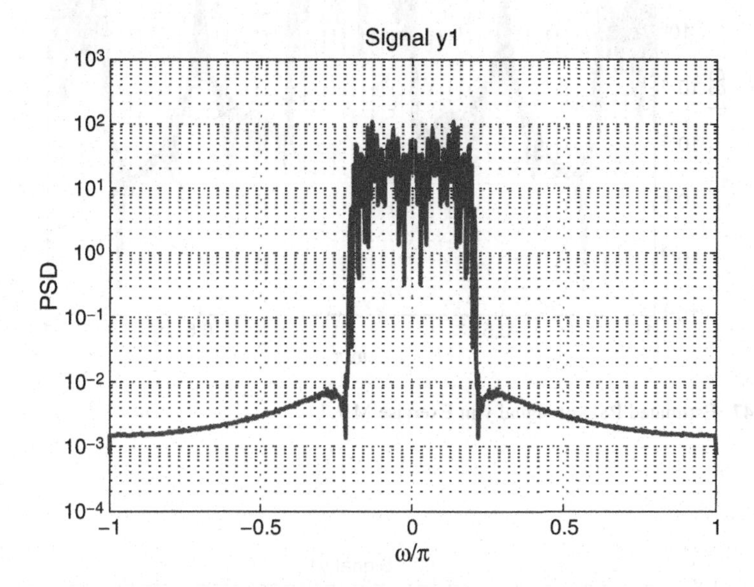

Fig. 7.45 Estimated PSD of signal y_1 in Exercise M.7.7

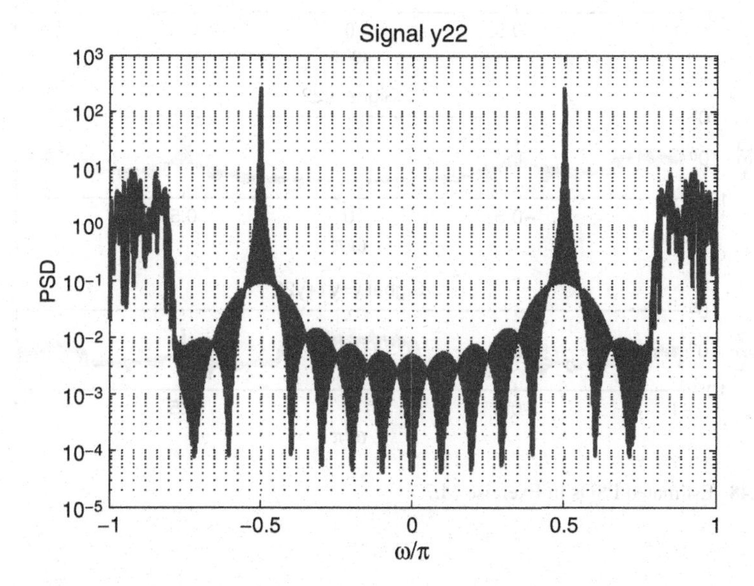

Fig. 7.46 Estimated PSD of signal y_{22} in Exercise M.7.7

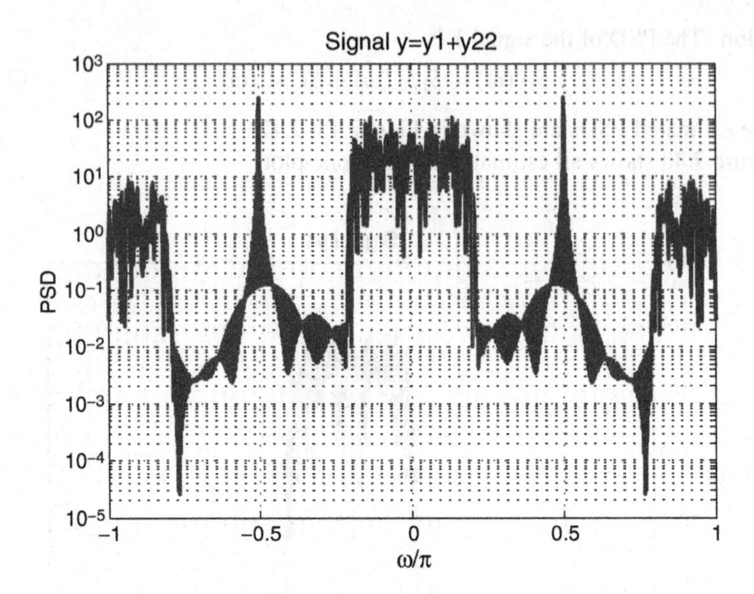

Fig. 7.47 Estimated PSD of signal y in Exercise M.7.7

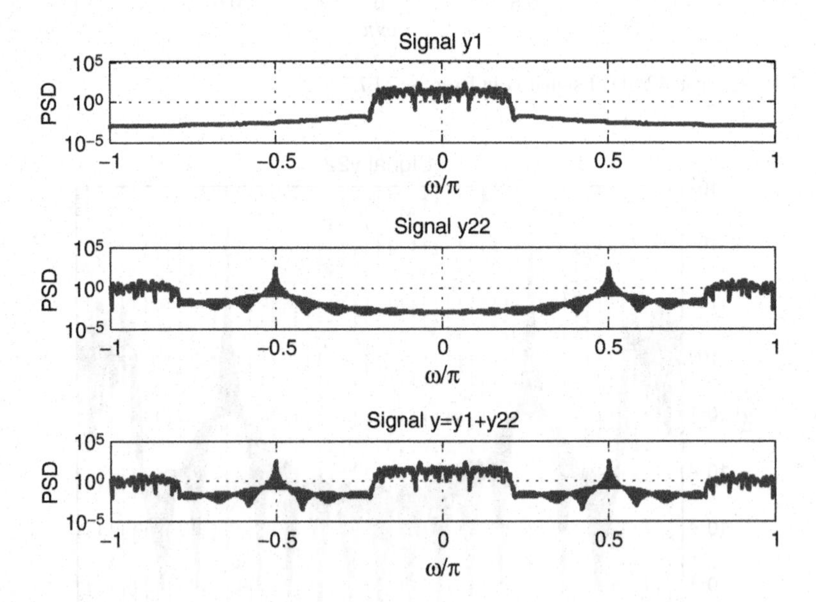

Fig. 7.48 Estimated PSDs in Exercise M.7.7

7.7 Questions

Q.7.1. Is it possible for a process *X*(*t*), which is at least WS stationary, to have an autocorrelation function given as:

$$R_{XX}(\tau) = \sin(\omega_0\tau), \tag{7.201}$$

where ω_0 is a constant.

Q.7.2. Does the Fourier transform of a random process give the phase information of a random process?

Q.7.3. Does the PSD represents a spectral characteristic of the process or only of a particular realization of a process?

Q.7.4. The Fourier transform of deterministic signals can be viewed as a decomposition of the signal into sinusoids of different frequencies with deterministic amplitudes and phases. Is there such decomposition in the case of a random process?

Q.7.5. Due to duality in time and frequency, the constant in frequency corresponds to a delta in time and vice versa, as shown in Table 7.1

Table 7.1 Duality of time and frequency

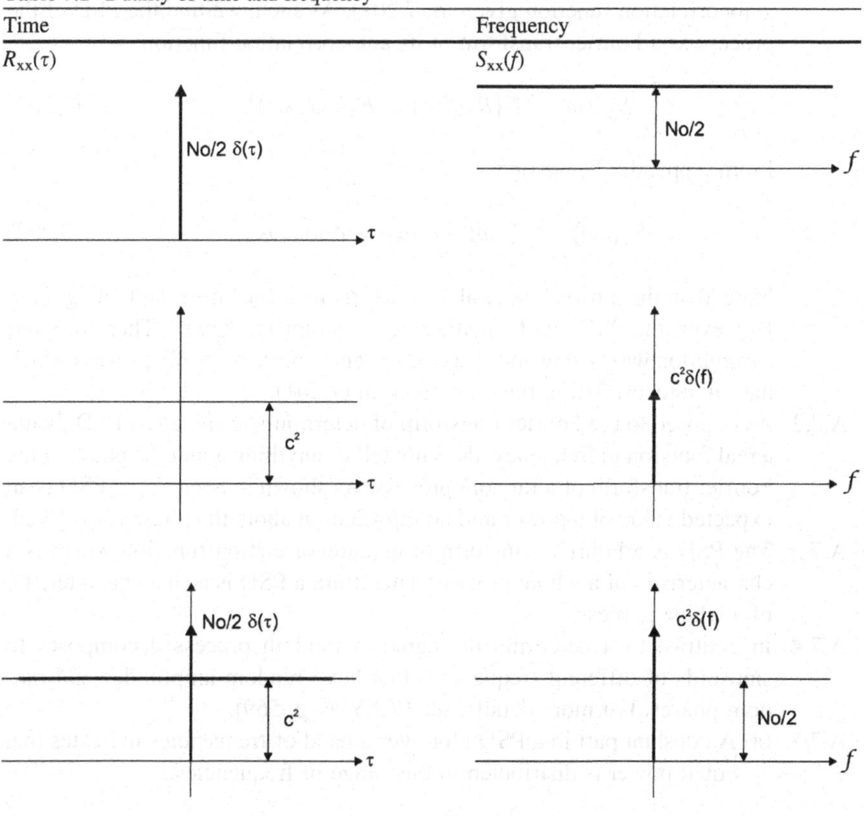

(a) What does a constant part in a PSD plot indicate?
(b) What does a constant part in an autocorrelation plot indicate?
(c) What does a delta function in a PSD plot indicate?
(d) What does a delta function in an autocorrelation plot indicate?

Q.7.6. Why must a process with a DC component have a delta function in its PSD?

Q.7.7. How we can detect from the PSD whether or not the process has a periodical component?

Q.7.8. Can a low-pass process have a zero average power at frequency $f = 0$?

Q.7.9. Why is the PSD of a white process usually presented as $N_0/2$ instead of N_0?

Q.7.10. Can a white process also be a non-Gaussian?

Q.7.11. Can a Gaussian process be a nonwhite process?

7.8 Answers

A.7.1. Let us suppose the contrary that there is a WS process $X(t)$ which has the autocorrelation function given in (7.201). As such, we find the PSD of the process as a Fourier transform of its autocorrelation function:

$$S_{XX}(\omega) = F\{R_{XX}(\tau)\} = F\{\sin(\omega_0\tau)\}. \tag{7.202}$$

From Appendix E, we find:

$$S_{XX}(\omega) = -\mathrm{j}\pi[\delta(\omega - \omega_0) - \delta(\omega + \omega_0)]. \tag{7.203}$$

Note that the power spectral characteristic is negative and imaginary. However, the PSD can be neither negative nor imaginary. Therefore, our assumption was wrong and as a consequence, there is no WS process which has an autocorrelation function given in (7.201).

A.7.2. As opposed to the Fourier transform of deterministic signals, a PSD, being a real function in frequency, does not tell us anything about the phase of the Fourier transform of a random process. As shown in Sect. 7.1, a PSD is an expected value of a power and no information about the phase is involved.

A.7.3. The PSD is a Fourier transform of an autocorrelation function which is a characteristic of a whole process. Therefore, a PSD is also a characteristic of a whole process.

A.7.4. In contrast to a deterministic signal, a random process decomposes to sinusoids of different frequencies that have random amplitudes and random phases. For more details, see [KAY06, p. 569].

A.7.5. (a) A constant part in a PSD plot over a band of frequencies indicates that equal power is distributed in this range of frequencies.

(b) A constant part in an autocorrelation function indicates that a process has a DC component.
(c) A delta function at the origin of a PSD plot indicates that there is a finite power at this frequency (i.e., that the process has a DC component).
(d) A delta function at the origin of an autocorrelation function indicates that the power is infinite.

A.7.6. The power is an area below a PSD. The power for the infinitesimal frequency band df is:

$$\mathrm{d}P_{XX} = \int_{0}^{\mathrm{d}f} S_{XX}(f)\mathrm{d}f. \tag{7.204}$$

In the frequency $f = 0$, this interval d$f \rightarrow 0$, and as consequence the power from (7.204) is zero. Therefore, the power which belongs to the frequency $f = 0$ is zero. In order for the signal to have a finite power in the frequency $f = 0$ (DC component), it is necessary that S_{XX} (f) in $f = 0$ is going to infinity, i.e., it must have a delta function at $f = 0$.

A.7.7. A PSD must have a delta function at the frequency of a periodic component.

A.7.8. A low-pass process has a zero power at the frequency $f = 0$ unless its PSD has a delta function at the frequency $f = 0$.

A.7.9. The PSD of a white process is presented as $N_0/2$ in order to include the negative frequencies, introduced for mathematical reasons, and which physically do not exist.

A.7.10. The term "white" is related with a PSD while the term "Gaussian" is related with a amplitude distribution. Therefore, a white process can have different amplitude distributions and, thus, not necessarily be a Gaussian process. However, in communication systems, the noise in the channel is usually presented as a Gaussian white noise (GWN) and because of this one can be made to think that a white process can be only a Gaussian process. (see [JOV11a]).

A.7.11. As explained in A.7.10., a Gaussian process can also be a nonwhite process. (see [JOV11a]).

Appendix A: Definite Integrals

1. $\int_0^\infty x^n \, \mathrm{e}^{-ax} \, \mathrm{d}x = \frac{\Gamma(n+1)}{a^{n+1}}; \quad a > 0, \quad n > -1$

2. $\int_0^\infty \frac{\cos ax}{1+x^2} \mathrm{d}x = \frac{\pi}{2} \mathrm{e}^{-|a|}$

3. $\int_0^\infty \mathrm{e}^{-a^2x^2} \cos bx \, \mathrm{d}x = \frac{\sqrt{\pi}}{2a} \mathrm{e}^{-\frac{b^2}{4a^2}}; \quad a > 0$

4. $\int_0^\infty x^n \, \mathrm{e}^{-ax^2} \, \mathrm{d}x = \frac{\Gamma\left(\frac{n+1}{2}\right)}{2a^{\frac{n+1}{2}}} = \begin{cases} \frac{1 \times 3 \times \cdots \times (2k-1)\sqrt{\pi}}{2^{k+1} a^{k+1/2}} & n = 2k \\ \frac{k!}{2a^{k+1}} & n = 2k+1 \end{cases}, \quad k = 0, 1, 2, \ldots$

5. $I_0(\beta) = \frac{1}{2\pi} \int_{-\pi}^{\pi} \mathrm{e}^{j\beta \sin x} \, \mathrm{d}x \approx 1 - \frac{\beta^2}{4}, \quad \text{for } \beta \ll 1$

6. $\int_{-\infty}^{\infty} \mathrm{e}^{-a^2x^2+bx} \, \mathrm{d}x = \frac{\sqrt{\pi}}{a} \mathrm{e}^{b^2/4a^2}, \quad a > 0$

7. $\int_0^\infty \frac{a}{a^2+x^2} \mathrm{d}x = \frac{\pi}{2}; \quad a > 0$

8. $\int_0^\infty \frac{\sin^2 ax}{x^2} \mathrm{d}x = |a|\pi/2; \quad a > 0$

G.J. Dolecek, *Random Signals and Processes Primer with MATLAB*,
DOI 10.1007/978-1-4614-2386-7,

Appendix B: Indefinite Integrals

1. $\int x\,\mathrm{e}^{-ax}\,\mathrm{d}x = \frac{\mathrm{e}^{-ax}}{a^2}(ax-1)$
2. $\int x^n\,\mathrm{d}x = \frac{x^{n+1}}{n+1}$
3. $\int \sin ax\,\mathrm{d}x = -(1/a)\cos x$
4. $\int \cos ax\,\mathrm{d}x = (1/a)\sin x$
5. $\int \sin^2 ax\,\mathrm{d}x = x/2 - \sin(2ax)/4a$
6. $\int \cos^2 ax\,\mathrm{d}x = x/2 + \sin(2ax)/4a$
7. $\int \ln x\,\mathrm{d}x = x\ln x - x$
8. $\int \mathrm{e}^{ax}\,\mathrm{d}x = \mathrm{e}^{ax}/a$
9. $\int a^x\,\mathrm{d}x = \frac{a^x}{\ln a}$
10. $\int \frac{\mathrm{d}x}{a^2+x^2} = (1/a)\tan^{-1}(x/a)$
11. $\int x\sin ax\,\mathrm{d}x = \frac{1}{a^2}(\sin ax - ax\cos ax)$
12. $\int x\cos ax\,\mathrm{d}x = \frac{1}{a^2}(\cos ax + ax\sin ax)$
13. $\int x^2\sin ax\,\mathrm{d}x = (2ax\sin ax + 2\cos ax - a^2x^2\cos ax)/a^3$
14. $\int x^2\cos ax\,\mathrm{d}x = (2ax\cos ax - 2\sin ax + a^2x^2\sin ax)/a^3$
15. $\int \frac{x\,\mathrm{d}x}{a^2+x^2} = \frac{1}{2}\ln(a^2+x^2)$

G.J. Dolecek, *Random Signals and Processes Primer with MATLAB*,
DOI 10.1007/978-1-4614-2386-7,

Appendix C: Gamma Function

Definition

$$\Gamma(x) = \int_0^\infty e^{-t} t^{x-1} \, dt, \quad x > 0$$

Properties

1.

$$\Gamma(1/2) = \sqrt{\pi}$$

2.

$$\Gamma(x+1) = x\Gamma(x), \quad x > 0$$

3.

$$\Gamma(n+1) = n!, \quad n \text{ is a nonnegative integer.}$$

G.J. Dolecek, *Random Signals and Processes Primer with MATLAB*,
DOI 10.1007/978-1-4614-2386-7,

Appendix D: Useful Mathematical Quantities

1. $\sin x = x - \frac{x^3}{3!} + \frac{x^5}{5!} - \cdots$
2. $\cos x = 1 - \frac{x^2}{2!} + \frac{x^4}{4!} - \frac{x^6}{6!} \cdots$
3. $e^x = 1 + x + \frac{x^2}{2!} + \frac{x^3}{3!} + \cdots = \sum_{n=0}^{\infty} \frac{x^n}{n!}$
4. $e^{jx} = \cos x + j \sin x$
5. $x! = 1 \times 2 \times 3 \times \cdots$, x positive integer
6. $(a+b)^n = C_n^0 + C_n^1 a^{n-1} b + C_n^2 a^{n-2} b^2 + C_n^3 a^{n-3} b^3 + \cdots + C_n^{n-1} a b^{n-1} + C_n^n b^n$
   $C_n^k = \frac{n!}{k!(n-k)!}$
7. $\cos(x \pm y) = \cos x \cos y \mp \sin x \sin y$
8. $\sin(x \pm y) = \sin x \cos y \pm \cos x \sin y$
9. $\cos 2x = \cos^2 x - \sin^2 x$
10. $\sin 2x = 2 \sin x \cos x$
11. $\cos x = \dfrac{e^{jx} + e^{-jx}}{2}$
12. $\sin x = \dfrac{e^{jx} - e^{-jx}}{2j}$
13. $\cos x \cos y = \frac{1}{2}[\cos(x-y) + \cos(x+y)]$
14. $\sin x \sin y = \frac{1}{2}[\cos(x-y) - \cos(x+y)]$
15. $\sin x \cos y = \frac{1}{2}[\sin(x-y) + \sin(x+y)]$
16. $\cos^2 x = \dfrac{1}{2}[1 + \cos 2x]$
17. $\sin^2 x = \dfrac{1}{2}[1 - \cos 2x]$
18. $\lim_{n \to \infty} \left(1 - \dfrac{x}{n}\right)^n = e^{-x}$
19. $\lim_{n \to \infty} \left(1 + \dfrac{1}{n}\right)^n = e$

G.J. Dolecek, *Random Signals and Processes Primer with MATLAB*,
DOI 10.1007/978-1-4614-2386-7,

Appendix E: Fourier Transform

Definition

$$X(\omega) = F\{x(t)\} = \int_{-\infty}^{\infty} x(t)\mathrm{e}^{-j\omega t}\,\mathrm{d}t$$

$$x(t) = F^{-1}\{X(\omega)\} = \frac{1}{2\pi}\int_{-\infty}^{\infty} X(\omega)\mathrm{e}^{j\omega t}\,\mathrm{d}t$$

Properties

1. Linearity:

$$F\left\{\sum_{i=1}^{N} a_i x_i(t)\right\} = \sum_{i=1}^{N} a_i F\{x_i(t)\}$$

2. Time shifting:

$$F\{x(t-t_0)\} = X(\omega)\mathrm{e}^{-j\omega t_0}$$

3. Frequency shifting:

$$F\{x(t)\mathrm{e}^{j\omega t_0}\} = X(\omega - \omega_0)$$

4. Time scaling:

$$F\{x(at)\} = \frac{1}{|a|}X(\omega/a)$$

DOI 10.1007/978-1-4614-2386-7,

5. Duality:

$$\text{If } X(\omega) = F\{x(t)\}, \quad \text{then} \quad F\{X(t)\} = 2\pi x(-\omega)$$

6. Multiplication in time:

$$F\{x_1(t)x_2(t)\} = X_1(\omega) * X_2(\omega) \quad \text{(convolution)}$$

7. Convolution in time:

$$F\{x_1(t) * x_2(t)\} = X_1(\omega)X_2(\omega)$$

8. Parseval's Theorem:

$$\int_{-\infty}^{\infty} |x(t)|^2 \, \mathrm{d}t = \frac{1}{2\pi} \int_{-\infty}^{\infty} |X(\omega)|^2 \, \mathrm{d}\omega$$

Table A1 Fourier transform pairs

$x(t)$	$X(\omega)$
$k\delta(t)$	k
k	$2\pi k\delta(\omega)$
$u(t)$	$\pi\delta(\omega) + \frac{1}{j\omega}$
$\frac{1}{2}\delta(t) - \frac{1}{j2\pi t}$	$u(\omega)$
$e^{j\omega t_0}$	$2\pi\delta(\omega - \omega_0)$
$\delta(t - t_0)$	$e^{-j\omega t_0}$
$\cos(\omega_0 t)$	$\pi[\delta(\omega - \omega_0) + \delta(\omega + \omega_0)]$
$\sin(\omega_0 t)$	$-j\pi[\delta(\omega - \omega_0) - \delta(\omega + \omega_0)]$
$u(t)\cos(\omega_0 t)$	$\frac{\pi}{2}[\delta(\omega - \omega_0) + \delta(\omega + \omega_0)] + \frac{j\omega}{\omega_0^2 - \omega^2}$
$u(t)\mathrm{e}^{-kt},\ k > 0$	$\frac{1}{k + j\omega}$
$u(t)t\,\mathrm{e}^{-kt}$	$\frac{1}{(k + j\omega)^2}$
$\mathrm{e}^{-k\lvert t\rvert}$	$\frac{2k}{k^2 + \omega^2}$
$\mathrm{e}^{-\frac{t^2}{2\sigma^2}}$	$\sigma\sqrt{2\pi}\mathrm{e}^{-\sigma^2\omega^2/2}$

References

[AGR07] A. Agresti, C. Franklin, "*Statistics, the Art and Science of Learning from Data,*" Pearson Prentice Hall, New Jersey, 2007.

[BRE69] L. Breiman, "*Probability and Stochastic Processes with a View Toward Applications,*" Houghton Mifflin Company, Boston, 1969.

[BRO97] R. G. Brown, P. Y. C. Hwang, "*Introduction to Random Signals and Applied Kalman Filtering,*" John Wiley & Sons, New York, 1997 (Third Edition).

[BURR94] C. S. Burrus, at all, "*Computer-Based Exercises for Signal Processing Using MATLAB,*" Prentice Hall, New Jersey, 1994.

[CAR75] B. Carlson, "*Communication Systems*", Mc Graw Hill Inc. New York, 1975.

[CHI97] D. G. Childers, "*Probability and Random Processes,*" The McGraw-Hill, Inc, New York, 1997.

[DAV87] W. B. Davenport, W. L. Root, "*An Introduction to the Theory of Random Signals and Noise,*" IEEE Press, New York, 1987.

[ENG07] S. Engelberg, "*Random Signals and Noise: A Mathematical Introduction,*" CRC Press, Boca Raton, 2007.

[GNE82] B. V. Gnedenko, "*The Theory of Probability,*" MIR Publishers, Moscow, 1982 (Fifth Edition).

[HAD06] A. H. Haddad, "*Probabilistic Systems and Random Signals,*" Pearson Education Inc., New Jersey, 2006.

[HAN96] D. Hanselman, B. Littlefield, "*Mastering MATLAB: A Comprehensive Tutorial and Reference,*" Prentice Hall, New Jersey 1996.

[HEL91] C. W. Helstrom, "*Probability and Stochastic Processes for Engineers,*" Macmillan Publishing Company, New York, 1991.

[HSU93] H. P. Hsu, "Theory and Problems of Analog and Digital Communications," McGraw-Hill, New York, 1993.

[JOV11a] G. Jovanovic Dolecek, "MATLAB-based Program for Teaching Autocorrelation Function and Noise Concepts," *IEEE Transactions on Education* Published on line 2011 (vol-pp, Issue 99).

[JOV11b] G. Jovanovic Dolecek, F. Harris, "On MATLAB Demonstrations of Narrowband Gaussian noise," *Computer Applications in Engineering Education.*, vol. 19, No. 3, pp. 598–603, Sept. 2011.

[JOV10] G. Jovanovic Dolecek, "Interactive MATLAB-based demo program for sum of independent random variables," *Computer Applications in Engineering Education*, published online 2010.

[JOV09] G. Jovanovic Dolecek, F. Harris, "Understanding Histograms, Probability and Probability Density Using MATLAB," *American Society for Engineering Education-Pacific Soithwest Conference, ASEE PSW-2009*, San Diego, March, 19–20, 2009.

DOI 10.1007/978-1-4614-2386-7,

(Proceedings, published by National University San Diego, edited by M. Amin and P. Dey, pp. 332–345).

[JOV97] G. Jovanovic Dolecek, "RANDEMO - Educational Software for Random Signal Analysis", *Computer Applications in Engineering Education*, vol. 5, No. 2, pp. 93–97, 1997.

[JOV87] G. Jovanovic Dolecek, "*Random Variables and Processes in Telecommunications*," (in Bosnian), Svjetlost, Sarajevo, 1987.

[JAFF00] R. C. Jaffe, "*Random Signals for Engineers Using MATLAB and MATHCAD*," Springer–Verlag, New York, 2000.

[KAY06] S. M. Kay, "*Intuitive Probability and Random Processes Using MATLAB*," Springer, New York, 2006.

[KLA89] K. B. Laassen, J. C. L. van Peppen, "*System Reliability: Concepts and Applications*," Edward Arnold, London, 1989.

[KLI86] G. Klimov, "*Probability Theory and Mathematical Statistics,*" Mir Publisher, Moscow, 1986.

[KOM87] J. J. Komo, "*Random Signal Analysis in Engineering Systems*," Academic Press, Orlando, Florida, 1987.

[LAT83] B. P. Lathi, "*Modern Digital and Analog Communication Systems*," CBS College Publ., New York, 1983.

[LEE60] Y. W. Lee, "*Statistical Theory of Communication*," John Wile & Sons Inc., New York, 1960.

[LEO94] A. Leon-Garcia, "*Probability and Random Processes for Electrical Engineering*," Addison-Wesley Publishing Company, Reading, Massachusetts, 1994 (Second Edition).

[MIL04] S. L. Miller, D. Childers, "*Probability and Random Processes with Applications to Signal Processing and Communications*," Elsevier Academic Press Inc., Burlington, MA, 2004.

[NEW93] D. E. Newland, "*An Introduction to Random Vibrations, Spectral & Wavelet Analysis,*" Dover Publications, Inc, Mineola, New York, 1993 (Third Edition).

[NGU09] H. H. Nguen, E. Shwedyk, "*A First Course in Digital Communications*," Cambridge University Press, Cambridge, UK, 2009.

[PAP65] A. Papoulis, "*Probability, Random Variables, and Stochastic Processes*," McGraw-Hill Inc., New York, 1965.

[PEE93] P. Z. Peebles, "*Probability, Random Variables, and Random Signal Principles*," McGraw-Hill Inc., New York, 1993.

[PRA99] R. Pratap, "*Getting Started with MATLAB A Quick Introduction for Scientists and Engineers*," Oxford University Press, 1999.

[POU09] A. R. Poularikis, "*Discrete Random Signal Processing and Filtering Primer with MATLAB,*" CRC Press Taylor & Francis Group, Boca Raton, 2009.

[ROSS10] S. Ross, "*A First Course in Probability*," Prentice Hall, New Jersey, 2010.

[SCH75] M. Schwartz, L. Shaw, "*Signal Processing: Discrete Spectral Analysis, Detection, and Estimation*," McGraw-Hill Inc., New York, 1975.

[THO71] J. B. Thomas, "*An Introduction to Applied Probability and Random Processes*," John Wile & Sons Inc., New York, 1971.

[THE04] C. W. Therrien, M. Tummala, "*Probability for Electrical and Computer Engineers*," CRC Press LLC, Boca Raton, Florida, 2004.

[URK83] H. Urkowitz, "*Signal Theory and Random Processes*," Artech House, Inc., Norwood, 1983.

[WEN86] E. Wentzel, L. Ovcharov, "*Applied Problems in Probability Theory*," Mir Publishers, Moscow, 1986.

[YAN09] W. Y. Yang at all, "*MATLAB/SIMULINK for Digital Communication*," The A-JIN Inc. 2009, Korea.

Index